WORLD ACADEMIC FRONTIERS
世界学术研究前沿丛书

Material Science

材料科学

"世界学术研究前沿丛书"编委会
THE EDITORIAL BOARD OF
WORLD ACADEMIC FRONTIERS

中国出版集团公司
世界图书出版公司
广州·上海·西安·北京

图书在版编目（CIP）数据

材料科学：英文 /"世界学术研究前沿丛书"编委会编.—广州：世界图书出版广东有限公司，2017.8
　ISBN 978-7-5192-2458-5

　Ⅰ.①材… Ⅱ.①世… Ⅲ.①材料科学－英文 Ⅳ.①TB3

中国版本图书馆 CIP 数据核字 (2017) 第 040980 号

Material Science © 2016 by Scientific Research Publishing

Published by arrangement with Scientific Research Publishing
Through Wuhan Irvine Culture Company

This Edition © 2017 World Publishing Guangdong Corporation
All Rights Reserved.
本书仅限中国大陆地区发行销售

书　　名：	材料科学 Cailiao Kexue
编　　者：	"世界学术研究前沿丛书"编委会
责任编辑：	康琬娟
出版发行：	世界图书出版广东有限公司
地　　址：	广州市海珠区新港西路大江冲25号
邮　　编：	510300
电　　话：	（020）84460408
网　　址：	http://www.gdst.com.cn/
邮　　箱：	wpc_gdst@163.com
经　　销：	新华书店
印　　刷：	广州市德佳彩色印刷有限公司
开　　本：	787 mm×1092 mm　1/16
印　　张：	42
插　　页：	4
字　　数：	1020千
版　　次：	2017年8月第1版　2017年8月第1次印刷
国际书号：	ISBN 978-7-5192-2458-5
定　　价：	598.00元

版权所有　翻印必究
（如有印装错误，请与出版社联系）

Preface

The interdisciplinary field of materials science, also commonly known as materials science and engineering, involves the discovery and design of new materials, with an emphasis on solids. The intellectual origins of materials science stem from the Enlightenment, when researchers began to use analytical thinking from chemistry, physics, and engineering to understand ancient, phenomenological observations in metallurgy and mineralogy. Materials science still incorporates elements of physics, chemistry, and engineering. In recent years, materials science has become more widely recognized as a specific and distinct field of science and engineering. Many of the most pressing scientific problems humans currently face are due to the limitations of the materials that are available and, as a result, breakthroughs in materials science are likely to have a significant impact on the future of technology.[1]

In the present book, numerous literatures about materials science published on international authoritative journals were selected to introduce the worldwide newest progress, which contains reviews or original researches on metallic materials, inorganic non-metallic materials, polymer materials and composite materials and so on. We hope this book can demonstrate advances in materials science as well as give references to the researchers, students and other related people.

编委会：
- ◇ 纳赛尔·巴迪教授，休斯顿大学，美国
- ◇ R·瓦森特·库玛教授，剑桥大学，英国
- ◇ 亚历山大·班格图亚兹教授，俄罗斯国家科学院，俄罗斯
- ◇ 陈昶乐教授，中国科技大学，中国

March 9，2017

---

[1] From Wikipedia: https://en.wikipedia.org/wiki/Materials_science

Selected Authors

D. L. Engelberg, Materials Performance Centre & Corrosion and Protection Centre, School of Materials, The University of Manchester, Sackville Street Campus, Manchester, UK.

M. Schreiber, Bio-products Discovery and Development Centre, Department of Plant Agriculture, Crop Science Building, University of Guelph, Guelph, Canada.

Ahmed S. J. Al-Zubaydi, Materials Research Group, Faculty of Engineering and the Environment, University of Southampton, Southampton, UK.

Izabella Rajzer, Division of Materials Engineering, Department of Mechanical Engineering Fundamentals, ATH University of Bielsko-Biala, Bielsko-Biała, Poland.

A. Keller, Department of Mechanical Engineering, Imperial College London, South Kensington Campus, London, UK.

Seung-Ryong Ha, Department of Dentistry, Ajou University School of Medicine, Suwon, Republic of Korea.

R. Eloirdi, European Commission, Joint Research Centre, Institute for Transuranium Elements, Karlsruhe, Germany.

Sarah H. Tolbert, Department of Chemistry, University of California, Los Angeles (UCLA), Los Angeles, USA.

Selected Authors

Nicholas W. M. Ritchie, Materials Measurement Science Division, National Institute of Standards and Technology, Gaithersburg, USA.

Peter I. Cowin, Department of Chemical and Process Engineering, University of Strathclyde, Glasgow, UK.

Contents

Chapter 1..1

Retardation of Plastic Instability via Damage-Enabled Microstrain Delocalization

by J. P. M. Hoefnagels, C. C. Tasan, F. Maresca, et al.

Chapter 2..31

Adsorption Characteristics of Noble Metals on the Strongly Basic Anion Exchanger Purolite A-400TL

by A. Wołowicz and Z. Hubicki

Chapter 3..53

A Facile Method for the Production of SnS Thin Films from Melt Reactions

by Mundher Al-Shakban, Zhiqiang Xie, Nicky Savjani, et al.

Chapter 4..67

Centrifugal Melt Spinning of Polyvinylpyrrolidone (PVP)/Triacontene Copolymer Fibres

by Tom O'Haire, Stephen. J. Russell, and Christopher M. Carr

Chapter 5..89

Conjugated Polymers as Robust Carriers for Controlled Delivery of Anti-Inflammatory Drugs

by Katarzyna Krukiewicz and Jerzy K. Zak

Chapter 6...107
Controlled-Release Formulation of Perindopril Erbumine Loaded PEG-Coated Magnetite Nanoparticles for Biomedical Applications

by Dena Dorniani, Aminu Umar Kura, Mohd Zobir Bin Hussein, et al.

Chapter 7...129
Correlative EBSD and SKPFM Characterization of Microstructure Development to Assist Determination of Corrosion Propensity in Grade 2205 Duplex Stainless Steel

by C. Örnek and D. L. Engelberg

Chapter 8...163
Determination of Bone Porosity Based on Histograms of 3D μCT Images

by M. Cieszko, Z. Szczepański and P. Gadzała

Chapter 9...189
Direct Reduction of Synthetic Rutile Using the FFC Process to Produce Low-Cost Novel Titanium Alloys

by L. L. Benson, I. Mellor and M. Jackson

Chapter 10...213
Effect of Interfacial Area on Densification and Microstructural Evolution in Silicon Carbide-Boron Carbide Particulate Composites

by Tom Williams, Julie Yeomans, Paul Smith, et al.

Chapter 11...231
Electrospun Green Fibres from Lignin and Chitosan: A Novel Polycomplexation Process for the Production of Lignin-Based Fibres

by Makoto Schreiber, Singaravelu Vivekanandhan, Peter Cooke, et al.

Chapter 12...251
Enhancement of Mechanical Properties of Biocompatible Ti-45Nb Alloy by Hydrostatic Extrusion

by K. Ozaltin, W. Chrominski, M. Kulczyk, et al.

Chapter 13..267
Evolution of Microstructure in AZ91 Alloy Processed by High-Pressure Torsion

by Ahmed S. J. Al-Zubaydi, Alexander P. Zhilyaev, Shun C. Wang, et al.

Chapter 14..287
Fabrication of Bioactive Polycaprolactone/Hydroxyapatite Scaffolds with Final Bilayer Nano-/Micro-Fibrous Structures for Tissue Engineering Application

by Izabella Rajzer

Chapter 15..305
Fast-Curing Epoxy Polymers with Silica Nanoparticles: Properties and Rheo-Kinetic Modelling

by A. Keller, K. Masania, A. C. Taylor, et al.

Chapter 16..343
Graphene as a Lubricant on Ag for Electrical Contact Applications

by Fang Mao, Urban Wiklund, Anna M. Andersson, et al.

Chapter 17..361
Growth of Carbon Nanofibers from Methane on a Hydroxyapatite-Supported Nickel Catalyst

by Ewa Miniach, Agata Śliwak, Adam Moyseowicz, et al.

Chapter 18..379
High-Temperature Reactivity and Wetting Characteristics of Al/ZnO System Related to the Zinc Oxide Single Crystal Orientation

by Joanna Wojewoda-Budka, Katarzyna Stan, Rafal Nowak, et al.

Chapter 19..397
Improving Shear Bond Strength of Temporary Crown and Fixed Dental Prosthesis Resins by Surface Treatments

by Seung-Ryong Ha, Sung-Hun Kim, Jai-Bong Lee, et al.

Chapter 20 ....**421**

In Vitro Biocompatibility of Anodized Titanium with Deposited Silver Nanodendrites

by Mariusz Kaczmarek, Karolina Jurczyk, Jeremiasz K. Koper, et al.

Chapter 21 ....**441**

Influence of Structural and Textural Parameters of Carbon Nanofibers on Their Capacitive Behavior

by Adam Moyseowicz, Agata Śliwak and Grażyna Gryglewicz

Chapter 22 ....**459**

Investigation of Ammonium Diuranate Calcination with High-Temperature X-Ray Diffraction

by R. Eloirdi, D. Ho Mer Lin, K. Mayer, et al.

Chapter 23 ....**475**

Metallic Muscles and Beyond: Nanofoams at Work

by Eric Detsi, Sarah H. Tolbert, S. Punzhin, et al.

Chapter 24 ....**517**

Microstructural Evolution in 316LN Austenitic Stainless Steel during Solidification Process under Different Cooling Rates

by Congfeng Wu, Shilei Li, Changhua Zhang, et al.

Chapter 25 ....**537**

Orientation of Cellulose Nanocrystals in Electrospun Polymer Fibres

by N. D. Wanasekara, R. P. O. Santos, C. Douch, et al.

Chapter 26 ....**557**

Oxide Dispersion-Strengthened Steel PM2000 after Dynamic Plastic Deformation: Nanostructure and Annealing Behaviour

by Z. B. Zhang, N. R. Tao, O. V. Mishin, et al.

Chapter 27..**575**

Early Root Development of Field-Grown Poplar: Effects of Planting Material and Genotype

by Grant B. Douglas, Ian R. McIvor, and Catherine M. Lloyd-West

Chapter 28..**607**

Preparation of (001) Preferentially Oriented Titanium Thin Films by Ion-Beam Sputtering Deposition on Thermal Silicon Dioxide

by Imrich Gablech, Vojtěch Svatoš, Ondřej Caha, et al.

Chapter 29..**623**

Probing Trace Levels of Prometryn Solutions: From Test Samples in the Lab toward Real Samples with Tap Water

by Rafael J. G. Rubira, Sabrina A. Camacho, Pedro H. B. Aoki, et al.

Chapter 30..**643**

Effect of Addition of BaO on Sintering of Glass–Ceramic Materials from SiO_2–Al_2O_2–Na_2O–K_2O–CaO/MgO System

by Janusz Partyka, Katarzyna Gasek, Katarzyna Pasiut, et al.

Chapter 1

Retardation of Plastic Instability via Damage-Enabled Microstrain Delocalization

J. P. M. Hoefnagels[1], C. C. Tasan[2], F. Maresca[1,3], F. J. Peters[1], V. G. Kouznetsova[1]

[1]Department of Mechanical Engineering, Eindhoven University of Technology (TU/e), P.O.Box 513, 5600MB Eindhoven, The Netherlands
[2]Max-Planck-Institut fur Eisenforschung, P.O.Box 140444, 40074 Dusseldorf, Germany
[3]Materials Innovation Institute (M2i), P.O.Box 5008, 2600GA Delft, The Netherlands

Abstract: Multi-phase microstructures with high mechanical contrast phases are prone to microscopic damage mechanisms. For ferrite-martensite dual-phase steel, for example, damage mechanisms such as martensite cracking or martensite-ferrite decohesion are activated with deformation, and discussed often in literature in relation to their detrimental role in triggering early failure in specific dual-phase steel grades. However, both the micromechanical processes involved and their direct influence on the macroscopic behavior are quite complex, and a deeper understanding thereof requires systematic analyses. To this end, an experimental-theoretical approach is employed here, focusing on three model dual-phase steel microstructures each deformed in three different strain paths. The micromechanical role of the observed damage mechanisms is investigated in detail by in-situ scanning electron microscopy tests, quantitative damage analyses, and finite element simulations. The comparative analysis reveals the unforeseen conclusion that damage nucleation may have a beneficial mechanical effect in ideally designed dual-phase steel microstructures (with effective crack-arrest mechanisms) through microscopic strain delocalization.

1. Introduction

In the last decades, novel advanced high-strength steels (AHSS) with more and more complex microstructures have been introduced (e.g., twinning-assisted plasticity steels[1][2], quench and partition steels[3][4], and carbide-free bainite steels[5][6]) to achieve superior mechanical performance compared to existing grades. Yet, the connection between the microstructure and the overall mechanical behavior is still not fully set even for the more established AHSS grades, such as dual-phase (DP) steels that have been present for decades[7]. The martensitic-ferritic microstructures of DP steels provide excellent combinations of high strength and good ductility[7]–[9] at low cost (*i.e.*, low alloying content) and relatively simple thermo-mechanical processing (*i.e.*, intercritical annealing). Thus, DP steels are nowadays being used or considered for different automotive components, e.g., for crash box structures.

The development of DP steels was triggered in the early 1970s and intensive research has been done since then. A huge experimental literature exists, which has shown the influence of martensite volume fraction[10][11], grain size of the constituents, and grain refinement[9][12][13], as well as carbon content[14], on the ultimate strength and ductility of DP steels. Models that account for such effects have been proposed and widely used, e.g.,[15]–[17]. The influence of the morphology of the constituents has also been extensively studied, both from experimental and computational points of view, e.g.,[18]–[20].

A wider application of DP steels is hampered by the limited understanding regarding their failure mechanisms. For example, it is beneficial for weight reduction purposes to employ higher strength DP grades in automotive body-in-white structures, as it would allow sheet thickness to be reduced. However, in such higher strength grades (with higher martensite content), activity of microstructural damage mechanisms may often lead to unpredicted failures during forming operations or upon crash[21][22]. The limited understanding of the macroscopic fracture processes in DP steel arises from the presence of multiple microstructural damage mechanisms that exhibit complex interactions[23]–[32]. As a consequence, the applicability of state-of-the-art damage models that aim at modeling multiple, interacting, damage nucleation mechanisms, e.g.,[33][34], is limited by the possibilities for experimental characterization, see e.g.,[29][31].

The challenge is thus clear: developing optimized martensite-ferrite microstructures that enable higher strengths in DP steels, while preserving good toughness. To this end, a vast variety of microstructure variations can be introduced in DP steels by small changes in the composition and/or thermomechanical processing[18][35]–[42]. To guide this microstructure design process, micromechanics-based foundations and design guidelines are needed that would ensure damage-prone microstructures. This research aims to provide an improved understanding in this direction.

There are many investigations in the literature on damage and failure mechanisms in DP steels[23]–[27][29]. These reports reveal three general observations:

- Aside from the rarely seen damage incidents at ferrite grain boundaries (DFGB), ferrite grain interiors (DFGI), or around inclusions (DINC), two main damage mechanisms are dominant in DP steel microstructures: martensite cracking (DMC) and martensite-ferrite interface damage (D_{MFI}).[1]

- The relative activity of these two mechanisms, their activation regimes, and their role on the overall mechanical response are strongly microstructure and strain path dependent.

- While its mechanical effect is critical, the overall damage fraction is difficult to detect as it is in the order of few percent even at high deformation levels.

Given these points, it is clear that generic microstructure design guidelines

---
[1]Note that the latter is often referred to in the literature as martensite-ferrite decohesion mechanism.

cannot be provided through qualitative analysis of a single microstructure deformed in a single strain path, as is done in most previous works. Therefore, in this research, we aim to improve on this by employing an experimental-numerical approach that has various novelties: (i) Experiments focus on quantitative characterization of ductile damage evolution up to failure, at different strain paths and strain levels; (ii) For these experiments, a recently designed miniaturized Marciniak setup[43] and a novel image post-processing methodology are employed for statistically sound quantification of damage evolution; (iii) Different model DP microstructures (with variation in only a single microstructural variable at a time) are investigated using these techniques; and (iv) For a deeper understanding of the most relevant damage nucleation mechanisms, follow-up in-situ scanning electron microscope (SEM) deformation experiments and finite element simulations are also carried out.

In what follows, first the employed methodology is introduced in detail. The results are presented, starting with the identification of active damage mechanisms and quantification of their activity, followed by focusing on the factors determining the relative activity of the damage mechanisms through a discussion of the numerical results and in-situ damage nucleation images. The report is finalized with the conclusions.

2. Methodology

2.1. Materials

To investigate systematically the influence of ferrite grain size and martensite volume fraction, different DP model microstructures are produced where a single microstructural parameter is changed at a time. These microstructures, referred in the text as fine-grained (μ_{FG}), coarse-grained (μ_{CG}), and high martensite (μ_{HM}) microstructures, are designed by thermal processing of non-commercial DP600 and DP800 steel grades of 1mm thickness from Tata steel IJmuiden (**Figure 1**). These base steels are chosen specifically, as they have almost equal (typical) concentrations of Mn, Si, and Cr, while differing only in C (0.092wt% vs. 0.147wt%, respectively). The μ_{CG} is produced by reaustenization of DP600 alloy at 960°C for 10min, followed by air cooling to room temperature, then intercritical

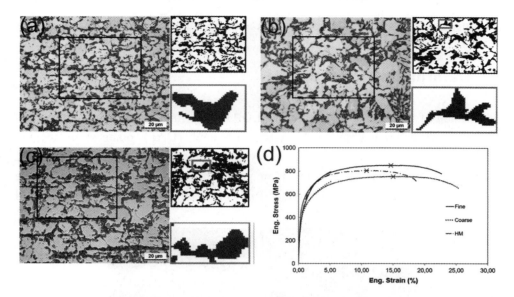

Figure 1. Optical microscopy images and SEM images (not shown) of a the µ$_{FG}$, b the µ$_{CG}$, and c the µ$_{HM}$ microstructures, recorded at the center cross section (sheet thickness in vertical direction) of the specimen; no dependence of the generated microstructures on the prior sheet rolling direction was observed. Each optical image a-c was first converted into black (martensite) and white (ferrite) images (top right subfigure of a-c), and then converted to a representative volume element (RVE) for FEM analysis (each SEM pixel is converted to a finite element to a total of ~1000 × 800 elements), of which a zoom with martensite colored red (bottom right subfigure of a-c) shows the fine mesh used. Shown in d are the global stress-strain curves under uniaxial tension of the µ$_{FG}$, µ$_{CG}$, and µ$_{HM}$ microstructures, with the point of plastic instability marked with a red cross. In dark red are shown the simulated stress-strain curves for each microstructure, which were fitted to the experimental curves by adapting the plastic model parameters (given in **Table 1**) (Color figure online).

annealing at 775°C for 30min, and finally quenching to room temperature. To produce the µ$_{FG}$, the re-austenization duration of the same alloy is decreased to 1min,[2], keeping the other conditions of the treatment identical. Decrease in austenization duration limits the growth of austenite grains but identical intercritical annealing treatment ensures largely unaffected martensite volume fraction (~33%) and morphology. To produce the µ$_{HM}$ microstructure, DP800 steel is heat treated in the same manner as the µ$_{CG}$. For the same intercritical annealing temperature, the DP800 steel with higher carbon content produces a higher martensite volume percentage (~41%) compared to the DP600 steel with lower carbon content, while the

---

[2]Note that accompanying dilatometry experiments reveals that the austenite transformation is completed within this duration.

martensite carbon contents in both are, on average, identical (**Figure 1**).

2.2. Deformation Experiments

Each of the three above-mentioned DP microstructures is deformed to fracture in three different strain paths: uniaxial tension (UAT), plane strain tension (PST), and biaxial tension (BAT). To carry out these deformation experiments, the miniaturized Marciniak setup with a punch diameter of 40mm[43], shown in **Figure 2(a)**, is employed. A finite element analysis of this Marciniak test showed

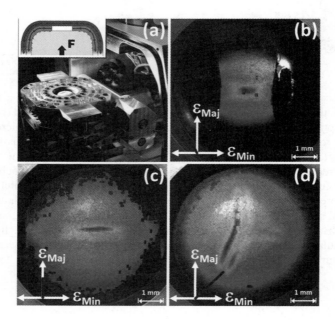

Figure 2. A Photograph of the miniaturized Marciniak setup[43], mounted in the door of the SEM, and used to perform all deformation experiments. The inset shows a schematic representation of the working principle of a Marciniak test on a specimen and so-called 'washer' (with central hole), where the red arrows show the in-plane displacement and the blue arrows the friction direction[43]. b-d The Real-time recorded optical images of the DP steel with fine-grained microstructure loaded under b UAT, c PST, and d BAT up to the first point of failure, and overlaid with the von Mises strain field obtained through digital image correlation. For each of the 3 strain paths, the strain fields have been used to calculate the evolution of the major strain, ε_{Maj}, as a function of the minor strain, ε_{Min}, which is shown in the inserts by the red curves. Note that at these critical strains the correlation of some subsets was lost due to large out-of-plane rotations at the specimen edge and detachment of the spray paint pattern at the specimen center, however, this did not interfere with the analysis (Color figure online).

that the stress in the thickness direction is negligible and that indeed a UAT, PST, or BAT stress state is achieved[43]. **Figures 2(b)-(d)** shows digital image correlation (DIC) overlays of the von Mises strain fields measured in situ under optical microscopy, obtained in the three considered strain paths. Aramis software (GOM Gmbh.) is employed for the DIC analysis. These samples are further characterized for the quantitative damage analysis which is described next. Furthermore, for a detailed analysis of the damage nucleation and growth mechanisms, in-situ scanning electron microscope deformation experiments are carried out in an FEI Quanta 600F microscope.

2.3. Quantitative Damage Analysis

For a systematic quantitative analysis of the deformation-induced evolution of the damage mechanisms, a semi-automatic Statistical Damage Identification program is developed (in MATLAB) and employed in this study. Within this methodology, five cross sections representing five different strain levels (measured using DIC) are metallographically prepared in each sample that is deformed to fracture. Per each strain level, five images are taken at an optimized magnification of 456× that ensures a large (*i.e.*, representative) field-of-view and sufficient resolution. Following inter-image contrast/brightness homogenization, each image is analyzed in the gray value thresholding-based image analysis algorithm (**Figure 3**). Each detected damage incident is also confirmed by the operator, and classified regarding the mechanism. Note that the damage incident density, *i.e.*, the number of damage sites per area, is recorded instead of the more commonly used damage area fraction in order to reduce the otherwise large influence of a few large damage sites on the damage statistics. Note also that during the calculation of the damage incident density for a given strain level, a correction is applied to take into account the change in reference area due to the evolving in-plane strain and cross contraction along the thickness direction.

2.4. Modeling Methodology

Optical microscopy images from μ_{FG}, μ_{CG}, and μ_{HM} specimens are binarized in MATLAB for clear classification of the martensite and ferrite regions. A 2D

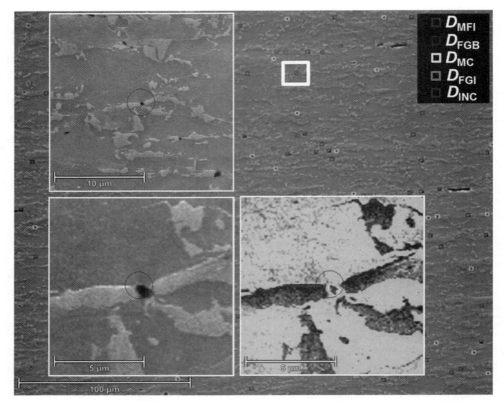

Figure 3. Screen capture of the semi-automatic statistical damage identification algorithm, which automatically identifies all damage incident areas and sequentially prompts each damage incident for classification by the user (through a pop-up selection box, not shown here). To assist in the assessment of the type of damage incident, the damage site is simultaneously shown at low, intermediate, and high magnification (respectively, background, top-left, and bottom-left image) and at high magnification with rainbow color map (bottom right image) (Color figure online).

finite element mesh with bilinear square finite elements is generated on a representative portion of the image, such that the global martensite volume fraction is preserved. Periodic boundary conditions are applied to all representative volume elements (RVE).

The elastic phase parameters are adopted from[44], *i.e.*, a Young's modulus of 220 and 195GPa for ferrite and martensite, respectively, and a Poisson's ratio of 0.3 for both phases. For each phase, the plastic deformation is modeled with a Ludwik-type stress-strain relationship $\left(\sigma = \sigma_y + K\varepsilon_p^n\right)$. Note that the effect of the crystal lattice misorientation of neighboring ferrite grains is not considered in such models. The Ludwik's model parameters, which are given in **Table 1**, were fitted on the experimental data of **Figure 1(d)**, where it is shown that a reasonable fit is achieved in the regime where the simulations are used in this work (below 5% major strain). Interestingly, the ferrite yield strength increases from μ_{CG} to μ_{FG}, as expected from the Hall-Petch effect, and the martensite yield strength for the lHM is lower as might be expected from the larger martensite island size.

A commercial finite element software package (MSC Marc) is used to perform the simulations. For each microstructure (μ_{FG}, μ_{CG}, and μ_{HM}), three strain paths UAT, PST, and BAT are considered. The UAT is simulated by employing plane stress finite elements (free out-of-plane contraction) and by assigning displacement along the rolling direction, while keeping the other directions free. The PST condition is simulated employing plane strain finite elements (fixed thickness), by assigning displacement along the rolling direction and free transverse displacement. Finally, the BAT condition is simulated using generalized plane strain finite elements, which allow prescription of constant thickness change together with the usual displacement along the rolling direction (transverse direction is free to contract).

Table 1. Model parameters of the Ludwik's yield strength for each phase and microstructure, as fitted to the data of **Figure 1(d)**.

Microstructures	Yield strength (MPa)	Hardening coefficient (MPa)	Ludwik coefficient
Fine (μ_{FG})			
Ferrite	220	1300	0.33
Martensite	800	6000	0.70
Coarse (μ_{CG})			
Ferrite	200	1450	0.43
Martensite	800	5000	0.70
High martensite (μ_{HM})			
Ferrite	180	1150	0.32
Martensite	650	4300	0.60

3. Results and Discussion

3.1. Variation of Strain Path

3.1.1. Quantitative Damage Analysis

As a first step toward the goal of statistically relevant characterization of ductile damage evolution up to failure, all possible damage mechanisms in the three DP microstructures (μ_{FG}, μ_{CG}, and μ_{HM}) and three strain paths (UAT, PST, and BAT) were extensively studied by exploiting the in-situ SEM capabilities of the miniaturized Marciniak setup. The five most relevant damage mechanisms are presented in **Figure 4**. These five mechanisms, which are also the dominant mechanisms observed in the literature[23]–[27][29], were chosen as categories in

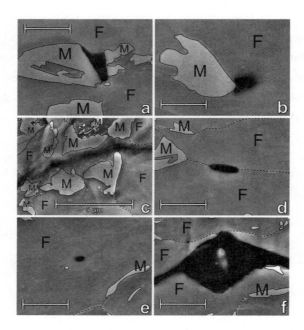

Figure 4. The most relevant damage mechanisms encountered in the DP microstructures, as observed with in-situ scanning electron microscopy: a Martensite cracking (MC), b, c Martensite-Ferrite Interface damage (MFI), d damage at a ferrite grain boundary (FGB), e damage at the ferrite grain interior (FGI), and f damage around an INClusion (INC). For all images, except image (c), the scale bar indicates a length of 1μm. The 'M' and 'F' symbols and solid and dashed guidelines denote, respectively, martensite, ferrite, a martensite-ferrite interface, and a ferrite-ferrite grain boundary, which have been identified by careful investigation at the highest magnification level.

the semi-automatic statistical damage identification algorithm (**Figure 3**) as a starting point for the quantitative damage analysis, discussed next.

The analysis starts with the fine-grained (lFG) microstructure, for which the different damage mechanisms were quantified for the three loading states (UAT, PST, and BAT). The damage incident densities of the five different damage mechanisms (D_{MC}, D_{MFI}, D_{FGB}, D_{FGI}, and D_{INC}) are shown in **Figure 5** as function of the von Mises strain, with the vertical dashed lines denoting the strain level at the point of necking (*i.e.*, global localization). Each data point was obtained by quantifying all damage incidents over five large-area ($300 \times 300\mu m^2$) SEM images, *i.e.*, a total area of $450,000\mu m^2$. This large amount of data allows for a very accurate determination of the averaged damage incident density. It should be noted, however, that the damage incident density inherently shows large variability due to the strong heterogeneity of the DP microstructure even in commercial grades, as can be observed by the wide error bands in **Figure 5**. Perhaps this inherent variability may also explain why, to our knowledge, such an extensive quantification of the relevant damage mechanisms as a function of strain level and for different strain paths and microstructures has not been carried out before.

The first aspect to note from **Figure 5** is that D_{FGB}, D_{FGI}, and D_{INC} damage incidents are all clearly present, however, only to a limited extent; therefore, these mechanisms most probably do not play a critical role in controlling the necking and failure behavior. For this reason, the investigation will focus on the D_{MFI} and D_{MC} damage mechanisms, for which a number of interesting observations can be made as follows:

(1) D_{MFI} is the dominant damage mechanism and its incident density increases from UAT to PST to BAT, whereas D_{MC} is negligible at UAT, increases slightly at PST, but becomes important for BAT.

(2) The necking strain is lowest for PST, which corresponds to the minimum that is typically found in forming limit diagrams. It may be surprising to see, however, that the BAT necking strain is much larger than that of UAT.

(3) Whereas damage versus strain measurements typically show damage initiation only after a minimum strain threshold, followed by an exponential damage

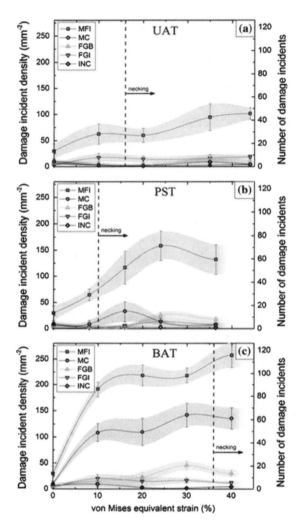

Figure 5. Damage incident areal density versus von Mises equivalent strain, measured using the damage quantification methodology for the fine-grained microstructure (μ_{FG}) loaded under a uniaxial tension (UAT), b plane strain tension (PST), and c biaxial tension (BAT). For each strain level, 5 large-view SEM scans were analyzed for a total area of 450,000μm². Data points and error bars represent, respectively, average values and their standard deviation of the five different damage mechanisms that are explained in **Figure 4**. Dashed vertical lines denote the point of global localization (deduced from the DIC strain maps).

increase[45], here all three load cases show that damage incidents are predominantly initiated at low strain levels, after which the total number of damage incidents saturates. This initial damage burst is particularly evident for BAT.

(4) It is remarkable that the BAT damage evolution trend of D_{MFI} and D_{MC} looks very similar, which is also true for the coarse-grained and high martensite microstructures (shown below in **Figure 9**). This suggests that both mechanisms are somehow linked.

Interestingly, the first three observations are in agreement with those of Tasan et al.[46], where the total number of damage incidents was measured (only) at the point of necking and failure, for the commercial (parent) DP600 microstructure with the same chemical composition (note that no comparison with observation 4 could be made).

In order to understand these observations, a thorough experimental and numerical analysis, discussed below, was initiated, which led to the following hypothesis on a chain of events that links D_{MC} to D_{MFI}:

(a) Plastic straining in F: upon deformation, due to the lower yield strength of ferrite compared to martensite, the ferrite matrix quickly strains plastically.

(b) Fracture of M: especially under biaxial loading, a large hydrostatic stress develops, even at early stage of deformation, causing the smallest or weakest cross section of the typically irregularly shaped martensite islands (or thin martensite bridges) to fracture.

(c) Extreme local straining in F and D_{MFI}: when a martensite island fractures, the surrounding ferrite must carry the released load. This results in extreme local plastic straining, stopped only by the increase in flow stress due to strain hardening. This extreme local straining in ferrite may trigger microdamage, *i.e.*, DMFI damage.

(d) Diffuse straining in F: a larger area around the damage site needs to increase in strain to accommodate the extreme local strains and to carry the increase in stress due to D_{MFI}.

One can easily see that this hypothesis, in which D_{MFI} is caused by D_{MC}, can explain the peculiar similarity in BAT trend for D_{MFI} and D_{MC} (observation 4). It may also explain why most damage incidents initiate at low strain (observation 3), while at the same time the built up of stress in M explains the relatively high yield strength of DP steels. Moreover, the critical role of hydrostatic stress can explain why D_{MC} primarily occurs at BAT [D_{MC} is negligible for UAT [**Figure 5(a)**] and small for PST [**Figure 5(b)**]. Furthermore, the coupling of D_{MFI} to D_{MC} can explain that D_{MFI} also increases from UAT to PST to BAT (observation 1). Lastly, the diffuse straining in combination with strain hardening may prevent the formation of percolation paths, and thus delaying global localization; such a necking retardation mechanism may explain the large necking strain at BAT (observation 2). Nevertheless, to test the validity of this D_{MC}-D_{MFI} hypothesis, additional numerical and experimental studies were conducted, which are presented next.

3.1.2. Microstructural Simulations

First, numerical simulations of the (measured) fine-grained microstructure loaded at UAT, PST, and BAT to 5% strain are investigated. To this end, **Figure 6** shows the hydrostatic stress and plastic strain fields. Note that the deviatoric stress (or von Mises stress) and volumetric strain are not shown as they scale with the plastic strain and hydrostatic stress, respectively, in the isotropic elasto-plastic model used ("Methodology" section). Also no damage mechanisms were included in these simulations, as they would require the measurement of constitutive laws for damage initiation and growth; the fundamental challenges in obtaining such laws have been described in detail in[31]. Since these simulations do not include damage-induced strain relaxation and stress redistributions, care should be taken when comparing to experimental results. Nevertheless, the simulations do provide qualitative insight in the differences in stress and strain state for the different strain paths.

Figures 6(a)-(c) shows that the equivalent plastic strain is higher in the ferrite matrix than the martensite islands and shows strain bands between 45° and 60° to the main loading direction, in agreement with[30]. Regarding the plastic strain magnitude and distribution in the ferrite, it is observed that, from BAT to PST to UAT, the strain localizes increasingly into peaks. Based on this trend, a decrease in DMFI from UAT to BAT would be expected; however, the opposite is observed in **Figure 5**, which indicates that another mechanism for damage in ferrite becomes active at PST and especially BAT.

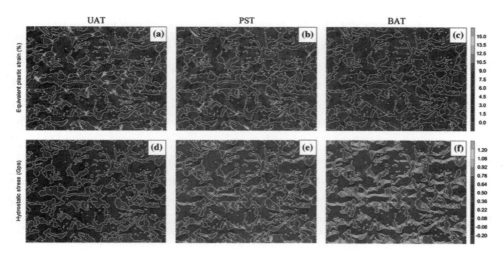

Figure 6. FEM simulation results for the fine microstructure (μ_{FG}), deformed to a global von Mises equivalent strain of 5%, for UAT, PST, and BAT. a-c The local equivalent plastic strain (in %). a-f The hydrostatic stress (in GPa). A white line demarks the martensite-ferrite phase boundaries and a fine white speckled pattern was added on the martensite phase to make it distinguishable from the ferrite phase. A map of the martensite-ferrite microstructural distribution is shown in **Figure 1(a)**.

The plastic straining releases the deviatoric stress in the ferrite matrix and, through stress redistribution (bounded by stress equilibrium at the phase boundaries), also the hydrostatic stress. This is seen in **Figures 6(d)-(f)**, which shows that the hydrostatic stress is (much) higher in the martensite islands. Naturally, the hydrostatic stress increases with the change of loading from UAT to PST to BAT. This increase in hydrostatic stress explains the observed increase in fracture of martensite (*i.e.*, DMC) from UAT to PST to BAT (**Figure 5**).

The simulations thus support the first two steps of the DMC-DMFI hypothesis; however, because of the absence of damage mechanisms, the last two steps (regarding the coupling between DMC and DMFI) cannot be investigated. Hence, two additional experiments were performed to examine the connection between DMC and DMFI.

3.1.3. In-Situ SEM Study

In the first additional experiment to study the evolution of individual damage incidents during the deformation, biaxial tension tests up to failure were performed in situ under SEM (SE-mode) observation using home-built miniaturized Marciniak setup, shown in **Figure 2(a)**. The measured large-area (300 × 300μm^2) in-situ SEM movies were analyzed in detail with respect to martensite cracking incidents and further deformation around these DMC sites. First of all, it was found that the areal density of DMC incidents at the surface was significantly lower than in the bulk, which is attributed to the lower hydrostatic stress at the surface. Still, many DMC incidents could be observed under biaxial loading, of which seven examples are given in **Figure 7**. It was found that most DMC incidents occurred in the smallest cross section of the irregularly shaped martensite islands, *i.e.*, the thin martensite bridges. Moreover, it was observed that almost all DMC incidents initiated at the early stages of deformation, see **Figure 7(b)**, and that DMC incidents were typically accompanied by one or more location of extreme plasticity in the surrounding ferrite, see **Figure 7(c)**. This would be counted as DMFI damage in the damage quantification methodology, giving direct evidence for the hypothesis that DMC triggers DMFI. Finally, it should be noted that around most D_{MC}-D_{MFI} locations the localized extreme plastic straining spreads out into the neighboring ferrite grains resulting in diffuse deformation zones that can cover the complete ferrite grain, see **Figure 7(d)**, thus supporting the necking retardation mechanism of the hypothesis. This mechanism of ferrite damage (*i.e.*, highly localized ferrite deformation) activating diffuse deformation zones in the adjacent ferrite grains was also observed in situ in the microstructural martensite bands observed in commercial DP600 sheet[30]. Combining **Figure 5** and **Figure 7**, it can be concluded that the early-initiated martensite cracking incidents are well enough dispersed to postpone the formation of percolation paths, which explains the late global localization.

3.1.4. 3D Depth Profiling

In the second additional experiment to investigate whether the coupling between DMC and DMFI damage initiation is also present in the specimen interior, high-resolution 3D depth profiling is performed on the cross section of a 16% biaxially strained fine-grained specimen. To this end, a series of flat profiles are made approximately 300nm apart. Note that the high requirements on surface roughness rule out the (Nital) surface etching, used before to distinguish between martensite and ferrite phases. Instead, precision polishing is used to reproducibly remove a ~300nm surface layer, while SEM imaging in backscatter electron (BSE)

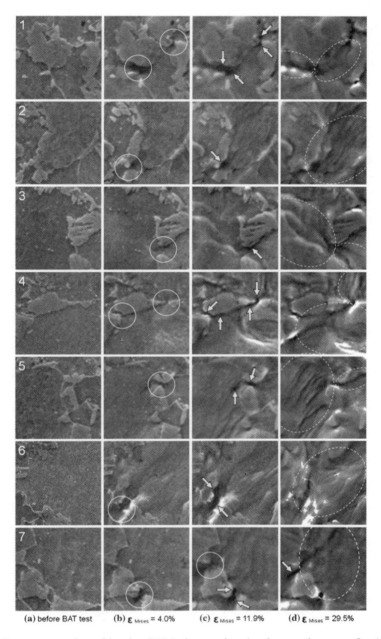

Figure 7. Seven examples of in-situ SEM observation (at the specimen surface) of the finegrained microstructure (a), which exhibits damage evolution under biaxial loading initiated by martensite cracking [solid circles in images (b)] at the early stages of deformation, followed by extreme localized plasticity in the surrounding ferrite [arrows in images (c)], followed by large deformation zones [dashed ellipses in images (d)]. All images are sized $10 \times 10 \mu m^2$.

imaging mode is used to identify the martensite and ferrite phases by the difference in channeling contrast (note that martensite shows much finer spatial variations in channeling contrast due to its much finer substructure compared to that of the relatively coarse ferrite sub-grains). This identification procedure was verified in detail using electron backscatter diffraction analysis (not shown). Note also that, due to the channeling contrast, D_{MC} and especially D_{MFI} damage locations appear differently.

Three typical examples of the detailed 3D shape of a D_{MC} damage location are shown in **Figure 8**. A number of observations could be made from these and other depth profiles measured in the specimen interior.

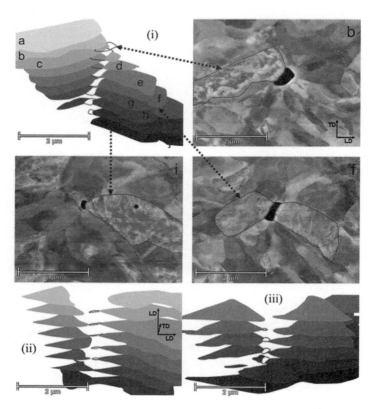

Figure 8. High-resolution 3D profiles of typical damage incidents in the specimen interior in BAT-strained (ε_{Mises} = 16%) fine-grained (μ_{FG}) microstructure, consistently showing a martensite crack at its center (e.g., image f of damage incident i) surrounded by severe plastic straining location (e.g., image b 'above' and image i 'below'). The depth profile layers are separated ~300nm (along one of the two loading directions, LD) and the SEM channeling contrast images were taken halfway through the sheet thickness direction (TD).

1) As expected, the 3D shape of the martensite islands is irregular and the fracture occurs always at the smallest cross section, or at least a small cross section. In other words, the microstructural configuration within the martensite islands seems to be play a secondary role, in agreement with[47].

2) The D_{MC} locations are typically surrounded on one or both sites by a DMFI location, see, e.g., micro-graphs b and i in **Figure 8**. This is a strong indication that martensite cracking triggers martensite-errite interface damage, because the force previously carried by the martensite island must be fully transferred to the neighboring ferrite matrix after the martensite cracking. Notice also that D_{MC}-to-D_{MFI} mechanism is activated already at the relatively low small strain of 16%, in agreement with **Figure 7(c)**.

3) The fact that the D_{MFI} location has opened up and has therefore become visible for micrographic observation in the SEM-BSE images also means that the surrounding ferrite must have strained heavily to accommodate the martensite crack opening displacement, which is typically in the order of hundreds of nanometers.

In addition, all recorded high-resolution SEM-BSE images (with a total area of 38200μm^2) were processed with the above-mentioned damage quantification methodology, *i.e.*, similar to **Figure 5**. A total of 202 damage incidents were automatically found by the software and identified as D_{MC}, D_{MFI}, D_{FGB}, D_{FGI}, or D_{INC}. Again D_{MFI} and D_{MC} damage dominated showing a mutual ratio of ~1.7 in good agreement with the ratio found in **Figure 5(c)** at 16% strain, especially when considering the differences in image contrast mode used. Detailed investigation of the 3D connections revealed that the 202 damage counts in these stacked images could be traced back to 81 3D damage zones and approximately half of the DMFI incidents originate from a martensite cracking event (D_{MC}), which may explain the increase in D_{MFI} from PST to BAT loading, observed in **Figure 5**.

Finally, it is noted that, with this insight in the 3D character of coupled D_{MC}-$_{MFI}$ damage incidents, it cannot be excluded that the damage incidents at a ferrite grain boundary or inside the grain interior (D_{FGB} and D_{FGI}) are in fact caused by a martensite island above or below the surface of observation, and thus should have been counted as DMFI. However, due to the relative unimportance of

DFGB compared to DFGI, this would not alter the conclusions.

3.1.5. Conclusions Part A

In all, it can be concluded that the $D_{MFI\text{-}MC}$ hypothesis is supported by many different forms of experimental and numerical evidence. Especially, the mechanism that spreads out the deformation over a larger ferrite area (the diffuse deformation zones) is interesting, as it seems to be the cause for the delay of global localization. For this necking retardation mechanism to be effective, however, the damage incidents need to be well enough dispersed, such that the early burst of DMC damage in BAT does not result in global localization by connection of DMC damage localizations. Therefore, next, the influence of microstructure features (grain size and martensite volume percentage) is investigated.

3.2. Variation of Microstructure

Figure 9 compares the BAT deformation of the fine-grained (μ_{FG}), coarse-grained (μ_{CG}), and high martensite (μ_{HM}) microstructures, with respect to the damage incident densities obtained with the damage quantification methodology [**Figures 9(a)-(c)**], the simulated hydrostatic stress fields [**Figures 9(d)-(f)**], and simulated plastic strain fields [**Figures 9(g)-(i)**]. All three microstructures show very similar damage density evolutions, with DMFI being approximately twice as much as D_{MC} and more than four times larger than the three other mechanisms (D_{FGB}, D_{FGI}, and D_{INC}), and D_{MFI} and D_{MC} showing roughly the same trend with a steep initial increase that reduces toward higher strains already before the point of necking. This suggests that the above-mentioned causal connection between D_{MFI} and D_{MC} is also active at larger grain size and higher martensite content. On a more subtle note, for μ_{CG}, the ratio of D_{MFI} to D_{MC} is slightly larger than those for the two other microstructures and the initial increase of D_{MC} is slightly less steeper. Perhaps, the number of "thin martensite bridges" is lower for the μ_{CG} microstructure which leads to fewer MC incidents.

3.2.1. Influence of Grain Size

The isolated influence of grain size is investigated by comparing the µFG

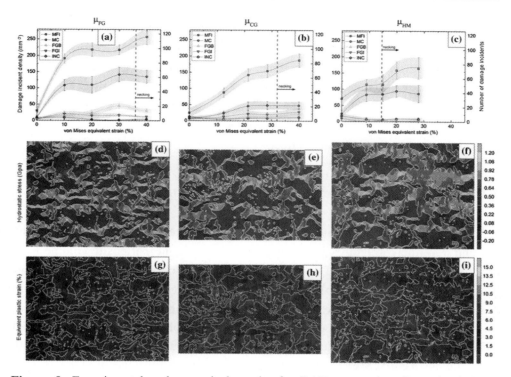

Figure 9. Experimental and numerical results for BAT, comparing finegrained (μ_{FG}), coarse-grained (μ_{CG}), and high martensite (μ_{HM}) microstructures. a-c Damage incident density versus equivalent strain, quantified from five 300 × 300µm² SEM scans for each data point (error bars represent standard deviation; dashed vertical lines show point of global localization). FEM simulations at a global von Mises equivalent strain of 5% of d-f the local equivalent plastic strain and g-i the hydrostatic stress. Same color scale bars as in **Figure 6** are used for easy comparison. A white line demarks the martensite-ferrite phase boundaries in d-i and a fine white speckled pattern was added on the martensite phase to make it distinguishable from the ferrite phase. A map of the martensite-ferrite microstructural distribution is shown in, respectively, **Figures 1(a)-(c)**.

and μ_{CG} microstructures: a reduction in grain size corresponds to an increase in DMFI and DMC densities and, especially, earlier damage initiation at low strains [**Figure 9(a)**, **Figure 9(b)**]. These effects could be caused by the same grain size effect underlying the well-known Hall-Petch relation between the yield (and flow) strength and the grain size, which is explained by the obstruction of plastic slip at the grain and/or phase boundaries causing dislocation pile-up, thereby locally increasing the stress level at the boundaries. Indeed, the experimental global stress-strain curves in **Figure 1(d)** show this increase in yield and flow strength. The D_{MFI}-D_{MC} hypothesis would predict that a faster rise of the stress level at the martensite-ferrite boundaries (due to a reduction in ferrite grain size) results in more and earlier DMC damage and, due to the D_{MC}-D_{MFI} causality, in more D_{MFI} damage, thus explaining the observed differences between **Figure 9(a)**, **Figure 9(b)**. The evolutions of the simulated hydrostatic stress also show significantly higher stress concentrations in the martensite islands of the μ_{FG} microstructure, but this is a direct result of the higher ferrite yield strength used, see **Table 1**, which indirectly takes into account the Hall-Petch effect.

3.2.2. Influence of Martensite Volume Fraction

To investigate the isolated influence of martensite volume fraction, next, the μ_{CG} and μ_{HM} microstructures are compared: an increase in martensite volume fraction results in an increase in D_{MC} damage, whereas it does not seem to significantly impact DMFI [**Figure 9(b)**, **Figure 9(c)**]. The increase in D_{MC} is attributed to the stress increase due to the reduction of plastically deforming ferrite phase resulting in a compact network of the harder martensite phase. Indeed, a pronounced increase in stress level (at equal global strain) is seen in the simulated hydrostatic stress fields [**Figure 9(e)** versus **Figure 9(f)**]. Interestingly, the increase in D_{MC} with increasing martensite volume fraction is not followed by an increase of D_{MFI}. This may be the result of the lower probability that a D_{MC} location is adjacent to an open ferrite area that is large enough (and thus the constraint by the surrounding martensite network low enough) to develop extreme localized plasticity, identified as D_{MFI}. As a direct consequence, the areal density of diffuse deformation zones, which are initiated from a D_{MFI} sites as shown in **Figure 7(d)**, will also be lower. This is precisely what is also seen in the simulated fields of the plastic strain, which for higher martensite volume fraction shows large regions with low ferrite strain, see, e.g., the lower left corner of **Figure 9(i)**. In other words, the compact martensite network in the μ_{HM} microstructure prevents the plastic straining around a D_{MC} location from spreading out to surrounding ferrite grains. Indeed, as a consequence of the fact that this spreading of plastic straining is hampered, **Figure 9(i)** also reveals a number of local spots where the plastic strain peaks to a level far above the maximum strain found in **Figure 9(g)**, **Figure 9(h)**.

3.2.3. Retardation of Plastic Instability

Let us next focus on the global localization behavior of these three microstructures. Comparing the necking behavior of μ_{CG} with μ_{HM}, a large reduction in global localization strain is observed, which can be related to the increase in martensite volume fraction. Global localization involves connection of the above-mentioned diffuse deformation zones into a global strain percolation path, which, for DP steel, will obviously run through the available ferrite grains. For μ_{HM}, less strain percolation paths form, and hence each percolation path must strain more to accommodate the same applied global strain, therefore earlier reaching the point of global localization. This reduction of the number of percolation paths is clearly seen in **Figure 9(i)**, which only shows one pronounced percolation path (running from upper left to lower right corner).

Figure 9 also shows that necking takes place at higher equivalent strain for μ_{FG} compared to μ_{CG}. Because the martensite volume fraction is the same for μ_{FG} and μ_{CG}, another mechanism must be at play, which may be explained as follows. Global localization is controlled by the weakest percolation path and, for μ_{CG} compared to μ_{FG}, less diffuse deformation zones need to be connected to complete a percolation path over the full sample thickness or width. Therefore, taking into account the large spread of grain properties and geometries, the percolation paths in μ_{CG} will exhibit a larger variability. As a result, the strength of the critical (weakest) percolation path will be smaller in μ_{CG}, which explains its lower global localization strain. The same mechanism was found to control the necking behavior observed in tensile tests of aluminum strips with very few grains over the specimen width[48], for which in-situ DIC strain maps showed direct evidence that weaker localized percolation paths develop when the grain size is increased, triggering earlier global localization. For our case, this possible explanation would indeed be supported by the strain fields in **Figure 9(g)**, **Figure 9(h)**, which shows that the number of percolation paths is higher in the μ_{FG} microstructure.

Finally, when the case of μ_{FG} is directly compared to that of μ_{HM}, it is interesting to note that the damage evolution at small strains looks quite similar, see **Figure 9(a)**, **Figure 9(c)**. However, there is a major difference, which exhibits itself in the observation of a higher flow stress as well as a higher fracture strain, see **Figure 1(d)**. Of course, the above-mentioned Hall-Petch effect could explain the increase in flow stress; however, there exists a well-known competition between high strength versus high elongation. Therefore, to explain the observed increase in fracture strain for μ_{FG} compared to μ_{HM} another mechanism is required. As was seen above, for μ_{HM}, the high hydrostatic stresses are a direct result from the limitation in the number of strain percolation paths, which also localizes the damage evolution causing earlier global localization and final fracture [**Figure 9(c)**]. For lFG, on the other hand, the damage is more dispersed due to its finer microstructure and more ferrite grains, which activates the necking retardation mechanism in which damage initiation triggers (many) diffuse deformations zones, as was seen in **Figure 7**, thereby spreading out plastic straining and thus postponing global localization. Hence, for μ_{FG}, the high hydrostatic stress does not seem to be detrimental, but actually beneficial as it increases the global flow strength compared to μ_{HM} [shown in **Figure 1(d)**] for the global stress-strain curves under uniaxial tension). This would mean that the well-known competition between high strength versus high elongation can be overcome by inserting many barriers in the microstructure that increase the hydrostatic stresses. It is crucial, however, that these barriers break open easily enough (as is the case in μ_{FG} and not in μ_{HM}) such that plasticity spreads out subsequently to the surrounding matrix in order to prevent early necking.

3.2.4. Microstructure Design

The role of the damage mechanisms in the localization and fracture behavior is critical. Without damage mechanisms, there is no stress release by diverging localized plasticity to non-local (diffuse) plasticity, thus the stress keeps on building up, leading to early necking. Of course, stress release can only activate a necking retardation mechanism when damage sets in before strain percolation paths have formed. In turn, early damage formation requires high hydrostatic stress built up at early stages of deformation, which can be achieved by microstructural refinement due to the grain size effect, while it also strongly depends on the loading conditions. For instance, for BAT, much higher hydrostatic stresses build up compared to UAT and PST, see **Figure 6**, which may explain the unusually high BAT necking strain (**Figure 5**) compared to typical forming limit diagrams which show the highest necking strain for UAT.

Based on these insights, it is anticipated that the ideal microstructure combining high strength with high ductility can be achieved through microstructural refinement, e.g., by careful design of a nano-grained DP. The hard phase (e.g., martensite) should be tailored to surround the softer grains with an approximately uniform layer that is strong enough to drive up the stress, but with enough weak spots that can lead to damage relatively easily, resulting in a high dispersion of damage locations, each activating a diffuse deformation zone, and thereby effectively retarding global localization. This mechanism may be the underlying reason for the recent success of nano-grain dual-phase steels[9]. The diffuse deformation and resulting strain hardening in the ferrite grains adjacent to the voids may also explain earlier observations that for DP steels the classical mechanism of ductile failure through void initiation, growth, and coalescence only becomes relevant close to the moment of final failure, *i.e.*, after global localization has set in[46].

4. General Conclusions

An extensive experimental-numerical campaign was set up to characterize, in a statistically relevant manner, the evolution of the key ductile damage mechanisms up to failure, for three strain paths and three well-controlled dual-phase microstructures. From the in-depth analysis, the following main conclusions can be drawn:

- A chain of damage events was hypothesized, in which plastic straining in ferrite grains triggers fracture of martensite islands and subsequently damage in neighboring ferrite, causing diffuse straining in a larger ferrite area. This hypothesis is supported by various direct and indirect evidence.

- An interesting necking retardation mechanism was elucidated, in which the diffuse straining in combination with strain hardening may postpone the formation of a global strain percolation path. This mechanism is enhanced for finer microstructures, in which the damage initiation sites as well as the resulting diffuse deformations zones are more dispersed.

Based on these new insights, a route to circumvent the well-known competition between high strength versus high elongation was proposed by exploiting the concept of microstructural refinement to greater depths.

Acknowledgements

The authors gratefully acknowledge the contributions of Marc van Maris,

Patrick Schoenmakers, Chaowei Du, Dingshun Yan, Carel ten Horn, and Henk Vegter.

Open Access: This article is distributed under the terms of the Creative Commons Attribution 4.0 International License (http://cre-ativecommons.org/licenses/by/4.0/), which permits unrestricted use, distribution, and reproduction in any medium, provided you give appropriate credit to the original author(s) and the source, provide a link to the Creative Commons license, and indicate if changes were made.

Source: Hoefnagels J P M, Tasan C C, Maresca F, *et al*. Retardation of plastic instability via damage-enabled microstrain delocalization[J]. Journal of Materials Science, 2015, 50(21):6882−6897.

References

[1] Grässel O, Kruger L, Frommeyer G, Meyer LW (2000) High strength Fe-Mn-(Al, Si) TRIP/TWIP steels development—properties—application. Int J Plast 16:1391–1409. doi:10.1016/ S0749-6419(00)00015-2.

[2] Barbier D, Gey N, Allain S, Bozzolo N, Humbert M (2009) Analysis of the tensile behavior of a TRIP steel based on the texture and microstructure evolutions. Mat Sci Eng A 500:196–206. doi:10.1016/j.msea.2008.09.031.

[3] Speer JG, Assuncao FCR, Matlock DK, Edmonds DV (2005) The "quenching and partitioning" process: background and recent progress. Mater Res 8:417–423. doi:10.1590/S1516-14392005000400010.

[4] Edmonds DV, He K, Rizzo FC, De Cooman BC, Matlock DK, Speer JG (2006) Quenching and partitioning martensite—a novel steel heat treatment. Mat Sci Eng A 440:25–34. doi:10.1016/j. msea.2006.02.133.

[5] Caballero FG, Bhadeshia HKDH (2004) Very strong bainite. Curr Opin Solid State Mater 8:251–257. doi:10.1016/j.cossms.2004. 09.005.

[6] Bhadeshia HKDH (2010) Nanostructured bainite. Proc R Soc A 466:3–18. doi:10.1098/rspa.2009.0407.

[7] Rashid MS (1981) Dual phase steels. Ann Rev Mater Sci 11:245–266. doi:10.1146/annurev.ms.11.080181.001333.

[8] Llewellyn DT, Hillis DJ (1996) Dual phase steels. IronmakSteelmak 23:278–471.

[9] Calcagnotto M, Ponge D, Raabe D (2010) Effect of grain refinement to 1µm on strength and toughness of dual-phase steels. Mat Sci Eng A 527:7832–7840.

doi:10.1016/j.msea.2010. 08.062.

[10] Calcagnotto M, Ponge D, Demir E, Raabe D (2010) Orientation gradients and geometrically necessary dislocations in ultrafine grained dual-phase steels studied by 2D and 3D EBSD. Mater Sci Eng A 527:2738–2746. doi:10.1016/j.msea.2010.01.004.

[11] Asadi M, De Cooman BC, Palkowski H (2012) Influence of martensite volume fraction and cooling rate on the properties of thermomechanically processed dual phase steel. Mater Sci Eng A 538:42–52. doi:10.1016/j.msea.2012.01.010.

[12] Son YI, Lee YK, Park K-T, Lee CS, Shin DH (2005) Ultrafine grained ferrite-martensite dual phase steels fabricated via equal channel angular pressing: microstructure and tensile properties. Acta Mater 53:3125–3134. doi:10.1016/j.actamat.2005.02.015.

[13] Delince' M, Jacques PJ, Pardoen T (2006) Separation of size-dependent strengthening contributions in fine-grained dual phase steels by nanoindentation. Acta Mater 54:339503404. doi:10. 1016/j.actamat.2006.03.031.

[14] Ahmad E, Manzoor T, Ali KL, Akhter JI (2000) Effect of microvoid formation on the tensile properties of dual-phase steel. J Mater Eng Perform 9:306–310. doi:10.1361/105994900770345962.

[15] Bouaziz O, Iung T, Kandel M, Lecomte C (2001) Physical modelling of microstructure and mechanical properties of dual-phase steel. J Phys IV France 11:223–231. doi:10.1051/jp4: 2001428.

[16] Rodriguez R-M, Gutie'rrez I (2003) Unified formulation to predict the tensile curves of steels with different microstructures. Mater Sci Forum 2003:426–432.

[17] Ramazani A, Mukherjee K, Schwedt A, Goravanchi P, Prahl U, Bleck W (2013) Quantification of the effect of transformation-induced geometrically necessary dislocations on the flow-curve modeling of dual-phase steels. Int J Plast 43:128–152. doi:10. 1016/j.ijplas.2012.11.003.

[18] Das D, Chattopadhyay PP (2009) Influence of martensite morphology on the work-hardening behavior of high strength ferrite– martensite dual-phase steel. J Mater Sci 44:2957–2965. doi:10. 1007/s10853-009-3392-0.

[19] Avramovic-Cingara G, Ososkov Y, Jain MK, Wilkinson DS (2009) Effect of martensite distribution on damage behaviour in DP600 dual phase steels. Mater Sci Eng A 516:7–16. doi:10. 1016/j.msea.2009.03.055.

[20] Sun X, Choi KS, Liu WN, Khaleel MA (2009) Predicting failure modes and ductility of dual phase steels using plastic strain localization. Int J Plast 25:1888–1909. doi:10.1016/j.ijplas.2008. 12.012.

[21] Davies RG (1978) Influence of martensite composition and content on the properties of dual phase steels. Metall Trans A 9:671–679. doi:10.1007/BF02659924.

[22] Marder AR (1982) Deformation characteristics of dual-phase steels. Metall Trans A 13:85–92. doi:10.1007/BF02642418.

[23] He XJ, Terao N, Berghezan A (1984) Influence of martensite morphology and its dispersion on mechanical properties and fracture mechanisms of Fe-Mn-C dual phase steels. Met Sci 18:367–373. doi:10.1179/030634584790419953.

[24] Ray RK (1984) Tensile fracture of a dual-phase steel. Scr Metall Mater 18:1205–1209. doi:10.1016/0036-9748(84)90106-6.

[25] Suh D, Kwon D, Lee S, Kim NJ (1997) Orientation dependence of microfracture behavior in a dual-phase high-strength low-alloy steel. Metall Mater Trans A 28:504–509. doi:10.1007/s11661-997-0152-0.

[26] Maire E, Bouaziz O, Di Michiel M, Verdu C (2008) Initiation and growth of damage in a dual-phase steel observed by X-ray microtomography. Acta Mater 56:4954–4964. doi:10.1016/j.acta mat.2008.06.015.

[27] Avramovic-Cingara G, Saleh C, Jain MK, Wilkinson DS (2009) Void nucleation and growth in dual-phase steel 600 during uniaxial tensile testing. Metall Mater Trans A 40:3117–3127. doi:10.1007/s11661-009-0030-z.

[28] Sun X, Choi KS, Soulami A, Liu WN, Khaleel MA (2009) On key factors influencing ductile fractures of dual phase (DP) steels. Mater Sci Eng A 526:140–149. doi:10.1016/j.msea.2009.08.010.

[29] Tasan CC, Hoefnagels JPM, Geers MGD (2009) A critical assessment of indentation-based ductile damage quantification. Acta Mater 57:4957–4966. doi:10.1016/j.actamat.2009.06.057.

[30] Tasan CC, Hoefnagels JPM, Geers MGD (2010) Microstructural banding effects clarified through micrographic digital image correlation. Scr Mater 62:835–838. doi:10.1016/j.scriptamat. 2010.02.014.

[31] Tasan CC, Hoefnagels JPM, Geers MGD (2012) Identification of the continuum damage parameter: an experimental challenge in modeling damage evolution. Acta Mater 60:3581–3589. doi:10. 1016/j.actamat.2012.03.017.

[32] Kadkhodapour J, Butz A, Ziaei-Rad S (2011) Mechanisms of void formation during tensile testing in a commercial, dual-phase steel. Acta Mater 59:2575–2588. doi:10.1016/j.actamat.2010.12. 039.

[33] Al-Abbasi FM, Nemes JA (2008) Predicting the ductile failure of DP-steels using micromechanical modeling of cells. Int J Damage Mech 17:447–472. doi:10.1177/1056789507077441.

[34] Uthaisangsuk V, Prahl U, Bleck W (2008) Micromechanical modelling of damage behaviour of multiphase steels. Comput Mater Sci 43:27–35. doi:10.1016/j.commatsci.2007.07.035.

[35] Marder AR (1981) The effect of heat treatment on the properties and structure of molybdenum and vanadium dual-phase steels. Metall Trans A 12:1569–1579. doi:10.1007/BF02643562.

[36] Goel NC, Chakravarty JP, Tangri K (1985) The influence of starting microstructure

on the retention and mechanical stability of austenite in an intercritically annealed-low alloy dual-phase steel. Metall Trans A 18:5–9. doi:10.1007/BF02646215.

[37] Huppi GS, Matlock DK, Krauss G (1980) An evaluation of the importance of epitaxial ferrite in dual-phase steel microstructures. Scr Metall Mater 14:1239–1243. doi:10.1016/0036-9748(80) 90264-1.

[38] Ramos LF, Matlock DK, Krauss G (1979) On the deformation behavior of dual-phase steels. Metall Trans A10:259–261. doi:10. 1007/BF02817636.

[39] Cai X-L, Feng J, Owen WS (1984) The dependence of some tensile and fatigue properties of a dual-phase steel on its microstructure. Metall Mater Trans A 16:1405–1415. doi:10. 1007/BF02658673.

[40] Sarwar M, Manzoor T, Ahmad E, Hussain N (2007) The role of connectivity of martensite on the tensile properties of a low alloy steel. Mater Des 28:1928–1933. doi:10.1016/j.matdes.2006.05. 010.

[41] Sarwar M, Ahmad E, Qureshi KA, Manzoor T (2007) Influence of epitaxial ferrite on tensile properties of dual phase steel. Mater Des 28:335–340. doi:10.1016/j.matdes.2005.05.019.

[42] Azizi-Alizamini H, Militzer M, Poole WJ (2007) A novel technique for developing bimodal grain size distributions in low carbon steels. Scr Mater 57:1065–1068. doi:10.1016/j.scriptamat. 2007.08.035.

[43] Tasan CC, Hoefnagels JPM, Dekkers ECA, Geers MGD (2012) Multi-axial deformation setup for microscopic testing of sheet metal to fracture. Exp Mech 52:669–678. doi:10.1007/s11340-011-9532-x.

[44] Furnémont Q (2003) The micromechanics of TRIP-assisted multi-phase steels, Ph.D. thesis, Universite´ Catholique de Louvain.

[45] Lemaitre J, Dufailly J (1987) Damage measurements. J. Eng Fract Mech 28:643–661. doi:10.1016/0013-7944(87)90059-2.

[46] Tasan CC, Hoefnagels JPM, ten Horn CHLJ, Geers MGD (2009). Experimental analysis of strain path dependent ductile damage mechanics and forming limits. Mech Mater 41:1264–1276. doi:10.1016/j.mechmat.2009.08.003.

[47] Ghadbeigi H, Pinna C, Celotto S, Yates JR (2010) Local plastic strain evolution in a high strength dual-phase steel. Mat Sci Eng A 527:5026–5032. doi:10.1016/j.msea.2010.04.052.

[48] Hoefnagels JPM, Janssen PJM, de Keijser TH, Geers MGD (2008) First-order size effects in the mechanics of miniaturized components. In: Dilieu-Barton JM, Lord JD, Greene RJ (eds) Applied mechanics and materials: advanced in experimental mechanics VI, Vol. 13–14. Trans Tech Publications, Switzerland, pp 183–192.

Chapter 2

Adsorption Characteristics of Noble Metals on the Strongly Basic Anion Exchanger Purolite A-400TL

A. Wołowicz[1], Z. Hubicki[2]

[1]Department of Inorganic Chemistry, Faculty of Chemistry, Maria Curie-Skłodowska University, Maria Curie-Skłodowska Square 2, 20-031 Lublin, Poland
[2]Fertilizer Research Institute, 24-100 Puławy, Poland

Abstract: Ion exchange is an alternative process for uptake of noble metals from aqueous solutions. In the present study, the sorption of Pd(II), Pt(IV), and Au(III) ions from aqueous solution was investigated by using Purolite A-400TL (strongly basic anion exchanger, gel, type I) in a batch adsorption system as a function of time (1min−4h). Initial Pd(II) concentration (100mg/L−1000mg/L), beads size (0.425−0.85mm), rate of phases mixing (0−180rpm), and temperatures (ambient, 313K) were taken into account during the Pd(II) sorption process. Moreover, the column flow adsorption study was carried out, and the breakthrough curves were obtained for Pd(II) ions. The equilibrium, kinetic, desorption, and ion-exchange resin reuse studies were carried out. The experimental results showed that Purolite A-400TL—the strongly basic anion-exchange resin could be used effectively for the removal of noble metal ions from the aqueous medium. The kinetics of sorption process is fast and the resin could be reused without reduction of capacity (three cycles of sorption-desorption, the reduction of capacity is smaller than 1%). The column studies indicated that in the dilute acidic solution (0.1M HCl) the working anion exchange capacity is high (0.0685mg/cm^3) in comparison with the other SBA resins examined under the same experimental conditions, e.g., Amberlite IRA-458 (0.0510mg/cm^3), Amberlyst A-29 (0.0490mg/cm^3), Dowex MSA-1 (0.0616mg/cm^3), Dowex MSA-2 (0.0563mg/cm^3), Varion ADM (0.0480mg/cm^3), and Varion ATM (0.0490mg/cm^3) etc. The highest % of Pd(II) desorption was obtained using thiourea, acidic thiourea, sodium hydroxide, and ammonium hydroxide as eluting agents (%D1 was in the range of 23.9−46.9mg/g).

1. Introduction

Ion-exchange resins have played a very significant role in many branches of industry. As was reported by "Global Industry Analysts, Inc." in 2010, the global market for ionexchange resins is projected to exceed $535 million by the year 2015. This tendency results from the growing demand for pure water and its lack in the world, increasing population growth, urbanization, industrialization, and pollution, etc.[1]. At present, the main ion-exchange resins manufacturing companies such as Dow Corporation, Rohm and Hass Company, Purolite Corporation, Lanxess, etc., produce a wide selection of ion-exchange resin types[2]. Among the ion-exchange resins available on market strongly basic anion-exchange resins play a significant role. Many examples of the commercial ion-exchange resins of different types for recovery of precious metals from solutions of different composition can be found in[1]–[6].

Among the recovery methods of valuable metal ions, the hydrometallurgical method is one of the most effective, whereas ion-exchange methods (sorption on ion-exchange resins) are the only economical methods for removal of gold and Platinum Group Metals from the diluted solution obtained after leaching the scrap materials containing the above mentioned metals. The advantage of the hydrometallurgical processes using the ion-exchange resins and ionexchange method applied for recovery of noble metal ions is their possible recovery even from solutions in which noble metal ions are in trace concentration. There are no other methods which can be so effective as ion exchange used for such application. Based on the above facts, the ion-exchange method was applied for noble metal ions recovery from the diluted solutions using Purolite A-400TL. Moreover, a lack of consistent knowledge related to the behavior of noble and base metals on Purolite A-400TL has lead us to cope with the resin for more complete understanding of their sorption and desorption properties. The SBA resin of similar matrix (PS-DVB) and structure-type (gel) Lewatite MonoPlus M-600 but type 2 was applied previously by us in Pd(II) recovery both in the batch and column studies[7].

This article reported the efficiency of the commercially available strongly basic anion-exchange resin of type 1[8] Purolite A-400 TL for noble [Pd(II), Pt(IV), Au(III)] metal ions removal from acidic solutions of different composition (HCl; HCl–HNO$_3$). Various parameters such as effect of phases contact time, initial metal ions concentration, agitation speed, temperature, and beads size distribution were considered in the batch mode to optimize conditions for the effective removal of these metal ions. Additionally, to characterize the loading processes of Pd(II) onto the SBA resin, a column system was applied. Equilibrium, kinetic, desorption, and SBA resin reuse studies were also carried out.

2. Experimental

2.1. Reagents and Solutions

The single stock solutions containing metal ions such as Pd(II), Pt(IV), and Au(III) were prepared from solids: PdCl$_2$ or liquid: H$_2$PtCl$_6$, HAuCl$_4$ in 0.1M HCl. In the case of palladium stock solution, a weighed amount of salt was dissolved in 1.0M HCl solutions (temperature: 333K, microwaves (Inter Sonic, type IS-1 with a thermoregulator), digestion time = 1h). The concentration and composition of acidic solutions containing selected metal ions were following: 0.1–6.0M HCl—100mg/L M(II), M(III) or M(IV) and 0.1–0.9M HCl—0.9–0.1M HNO$_3$—100mg/L M(II), M(III) or M(IV).

All the reagents were of analytical purification grade (POCh, Poland).

2.2. Purolite A-400TL Characteristics

Purolite A-400TL is a strongly basic polystyrene-divinylbenzene anion-exchange resin of –N$^+$(CH$_3$)$_3$ (type 1) functional groups. It has a gel structure; the total exchange capacity is 1.3eq/L (in the Cl$^-$ form). Harmonic mean sizes are equal to 0.425–0.85mm (uniform coefficient max. 1.3). The maximum operating temperature is 373K (in the Cl⁻ form) and 333K (in the OH⁻ form). The pH limits are following: 0–14 (stability) and 1–10 (operating; the OH⁻ form). Moisture retentions and reversible swelling are equal to 48%–54% and 20% (Cl⁻ to OH⁻), respectively. Purolite A-400TL has excellent physical stability which permits a long life without the development of excessive pressure drop, even when operating at high flow rates. It also shows good kinetics of exchange. Purolite A-400TL is advisable for the demineralization of water and silica removal.

From the physical structure point of view, gel-type ionexchange resin beads are homogeneous, whereas in the case of macroporous (macroreticular) one of the resin beads are heterogeneous and consist of interconnected macrospores surrounded by gel-type microbeads agglomerated together. The macrospores have sizes ranging from several angströms up to many hundreds of angström, while microbeads give the resin a large internal surface area which depends on the size of these microbeads.

2.3. Methods and Measurements

The sorption, equilibrium, kinetic, desorption, and reuse studies were carried out by means of the batch method using a thermostated shaker (Elpin⁺, 358S, Lubawa, Poland). Sorption studies: The sorption studies were conducted in Erlenmeyer 100mL flasks by adding 0.5 (±0.0005)g of Purolite A-400TL to 50mL of metal solutions of 100mg/L concentration [0.1–6.0M HCl—100mg/L M(II), M(III) or M(IV) and 0.1–0.9M HCl—0.9–0.1M HNO₃—100mg/L. M(II), M(III) or M(IV)]. After shaking the flasks at 180rpm and ambient temperature and time interval from 1min to 4 h, the SBA resin was separated from aqueous solution by filtration. The final concentration of metal ions in the solution after sorption processes was determined using The Fast Sequential Atomic Absorption Spectrometer, VarianAA240FS, equipped with the appropriate hollow cathode lamps and SIPS autosampler and then calculated using the following equation:

$$q_e = (C_o - C_e)V/W \tag{1}$$

where C_o and C_e are the initial and equilibrium concentrations of M(II), M(III), or M(IV), respectively (in mg/L); V (in L) is the volume of M(II), M(III), or M(IV) contacting solution; W (in g) is the SBA resin mass.

2.3.1. Kinetic Studies

Kinetic experiments were identical to those of the sorption test, but some experimental conditions were changed to examine the influences of Pd(II) initial metal concentration, temperature, beads size distribution, and agitation speed on the sorption processes. The experimental conditions were following (mass of the ion exchanger, m_j; volume of the solution, V; initial Pd(II) concentration, C_o; amplitude, A; agitation speed, V_{as}; temperature T; phases contact time, t; beads size distribution, bs):

- m_j = 0.5g, V = 50mL, C_o = 100, 500, 1000mg Pd(II)/L, A = 8, V_{as} = 180rpm, T—ambient, t = 1min to 4 h, bs—0.425mm–0.85mm—effect of the initial Pd(II) concentration,

- m_j = 0.5g, V = 50mL, C_o = 500mg Pd(II)/L, A = 8, V_{as} = 120, 150, 180rpm, T—ambient, t = 1min to 4h, bs—0.425mm–0.85mm—effect of agitation speed,

- m_j = 0.5g, V = 50mL, C_o = 500mg Pd(II)/L, A = 8, V_{as} = 180rpm, T—ambient, 313, 333K, t = 1min to 4h, bs—0.425mm–0.85mm—effect of temperature,

- m_j = 0.5g, V = 50mL, C_o = 500mg Pd(II)/L, A = 8, V_{as} = 180rpm, T—ambient, t = 1min to 4, beads size— f_1, f_2, f_3, f_4 (0.85 > f_1 ≥ 0.6mm; 0.6 > f_2 ≥ 0.5mm; 0.5 > f_3 ≥ 0.43mm, 0.43 > f_4 ≥ 0.425mm)—effect of beads size distribution.

The kinetic studies were carried out from the solutions containing Pd(II) ions (single solutions) of the following composition: 0.1M HCl—100, 500 or 1000mg Pd(II)/L and 0.1M HCl—0.9 M HNO_3—100, 500 or 1000mg Pd(II)/L (effect of initial Pd(II) concentration) and 0.1M HCl—500mg Pd(II)/L and 0.1M

HCl—0.9 M HNO$_3$—500mg Pd(II)/L (effect of agitation speed, effect of temperature, and effect of beads size distribution).

The saturation degree (or fractional attainment of equilibrium) (F) is calculated from Equation (2):

$$F = q_t / q_e \qquad (2)$$

where q_t and q_e are the amount of Pd(II) sorbed (in mg/g) at time t and equilibrium, respectively. Based on the kinetic curves (plot of F vs t) the half-exchange times, $t_{1/2}$ (in s), were determined at F = 0.5[9].

2.3.2. Equilibrium Studies

The equilibrium studies were carried out under the experimental conditions: m_j = 0.5g, V = 50mL, C_o = 100–6000mg/L in 0.1M HCl, A = 8, V_{as} = 180rpm, T—ambient, t = 24h, bs—0.425–0.85mm, and the procedure was identical to those applied during the sorption studies.

2.3.3. Desorption Studies

Different eluting agents were prepared by dilution of concentrated hydrochloric acid (0.1–6.0M HCl), nitric acid (0.1–4.0 M HNO$_3$), sulfuric acid (0.5–4.0M H$_2$SO$_4$), ammonia (0.5–2.0M NH$_4$OH), sodium hydroxide (0.1–3.0M NaOH), 1.0 M TU (thiourea), 1.0M TU—1.0M HCl; 1.0M TU—1.0M HNO$_3$ in order to elute the retained noble metal ions from Purolite A-400TL. The desorption experimental conditions were: m_j = 0.5g, V = 50mL, A = 8, V_{as} = 180rpm, T—ambient, t = 2h and changeable concentration and types of eluting agents. The sorption-desorption cycle was repeated three times to obtain information about Purolite A-400TL reuse possibility.

2.3.4. Breakthrough Capacities

In characterization of SBA resin applicability for metal sorption purposes, loading capacity is one of the main parameters commonly used. It is defined as a number of metal ion equivalent per mass of resin that can be removed from the solution containing the ion (in equilibrium state)[2][7].

The dynamic procedures were applied to obtain the breakthrough curves and capacity. The one-centimeter diameter columns were filled with swollen Puolite A-400TL in the amount of 10mL. Then, the solution of 100mg/L was passed through the anion-exchange resin bed at the rate of 0.4mL/min. The eluate was collected in the fractions, and the metal concentrations were determined.

2.3.5. Analytical Procedure

The concentration in the solution after the sorption, desorption, kinetic, equilibrium, and reuse studies was obtained by the AAS method. Standard solutions were prepared by dilution of the standard stock solutions (1000mg/dm^3 in 0.1M HNO$_3$) with acids and distilled water. The concentration of the standard solutions was changeable depending on the metal ions concentration determination. An oxidizing air-acetylene flame was used for atomization. The other parameters were following: lamp current 10mA—Pd(II); 4.0mA—Au(III); 7mA—Pt(IV), slight width 0.2nm—Pd(II), Pt(IV); 1.0nm—Au(III), acetylene flow 2dm^3/min, air flow 13.5dm^3/min, and the analytical wavelength 247.6nm—Pd(II); 242.8nm—Au(III); 265.9nm—Pt(IV).

3. Results and Discussion

3.1. Sorption Capacity—Effect of Experimental Conditions

The complexation chemistry and ionic state of Platinum Group Metals and gold in the chloride solutions with varying chloride ion concentration have been described by other researchers[2][10][11]. In the chloride solutions the palladium(II), platinum(IV), and gold(III) metal ions can exist in different forms of their complexes. Depending on the solution pH and total concentration of chloride anions, these metals form cationic, non-anionic, and anionic complexes. More information about the Pd(II), Pt(IV), and Au(III) complexes species can be found in[2][12][13].

In our studies [metal ions sorption from the chloride solutions: 0.1–6.0M HCl—100mg/L M(II), M(III), or M(IV)] Pd(II), Pt(IV), and Au(III) metal ions exist in the forms of anionic complexes of different ionic structure and properties (see **Table 1**)[10][11]. The effect of acids concentrations (chloride concentrations) and phases contact time on the metal ions sorption on Purolite A-400TL is presented in **Figures 1(a)-(c)** for the chloride and **Figure 1(d)** chloride-nitrate solutions (in this case only the equilibrium sorption capacities for Pd(II), Pt(IV), and Au(III) were compared). In the dilute chloride solutions such as 0.1M HCl, the equilibrium sorption capacities achieved the highest possible values for Pd(II) and Pt(IV) (removal is quantitative) 10mg/g and 9.99mg/g for Au(III). With the hydrochloric acid concentration increase, the equilibrium sorption capacities decrease. The reduction of sorption capacity is the highest for Pd(II) and is equal to 29%. The sorption capacities drop from 10mg/g (0.1M HCl) to 7.10mg/g (6.0M HCl); from 10mg/g (0.1M HCl) to 9.31mg/g (6.0M HCl), and from 9.99mg/g (0.1M HCl) to 9.72mg/g (6.0M HCl) for Pd(II), Pt(IV), and Au(III), respectively. These sorption capacities changes indicate only 6.9% and 2.7% reduction of their values for Pt(IV) and Au(III). Phases contact time also influences on the sorption capacities. The amount of the metal ions uptake increased with the increasing phases contact time. At the beginning of the sorption process (for short phases contact time), the q_t values increase is high and they decrease with the phases contact time increase. When the system reached equilibrium, further increase of the

Table 1. Noble metals oxidation and chloro-complexes[10][11].

Metal	Electron configuration	LK	Ionic structure		Complexes formed low—[Cl⁻]—high		Redox stability	Kinetic stability	Thermal stability
Pd(II)	d^8	4	Square planar		$PdCl_4^{2-}$	$PdCl_4^{2-}$	Stable	Very stable	–
Pd(IV)	d^6	6	Octahedral		$PdCl_6^{2-}$	$PdCl_6^{2-}$	Unstable	Stable	Unstable
				$PdCl_4^{2-}$					
Pt(II)	d^8	4	Square planar		$PtCl_4^{2-}$	$PtCl_4^{2-}$	Unstable	Unstable	–
Pt(IV)	d^6	6	Octahedral		$PtCl_6^{2-}$	$PtCl_6^{2-}$	Stable	Very stable	Very stable
				$PtCl_6^{2-}$					
Au(III)	d^8	4	Square planar		$AuCl_4^-$	$AuCl_4^-$	–	–	–
				$AuCl_4^-$					

LK coordination number, – data not available

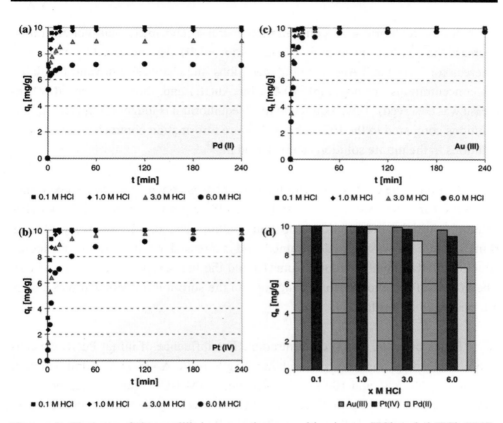

Figure 1. Changes of the equilibrium sorption capacities in a-c HCl and d HCl−HNO$_3$ systems obtained during the a, d Pd(II), b, d Pt(IV), c, d Au(III) metal ions sorption on Purolite A-400TL (experimental conditions: m_j = 0.5g, V = 50mL, C_o = 100mg/L in 0.1M HCl, A = 8, V_{as} = 180rpm, T—ambient, t = 1min to 4h, bs—0.425−0.85mm).

phases contact time does not result in the increase of q_t values because they remain unchanged. The rapid sorption observed during the first 30min is probably due to the abundant availability of active sites on the anion exchanger surface (the solute concentration gradient was relatively high) and with the gradual occupancy of these sites, the sorption becomes less efficient. With the hydrochloric acid concentrations increase, the time necessary to attain equilibrium is longer. In 0.1M HCl solution quantitative removal of Pd(II) is obtained after only 15min, whereas for Pt(IV) and Au(III) longer time is needed. Compared to other studies this time is very short and similar to that obtained for the Dowex MSA-1[14], Lewatite TP-220[12], and Purolite A-830[15] ion exchange resins.

For the chloride-nitrate(V) solutions, the changes of q_t values are different.

With the hydrochloric acid concentration increase and nitric acid concentration decrease the equilibrium sorption capacities for Pd(II) also increased by about 8%, whereas for Pt(IV) and Au(III) noble metal ions, the changes of q_t values with acids concentrations are negligible (0.4% for Au(III) and about 1% for Pt(IV) increase was observed). Time required to reach equilibrium is longer for the HCl-HNO$_3$ solutions compared to the HCl solutions. More information about changes of Pd(II) solutions in the nitrate solution can be found in[16].

Not only the acids concentrations and phases contact time were taken into account during the recovery process but also the agitating speed, beads size distribution, initial concentration, and temperature. The rate of recovery process (kinetic studies) was also determined and expressed by the saturation degree (fractional attainment of equilibrium, F) and the half-exchange times, $t_{1/2}$ (in s). The effects of the experimental condition on the sorption process and its kinetics are presented in **Table 2**.

Effect of initial Pd(II) concentration: The influence of initial Pd(II) concentration on loading was studied by adding 0.5g Purolite A-400TL to 50mL solution containing 100, 500, and 1000mg/L Pd(II) and 0.1M HCl at ambient temperature, with the results showing that the sorption rate was very fast at the initial stage: after 1min of sorption 71%, 81%, and 91% of Pd(II) were removed from the solutions containing 100, 500, and 1000mg/L Pd(II), respectively. F reached very high values after 1min (higher than 0.5), therefore, $t_{1/2}$ cannot be determined with high quality (for F = 0.5). Half-exchange time is smaller than 60 s for all presented cases. After 10min (100mg Pd(II)/L) and 15min (500 and 1000mg Pd(II)/L), F reached the constant values. Similar observation was also made for the HCl-HNO$_3$ solutions (F reached 0.55−0.76 values after 1min). As expected, increasing the initial concentration caused an increase in equilibrium sorption capacity. As was pointed out previously[16][17], the rate of ion exchange is affected by the initial metal concentration when the controlling step is film diffusion, whereas the system is governed by intraparticle diffusion, the sorption rate is not influenced by metal concentration. In our cases at the beginning of the sorption process, an insignificant effect of metal ion concentrations on the Pd(II) sorption is observed. For the solutions of 500 and 1000mg Pd(II)/L concentration, the metal recovery is not quantitative, the sorption capacities are 49.97mg/g (max. 50mg/L) and 94.95mg/g (max. 100mg/L). This is explained by the fact that as the concentration of

Table 2. Comparison of F values depending of the experimental conditions applied.

Time	F						Plot q_t versus t	Experimental conditions
	Effect of initial Pd(II) concentration (mg/L)							
	HCl system			HCl–HNO$_3$ system			$t_{1/2}$ (s)	
t (min)	100	500	1000	100	500	1000		
1	0.71	0.81	0.91	0.68	0.55	0.76	<60 s	$m_j = 0.5$ g,
3	0.90	0.93	0.97	0.85	0.77	0.78		$V = 50$ mL,
5	0.95	0.97	0.99	0.93	0.88	0.85		$C_o = 100, 500, 1000$ mg/L in 0.1 M HCl,
10	1.00	0.99	0.99	0.98	0.99	0.88		$A = 8$, $V_{as} = 180$ rpm,
15	1.00	1.00	1.00	0.99	0.99	1.00		T—ambient,
30	1.00	1.00	0.99	1.00	1.00	1.00		$t = 1$ min to 4 h,
60	1.00	1.00	1.00	1.00	1.00	1.00		bs—0.425–0.85 mm
120	1.00	1.00	1.00	1.00	1.00	1.00		
180	1.00	1.00	0.99	1.00	1.00	1.00		
240	1.00	1.00	1.00	1.00	1.00	1.00		

Time	Effect of agitation speed (rpm)						Plot q_t versus t	Experimental conditions
	HCl system			HCl–HNO$_3$ system			$t_{1/2}$ (s)	
t (min)	120	150	180	120	150	180		
1	0.72	0.77	0.81	0.71	0.56	0.55	<60 s	$m_j = 0.5$ g,
3	0.73	0.87	0.93	0.76	0.64	0.77		$V = 50$ cm^3, $C_o = 500$ mg/L,
5	0.76	0.92	0.97	0.72	0.78	0.88		$A = 8$,
10	0.80	0.98	0.99	0.82	0.95	0.99		$V_{as} = 120, 150, 180$ rpm, T—ambient,
15	0.80	1.00	1.00	0.83	0.95	0.99		$t = 1$ min to 4 h,
30	0.89	1.00	1.00	0.94	1.00	1.00		bs—0.425–0.85 mm
60	0.99	1.00	1.00	0.98	1.00	1.00		
120	1.00	1.00	1.00	1.00	1.00	1.00		
180	1.00	1.00	1.00	1.00	1.00	1.00		
240	1.00	1.00	1.00	1.00	1.00	1.00		

Time	Effect of temperature (K)						Plot q_t versus t	Experimental conditions
	HCl system			HCl–HNO$_3$ system			$t_{1/2}$ (s)	
t (min)	298	313	333	298	313	333		
1	0.81	0.86	*	0.55	0.83	*	<60 s	$m_j = 0.5$ g,
3	0.93	0.98		0.77	0.94			$V = 50$ cm^3, $C_o = 500$ mg/L,
5	0.97	0.99		0.88	0.97			$A = 8$, $V_{as} = 180$ rpm, T—ambient (about 298), 313 K.
10	0.99	1.00		0.82	0.95			$t = 1$ min to 4 h,
15	1.00	1.00		0.83	0.95			bs—0.425–0.85 mm
30	1.00	1.00		0.94	1.00			
60	1.00	1.00		0.98	1.00			
120	1.00	1.00		1.00	1.00			
180	1.00	1.00		1.00	1.00			
240	1.00	1.00		1.00	1.00			

Time	Effect of beads size						Plot q_t versus t	Experimental conditions
	HCl system			HCl–HNO$_3$ system			$t_{1/2}$ (s)	
t (min)	f_3	f_4	f_1, f_2	f_3	f_4	f_1, f_2		
1	0.80	0.82	**	0.78	0.80	**	<60 s	$m_j = 0.5$ g,
3	0.92	0.93		0.87	0.89			$V = 50$ cm^3, $C_o = 500$ mg/L,
5	0.96	0.98		0.90	0.91			$A = 8$, $V_{as} = 180$ rpm,
10	1.00	1.00		0.96	0.96			T—ambient,
15	1.00	1.00		0.99	0.97			$t = 1$ min to 4,
30	1.00	1.00		0.99	0.99			beads size—f_1, f_2, f_3, f_4
60	1.00	1.00		0.99	0.95			$0.85 > f_1 \geq 0.6$ mm,
120	1.00	1.00		0.99	0.99			$0.6 > f_2 \geq 0.5$ mm,
180	1.00	1.00		0.99	0.99			$0.5 > f_3 \geq 0.43$ mm,
240	1.00	1.00		1.00	1.00			$0.43 > f_4 \geq 0.425$ mm

* 333 K was the maximum PuroliteA-400TL temperature, therefore, to avoid possible anion-exchange resin decomposition this temperature was not applied

** Mass of the beads from this population was not enough to carry out the batch sorption for all phases contact time

metal ion increases, more and more surface sites are covered, and hence at higher concentration of metal ions, the capacity of the anion-exchange resin is exhausted due to non-availability of the surface sites. It is, therefore, evident that in low concentration ranges the percentage of sorption is high (usually quantitative) because of availability of more active sites on the surface of sorbent.

3.1.1. Effect of Agitation Speed

The kinetic profiles obtained for different agitation speeds are similar after 60min–120min of phase contact time (in all cases the applied speed was 120, 150, and 180rpm). This time is enough to reach equilibrium by the system even for the cases of 120rpm. When the agitation speed increases, the time required to reach equilibrium decreases. For 180rpm agitation speed, this time is reduced to 30min (HCl and HCl–HNO$_3$ systems). As expected, the sorption capacities at equilibrium are very close (varying by less than 0.5% for HCl and 3% for the HCl–HNO$_3$ systems). The effect of agitation speed on the sorption process is observed at the beginning of the process. With the agitation speed increase from 120 to 180rpm, the F values also increase and exceed 0.5 after 1min (F = 0.72–120rpm; F = 0.77–150rpm; F = 0.81–180rpm—HCl system, F = 0.71rpm–120rpm; F = 0.56–150rpm; F = 0.55–180rpm for HCl–HNO$_3$ system). Exception to the rule are the F values for 1 and 3min phases contact time for the HCl–HNO$_3$ system where the values decrease or change in some way. Faster phases mixing assure that all the surface binding sites are made readily available for metal uptake. Additionally, changes of agitation speed cause changes of the external boundary film surrounding the resin beads (lower mixing—the external boundary film is thicker). A similar effect of sorption capacities changes with the agitation speed was observed by other researchers e.g. Sepideh et al.[18] half-exchange time is smaller than 60 s for all presented cases and it decreases with the agitation speed increase (the HCl system). As was pointed out previously[16][17] when the rate of ion exchange increases with the agitation speed, the process is controlled by resistance to film diffusion, otherwise if the agitation speed does not influence on the rate of sorption the process is controlled by intraparticle diffusion.

3.1.2. Effect of Temperature

The purpose of this research is to study the effect of temperature on the sorption of Pd(II) ions by Purolite A-400TL. The effect of temperature on the removal of Pd(II) in the acidic solution by Purolite A-400TL was studied at ambient temperature and 313K. Due to the fact that 333K is the maximum temperature of thermal stability of anionexchange resin, the temperature study was neglected to avoid anion-exchange resin decomposition. The data are presented in **Table 2** which shows that sorption of Pd(II) ions by Purolite A-400TL negligibly increased with the increase in temperature [HCl system; q_t values increased from 40.5mg/g (298K) to 43.0mg/g (333K)—1min phases contact time, from 46.5mg/g (298K) to 48.7mg/g (333K)—3min phases contact time, from 48.5mg/g (298K) to 49.7mg/g (333K)—5min phases contact time, from 49.7mg/g (298K) to 49.9mg/g (333K)—10min phases contact time]. A similar temperature effect is observed for 1–3min phases contact time for the HCl–HNO$_3$ system [HCl–HNO$_3$ system; q_t values increased from 27.1mg/g (298K) to 40.1mg/g (333K)—1min phases contact time, from 38.4mg/g (298K) to 45.3mg/g (333K)—3min phases contact time, from 43.6mg/g (298K) to 46.6mg/g (333K)—5min phases contact time]. After 10min of phases contact time for the HCl–HNO$_3$ system, the obtained q_t values are higher at ambient temperature than at 333K. The F values increased with the phases contact time significantly and also negligibly with the increase of temperature. Usually after 15min of phases contact time, the F values remain unchanged. At high temperature, the thickness of the boundary layer decreases, due to the increased tendency of the metal ions to escape from the anion-exchange resin surface to the solution phase, which results in a decrease in sorption as temperature increases.

3.1.3. Effect of Bead Size Distribution

200g of the anion-exchange resin was divided into fractions of different beads size by the classical sieve analysis. The fractions were following: $0.85 > f_1 \geq 0.6$mm; $0.6 > f_2 \geq 0.5$mm; $0.5 > f_3 \geq 0.43$mm, and $0.43 > f_4 \geq 0.425$mm.

Due to the fact that the most of the beads belong to f_3 and f_4 fractions and the first and second ones were not enough abundant in this case the sorption of Pd(II) was checked only at equilibrium time for comparison. Based on the values of the equilibrium sorption capacities, the effect of beads size distributions on the sorption process of Pd(II) on Purolite A-400TL is not marked. The differences of the sorption capacity values at equilibrium time are so small that they can be neglected: q_t = 49.95mg/L (f1), q_t = 49.95mg/L (f$_2$), q_t = 49.96mg/L (f$_3$), q_t = 49.96mg/L (f$_4$)—HCl system; q_t = 48.53mg/L (f$_1$), q_t = 48.49mg/L (f$_2$), q_t = 48.52mg/L (f$_3$), q_t = 48.66mg/L (f$_4$)—HCl–HNO$_3$ system. At the beginning of the sorption process, this effect is also not marked enough compared to the other experimental parameters described above for which at the beginning the changes of q_t values were more marked. As expected, the sorption rate and the time required to reach equilibrium do not change with the beds size increase. Our previous studies of the effect of beads size distribution on the Pd(II) sorption process indicated that the sorption process can be effected by ionexchange resin beads size much more than in this paper (Purolite A-400TL—polystyrene-divinylbenzene, gel)[15]. Kinetic studies of Purolite A-830 (polyacrylic, macroporous)[15] indicated that the equilibrium sorption capacities are not changed but the rate of sorption increases with the beads size decrease. Similar observation was made by Wawrzkiewicz[19] who applied Amberlite IRA-458 and Am-berlite IRA-67 (polyacrylic, gel) for Direct Red 75 sorption. This change of rate sorption was explained by the fact that for the small beads the diffusion path lengths of the exchanging ions to and from the active sites are shorter[19]. The fast Pd(II) sorption in this case is confirmed also by the F values close to 1 after 5–10min of phases contact time and by small values of $t_{1/2}$. Due to the fact that the bead size effect can be neglected, it can be stated that in this case the diffusion path lengths of the exchanging ions did not play a significant role. Based also on the other experimental parameters examined here (agitation speed, temperature, and initial concentration), it can be concluded that the film diffusion plays a more significant role in the sorption process of Pd(II) on Puroite A-400TL.

3.2. Equilibrium Studies

The equilibrium isotherm equations are used to describe the experimental sorption data. The equation parameters and the underlying thermodynamic assumptions of these equilibrium models often provide some insight into both the sorption mechanism and the surface properties and affinity of the sorbent.

Equilibrium studies were carried out using the batch method. The initial Pd(II) concentrations were in the range from 100 to 6000mg/L (in 0.1M HCl). The Langmuir and Freundlich isotherms described in[12][15] were applied. The Langmuir and Freundlich parameters were obtained by plotting C_e/q_e versus C_e and ln q_e

versus ln C_e, respectively, and the data are provided in **Table 3**. **Table 3** shows that the Langmuir model is more suitable than the Freundlich adsorption isotherm.

The Freundlich isotherm plot of Pd(II) ions sorption provides a correlation coefficient of 0.9266, the values of kF and 1/n obtained from the plot were 51.25mg/g and 0.2745, respectively. The values of R^2 are smaller than 0.99 which indicate that the relationship C_e/q_e versus Ce is not linear. The Langmuir plot gives a better correlation coefficient than the Freundlich one, and the correlation coefficient is equal to 0.9448 but it is still not satisfactory. The sorption capacity value calculated from the Langmuir isotherm equation is equal to 404.15mg/g, whereas the experimental qe is higher by about 3% (413.93mg/g). The difference between the calculated and the experimental sorption capacity values is not high. The essential characteristics of a Langmuir isotherm can be expressed in terms of a dimensionless constant separation factor or the equilibrium parameter, R_L [R_L = 1/(1 + bC_o)]. The parameter indicates the isotherm shape as follows: R_L > 1—unfavorable, R_L = 1—linear, 0 < R_L < 1—favorable, R_L = 0—irreversible. The R_L values are 0.4868, and it is a typical behavior of the favorable isotherm.

Table 3. Equilibrium results—parameters and fitting plot.

Isotherm model applied					Parameters	
	Equation	No.	Plot	Symbols		
Langmuir	$\frac{C_e}{q_e} = \frac{1}{Q_o b} + \frac{C_e}{Q_o}$	(3)	C_e/q_e vs C_e	Q_0—the Langmuir monolayer sorption capacity (mg/g),	Q_o (mg/g)	404.15
				b—the Langmuir constant related to the free energy of sorption (dm³/mg),	b (dm³/mg)	0.0106
					R_L	0.4868
					R^2	0.9448
Freundlich	$\log q_e = \log k_F + \frac{1}{n} \log C_e$	(4)	$\log q_e$ vs $\log C_e$	R_L—separation factor or equilibrium parameter,	k_f (mg/g)	51.25
				k_F—the Freundlich adsorption capacity (mg/g),	$1/n$	0.2745
				$1/n$—the Freundlich constant related to the surface heterogeneity	R^2	0.9266
Fitting						

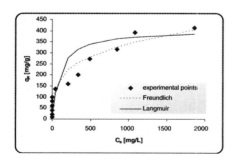

3.3. Column Experiment

Column experiments were conducted using a glass tube of 1 cm diameter by passing the initial solution concentration of 100mg Pd(II)/L through the Purolite A-400TL beads. The breakthrough curve (plot C/C$_o$ vs V) and the working ion exchange capacities [Equation (5)], the weight [Equation (6)] and bed [Equation (7)] distribution coefficients were calculated using the following equations:

$$C_r = (V_p \cdot C_o)/V_j \qquad (5)$$

$$D_w = (U - U_o - V)/m_j \qquad (6)$$

$$D_v = D_w d_z \qquad (7)$$

where C$_r$ is the working ion exchange capacity, V$_p$ is the collected volume of effluent between the first fraction and that to the breakthrough point (mL), C$_o$ is the initial Pd(II) concentration, V$_j$ is the volume of ion exchanger bed put into the columns, Dw is the weight distribution coefficient, U is the effluent volume at C = 0.5 C/C$_o$ (mL), U$_o$ is the dead volume in the column (mL), V is the void (interparticle) ion exchanger bed volume (which amounts to ca. 0.4), m$_j$ is the dry ion exchanger weight (g), D$_v$ is the bed distribution coefficient, d$_z$ is the ion exchanger bed density. The breakthrough curves are presented in **Figure 2**. whereas the calculated parameters in **Table 4**. As can be seen the breakthrough curves possess at the beginning the S-shape but then the end of the curves is not typical. **Figure 2** indicates that the ratio of C to C$_o$ achieved a constant value. This tendency was not observed previously, only for Purolite S-984 the breakthrough curves possess also unusual shape[20]. The explanation of such tendency has not been found yet. Probably, a gel structure as well as the resin capacity can play a main role here. The SBA resin capacities values change with the concentration of acids. For the HCl system, the values of the working ion exchange capacities are much higher for dilute acidic solutions compared to the capacities obtained for the HCl–HNO$_3$ system. Additionally, typical reduction of C$_r$ values was observed with the increasing HCl acid concentrations. The competitive effect of Cl$^-$ anion and Pd(II) complexes is marked here[2]. For the HCl–HNO$_3$ system, the capacities increase with the HCl concentration increase and HNO$_3$ concentration decreases. Such

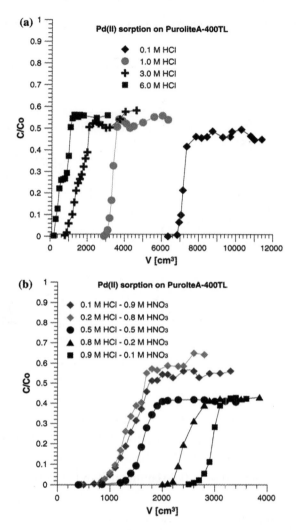

Figure 2. Breakthrough curves of Pd(II) sorption on Purolite A-400TL from a HCl and b HCl–HNO$_3$ systems.

behavior was observed previously for about 30 examined ion-exchange resins[4][5][8][12][14][15][20] etc. The mechanism of Pd(II) can be presented as follows:

Table 4. Comparison of the sorption parameters—column studies of Pd(II) recovery on Purolite A-400TL.

System	Dw	Dv
0.1 M HCl	-	-
1.0 M HCl	1057.9	358.4
3.0 M HCl	615.1	208.4
6.0 M HCl	320.8	108.7
0.1 M HCl - 0.9 M HNO$_3$	518.9	175.8
0.2 M HCl - 0.8 M HNO$_3$	489.7	165.9
0.5 M HCl - 0.5 M HNO$_3$	-	-
0.8 M HCl - 0.2 M HNO$_3$	-	-
0.9 M HCl - 0.1 M HNO$_3$	-	-

– values were not obtained

Purolite A-400TL shows high sorption capacity (0.0685 mg/cm^3) compared to the other SBA resins such as Amberlite IRA-458, Amberlyst A-29, Dowex MSA-1, Dowex MSA-2, Varion ADM, Varion ATM (capacities for other SBA resins are in the range from 0.048 to 0.0616 mg/cm^3) in 0.1M HCl.

Desorption and reusable properties of purolite A-400TL

The desorption studies and possibilities of Purolite A-400TL reuse were carried out. In the desorption studies, many eluting agents of different concentrations were applied such as 0.1–4.0M HNO$_3$, 0.1–6.0M HCl, 0.5–4.0M NH$_4$OH, 0.5–3.0M NaOH, 0.5–4.0M H$_2$SO$_4$, and 1.0M TU (thiourea) or acidic TU (1.0M TU—1.0M HCl; 1.0M TU—1.0M HNO$_3$). The desorption results are shown in **Table 5**. Based on **Table 5** the following observations can be made:

- sorption SBA resin capacities are high and even after three cycles of sorption-desorption these values remain almost unchanged—the capacity

Table 5. Effectiveness of sorption and desorption processes.

0.1 M HCl—100 mg Pd (II)/L							
	Eluting agent	S_1	D_1	S_2	D_2	S_3	D_3
1	0.1 M HNO$_3$	100.0	0.5	99.8	0.6	99.8	0.6
2	1.0 M HNO$_3$	100.0	10.5	99.8	8.0	99.8	8.2
3	2.0 M HNO$_3$	100.0	19.4	99.7	13.1	99.8	12.9
4	3.0 M HNO$_3$	100.0	25.9	99.8	16.1	99.8	15.6
5	4.0 M HNO$_3$	100.0	30.5	99.8	20.3	99.9	17.4
6	0.1 M HCl	100.0	0.1	99.8	0.0	99.9	0.1
7	1.0 M HCl	100.0	1.9	99.9	2.0	99.8	2.5
8	2.0 M HCl	100.0	5.4	99.9	4.7	99.9	5.8
9	3.0 M HCl	100.0	8.0	99.9	7.4	99.9	8.6
10	6.0 M HCl	100.0	24.3	99.9	19.1	99.9	18.1
11	0.5 M NH$_4$OH	100.0	44.9	99.8	22.4	99.6	19.1
12	1.0 M NH$_4$OH	100.0	42.9	99.8	23.3	99.6	17.5
13	2.0 M NH$_4$OH	100.0	46.9	99.8	23.2	99.8	18.3
14	3.0 M NH$_4$OH	100.0	44.8	99.8	21.4	99.8	17.7
15	4.0 M NH$_4$OH	100.0	43.8	99.7	24.5	99.8	18.9
16	0.5 NaOH	100.0	23.9	99.9	14.3	99.9	5.1
17	1.0 M NaOH	100.0	27.3	99.9	20.6	99.9	16.8
18	2.0 M NaOH	100.0	35.2	99.9	20.8	99.9	17.4
19	3.0 M NaOH	100.0	37.5	99.9	22.4	99.9	18.5
20	0.5 M H$_2$SO$_4$	100.0	0.2	99.9	0.2	99.9	0.2
0.1 M HCl—100 mg Pd (II)/L							
	Eluting agent	S_1	D_1	S_2	D_2	S_3	D_3
21	2 M H$_2$SO$_4$	100.0	3.3	99.9	2.8	99.8	2.6
22	4 M H$_2$SO$_4$	100.0	5.2	99.9	3.9	99.8	3.4
23	1.0 M TU	100.0	40.0	99.1	23.3	99.3	14.4
24	1.0 M TU—0.1 HCl	100.0	41.7	99.7	25.0	99.0	16.2
25	1.0 M TU—0.1 HNO$_3$	100.0	38.0	99.4	14.1	99.6	15.0

S sorption, D desorption, *1,2,3* number of cycle

reduction is smaller than 1%. The advantage of this resin is the fact that it can be used many times without significant reduction of capacity.

- desorption effectiveness of desorption studies is not satisfactory enough. Application of acids gives desorption yield in the range from 0.1% to 30.5% (D1), from 0 to 20.3% (D2), and from 0.1% to 18.5% (D3). Better % of desorption was obtained by using basic solution but the desorption yield did not exceed 47%. Acidic solutions of TU and those without acid in this case did not give satisfactory results either. The % of desorption usually decreases with the next cycle of sorption-desorption. In the calculation of % D, e.g., in the second step, the amount of Pd(II) not desorbed in the first cycle was taken into account and added to the amount of Pd(II) retained in the second step of sorption.

The comparison of eluting agents applied in Pd(II) desorption from other (bio)sorbents and the effectiveness of ion-exchange resin regeneration was presented previously in Table S7 in[15]. As follows from the table acids, bases and TU solutions were usually applied as eluting agents. The effectiveness of % D is different but acidic thiourea solution seems to be the most appropriate for this purpose but in many cases the use of such solutions did not give a quantitative Pd(II) recovery. Sometimes changes of volume of the eluting agents, concentrations, and temperature make the elution more quantitative.

4. Conclusions

Based on the present study, it is clearly shown that Purolite A-400TL (polystyrene-divinylbenzene anion-exchange resin of $-N^+(CH_3)_3$ (type 1) functional groups and gel-type) is found to be an effective sorbent for removal of Pd(II), Pt(IV), and Au(III) ions from aqueous solution. The experimental sorption capacity of Purolite A-400TL, strongly basic anion-exchange resin, was found to be 413.93mg Pd(II)/g, and the calculated one (Langmuir capacity) was equal to 404.15mg Pd(II)/g. The sorption process is slightly affected by such parameters as agitation rate, bead size distribution, and temperature. The initial Pd(II) concentration and phases contact time play a significant role in Pd(II) recovery. The percentage removal of Pd(II) increased with the increase in contact time. The rate of Pd(II) removal and the kinetics of sorption process are very fast, therefore, the time required to reach equilibrium is very short and the saturation degree (or fractional attainment of equilibrium) reached the constant values after short phases contact time. Half-exchange time is smaller than 60 s for all presented cases. Desorption studies showed that the strongly basic anion-exchange resin can be regenerated (but not quantitative Pd(II) desorption is observed) and reused. Moreover, the reduction of capacity after three cycles of sorption-desorption is negligible (smaller than 1%). Column studies indicated that in the dilute acidic solution (0.1M HCl) the working anion exchange capacity is high (0.0685mg/cm^3) in comparison with other SBA resins examined under the same experimental conditions.

Open Access This article is distributed under the terms of the Creative Commons Attribution License which permits any use, distribution, and reproduction in any medium, provided the original author(s) and the source are credited.

Source: A. Wolowicz, Z. Hubicki. Adsorption characteristics of noble metals on the strongly basic anion exchanger Purolite A-400TL[J]. Journal of Materials Science, 2014, 49(18):6191–6202.

References

[1] Jermakowicz-Bartkowiak D, Kolarz BN (2013) Anionity polimerowe do odzyskiwania metali szlachetnych. Polimery 58:524–532 (in Polish).

[2] Nikolski AN, Ang KL (2014) Review of the application of ion exchange resins for the recovery of platinum-group metals from hydrochloric acid solutions. Miner Process Extr Metall Rev 35:369–389.

[3] Mladenova E, Karadjova I, Tsalev DL (2012) Solid-phase extraction in the determination of gold, palladium, and platinum. J Sep Sci 35:1249–1265.

[4] Hubicki Z, Wawrzkiewicz M, Wołowicz A (2008) Application of ion exchange methods in recovery of Pd(II) ions—a review. Chem Anal 53:759–784.

[5] Wołowicz A (2011) Zastosowanie włókien i jonitów chela-tuja̧cych w procesie sorpcji i separacji jonów Pd(II). Przem Chem 90:1001–1015.

[6] Syed S (2012) Recovery of gold from secondary sources—a review. Hydrometallurgy 30:115–116.

[7] Wołowicz A, Hubicki Z (2011) Comparison of strongly basic anion exchange resins applicability for the removal of palladium(II) ions from acidic solutions. Chem Eng J 170:206–215.

[8] Harland CE (1994) Ion exchange: theory and practice, 2nd edn. Royal Society of Chemistry, Cambridge.

[9] Alguail FJ, Adeva P, Alonso M (2005) Processing of residual gold(III) solutions via ion exchange. Gold Bull 38:9–13.

[10] Renner H (1997) Platinum Group Metals. In: Habashi F (ed) Handbook of extractive metallurgy. Wiley-VCH, Weinheim, Chichester.

[11] Harris GB (1993) A review of precious metals refining. In: Mishra RK (ed) Precious metals. IPMI, Allentown.

[12] Wołowicz A, Hubicki Z (2012) The use of the chelating resin of a new generation Lewatit MonoPlus TP-220 with the bis-picolylamine functional groups in the removal of selected metal ions from acidic solutions. Chem Eng J 197:493–508.

[13] Parodi A, Vincent T, Pilsniak M, Trochimczuk AW, Guibal E (2008) Palladium and

platinum binding on an amidazol containing resin. Hydrometallurgy 92:1–10.

[14] Wołowicz A, Hubicki Z (2011) Investigation of macroporous weakly basic anion exchangers applicability in palladium(II) removal from acidic solutions—batch and column studies. Chem Eng J 174:510–521.

[15] Wołowicz A, Hubicki Z (2012) Applicability of new acrylic, weakly basic anion exchanger Purolite A-830 of very high capacity in removal of palladium(II) chloro-complexes. Ind Eng Chem Res 51:7223–7230.

[16] Kononova ON, Goryaeva NG, Dychko OV (2009) Ion exchange recovery of palladium(II) from nitrate weak acidic solutions. Nat Sci 1:166–175.

[17] Helfferich F (1995) Ion exchange. Dover Publications Inc., Mineola.

[18] Sepideh J, Mahmoud A, Hossein A, Ahmad KD (2012) The effect of kinetics parameters on gold extraction by Lewis cell: comparison between synthetic and leach solutions. Iran J Chem Eng 31:59–67.

[19] Wawrzkiewicz M (2011) Comparison of gel anion exchangers of various basicity in direct dye removal from aqueous solutions and wastewaters. Chem Eng J 173:773–781.

[20] Wołowicz A, Hubicki Z (2014) Polyacrylate ion exchangers in sorption of noble and base metal ions from single and tertiary component solutions. Solv Extr Ion Exch 32:189–205.

Chapter 3
A Facile Method for the Production of SnS Thin Films from Melt Reactions

Mundher Al-Shakban[1], Zhiqiang Xie[1], Nicky Savjani[2], M. Azad Malik[1], Paul O'Brien[1,2*]

[1]School of Materials, University of Manchester, Oxford Road, Manchester M13 9PL, UK

[2]School of Chemistry, University of Manchester, Oxford Road, Manchester M13 9PL, UK

Abstract: Tin(II)O-ethylxanthate [$Sn(S_2COEt)_2$] was prepared and used as a single-source precursor for the deposition of SnS thin films by a melt method. Polycrystalline, (111)-orientated, orthorhombic SnS films with controllable elemental stoichiometries (of between $Sn_{1.3}S$ and SnS) were reliably produced by selecting heating temperatures between 200°C and 400°C. The direct optical band gaps of the SnS films ranged from 1.26eV to 1.88eV and were strongly influenced by its Sn/S ratio. The precursor [$Sn(S_2COEt)_2$] was characterized by thermogravimetric analysis and attenuated total reflection Fourier-transform infrared spectroscopy. The as-prepared SnS films were characterized by scanning electron microscopy, energy-dispersive X-ray spectroscopy, powder X-ray diffractometry, Raman spectroscopy, and UV-Vis spectroscopy.

1. Introduction

Tin sulfides (SnS_2, Sn_2S_3, and SnS) are members of the IV-VI family of semiconductors that have shown promise in photovoltaic and optoelectronic applications[1]–[4]. Tin(II) sulfide (SnS) in particular has been seen as a potential candidate as an absorber layer in photovoltaic cells due to its 1.4eV direct band gap that can harvest the visible and near-IR regions of the EM spectrum, the lower costs and toxicity of the constituent elements as compared to other potential materials (e.g., PbS and CdS), and the simplicity of the binary system compared to multicomponent materials such as copper zinc tin sulfide (CZTS) and copper indium gallium sulfide (CIGS)[5]–[8].

Single-source precursors (SSPs) are compounds that are designed to decompose to materials of specific compositions, by containing the desired elements. In many cases, the uses of SSPs have granted control of both its physical and optical properties that dualsource precursors cannot[9]–[11]. In the last 20 years[12]–[18], many SSPs comprised metal (N,N-dialkyldithiocarbamates) [$M(S_2CNR_2)n$] have been used to synthesize metal sulfide nanocrystals. More recently, complexes containing (O-alkyl)xanthate ($-S_2COAk$) ligands have been viewed as a potentially useful class of SSPs for the production of metal sulfide nanomaterials. The decomposition of metallo-organic xanthates is known to take place via the relatively low-temperature and clean Chugaev elimination reaction[19]. The use of the xanthate ligand in SSPs has permitted the formation of many metal sulfides, including, but not limited to, MoS_2[20], CdS[21], NiS, PdS[22], and CZTS[23], at lower temperatures than those needed by their respective (N,N-dialklydithio-carbamato-) analogs. Recently, we have reported the preparations of PbS/polymer composites from both lead(II)xanthate and lead(II)dithiocar-bamate complexes by a melt process[18][24], finding that the decomposition of $Pb(S_2COnBu)_2$ in a polymer matrix produced pure cubic PbS nanocrystals at 150°C; significantly lower temperatures than 275°C are needed to decompose $Pb(S_2CNnBu_2)_2$. As a result, the xan-thate-containing SSP can be used in a wider temperature window, giving greater control over nanocrystal size, shape variation, and orientation preference of the PbS crystals. Among the other known methods to SnS nanomaterials[18][25], explorations of the syntheses of orthorhombic SnS nanoparticles [26]–[28] and films[13][29][30], using Sn-SSPs such as [$Sn^{II}(S_2CNR2)_2$] and [$R'_2Sn^{IV}(S_2CNR_2)_2$], have been reported. To date, however, no studies on the uses of tin(O-alkylxan-thate) complexes have been documented.

In this report, we investigate the use of the SSP [Sn(S$_2$COEt)$_2$] as a coating material for the production of herzenbergite SnS films on glass. We focus on both the annealing temperature and the role of the xan-thate ligand during the decomposition process for the potential in controlling the structural and optoelectronic properties of the SnS films produced

2. Experimental

2.1. Materials and Methods

Potassium ethyl xanthate, chloroform and tetrahydrofuran were purchased from Sigma-Aldrich. Tin(II) chloride was purchased from Alfa Aesar. All chemicals were used as received. Elemental (EA) and thermogravimetric (TGA) analyses were carried out by the Microelemental Analysis service at University of Manchester. EA was performed using a Flash 2000 Thermo Scientific elemental analyzer and TGA data obtained with Mettler Toledo TGA/DSC1 stare system between the ranges of 30°C–600°C at a heating rate of 10°C min^{-1} under nitrogen flow. Scanning electron microscopy (SEM) analysis was performed using a Philips XL30 FEG microscope, with energy-dispersive X-ray spectroscopy (EDX) data obtained using a DX4 instrument. Thin-film X-ray diffraction (XRD) analyses were carried out using an X-Pert diffractometer with a Cu-K$_{\alpha 1}$ source (λ = 1.54059Å), the samples were scanned between 20° and 75°, the applied voltage was 40kV, and the current was 30mA. Raman spectra were measured using a Renishaw 1000 Micro-Raman System equipped with a 514nm laser. UV-Vis measurements were made using a Shimadzu UV-1800 spectrophotometer.

2.2. Synthesis of Tin(II)(O-Ethylxanthate)

[Sn(S$_2$COEt)$_2$] was prepared by a procedure that was modified for that described in literature[19][31]. An aqueous solution of potassium ethylxanthate (10.0g, 12.5mmol) was added to a stirred solution of tin(II) chloride (5.9g, 6.2mmol) in distilled water (100ml) and stirred for a further 30min. The yellow precipitate produced was filtered by vacuum filtration, washed three times with water, and finally dried in a vacuum oven at room temperature for 2h. Yield = 7.2g (67%). Melting point = 44°C–53°C. Anal. Calcd for [Sn(S$_2$COEt)$_2$]: C, 19.98; H, 2.79; S, 35.45; Sn, 32.91 Found: C, 19.67; H, 2.74; S, 35.45; Sn, 32.17. FTIR data (cm); 2986.8 (w), 2930.7 (w), 1457 (w), 1355 (w), 1195.6 (s), 1108.1 (s), 1020.6 (s), 852.0 (w), 801.3 (w), 563.4 (w).

2.3. Preparation SnS Thin Films by Spin Coating and Heating

Glass slides were cut to 20mm × 15mm, cleaned by sonication in acetone (twice) and water, and allowed to dry. Three cycles of coating was performed; in each cycle, 300μL of a 3M [Sn(S$_2$COEt)$_2$] solution in THF was coated onto the glass slide by spin coating at 700 rpm for 60 s and allowed to dry. The resulting films were loaded into a glass tube for decomposition in a dry nitrogen environment. The tube was then heated in the furnace to the desired temperature (150°C–400°C) at a rate of ~3°C min^{-1} and held at that temperature for 60min; after this time had elapsed, the furnace was turned off and the tube allowed to cool to room temperature.

3. Results and Discussion

The tin xanthate precursor [Sn(S$_2$COEt)$_2$] was synthesized by the literature procedure[19][31] and elemental analyses confirming its purity. The yellow powder is readily soluble in THF and many other common organic solvents. It was found that storage at −20°C was necessary to limit decomposition. The thermal decomposition of [Sn(S$_2$COEt)$_2$] was studied using thermogravimetric analysis (TGA). The thermogram showed rapid single decomposition step between 80 and 130°C (Figure S1). The final weight of the residue (44.1%) is close to the predicted value for residual SnS (41.8%). The precursor is predicted to break down via the Chugaev elimination mechanism[19], as shown in Scheme S1. The IR spectrum of [Sn(S$_2$COEt)$_2$] shows bands corresponding to m(C-O) (1196 and 1224cm^{-1}) and m(C-S) (1021 and 1108cm^{-1}). A peak at 563cm^{-1} is also observed consistent with a m(Sn-S) mode (Figure S2).

SnS films were prepared by coating glass slides with the [Sn(S$_2$COEt)$_2$] precursor, followed by heating step in an N$_2$ environment, at 150°C, 200°C, 250°C,

300°C and 400°C for 1h. The resulting films were gray and uniform at all the heating temperatures (Figure S3); the films were found to be between 2.2 and 2.9lm thick. Both the morphology and composition of the SnS films produced showed a dependence on the heating temperature. The films contained nearly spherical structures with some flakes [**Figures 1(a)-(d)**]. Elemental maps for the film produced at all temperatures (elemental maps for 300°C shown in **Figure 1(e)**, **Figure 1(f)**; maps for the other films in Figure S4) demonstrate the uniform distribution of these elements among the films. The Sn/S ratio within the films steadily decreased when higher reaction temperatures were used (**Table 1**; **Figure 2**): Sn/S ratios of 1.31 and 1.32 ($Sn_{1.31}S$ and $Sn_{1.32}S$) were seen in the films produced at 150°C and 200°C (sulfur deficient). Increasing the heating temperature also increased the sulfur content in the film, reactions performed at 250°C and 300°C produced SnS films with reduced sulfur deficiency [Sn/S ratios of 1.15:1 ($Sn_{1.15}S$) and 1.11:1 ($Sn_{1.11}S$), respectively], and a stoichiometric SnS film was obtained upon heating at 400°C [Sn/S ratio 1.03:1; ($Sn_{1.03}S$)]. Such control of the stoichiometry in SnS nanomaterials has previously been observed[32][33].

The p-XRD patterns of the SnS thin films produced at temperatures between 200°C and 400°C gave peaks that can be indexed to herzenbergite-SnS with the expected orthorhombic crystal structure [matches ICCD pattern No. 00-039-0354; see **Figure 3(a)**]; no other peaks are observed that correspond to other tin oxide or sulfide species. The calculated unit cell parameters of the herzenbergite SnS films (shown in **Table 1**) match the expected (Pbnm) space group, with lattice

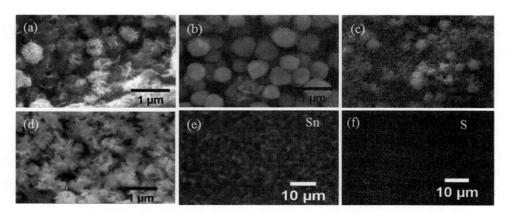

Figure 1. SEM images of SnS films grown on glass substrates from [$Sn(S_2COEt)_2$] at a 200°C, b 250°C, c 300°C, and d 400°C (scale bars represent 1μm); e and f the elemental maps of the SnS film produced at 300°C (scale bars represent 10μm).

Table 1. Thicknesses, compositions, and unit cell parameters of the SnS films produced by the melt method.

Heating Temp. (°C)	Thickness (μm)	Sn atomic (%)[a]	S atomic (%)[a]	Sn/S ratio	Unit cell parameters (orthorhombic, Å)[b,c]
150	–	56.8	43.2	1.31:1 (± 0.07)	$a = 4.355, b = 11.232, c = 3.985$
200	2.2	56.99	43.0	1.32:1 (± 0.07)	$a = 4.324, b = 11.237, c = 3.980$
250	2.9	53.4	46.4	1.15:1 (± 0.06)	$a = 4.324, b = 11.231, c = 3.984$
300	2.8	52.5	47.5	1.11:1 (± 0.06)	$a = 4.324, b = 11.236, c = 3.986$
400	2.5	50.7	49.3	1.03:1 (± 0.05)	$a = 4.324, b = 11.219, c = 3.986$

[a] Determined by SEM–EDX
[b] Determined by p-XRD
[c] Rock-salt SnS phase also observed in the films produced at 150 °C (unit cell parameter: $a = 5.801$ Å)

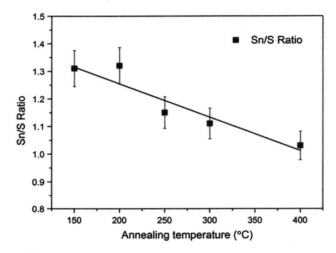

Figure 2. The Sn/S ratio by EDX for samples heated for 60min at temperatures between 150°C and 400°C.

parameters that closely match with those reported in literature[34][35]. The films heated at 400°C have strong, well-defined diffraction peaks. At lower temperatures, however, the films were found to exhibit broader peaks, possibly due to the increasing sulfur deficiency within the crystalline film. The temperature during [Sn(S$_2$COEt)$_2$] decomposition seems crucial; at lower temperatures, the rate at which the precursor decomposes will be slowed considerably, with the resulting intermediate species exposed to temperatures that may promote evaporation. In contrast, the Sn$_{1.31}$S film produced at 150°C was found to consist of a mixture of orthorhombic and cubic phases, with the latter phase matching well with a rock-salt SnS phase (space group Fm-3 m; ICCD pattern No. 04-004-8426) recently discussed by first-principle calculations[36][37]. In addition, Raman spectroscopy of all of the SnS films [**Figure 3(b)**] revealed Raman bands at 94cm^{-1}, 160cm^{-1}, 188cm^{-1}, and 218cm^{-1}, in good

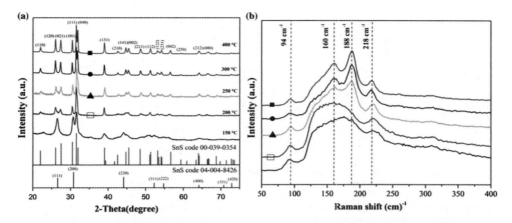

Figure 3. A p-XRD patterns of SnS films grown on glass substrate at different temperatures, accompanied by reference patterns of herzenbergite SnS (ICCD pattern No. 00-039-0354) and rock-salt SnS (ICCD pattern No. 04-004-8426). b Raman spectra for the SnS films grown on glass substrates from [Sn(S$_2$COEt)$_2$] at 150°C, 200°C, 250°C, 300°C, and 400°C.

agreement with the herzenbergite SnS phase reported previously[38]-[40].

The optical band gaps (Eg) of the SnS films (produced at temperatures between 200°C and 400°C) were determined from optical absorption measurements by the Tauc method [**Figure 4(a)**][41][42]. All of the films analyzed have high absorption coefficients ($\alpha > 104 cm^{-1}$ above the fundamental absorption). The band gaps were evaluated by extending linear part of the plots of $(\alpha h\upsilon)^2$ versus $h\upsilon$[43]. The band gap of the Sn$_{1.32}$S films formed at 200°C is 1.88eV. Increased decomposition temperatures gave lower band gaps; 250 (Sn$_{1.15}$S), 300 (Sn$_{1.11}$S), and 400°C (Sn$_{1.03}$S) gave band gaps at 1.75, 1.49, and 1.26eV, respectively. It is clear that the control of the stoichiometry that we have achieved in the syntheses allows for tuning of optical band gaps. The band gaps of SnS films previously reported[44]–[48] show similar change with sulfur deficiency [**Figure 4(b)**]. However, care needs to be taken when comparing the results obtained from literature to our dataset, as the films produced by each citation varies from both our work and each other. These experimental variations will introduce variations in the macrostructures, elemental stoichiometries, film thicknesses, and concentrations of Sn$_x$S$_y$-based impurities contained within the SnS films documented. The physical properties of the materials and the potential of quantum confined materials may also need to be considered.

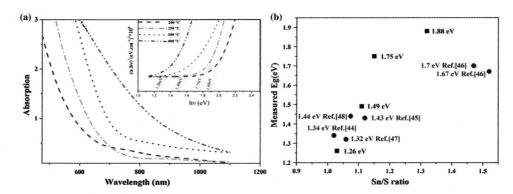

Figure 4. A UV-Vis and Tauc plots (inset) for the SnS films grown on glass substrates from [Sn(S$_2$COEt)$_2$] at 200°C, 250°C, 300°C, and 400°C. b Graph showing the relationship between the Sn/S ratio of SnS$_{1-x}$ materials with its measured band gap. Data in black represents the findings in this report, whereas data in red were obtained from literature.

4. Conclusions

A simple process has been described for the growth of SnS films. Heating of substrates spin-coated with tin(II)O-ethylxanthate at different temperatures between 150°C and 400°C produced SnS films mainly in orthorhombic phase with good crystallinity. Analyses of the films reveal them to be sulfur deficient with Sn/S ratio controlled by selecting the heating temperatures. In addition, a correlation was found between the optical band gap of the SnS films (as determined by UV-Vis spectroscopy) and the stoichiometry. The measured optical band gaps were lowered from 1.88 to 1.26eV. We believe that the process described in this report could be used in the production of SnS films with tuneable band gaps for solar cell applications.

Acknowledgements

The authors would like to acknowledge the EPSRC Core Capability in Chemistry (CCC), Grant Number EP/K039547/1 (Director: Prof. Gareth Morris), for access to numerous analytical equipment. The authors would also like to thank Dr. Christopher Wilkins at School of Materials University of Manchester for helpful discussions on SEM and EDX. MAS acknowledges the Iraqi Culture Attaché in London for financial support. NS thanks the Parker family for funding his position.

Compliance with Ethical Standards

Conflict of Interest: The authors declare that they have no conflict of interest.

Open Access: This article is distributed under the terms of the Creative Commons Attribution 4.0 International License (http://creativecommons.org/licenses/by/4.0/), which permits unrestricted use, distribution, and reproduction in any medium, provided you give appropriate credit to the original author(s) and the source, provide a link to the Creative Commons license, and indicate if changes were made.

Electronic supplementary material: The online version of this article (doi:10.1007/s10853-016-9906-7) contains supplementary material, which is available to authorized users.

Source: Al-Shakban M, Xie Z, Savjani N, *et al.* A facile method for the production of SnS thin films from melt reactions[J]. Journal of Materials Science, 2016, 51(13):6166–6172.

References

[1] Reddy KTR, Reddy NK, Miles RW (2006) Photovoltaic properties of SnS based solar cells. Sol Energ Mater Sol Cells 90:3041–3046.

[2] Lei Y, Song S, Fan W, Xing Y, Zhang H (2009) Facile synthesis and assemblies of flowerlike SnS_2 and In^{3+}-doped SnS_2: hierarchical structures and their enhanced photocatalytic property. J Phys Chem C 113:1280–1285.

[3] Motevalizadeh L, Khorshidifar M, Abrishami ME, Mohagheghi MMB (2013) Nanocrystalline ITO-Sn2S3 transparent thin films for photoconductive sensor applications. J Mater Sci 24:3694–3700.

[4] Zhu H, Yang D, Ji Y, Zhang H, Shen X (2005) Two-dimensional SnS nanosheets fabricated by a novel hydrothermal method. J Mater Sci 40:591–595. doi:10.1007/s10853-005-6293-x.

[5] Ichimura M (2009) Calculation of band offsets at the CdS/ SnS heterojunction. Sol Energy Mater Sol Cells 93:375–378.

[6] Ghosh B, Das M, Banerjee P, Das S (2009) Fabrication of the SnS/ZnO heterojunction for PV applications using electrodeposited ZnO films. Semicond Sci Technol 24:025024.

[7] Dussan A, Mesa F, Gordillo G (2010) Effect of substitution of Sn for Bi on structural and electrical transport properties of SnS thin films. J Mater Sci 45:2403. doi:10.1007/s10853-010-4207-z.

[8] Robles V, Trigo JF, Guillén C, Herrero J (2013) Structural, chemical, and optical properties of tin sulfide thin films as controlled by the growth temperature during co-evaporation and subsequent annealing. J Mater Sci 48:3943–3949. doi:10.1007/s10853-013-7198-8.

[9] Lazell M, O'Brien P, Otway D, Park J-H (2000) Single source molecular precursors for the deposition of III/VI chalcogenide semiconductors by MOCVD and related techniques. J Chem Soc Dalton Trans 24:4479–4486.

[10] Castro SL, Bailey SG, Raffaelle RP, Banger KK, Hepp AF (2004) Synthesis and characterization of colloidal CuInS2 nanoparticles from a molecular single-source precursor. J Phys Chem 108:12429–12435.

[11] Tian L, Tan HY, Vittal JJ (2007) Morphology-controlled synthesis of Bi2S3 nanomaterials via single-and multiple-source approaches. Cryst Growth Des 8:734–738.

[12] Trindade T, O'Brien P, Zhang X-M (1997) Synthesis of CdS and CdSe nanocrystallites using a novel single-molecule precursors approach. Chem Mater 9:523–530.

[13] Kevin P, Lewis DJ, Raftery J, Malik MA, O'Brien P (2015) Thin films of tin(II) sulphide (SnS) by aerosol-assisted chemical vapour deposition (AACVD) using tin(II) dithio-carbamates as single-source precursors. J Cryst Growth 415:93–99.

[14] Trindade T, O'Brien P, Pickett NL (2001) Nanocrystalline semiconductors: synthesis, properties, and perspectives. Chem Mater 13:3843–3858.

[15] O'Brien P, Nomura R (1995) Single-molecule precursor chemistry for the deposition of chalcogenide (S or Se) containing compound semiconductors by MOCVD and related methods. J Mater Chem 5:1761–1773.

[16] Malik MA, Afzaal M, O'Brien P (2010) Precursor chemistry for main group elements in semiconducting materials. Chem Rev 110:4417–4446.

[17] Ramasamy K, Malik MA, Revaprasadu N, O'Brien P (2013) Routes to nanostructured inorganic materials with potential for solar energy applications. Chem Mater 25:3551–3569.

[18] Lewis DJ, Kevin P, Bakr O, Muryn CA, Malik MA, O'Brien P (2014) Routes to tin chalcogenide materials as thin films or nanoparticles: a potentially important class of semiconductor for sustainable solar energy conversion. Inorg Chem Front 1:577–598.

[19] Kociok-Köhn G, Molloy KC, Sudlow AL (2014) Molecular routes to Cu2ZnSnS4: a comparison of approaches to bulk and thin film materials. Can J Chem 92:514–524.

[20] Savjani N, Brent JR, O'Brien P (2015) AACVD of molyb-denum sulfide and oxide thin films from molybdenum (V) based single-source precursors. Chem Vap Depos 21:71–77.

[21] Pradhan N, Efrima S (2003) Single-precursor, one-pot versatile synthesis under near ambient conditions of tunable, single and dual band fluorescing metal sulfide nanoparticles. J Am Chem Soc 125:2050–2051.

[22] Cheon J, Talaga DS, Zink JI (1997) Laser and thermal vapor deposition of metal sulfide (NiS, PdS) films and in situ gasphase luminescence of photofragments from M(S2COCHMe2)2. Chem Mater 9:1208–1212.

[23] Fischereder A, Schenk A, Rath T, Haas W, Delbos S, Gougaud C, Naghavi N, Pateter A, Saf R, Schenk D (2013) Solution-processed copper zinc tin sulfide thin films from metal xanthate precursors. Monatsh Chem 144:273–283.

[24] Lewis EA, Mcnaughter PD, Yin Z, Chen Y, Brent JR, Saah SA, Raftery J, Awudza JAM, Malik MA, O'Brien P, Haigh S (2015) In situ synthesis of PbS nanocrystals in polymer thin films from lead(II) xanthate and dithiocarbamate complexes: evidence for size and morphology control. Chem Mater 27:2127–2136.

[25] Brent JR, Lewis DJ, Lorenz T, Lewis EA, Savjani N, Haigh SJ, Seifert G, Derby B, O'Brien P (2015) Tin(II) sulfide (SnS) nanosheets by liquid-phase exfoliation of herzenbergite: IV–VI main group two-dimensional atomic crystals. J Am Chem Soc 137:12689–12696.

[26] Petkov N, Xu J, Morris MA, Holmes JD (2008) Confined growth and crystallography of one-dimensional Bi2S3, CdS, and SnSx nanostructures within channeled substrates. J Phys Chem C 112:7345–7355.

[27] Ning J, Men K, Xiao G, Wang L, Dai Q, Zou B, Liu B, Zou G (2010) Facile synthesis of IV–VI SnS nanocrystals with shape and size control: nanoparticles, nanoflowers and amorphous nanosheets. Nanoscale 2:1699–1703.

[28] Hong SY, Popovitz-Biro R, Prior Y, Tenne R (2003) Synthesis of SnS2/SnS fullerene-like nanoparticles: a superlattice with polyhedral shape. J Am Chem Soc 125:10470–10474.

[29] Ramasamy K, Kuznetsov VL, Gopal K, Malik MA, Raftery J, Edwards PP, O'Brien P (2013) Organotin dithiocarbamates: single-source precursors for tin sulfide thin films by aerosol-assisted chemical vapor deposition (AACVD). Chem Mater 25:266–276.

[30] Xu Z, Chen Y (2012) Fabrication of SnS thin films by a novel multilayer-based solid-state reaction method. Semicond Sci Technol 27:035007.

[31] Raston C, Tennant PR, White AH, Winter G (1978) Reactions of tin(II) and tin(IV) xanthates: crystal structure of Tetrakis(O-ethylxanthato)tin(IV). Aust J Chem 31:1493–1500.

[32] Robles V, Trigo JF, Guillén C, Herrero J (2015) SnS absorber thin films by co-evaporation: optimization of the growth rate and influence of the annealing. Thin Solid Films 582:249–252.

[33] Ichimura M, Takeuchib K, Onob Y, Arai E (2000) Electrochemical deposition of SnS thin films. Thin Solid Films 361:98–101.

[34] Wiedemeier H, Schnering HGV (1978) Refinement of the structures of GeS, GeSe, SnS and SnSe. Z Kristallogr 148:295–303.

[35] El-Nahass MM, Zeyada HM, Aziz MS, El-Ghamaz NA (2002) Optical properties of thermally evaporated SnS thin films. Opt Mater 20:159–170.

[36] Burton LA, Walsh A (2012) Phase stability of the earth-abundant tin sulfides SnS, SnS_2, and Sn2S3. J Phys Chem C 116:24262–24267.

[37] Sun Y, Zhong Z, Shirakawa T, Franchini C, Li D, Li Y, Yunoki S, Chen X-Q (2013) Rock-salt SnS and SnSe: native topological crystalline insulators. Phys Rev B 88:235122.

[38] Chandrasekhar HR, Humphreys RG, Zwick U, Cardona M (1977) Infrared and Raman spectra of the IV-VI compounds SnS and SnSe. Phys Rev B 15:2177.

[39] Sinsermsuksakul P, Heo J, Noh W, Hock AS, Gordon RG (2011) Atomic layer deposition of tin monosulfide thin films. Adv Energy Mater 1:1116–1125.

[40] Price LS, Parkin IP, Hardy AME, Clark RJH, Hibbert TG, Molloy KC (1999) Atmospheric pressure chemical vapor deposition of tin sulfides (SnS, Sn2S3, and SnS2) on glass. Chem Mater 11:1792–1799.

[41] Jain P, Arun P (2013) Influence of grain size on the band-gap of annealed SnS thin films. Thin Solid Films 548:241–246.

[42] Koktysh DS, McBride JR, Rosenthal SJ (2007) Synthesis of SnS nanocrystals by the solvothermal decomposition of a single source precursor. Nanoscale Res Lett 2:144–148.

[43] Tanuševski A, Poelman D (2003) Optical and photoconductive properties of SnS thin films prepared by electron beam evaporation. Sol Energy Mater Sol Cells 80:297–303.

[44] Yue GH, Peng DL, Yan PX, Wang LS, Wang W, Luoa XH (2009) Structure and optical properties of SnS thin film prepared by pulse electrodeposition. J Alloys Compd 468:254–257.

[45] Ghosh B, Das M, Banerjee P, Das S (2008) Fabrication and optical properties of SnS thin films by SILAR method. App Surf Sci 254:6436–6440.

[46] Calixto-Rodriguez M, Martinez H, Sanchez-Juarez A, Campos-Alvarez J, Tiburcio-Silver A, Calixto ME (2009) Structural, optical, and electrical properties of tin sulfide thin films grown by spray pyrolysis. Thin Solid Films 517:2497–2499.

[47] Reddy NK, Reddy KTR (1998) Growth of polycrystalline SnS films by spray pyrolysis. Thin Solid Films 325:4–6.

[48] Xu Z, Chen Y (2011) Synthesis of SnS thin films from nano-multilayer technique. Energy Procedia 10:238–242.

Chapter 4

Centrifugal Melt Spinning of Polyvinylpyrrolidone (PVP)/Triacontene Copolymer Fibres

Tom O'Haire[*], Stephen. J. Russell, Christopher M. Carr

School of Design, University of Leeds, Leeds LS2 9JT, UK

Abstract: Polyvinylpyrrolidone/1-triacontene (PVP/TA) copolymer fibre webs produced by centrifugal melt spinning were studied to determine the influence of jet rotation speed on morphology and internal structure as well as their potential utility as adsorbent capture media for disperse dye effluents. Fibres were produced at 72°C with jet head rotation speeds from 7000r min^{-1} to 15,000r min^{-1}. The fibres were characterised by means of SEM, XRD and DSC. Adsorption behaviour was investigated by means of an isothermal bottle point adsorption study using a commercial disperse dye, Dianix AC-E. Through centrifugal spinning nanofibers and microfibers could be produced with individual fibres as fine as 200–300nm and mean fibre diameters of ca. 1–2μm. The PVP/TA fibres were mechanically brittle with characteristic brittle tensile fracture regions observed at the fibre ends. DSC and XRD analyses suggested that this brittleness was linked to the graft chain crystallisation where the PVP/TA was in the form of a radial brush copolymer. In this structure, the triacontene branches interlock and form small lateral crystals around an amorphous backbone. As an adsorbent, the

PVP/TA fibres were found to adsorb 35.4mg g^{-1} compared to a benchmark figure of 30.0mg g^{-1} for a granular-activated carbon adsorbent under the same application conditions. PVP/TA is highly hydrophobic and adsorbs disperse dyes through the strong "hydrophobic bonding" interaction. Such fibrous assemblies may have applications in the targeted adsorption and separation of non-polar species from aqueous or polar environments.

1. Introduction

The ability to influence the interaction between multicomponent liquids and fibre surfaces to control behaviour such as mutual hydrophobicity and oleophilicity is important in numerous applications such as filter media, chemical sorbents and protective clothing. An attractive development strategy is to identify novel polymer systems for fibre production for use in such products. However, the limitations and difficulties when processing such polymers on conventional spinning lines continue to restrict feasibility. Polyvinylpyrrolidone/1-triacontene (PVP/TA) consists of a polyvinylpyrrolidone (PVP) backbone with 1-triacontene (C$_{30}$) side chains[1], **Figure 1**, and whilst homogenous PVP is hydrophilic and water soluble, PVP/TA is non-polar and highly hydrophobic. The 1-triacontene is densely grafted and forms a brush copolymer structure[2] and in this form, the inherent hydrophilicity of the PVP is modified by the hydrophobicity of the non-polar C$_{30}$ chains[3][4]. Currently, PVP/TA is used as a water-proofing agent in cosmetics and sun-screens as well as in printing inks and other liquid formulations[1][4]–[7]. It is

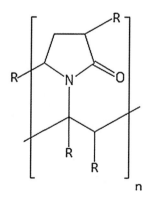

Figure 1. Typical structure of alkyated PVP brush copolymers, where R is either hydrogen or a long chain hydrocarbon such as C$_{30}$H$_{61}$ in the case of PVP/TA[1][2].

normally used as a film or in the form of a liquid dispersion. Given the hydrophobicity, water insolubility and oleophilic behaviour of PVP/TA, the production of fibres from this copolymer provides significant opportunities to extend the range of industrial applications.

Spinning of homogenous PVP homopolymer and in blends with other polymers has been previously demonstrated through electrospinning[8][9] and needleless centrifugal spinning[10]. Fibres have been formed using PVP dissolved in volatile solvents in conjunction with: a CO_2 atmosphere[9]; a saline sheath[8]; and through centrifugal electrospinning, a hybrid technique[11]. The needleless centrifugal spinning of PVP used a 20% aqueous solution to form fibres as 400 ± 100nm in diameter at 4000r min^{-1}. However, there is little information on the melt processing of PVP into fibres[12] or the utilisation of PVP/TA or similar alkylated PVP materials in fibre spinning studies. Although electrospinning is a versatile spinning technique, it is limited by many factors such as the need to form an electrified jet. PVP/TA is insoluble in aqueous solvents and has a depressed melting point compared to PVP, creating the possibility of low temperature melt processing. Recently, centrifugal spinning has been increasingly applied to form non-ideal materials into ultrafine fibres from either polymer melts or solutions[13]. This technique involves rotating polymer fluids at high speeds to induce jet formation and elongation without the need for external drawing, electrostatic forces or high velocity hot air[14]. It has been demonstrated that centrifugal spinning can readily produce fibres finer than 1μm from a variety of polymer materials[14][15] from either thermoplastic liquid melts or solutions. Fibres have been produced from conventional thermoplastics such as nylon and polyester; speciality polymers such as polycaprolactone and poly(lactic acid); bismuth; ceramic materials; compounds of polypropylene and carbon nanotubes; compounds of poly(ethylene terephthalate) with graphene[13][16][17]. In centrifugal spinning, the operating conditions can have a significant influence on fibre diameter and investigation is often necessary to establish optimum parameters, balancing fibre fineness and the level of beading[15].

Forming PVP/TA into submicron fibres using centrifugal spinning has the potential to significantly increase the available surface area of the material compared to films or granulate. This would usefully facilitate the "capture" of molecules through mutual hydrophobic and oleophilic adsorption[4][18]. Specifically, this could assist in the chemical adsorption of hydrophobic disperse dyes from textile waste effluent with improved efficiency compared to traditional sorbents. Disperse dyes are used in the colouration of polyester and other synthetic fibres and are typically non-polar and hydrophobic. Disperse dyes are not always decolourised or decomposed by biological or reductive treatments and may therefore persist in the environment and in aquatic systems[19][20]. Adsorption of disperse dyes on to a capture media has been proposed as one solution for removal, possible recycling and minimising potential accumulation in the environment[21]. In addition to the treatment of dyehouse effluent, such hydrophobic/oleophilic fibres could also be used as sorbents and capture agents in domestic laundering, potentially preventing the unwanted cross-staining of dyes[22].

The purpose of this investigation was to study the centrifugal spinning of submicron PVP/TA copolymer fibres and to evaluate the structure and physical properties of resulting fibres and fabrics. In addition to studying as-spun fibre morphology by scanning electron microscopy (SEM), the thermal behaviour and fine structure of the fibres were elucidated by differential scanning calorimetry (DSC) and X-ray diffraction (XRD), respectively. Finally, the potential applications of PVP/TA fibre in porous sorption media were explored by comparing the adsorption of a disperse dye against an activated carbon benchmark.

2. Materials and Methods

2.1. Materials

The polyvinylpyrrolidone/1-triacontene polymer was supplied as a flake by Sigma Aldrich, UK, under the commercial name Antaron WP660, which is also known as triacontyl polyvinylpyrrolidone, CAS registered as 2-pyrrolidinone, 1-ethenyl with 1-triacon-tene (CAS number 136,445-69-7). A commercial anthraquinone-based disperse dye (Dianix Blue AC-E, CAS number 98725-74-7) was supplied by Dystar Textilfarben, Germany, and was used without further treatment or purification. Chromatography grade acetone, (99.8%) and laboratory grade Triton X100 (CAS 9002-93-1), both via Sigma Aldrich, UK, were used for the dye adsorption study along with an activated carbon adsorbent, 1mm granular Norit®supplied by Sigma Aldrich, UK. The as-delivered activated carbon was rinsed in deionised water and then dried at 50°C for 24h in a laboratory oven prior to use.

2.2. Capillary Rheometry

Apparent melt viscosity of PVP/TA was assessed using a RH2000 capillary rheometer (Bohlin Instruments, UK) configured with a capillary die 1mm in diameter and 16mm in length. Rheology measurement was conducted at temperatures of 65°C, 70°C, 75°C, and 80°C with piston speeds of 50mm min^{-1}, 70mm min^{-1}, and 120mm min^{-1}.

2.3. Centrifugal Spinning

Fibre production was carried out using the Force-spinningTM L1000 M centrifugal spinner (Fiberio, USA) in a melt spinning configuration using tri-orifice spinnerets with two different orifice diameters available: 159μm (fine) and 602μm (coarse). A static arrangement of posts positioned 115mm circumferentially from the spinneret was used as a supporting collector. The influence of rotational speed was assessed by varying the rotational speed from 6000r min^{-1} to 15,000r min^{-1}. A mass of 300mg of PVP/TA was added to the spinneret and external heat was applied until a stable polymer temperature of 72°C ± 1°C was achieved. The spinneret was then accelerated and held constant at a given speed for 45 s before decelerating. A minimum of two fibrous webs was formed at each operating condition for assessment. For the dye adsorption investigations, the PVP/TA fibres were produced at a rotational speed of PVP/TA at 11,000r min^{-1} using the coarser spinneret (602μm) only.

2.4. Scanning Electron Microscopy (SEM) Analysis

The fibres were examined using a Jeol JSM-6610LV scanning electron microscope (Japan). The fibrous samples were mounted on conductive carbon tape and gold sputter coated prior to analysis. An acceleration voltage of 5kV–15kV was used with a typical working distance of 100mm. Images were captured at magnifications of ×1000 for fibre diameter measurements. Higher magnifications were used to image specific areas and fibres of interest. Image analysis software,

Image J, was used to assess the average fibre diameters by evaluating a minimum of 200 fibres from 5 different regions of the sample stub.

2.5. Thermal Analysis

The thermal properties of the PVP/TA fibres and flake were assessed using a Perkin Elmer Jade differential scanning calorimeter. Samples were analysed in aluminium pans from −10°C to 250°C at a rate of 20°C min^{-1} with a nitrogen gas supply of 20mL min^{-1}. A heat-cool-heat cycle was conducted on a fibrous sample from 25°C to 120°C at the same heating/cooling rate under identical conditions.

2.6. X-Ray Diffraction (XRD) Analysis

The internal fine structure of the PVP/TA raw material and as-spun PVP/TA fibres were investigated using a P'ANanalytical X'Pert MPD X-ray diffractometer. The as-supplied PVP/TA flake was analysed without modification and the fibrous webs were cold pressed flat into a thin sample prior to observation. The radiation source was CuKα (λ = 1.540Å) and the scans were taken in the theta:theta orientation through 4°−60° with a step size of 0.066°.

2.7. Dye Adsorption Study

The level of dye removal by the fibres and the subsequent adsorption isotherm for the PVP/TA fibres and the disperse dye were determined using a bottle point adsorption method[23][24]. Dye dispersions of concentrations of 25mg dm^{-3}–300mg dm^{-3} were made using Dianix Blue AC-E and distilled water. The PVP/TA fibres were added to the dispersions at a ratio of 0.5g adsorbent to 0.1dm^3 of liquor. Triton ×100 was added at a concentration of 1g dm^{-3} to act as a wetting agent. A parallel set of dispersions were treated withgranulated activated carbon under the same conditions. The dispersions were stirred for 3 days in a sealed jar using a magnetic impeller. After this period, the dye liquors were passed through porosity 2 sintered class crucibles to separate the PVP/TA fibres and carbongranules from the solutions. The filtrates were mixed 50:50 with HPLCgrade acetone in order to dissolve the disperse dye molecules and the λ_{max}

was measured using UV-Vis spectrophotometry. The measured values then converted into concentrations using a calibration chart produced for Dianix Blue AC-E in 50:50 distilled water and acetone.

3. Results and Discussion

3.1. Capillary Rheometry

The capillary rheometry resultsgiven in **Figure 2** show that the melt viscosity of PVP/TA is sensitive to temperature changes, reducing from >25Pa s^{-1} at temperatures 65°C and 70°C to < 2Pa s^{-1} at 80°C.

3.2. Centrifugal Spinning of PVP/TA

PVP/TA was successfully formed into fibrous webs using melt-centrifugal spinning using both the coarse and fine spinnerets at a polymer temperature of 72°C ± 1°C. When using the fine melt spinneret, fibres were successfully formed at all rotational spinneret speeds from 12,000 to 15,000r min with all other conditions fixed. At 11,000r min^{-1} and below, the PVP/TA melt was unable to be ejected from the fine spinneret due to the relatively high melt viscosity. It is proposed that

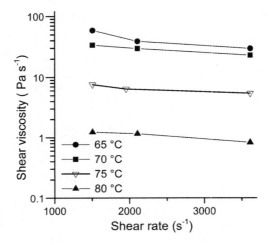

Figure 2. Apparent shear viscosity curves as a function of shear rate and temperature.

at low rotational speeds the inertia of the melt is insufficient to overcome the capillary resistance of the 159μm orifice; thus, the critical speed necessary for successful extrusion was in excess of 11,000r min for the combination of spinneret and temperature investigated[25]. At 16,000r min and above the level of beading it became excessive and the level of fibre produced decreased significantly. It is proposed that the inertial and aerodynamic forces encountered at such rotational speeds is excessively high, leading to jet breakup and bead formation as instabilities become critical[25][26]. For the coarse spinneret (602μm), the critical rotational speed was lower, with fibres being produced at speeds 7000r min^{-1} and above. This was due to coarser spinneret requiring a lower inertial pressure to overcome the capillary resistance. With this spinneret, the beading became excessive at 15,000r min^{-1}.

Figure 3 shows SEM micrographs from selected webs which show that the PVP/TA fibres produced were cylindrical and smooth with very little surface texture. A disparity in fibre diameters within a sample can also be observed. Variation in fibre diameter was evident with very fine fibres (>300nm) proximal with much

Figure 3. SEM micrographs of PVP/TA fibres produced using the coarse spinneret at a 8000, b 10,000, c 12,000 and d 14,000r min^{-1}, respectively.

coarser fibres (<5μm), resulting in a relatively high value of standard deviation, **Table 1**. A one-way ANOVA ($\alpha = 0.05$) analysis of the date on the centrifugal revealed that the means were statistically different for conditions measured [$p < 0.001$; $F (63.1) >$ Fcrit (1.79)] and different within the samples formed using the coarse spinneret [$p < 0.001$; $F (23.6) >$ Fcrit (2.02)] and the fine spinneret [$p < 0.05$; $F (4.29) >$ Fcrit (2.61)].

The formation of coarse fibres was attributed to the acceleration and deceleration phases in the spinning cycle; the turbulent attenuation of fibres; premature jet breakage; and variability in the throughput of polymer. It is proposed that a combination of these factors contributes to generating fibres with a range of fibre diameters significantly greater than the reported mean value. It has been previously reported that mean fibre diameter of centrifugal spun fibres decreases with spinning duration and it is proposed that a continuous machine would create a lower proportion of coarse fibres as the effects of acceleration and deceleration are minimised[15].

At low spinneret rotation speeds, the extensional forces on the polymer jet are lower relative to higher speeds and the level of attenuation is lower, resulting in larger fibre diameters. As with many polymers, the formation of PVP/TA into fibres is a balancing act between polymer throughput, initial jet diameter, jet attenuation and jet break-up[25][26].

Examination of the discontinuities or fibre ends resulting from breakage is shown in **Figure 4(a)**, **Figure 4(b)**, which shows that the fibre end breaks are relatively abrupt and smooth-faced with no indications of fibrillation or ductile fracture. Such fibre breakage was observed across all samples irrespective of the process conditions employed, and therefore appears to be a characteristic feature of centrifugally spun PVP/TA fibres and the bulk properties of the co-polymer. A

Table 1. Mean fibre diameters for PVP/TA fibres made through centrifugal spinning.

Spinneret	Fibre diameter (μm)	Rotational speed (r min^{-1})									
		6000	7000	8000	9000	10,000	11,000	12,000	13,000	14,000	15,000	16,000
Coarse	$\bar{x}$	x	7.88	5.02	5.38	4.23	4.01	3.43	2.81	2.33	b	x/b
	σ	x	6.16	3.12	4.31	3.04	3.00	2.20	1.69	1.47	b	x/b
Fine	$\bar{x}$	x	x	x	x	x	x	1.83	1.43	1.52	1.46	b
	σ	x	x	x	x	x	x	1.66	1.02	1.20	0.87	b

x web was not produced, b excessive beading/spraying occurred

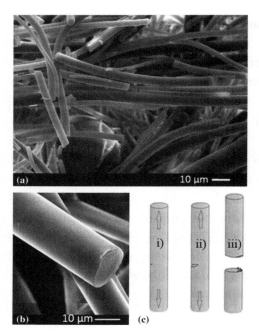

Figure 4. Fibre end breaks as observed in PVP/TA fibres formed in a 15,000rpm (fine spinneret); b 12,000r min^{-1} (coarse spinneret) with the mechanism of brittle fracture modelled in c.

slight imperfection frequently observed on one edge of the break cross section is strongly indicative of brittle tensile fracture, as previously detailed[27].

A schematic representation of the brittle tensile fracture in a PVP/TA fibre is shown in **Figure 4(c)**. Within a fibre with a surface flaw (i), the crack will propagate (ii) until eventually the fibre breaks (iii) leaving a characteristic clean break across most of the width of the fibre fracture face. The increased propensity of fibre breakage is not only thought to be a combination of disruption of polymer streams during spinning but also to fracture of fibres during handling of the PVP/TA webs, which although self-supporting, were relatively weak. Bending failure in brittle materials also produces the clean fracture as shown in **Figure 4**[28], and it is likely that load in bending also causes a significant number of breaks.

3.3. Internal Fine Structure and Thermal Analyses of Fibres

The DSC experiments revealed the thermal behaviour of PVP/TA flake and fibres when heated through the melting point, **Figure 5**.

The PVP/TA flake has a broad melting endotherm with $T_{\text{-onset}}$ at 55°C and a T_{melt} at 72°C with a profile typical of the melting of a semicrystalline polymer with one crystal form. The DSC thermograms of the PVP/TA fibres exhibit a significantly different thermal profile. The fibre endotherms have a double peak or shoulder trace (50°C–60°C), which is most clearly evident in the fibres produced from the coarse spinneret. This double endotherm indicates that PVP/TA fibres have a clear α and β peak profile not observed in the flake. The major α peak was the main melt transition and is found at around 76°C–80°C which was preceded by smaller β peak at 57°C–62°C that represented a pre-melt transition. Double melting peaks such as these in DSC can be attributed to the following conditions:

i. Two distinct crystal morphologies and configurations;

ii. Re-crystallisation behaviour during the DSC scan[29];

iii. Material is actually a binary blend of two differentgrade products[30];

A typical heat-cool-heat DSC scan for PVP/TA isgiven in **Figure 6** which

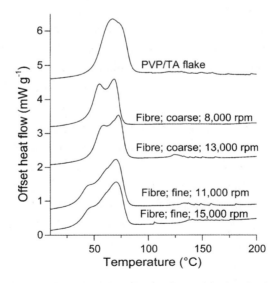

Figure 5. DSC thermograms of PVP/TA fibres formed using both the coarse and fine spinnerets at selected processing speeds.

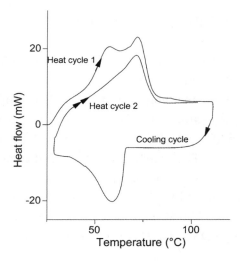

Figure 6. DSC thermogram showing a heat-cool-heat cycle of PVP/TA fibres formed at 8000r min^{-1} in the coarse spinneret.

shows that the double peak profile observed in fibres was erased by the first heating cycle and the melting profile for the second heat scan does not include this secondary peak. This secondary heating profile was observed in all the PVP/TA fibres regardless of processing conditions. The deletion of this early peak indicates that it is a feature of processing not inherent to PVP/TA. The absence of such a β-peak for the flake material suggests that conditions (i) and (ii) are not applicable in PVP/TA. It is thus concluded and appears that centrifugally spun PVP/TA has two crystal conformations: a primary α-crystal that is found in both the flake and the fibre; and a secondary β-crystal form, which melts at a lower temperature and is formed during the processing of PVP/TA into a fibre using melt processing.

The XRD patterns, **Figure 7**, for PVP/TA flake and the centrifugally spun fibres generated two dominant peaks superimposed onto a broad and weak halo that stretches from 10° to 30° 2θ. This broad shoulder is indicative of an amorphous region[31] and it has been demonstrated previously that homogenous PVP exhibits low levels of order in XRD analyses indicating that PVP is entirely amorphous[32][33]. However, the distinct peaks evident in the XRD patterns for PVP/TA fibres are a strong indicator of long range order and therefore semi-crystallinity. The PVP/TA diffraction pattern has peaks at 21.6°, 24.0°, and 36.4° 2θ along with additional secondary peaks between 25° and 40° 2θ. The peak locations and diffraction pattern shape are very similar to that of linear and branched polye-

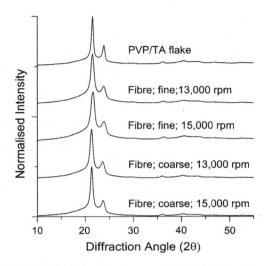

Figure 7. Selected XRD plots of PVP/TA flake and fibres.

thylene[34]. Polyethylene forms an orthorhombic crystal with peaks found around 21.4°, 24.2°, and 36.5° 2θ[35][36].

It is proposed that PVP/TA forms branched chain crystals where the C_{30} side chains falls and in crystallographic register with local C_{30} elements whilst the PVP backbone remains in an amorphous state. The crystallisation of a graft chain on an amorphous backbone is known as interdigitating packing[37]. Deviations in backbone chain registration and variations in chain length would allow for a single copolymer brush to enter both crystalline and amorphous regions and would incorporate further defects and disorder into the crystal system. Previous research has indicated that graft side chains will only crystallise at sufficient distance away from the backbone; this will further limit the size of the crystal formed by the C_{30} chains in at least one dimension[38][39]. A simple model of this interaction isgiven in **Figure 8**.

Based on this structure, there is a possible link between the laterally crystalline structure and the propensity for PVP/TA fibres to undergo brittle fracture. The net orientation of the brush polymer is proposed to be along the length of the fibre and the crystalline elements roughly perpendicular in net orientation. Therefore due to the relatively small dimensions of the crystals, the microfibrils will be capable of slipping more easily relative to chain folding high polymers[40]. This

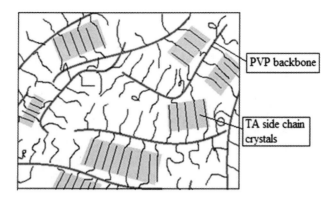

Figure 8. Schematic model for crystallisation of graft copolymers[37]–[39].

could explain the brittle tensile fractures observed in the SEM micrographs of the fibres. It is hypothesised that these lateral crystals do not contribute to the tensile strength and elasticity associated with crystallised regions in long linear chains[41].

3.4. Adsorbent Behaviour of PVP/TA Fibres

Having successfully produced PVP/TA fibres for the first time it was deemed useful to determine their feasibility as a fibrous sorption material and their relative performance against activated carbon. The UV-Vis adsorption values for the filtered solutions were converted into dye concentrations at equilibrium (C_e) at each dyebath concentration (C_0), using a calibration curve for the disperse dye, Dianix Blue AC-E, in 50:50 acetone. The adsorption at equilibrium (Q_e) for the PVP/TA fibres was calculated using $Q_e = (C_o - C_e)\dfrac{V}{m}$ where V is the volume of solution and m is the mass of adsorbent.

Table 2 provides the summary of the adsorption at equilibrium figures at each dye concentration along with the degree of removal, calculated using: Removal (%) = (C_o − C_e)/C_0 × 100. As indicated the PVP/TA fibres effectively adsorbed the disperse dye molecule, with removal rates of up to 97.1%, based on an original dye concentration of 25mg dm with an adsorption at equilibrium as high as 35.4mg g^{-1}.

Table 2. Proportion of disperse dye removed by PVP/TA fibres and activated carbon and amount of dye adsorbed onto the adsorbent fibres and carbon.

	Initial dyebath concentration, C_o (mg dm^{-3})					
	25	50	100	150	200	300
Dye/fibre parameters						
Equilibrium concentration C_e (mg dm^{-3})	0.7	8.4	24.7	38.5	61.3	122.8
Dye removal from solution (%)	97.1	83.2	75.3	74.3	70.4	59.1
Adsorption at equilibrium Qe (mg g^{-1})	4.9	8.3	15.1	22.3	27.7	35.4
Dye/activated carbon parameters						
Equilibrium concentration, C_e (mg dm^{-3})	1.0	5.9	21.5	40.2	73.4	150.4
Dye removal from Solution, (%)	95.9	88.3	78.5	73.2	63.3	49.9
Adsorption at equilibrium, Qe (mg g^{-1})	4.8	8.8	15.7	22.0	25.3	30.0

Comparison of the adsorption data, relative to granular activated carbon, indicated that the activated carbon had comparable adsorption of the disperse dye to the PVP/TA fibres but overall had a lower capacity for adsorption with a Q_e of 30.0mgg The isothermal plots for PVP/TA fibres and activated carbon show that the empirical data are well represented by the Langmuir isothermal model, **Figure 9**. The Langmuir model stipulates that adsorption occurs as a single monolayer limited to a finite number of dye sites. Therefore, the isothermal adsorption of disperse dyes on these materials is related to the surface area as this will determine the number of adsorption sites available.

Activated carbon is known for its high surface area due to a fine pore structure, enabling relatively high capture of particulate such as dyes on the surface. Numerous chemical bonding mechanisms can exist between a surface and a disperse dyes in aqueous media[42]. The literature shows that the adsorption of disperse dyes and other hydrophobic/insoluble compounds is linked to the prevalence of hydrophobic sites on the surface of the material[21][43]. It is proposed that the adsorption of non-ionic disperse dyes onto such materials occurs through a weak chemical bond such as the non-polar bonding of hydrophobic materials[44][45]. This "hydrophobic bonding" of non-polar materials in aqueous environments is shown to be unusually strong as it is a combination of interface bonding between the adsorbent and adsorbate and the mutual phobicity for water and other polar aqueous liquids[46]–[48]. Once two hydrophobic species are brought together, the energy required to dissociate the elements is a function of the energy required to disrupt

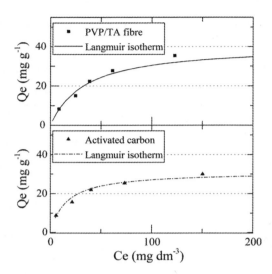

Figure 9. Recorded data and calculated Langmuir isothermal curves for PVP/TA fibres and activated carbon.

the hydrogen bonding of the surrounding water molecules, creating a bonding phenomenon that is long range and much stronger than the ionic and electrostatic interaction between the two molecules[46][47]. The Langmuir model assumes that there is a finite number of 'active sites' onto which adsorption can occur. For reactive dyeing, these sites can be related the number of functional-groups at the surface of the polymer; for disperse dyeing, onto hydrophobic materials these 'active sites' may be regions where hydrophobic bonds are most likely to occur The triacontene-grafted chains on the PVP polymer backbone provide the hydrophobic adsorption sites for the PVP/TA fibre. The hydrophilic PVP is masked by thegraft chains, creating an entirely hydrophobic surface. In contrast for activated carbon, there are a mixture of hydrophobic and hydrophilic regions, with disperse dyes being captured at hydrophobic regions[49]. As the interaction between PVP/TA and the dye molecules relies on mutual hydrophobicity and does not require elevated temperatures or secondary salts, there is potential to use such a material to collect not only dyes but other material such as hydrocarbons and fatty soils in either a laundry or industrial setting.

4. Conclusions

Polyvinylpyrrolidone/triacontene was successfully melt spun into ultrafine fibres through melt centrifugal spinning. Increasing the rotational speed reduced the average fibre diameter with typically the fibre average diameters observed being 1μm–2μm, although some individual fibres were observed as fine as 0.2μm. The fibres were found to be highly brittle, and examples of brittle elastic fracture were observed through SEM analysis of the fibres. DSC and XRD analyses suggest that this brittleness was linked to thegraft chain crystallisation where the PVP/TA was in the form of a radial brush copolymer which forms a semi-crystalline structure where the triacontene branches interlock and form small lateral crystals around an amorphous backbone. The radial triacontene branches also impart hydrophobicity and "mask" the hydrophilic nature of the PVP backbone.

This hydrophobic behaviour could potentially be useful in adsorption applications and this research assessed the effectiveness of PVP/TA fibres as a dye adsorbent. This was the first time that PVP/TA fibres have been considered as an adsorbent for disperse dyes. The PVP/TA fibres had an adsorption capacity of 35.4mg g^{-1} of Dianix Blue A-CE under neutral conditions, exceeding the capacity observed forgranular-activated carbon. The hydrophobic nature of the PVP/TA fibres and associated high surface area potentially offers a suitable adsorbent medium for disperse dyes, binding the colorants through nonpolar hydrophobic interactions.

Acknowledgements

The authors would like to thank The Clothworkers' Company for funding the PhD studies of Tom O'Haire.

Open Access: This article is distributed under the terms of the Creative Commons Attribution 4.0 International License (http://creativecommons.org/licenses/by/4.0/), which permits unrestricted use, distribution, and reproduction in any medium, provided yougive appropriate credit to the original author(s) and the source, provide a link to the Creative Commons license, and indicate if changes were made.

Source: O'Haire T, Russell S J, Carr C M. Centrifugal melt spinning of polyvinylpyrrolidone (PVP)/triacontene copolymer fibres[J]. Journal of Materials Science, 2016, 51(16):7512–7522.

References

[1] Login RB, Barabas ES (1996) Personal care application polymers (acetylene-derived). In: Salamone JC (ed) Polymeric materials encyclopedia. CRC Press, Boca Raton.

[2] Brittain WJ, Minko S (2007) A structural definition of polymer brushes. J Polym Sci Part A 45(16):3505–3512.

[3] Puccettig, Fares H (2014) A new approach for evaluating the water resistance of sunscreens on consumers: tap water vs. salt water vs. chlorine water. Int J Cosmet Sci 36(3):284–290.

[4] Brugnara M, Degasperi E, Volpe CD, Maniglio D, Penati A, Siboni S, Toniolo L, Poli T, Invernizzi S, Castelvetro V (2004) The application of the contact angle in monument protection: new materials and methods. Colloid Surf A 241(1–3):299–312.

[5] Liu KC, Helioff M, Rerek M, Davis LJ, grenner DE, Waldorf-Geber A, Koppel R (1997) Suncreen concentrate. United States Patent 5,916,544.

[6] Snyder F, Reinhartg, DiGirolamo D (1995) Cosmetic compositions for lengthening, coloring and curling eyelashes. United States Patent 5,389,363.

[7] Grinwald Y, Bar-Haimg, Berson Y (2015) Positively charged ink composition. United States Patent 8,927,635.

[8] Yu DG, White K, Yang JH, Wang X, Qian W, Li Y (2012) PVP nanofibers prepared using co-axial electrospinning with salt solution as sheath fluid. Mater Lett 67(1):78–80.

[9] Wahyudiono Machmudah S, Murakami K, Okubayashi S, goto M (2013) generation of PVP fibers by electrospinning in one-step process under high-pressure CO_2. Int J Ind Chem 4(1):1–6.

[10] Chen H, Xu H, Sun J, Liu C, Yang B (2015) Effective method for high-throughput manufacturing of ultrafine fibres via needleless centrifugal spinning. Micro Nano Lett 10(2):81–84.

[11] Liu SL, Long YZ, Zhang ZH, Zhang HD, Sun B, Zhang JC, Han WP (2013) Assembly of oriented ultrafine polymer fibers by centrifugal electrospinning. J Nanomater 2013:9.

[12] McCann JT, Marquez M, Xia Y (2006) Melt coaxial electrospinning: a versatile method for the encapsulation of solid materials and fabrication of phase change nanofibers. Nano Lett 6(12):2868–2872.

[13] O'Haire T, Rigout M, Russell S, Carr C (2014) Influence of nanotube dispersion and spinning conditions on nanofibre nanocomposites of polypropylene and multi-walled carbon nanotubes produced through ForcespinningTM. J Thermoplast Compos Mater 27(2):205–214.

[14] Sarkar K, gomez C, Zambrano S, Ramirez M, de Hoyos E, Vasquez H, Lozano K

(2010) Electrospinning to Force-spinningTM. Mater Today 13(11):12–14.

[15] McEachin Z, Lozano K (2011) Production and characterisation of polycaprolactone nanofibres via Forc-espinningTM technology. J Appl Polym Sci 126:473–479.

[16] Bandla S, Winarski R, Hanan J (2013) Nanotomography of polymer nanocomposite nanofibers. In: Jin H, Sciammarella C, Furlong C, Yoshida S (eds) Imaging methods for novel materials and challenging applications, vol 3. Springer, New York, pp 193–198.

[17] Gramley K (2012) Forcespinning ceramic nanofibres. Adv Ceram Rep 2012(2):6–7.

[18] Cavazzuti R, Mattox BK, Swanborough M (2002) Mixture of wax, ester and fatty alcohol. United States Patent 6,444,212.

[19] Christie R (2014) Colour chemistry. Royal Society of Chemistry, London.

[20] Carneiro PA, UmbuzeirogA, Oliveira DP, Zanoni MVB (2010) Assessment of water contamination caused by a mutagenic textile effluent/dyehouse effluent bearing disperse dyes. J Hazard Mater 174(1):694–699.

[21] Rai PB, Banerjee SS, Jayaram RV (2007) Removal of disperse dyes from aqueous solution using sawdust and BDTDA/sawdust. J Dispers Sci Technol 28(7): 1066–1071.

[22] Eschway H, Hackler L, Hartmann L, Kauschke M, Kumpel T, Kunkel HA, Nahe T, Ruzek I (1990) Absorbent body of nonwoven material and a method for the production thereof. United States Patent, 4,902,559.

[23] Allen S, Mckayg, Porter J (2004) Adsorption isotherm models for basic dye adsorption by peat in single and binary component systems. J Colloid Interface Sci 280(2):322–333.

[24] El-Geundi MS (1991) Colour removal from textile effluents by adsorption techniques. Water Res 25(3):271–273.

[25] O'Haire T (2015) The production of ultrafine fibres using variations of the centrifugal spinning technique. PhD Dissertation, University of Leeds.

[26] Ellison CJ, Phatak A,giles DW, Macosko CW, Bates FS (2007) Melt blown nanofibers: fiber diameter distributions and onset of fiber breakup. Polymer 48(11): 3306–3316.

[27] Hearle JW, Lomas B, Cooke WD (1998) Atlas of fibre fracture and damage to textiles. Elsevier, London.

[28] Hearle JW (2002) Forms of fibre fracture. In: Elices M, Llorca J (eds) Fiber fracture. Elsevier, Oxford, pp 57–71 .

[29] Barham P, Hill M, Keller A, Rosney CD (1988) Phase separation in polyethylene melts. J Mater Sci Lett 7(12):1271–1275.

[30] Blundell D (1987) On the interpretation of multiple melting peaks in poly (ether

ether ketone). Polymer 28(13):2248–2251.

[31] Camini R, Pandolfi L, Ballirano P (2000) Structure of polyethylene from X-ray powder diffraction: influence of the amorphous fraction on data analysis. J Macromol Sci B 39(4):481–492.

[32] Razzak MT, Dewi S, Lely H, Taty E (1999) The characterization of dressing component materials and radiation formation of PVA–PVP hydrogel. Radiat Phys Chem 55(2):153–165.

[33] Sethia S, Squillante E (2004) Solid dispersion of carbamazepine in PVP K30 by conventional solvent evaporation and supercritical methods. Int J Pharm 272(1–2):1–10.

[34] Ueno N, Seki K, Sugita K, Inokuchi H (1991) Nature of the temperature dependence of conduction bands in polyethylene. Phys Rev B 43(3):2384–2390.

[35] Smith P, Lemstra P (1980) Ultra-high-strength polyethylene filaments by solution spinning/drawing. J Mater Sci 15(2):505–514. doi:10.1007/BF00551705.

[36] Zheng L, Waddon AJ, Farris RJ, Coughlin EB (2002) X-ray characterizations of polyethylene polyhedral oligomeric silsesquioxane copolymers. Macromolecules 35(6):2375–2379.

[37] Neugebauer D, Theis M, Pakula T, Wegnerg, Matyjaszewski K (2006) Densely heterografted brush macromolecules with crystallizablegrafts: synthesis and bulk properties. Macromolecules 39(2):584–593.

[38] Schouten AJ, Wegnerg (1991) Langmuir-Blodgett mono- and multi-layers of preformed poly (octadecyl methacrylate)s, Surface-area isotherms of atactic and isotactic poly(octadecyl methacrylate)s and transfer into multilayers. Die Makromol Chem 192(10):2203–2213.

[39] Arndt T, Schouten AJ, SchmidtgF, Wegnerg (1991) Langmuir–Blodgett mono- and multilayers of preformed poly (octadecyl methacrylate)s, 2. Structural studies by IR spectroscopy and small-angle X-ray scattering. Die Makro-mol Chem 192(10):2215–2229.

[40] Peterlin A (1972) Morphology and properties of crystalline polymers with fiber structure. Text Res J 42(1):20–30.

[41] Sui K, Liang H, Zhao X, Ma Y, Zhang Y, Xia Y (2012) Synthesis of amphiphilic poly(ethylene oxide-co-glycidol)-graft-polyacrylonitrile brush copolymers and their self-assembly in aqueous media. Macromol Chem Phys 213(16):1717–1724.

[42] Giles CH (1961) Studies in adsorption, part XIV: the mechanism of adsorption of dispersedyes by cellulose acetates and other hydrophobic fibers. Tex Res J 31(2):141–151.

[43] Da̧browski A, Podkos'cielny P, Hubicki Z, Barczak M (2005). Adsorption of phenolic compounds by activated carbon—a critical review. Chemosphere 58(8):1049–1070.

[44] Derbyshire AN, Peters RH (1955) An explanation of dyeing mechanisms in terms of non-polar bonding. J Soc Dyers Colour 71(9):530–536.

[45] Moreno-Castilla C (2004) Adsorption of organic molecules from aqueous solutions on carbon materials. Carbon 42(1):83–94.

[46] Israelachvili J, Pashley R (1982) The hydrophobic interaction is long range, decaying exponentially with distance. Nature 300(5890):341–342.

[47] Franks F (2013) Water: a comprehensive treatise: volume 4: aqueous solutions of amphiphiles and macromolecules. Springer, London.

[48] Pashley R, McGuiggan P, Ninham B, Evans D (1985) Attractive forces between uncharged hydrophobic surfaces: direct measurements in aqueous solution. Science 229(4718):1088–1089.

[49] Groszek A, Partyka S (1993) Measurements of hydrophobic and hydrophilic surface sites by flow microcalorimetry. Langmuir 9(10):2721–2725.

Chapter 5

Conjugated Polymers as Robust Carriers for Controlled Delivery of Anti-Inflammatory Drugs

Katarzyna Krukiewicz, Jerzy K. Zak

Silesian University of Technology, Gliwice, Poland

Abstract: Conjugated polymers due to their reversible transition between the redox states are potentially able to immobilise and release ionic species. In this study, we have successfully developed a conducting polymer system based on poly(3,4-ethylenedioxythiophene) (PEDOT) for electrically triggered, local delivery of an ionic form of ibuprofen (IBU), a non-steroidal anti-inflammatory, and analgesic drug. It was shown that by changing the electropolymerisation conditions, the polymer matrix of specified IBU content can be synthesised. The electrochemical synthesis has been optimised to obtain the conducting matrix with the highest possible drug content. The process of electrically stimulated drug release has been extensively studied in terms of the dynamics of the controlled IBU release under varying conditions. The maximum concentration of the released IBU, 0.66 ($\pm$0.10)mM, was observed at the applied potential E = -0.5V (vs. Ag/AgCl). It was demonstrated that the immobilisation-release procedure can be repeated several times making the PE-DOT matrix promising materials for controlled drug release systems applied e.g. in neuroprosthetics.

1. Introduction

Conjugated polymers are widely known for their electrical properties and have found numerous applications in areas such as organic solar cells, organic light-emitting diodes, supercapacitors, actuators, etc.[1]–[4]. Nowadays, conducting polymers have appeared also as materials suitable for biomedical engineering. For the last decade, numerous biomedical applications based on biocompatible conducting polymers have been investigated, *i.e.* biosensors, molecular scaffolds, coatings for neuroprosthetics, and drug delivery systems[5]–[8]. Conjugated polymers, especially polypyrrole (PPy) and poly(3,4-ethylenedioxythiophene) (PEDOT), are now materials that are in a centre of attention for biomedical engineering; they have been proven to be biocompatible and may be introduced into a human body without any harmful effect on health[9]–[14]. Besides, their ability to conduct electricity is a significant advantage that enables efficient ways of processing and opens new potentially commercial applications, e.g. in neuroprosthetics[15]–[17]. It is known that implantable neural electrodes made of metal cause allergic reactions and inflammation, moreover, the introduction of implants carries the risk of infections[18]. Due to their biocompatibility, PPy and PEDOT films may act as an interface between the metal surface of implant and neural tissue reducing the risk of inflammation and glial scar formation[19]–[21].

The immobilisation of biologically active molecules in conjugated polymer matrices may lead to further increase in polymer biocompatibility[20][22]. Through an intentional choice of immobilised biomolecule, it is possible to prevent inflammation (immobilisation of anti-inflammatory drugs), infection (immobilisation of antibiotics), or brain oedema (immobilisation of steroid drugs)[23]–[25]. The immobilisation of neurotransmitters such as glutamate may be a step toward the development of retinal prosthesis[26]. The ion-exchangeable properties of intrinsically conducting polymers make them promising materials for controlled drug delivery systems. Biologically active compounds, which typically include ionic bonds in their structure, can be therefore immobilised in a conducting polymer matrix and released in a controlled way when stimulated electrically[23][24].

Numerous procedures of electrochemical immobilisation followed by electrically triggered drug release have been already presented in the literature. Piro *et al.*[27] demonstrated the electrochemical method of incorporation of oligonucleotides into PEDOT films followed by their release in a simple ion exchange process. Wadhwa et al.[18] developed PPy matrix for electrically controlled and local delivery of dexamethasone—a synthetic glucocorticoid anti-inflammatory drug. Esrafilzadeh et al.[28] presented conducting polymer fibres loaded with ciprofloxacin—an antibiotic that can be released or sustained in response to electrical stimulation. None of them, however, studied the stability of polymer matrix during stimulated drug release. This is an important question when considering medical applications of such matrices and their implementation into a human body[29]–[32].

In this study, we have described the synthesis of the conducting polymer matrix based on PEDOT as an example of biocompatible, conjugated polymer that recently has gained much interest in biomedical engineering[33]–[36]. The superior chemical and electrochemical stability is the main advantage of PEDOT compared to other currently applied conjugated polymers[37]. As a drug of interest, α-methyl-4-(isobutyl)phenylacetic acid (ibuprofen, IBU) has been chosen, which is known as a non-steroidal anti-inflammatory and analgesic drug showing an ability to enhance wound healing[38]–[40]. The electrochemical immobilisation of IBU in PEDOT matrix has been investigated together with the study of electrically triggered drug release and the examination of the stability of polymer matrix against applied potential.

2. Materials and Methods

2.1. Materials

3,4-Ethylenedioxythiophene (EDOT), lithium perchlorate of A.C.S. grade and IBU sodium salt (sodium α-methyl-4-(isobutyl)phenylacetate, IBU, MW = 228.26 g mol^{-1}) were obtained from Sigma Aldrich. Potassium chloride, dipotassium phosphate and monopotassium phosphate were obtained from POCh. All aqueous solutions were prepared with the use of deionised water.

2.2. Instrumentation

Electrochemical measurements and conditioning were performed in a three-electrode cell by use of CH Instruments 620 electrochemical workstation. The electrochemical synthesis of polymers was performed by means of cyclic voltammetry; the process of conditioning was carried out potentiostatically. The concentration of released IBU was monitored by means of Hewlett Packard 8453 UV-Vis Diode Array Spectrophotometer.

2.3. One-Step Synthesis of PEDOT/LiClO$_4$/IBU

A three-electrode electrochemical cell was set up in a 3-ml electrochemical glass cell with 1cm^2 platinum foil working electrode, Ag/AgCl as a reference electrode and glassy carbon rod as an auxiliary electrode. The polymer films containing IBU (PEDOT/LiClO$_4$/IBU) were obtained by means of cyclic voltammetry (CV) in aqueous solution containing 10mM EDOT, 10mM IBU, 0.1M LiClO$_4$, which was sonicated prior to the experiment for 30min. Typical polymer film was formed on Pt electrode within the potential range of 0–1.1V (vs. Ag/AgCl) at the scan rate v = 100mV s^{-1} for 25 potential cycles.

2.4. Three-Step Synthesis of PEDOT/LiClO$_4$/IBU

In the three-step synthesis, the same three-electrode electrochemical cell was used as that described in the previous section. In the first step, the polymer film was obtained from aqueous solution containing 10mM EDOT and 0.1M LiClO$_4$ in the similar procedure employing cyclic voltammetry with a narrower potential range of 0–1.0V (vs. Ag/AgCl), scan rate v = 100mV s^{-1}, and number of potential cycles equal to 50.

In the second step, the original ClO$_4^-$ dopant was removed from the polymer matrix by conditioning it at a reduction potential E_{red} = −0.7V (vs. Ag/AgCl) for 10min in 0.1M LiClO$_4$ aqueous solution. The IBU immobilisation occurred in the third step, when the polymer matrix after rinsing with deionised water was oxidised at a potential E_{ox} = 0.8V (vs. Ag/AgCl) in 0.4M IBU aqueous solution for 10min.

2.5. Release of IBU from PEDOT/LiClO₄/IBU Matrix

The concentration of IBU released from PEDOT/LiClO$_4$/IBU matrix was determined using UV-Vis spectroscopic measurements under potentiostatic conditions. The three-electrode electrochemical cell was set up in a 2-mm quartz cuvette, in which the PEDOT/LiClO$_4$/IBU-modified platinum-working electrode was mounted together with Ag wire used as a pseudoreference electrode, and graphite rod as an auxiliary electrode. Before measurements, the electrode covered with PEDOT/LiClO$_4$/IBU was soaked in PBS solution for 10min to wash off any loosely attached IBU and unreacted monomer. In order to release IBU from PEDOT/LiClO$_4$/IBU matrix, a constant potential in the range −0.8 to 0.8V (vs. Ag/AgCl) was stepped for a specified period of time. After each period of time at the given potential, the UV-Vis spectra were recorded for IBU released to the solution. All these measurements of the controlled release of the immobilised IBU were performed in the pH 6.5 phosphate buffer solution, PBS, containing 0.15 M KCl, 0.006 M K$_2$HPO$_4$, 0.001 M KH$_2$PO$_4$.

2.6. Stability of PEDOT/LiClO₄/IBU Matrix at Variable Potential

The stability of PEDOT/LiClO$_4$/IBU matrix exposed to a given electrode potential was determined in terms of the charge storage capacity, CSC, of the matrix. That factor was calculated as the electric charge integrated under corresponding CV curve during one CV cycle

$$CSC = \int_{t_1}^{t_2} I(t)\, dt$$

where t_1 is the beginning of CV cycle, t_2 is the end of CV cycle.

The CVs were recorded for the modified platinum electrode in PBS after 20min exposition to a given potential.

3. Results and Discussion

3.1. One-Step Synthesis of PEDOT/LiClO$_4$/IBU

As many other anions, IBU in its ionic form (**Scheme 1**) is able to participate in the polymer doping process, which is a part of the electropolymerisation reaction. Molecules like EDOT when oxidised (chemically or electrochemically) extend their conjugated bonding system, forming positively charged polymer chain that is stabilised by doping of anions from the reaction environment. Under the electro-chemical conditions, the polymerisation/doping process occurs at positive potentials applied to the electrode in the electrolyte solution containing monomer. Thus, a presence of IBU-in the electrolyte solution may lead to its immobilisation in the polymer matrix, which is created on the electrode surface. This process is shown in **Scheme 2**.

The current versus potential curves recorded in subsequent scans make it possible to observe the progress of electropolymerisation process, as shown in **Figure 1**. Two sets of CV curves were collected, the first for the solution containing EDOT, LiClO$_4$ and IBU [**Figure 1(a)**] and the second for the case, when IBU was not present in the solution [**Figure 1(b)**]. In both cases, the increase of current indicates that the monomer is irreversibly oxidised near 1.1V (vs. Ag/AgCl). Also

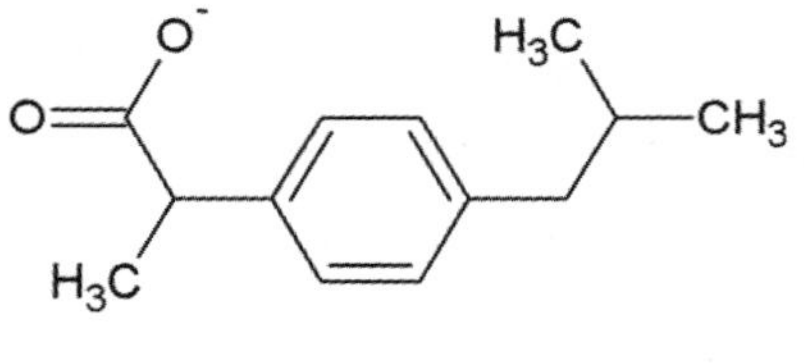

Scheme 1. The chemical structure of anionic form of IBU.

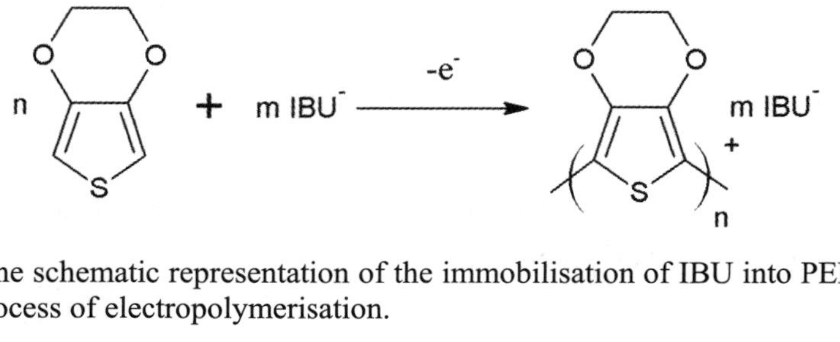

Scheme 2. The schematic representation of the immobilisation of IBU into PEDOT matrix during the process of electropolymerisation.

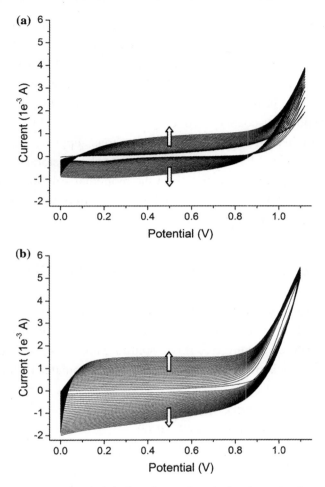

Figure 1. The CV curves recorded during electrochemical polymerization of 10mM EDOT in 0.1M LiClO$_4$ aqueous solution in the presence of 10mM IBU (a), and without IBU (b); potential range 0–1.1V (vs. Ag/AgCl), scan rate v = 100mV/s, 25 potential cycles.

for both data sets, the gradually increasing currents confirm formation of conductive deposits, typical for the electropolymerisation process. At the same number of potential scans, the final level of the resulting currents and corresponding electric charge are evidently different, indicating a slower process of electrochemical polymerisation of EDOT in case of IBU presence in solution [**Figure 1(a)**]. This may be due to the fact that the molecule of IBU is significantly larger than the molecule of electrolyte, thus the incorporation of drug into the structure of the polymer is less spatially favourable. The preliminary results indicated that the molar ratio of EDOT to IBU equal to 1:1 produces a polymer with the highest drug content, approximately 60μg cm^{-2} (mass of IBU immobilised in the unit surface area of Pt electrode). The charge storage capacity, CSC, calculated for PEDOT/LiClO$_4$/IBU (22.3mC) was however not as large as for PEDOT/LiClO$_4$ film (43.6mC) polymerised under analogous conditions as indicate the CV curves shown in **Figure 2**. The CSC factor used in here is simply current integrated along the potential axis recorded for the polymer-coated electrode placed in the pure electrolyte solution. It was found that the use of higher IBU concentrations suppresses the formation of polymer film, therefore it has adverse effects on its charge storage capacity. When IBU was present at lower concentration in the monomer solution (1mM, 2mM, or 5mM), the amount of IBU immobilised in the polymer film was as low as 5μg cm^{-2}, 6μg cm^{-2} and 11μg cm^{-2}, respectively.

3.2. Three-Step Synthesis of PEDOT/LiClO$_4$/IBU

The three-step synthesis of PEDOT/LiClO$_4$/IBU matrix allows to separate the processes of electropolymerisation and immobilisation. In this procedure, the polymer film is formed in pure electrolyte solution, thus it is doped only with perchlorate ions. Electrochemical reduction of the film eliminates to some extent that

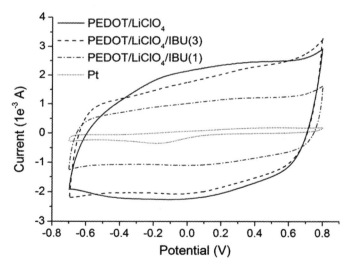

Figure 2. The CVs recorded on PEDOT/LiClO$_4$ (red line) and PEDOT/LiClO$_4$/IBU working electrodes synthesised in one-step (1) (green line and dots) and three-step(3) (blue dotted line) procedure, and CV recorded on bare Pt-working electrode in 0.1M LiClO$_4$ solution (grey dots); scan rate v = 100mV/s.

dopant from the polymer. Oxidation of the polymer is then carried out in the aqueous solution of IBU only. The CVs recorded on Pt electrodes modified with PEDOT/LiClO$_4$ and PEDOT/LiClO$_4$/IBU obtained via one-step (1) and three-step (3) procedures are compared in **Figure 2**. The average current magnitude of PEDOT/LiClO$_4$/IBU(3) is larger than for PEDOT/LiClO4/IBU(1) and only slightly smaller than that for PEDOT/LiClO$_4$. Hence, it may be concluded that a three-step synthesis route results in a matrix with a higher CSC (40.9mC) than matrix synthesised according to the one-step procedure (22.3mC), only slightly smaller than CSC of PEDOT/LiClO$_4$ film (43.6mC).

3.3. Release of IBU from PEDOT/LiClO₄/IBU

During the synthesis, anionic compounds are immobilised in a positively charged conducting polymer matrix to maintain a neutral charge of the polymer backbone. When the sufficiently negative voltage is applied to the matrix, the anionic dopant becomes unnecessary, therefore it is released to the solution. The schematic representation of the electrically triggered drug release from conducting polymer matrix is shown in **Scheme 3**.

The process of IBU release from PEDOT/LiClO$_4$/IBU matrix was monitored under potentiostatic conditions using UV-Vis spectroscopy. Calibration curve was plotted for absorbance versus IBU concentration (where y is the absorbance and x is the IBU concentration inmM). A linear relationship was observed between 0.01 and 1mM IBU concentration satisfying the equation y = 0.0176x + 1.9365 (R^2 = 0.9992).

In order to release IBU from PEDOT/LiClO$_4$/IBU matrix, a constant potential in the range between −0.8V and 0.8V (vs. Ag/AgCl) was applied to the matrix

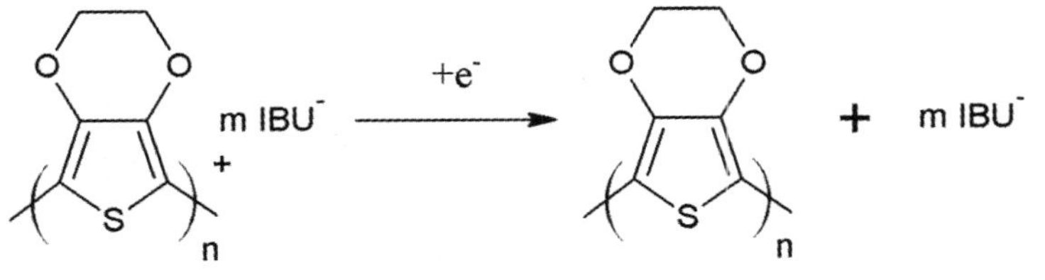

Scheme 3. The schematic representation of electrically triggered drug release from PEDOT/LiClO$_4$/IBU matrix.

placed in the cuvette. The absorbance at k = 222nm, which is characteristic for IBU, was recorded versus time at a given potential. The changes of absorbance spectra in time for the matrix stimulated with E = −0.5V (vs. Ag/AgCl) are presented in **Figure 3**.

The time profiles of IBU release from the matrix stimulated with E = −0.5V (vs. Ag/AgCl) and matrix maintained under open circuit conditions are presented in **Figure 4**. As it can be seen, for the non-stimulated matrix (open circuit potential), IBU was released in small concentration, six times lower than for the matrix stimulated with the reducing potential E_{red} = −0.5V (vs. Ag/AgCl). This result evidently confirms the effect of the applied potential on the process of the drug release. The greatest quantity of IBU appears to be released during the first 2min of the electric stimulation, as compared to longer times. It suggests that only a small portion of IBU molecules is imbedded deeply within the polymeric structure of the matrix; also the process of full reduction of PEDOT is practically completed within the applied stimulation time. Under the described conditions, a total time below 10min is required to reach equilibrium (tequil) in the system, then the solution concentration of IBU is stable.

Since the ratio of the amounts of oxidised and reduced forms of the conjugated polymer is strictly determined by the potential applied to the polymer film, therefore one can expect variation in the concentration of IBU released to the

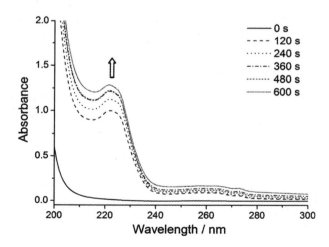

Figure 3. The change of absorbance spectra in time for IBU released from the matrix stimulated with E = −0.5V (vs. Ag/AgCl).

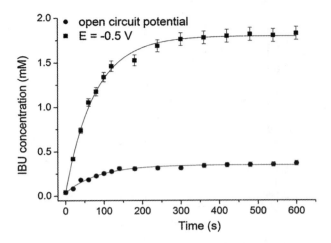

Figure 4. The concentration of the released IBU calculated from UV-Vis data as a function of time of the electric stimulation of PEDOT/LiClO$_4$/IBU matrices with potential E = −0.5V (black square) and under open circuit conditions (red circle). The plain line is only a visual guide.

solution as a function of the value of stimulating potential. These data were collected for t_{equil} = 10min for polymer films of approximately the same amount of immobilised IBU. The resulting concentration of IBU released as a function of applied potential is shown in **Figure 5**. The maximum concentration of IBU released, 0.66 (±0.10)mM, was observed when the potential E = −0.5V (vs. Ag/AgCl) was applied. When PEDOT/LiClO$_4$/IBU was subjected to more negative potentials, the concentration of released IBU decreased. This is caused by the effect of n-doping of negatively charged polymer matrix with cations present in PBS solution: the incorporation of potassium cations occurs instead of the release of the anionic form of IBU. The concentration of IBU released decreases as the value of applied potential increases since the positively charged film holds the anionic drug; the concentration of IBU released at E = 0.8V (vs. Ag/AgCl) was 0.21 (±0.05)mM, which is one-third of the concentration released at E = −0.5V (vs. Ag/AgCl). These results are consistent with the mentioned above effects of the applied potential on the redox state of conjugated polymer. The application of negative potentials results in facilitated release of anionic co-dopant, IBU, while the application of positive potential results in the drug retention.

The capacity of the polymer matrix towards the amount of immobilised IBU generally depends on the amount of polymer; the last in turn is easy to control in

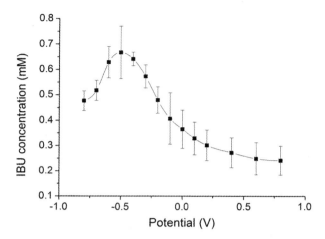

Figure 5. The concentration of IBU released as a function of the potential applied to PEDOT/LiClO$_4$/IBU matrix; the plain line is only a visual guide. The error bars represent standard deviations for five independent measurements.

the process of electropolymerisation by variation of the number of CV cycles. The relation between the drug capacity of the matrix and the conditions of electrochemical polymerisation, in terms of number of CV cycles, is presented in **Figure 6**. The data indicate nearly linear relation between the electropolymerisation time and the drug capacity; the first is represented by the number of CV scans, whereas the other is expressed in terms of the concentration of released IBU. These results demonstrate that it is possible to synthesise PEDOT/LiClO$_4$/IBU matrix of a specified IBU content that can be further effectively released.

Since the three-step synthesis of PEDOT/LiClO$_4$/IBU allows to separate the processes of electropolymerisation and immobilisation, it should be possible to apply the immobilisation-release procedure several times for the same matrix. Again, the curves representing the concentration of IBU released versus time were plotted for the PEDOT/LiClO$_4$/IBU matrix that was subjected to four consecutive immobilisation-release cycles. The plot of the concentration of released IBU calculated from the corresponding UV–Vis data is shown in **Figure 7**. After the first cycle, the concentration of released IBU was equal to 0.70 (±0.05)mM, whereas each subsequent immobilisation/release cycle appeared to be less efficient; in the forth cycle the concentration of IBU released was reduced to 0.34 (±0.02)mM. The observed effect is caused by the gradual degradation of PEDOT/LiClO$_4$/IBU matrix that usually occurs when the polymer is exposed to oxidative potential

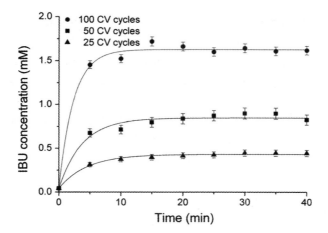

Figure 6. The concentration of IBU released versus stimulation time of polymer matrix for different film thicknesses measured as electropolymerisation number of CV cycles: 25 CV cycles (black triangle), 50 CV cycles (blue square), 100 CV cycles (red circle). PEDOT/LiClO$_4$/IBU matrices were electrically stimulated at potential E = −0.5V (vs. Ag/AgCl). The plain line is only a visual guide.

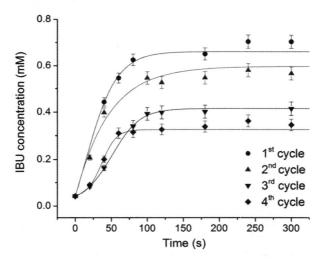

Figure 7. The concentration of IBU released versus time for PEDOT/LiClO$_4$/IBU matrix subjected to four immobilisation-release cycles: first cycle (red circle), second cycle (green triangle), third cycle (blue inverted triangle), forth cycle (pink diamond). The plain line is only a visual guide.

exceeding its optimal value (overoxidation). If, however, the potential is carefully chosen, the polymer degradation during its destructive oxidation can be optimised. On the other side, the matrix exposed to negative potentials for longer times (or at no potential applied) keeps its CSC factor practically unchanged.

The gradual loss of the matrix loading effectiveness, as observed in the spectral measurements, is likely caused by a degradation of the polymer film. The CV curves recorded after each IBU load/release cycle reveal decreasing electric charge. These data shown in terms of CSC values (**Table 1**) are consistent with the spectral results, which all together give a proof of a limited durability of the PEDOT matrix when multiple IBU-loadings are attempted.

4. Conclusions

In this study, a PEDOT-based conducting polymer system for electrically triggered delivery of an ionic form of IBU has been developed. The most favourable electropolymerisation conditions for achieving good electrochemical properties along with the highest drug content have been found and the process of electrically stimulated drug release has been extensively studied. The time profiles have provided the information about the dynamics of drug release, while application of different potentials has shown that the process of the drug release may be effectively controlled. The concentration of the released IBU has been shown to be dependent on the redox state of the polymer matrix, therefore application of negative potential resulted in the drug release, while the positive potentials resulted in the drug retention.

The proposed three-step synthesis of PEDOT/LiClO$_4$/IBU matrix allowed

Table 1. The charge storage capacity (CSC) of PEDOT/LiClO$_4$/IBU matrix calculated from CV data before and after the first, second, third and forth immobilisation-release cycle as compared to CSC of bare Pt electrode.

Type of surface	CSC (mC)
Bare Pt	3.77
Modified Pt before 1st load/release cycle	53.26
Modified Pt after 1st load/release cycle	39.48
Modified Pt after 2nd load/release cycle	25.00
Modified Pt after 3rd load/release cycle	15.53
Modified Pt after 4th load/release cycle	8.93

the separation of the electropolymerisation and immobilisation processes, which provided the matrix of a superior charge storage capacity. It was demonstrated that the immobilisation-release procedure can be repeated several times making the drug delivery system able to be reloaded. However, decreasing electric charge and gradual loss of the matrix loading efficiency after each IBU load/release cycle show the limited durability of the PEDOT matrix when multiple IBU-loadings are attempted. That is why further research should be focused on the optimisation of the loading-reloading conditions, especially on the careful choice of the oxidation and reduction potentials. If, however, the aforementioned conditions are optimised and the durability of PEDOT is improved, the PEDOT/LiClO$_4$/IBU system is expected to find application as a novel carrier for controlled-release local delivery of anti-inflammatory drug for biomedical engineering.

Acknowledgements

This work was supported by the Polish National Science Centre (Preludium 2012/07/N/ST5/01878).

Open Access: This article is distributed under the terms of the Creative Commons Attribution License which permits any use, distribution, and reproduction in any medium, provided the original author(s) and the source are credited.

Source: Krukiewicz K, Zak J K. Conjugated polymers as robust carriers for controlled delivery of anti-inflammatory drugs[J]. Journal of Materials Science, 2014, 49(16):5738–5745.

References

[1] Kim WH, Mäkinen AJ, Nikolov N *et al* (2002) Molecular organic light-emitting diodes using highly conducting polymers as anodes. Appl Phys Lett 80:3844–3846.

[2] Matoetoe MC (2010) A review of dye incorporated conducting polymers application as sensors and in solar cells. Mater Sci Forum 657:208–230.

[3] Mastragostino M, Arbizzani C, Soavi F (2002) Conducting polymers as electrode materials in supercapacitors. Solid State Ionics 148:493–498.

[4] Coskun Y, Cirpan A, Toppare L (2007) Construction of electrochromic devices using thiophene based conducting polymers. J Mater Sci 42:368–372. doi:10.1007/s10853-006-1076-6.

[5] Guimard NK, Gomez N, Schmidt CE (2007) Conducting polymers in biomedical engineering. Prog Polym Sci 32:876–921.

[6] Ravichandran R, Sundarrajan S, Venugopal JR et al. (2010) Applications of conducting polymers and their issues in biomedical engineering. J R Soc Interface 7:559–579.

[7] Svennersten K, Larsson KC, Berggren M et al (2011) Organic bioelectronics in nanomedicine. Biochim Biophys Acta 1810: 276–285.

[8] Subramanian A, Krishnan UM, Sethuraman S (2009) Development of biomaterial scaffold for nerve tissue engineering: biomaterial mediated neural regeneration. J Biomed Sci 16:108–119.

[9] Geetha S, Rao CRK, Vijayan M et al. (2006) Biosensing and drug delivery by polypyrrole. Anal Chim Acta 568:119–125.

[10] George PM, Lyckman AW, LaVan DA et al. (2005) Fabrication and biocompatibility of polypyrrole implants suitable for neural prosthetics. Biomaterials 26:3511–3519.

[11] Asplund M, Thaning E, Lundberg J et al. (2009) Toxicity evaluation of PEDOT/biomolecular composites intended for neural communication electrodes. Biomed Mater 4:1–12.

[12] Bennet D, Kim S (2011) Implantable microdevice for peripheral nerve regeneration: materials and fabrications. J Mater Sci 46:4723–4740. doi:10.1007/s10853-011-5510-z.

[13] Ateh DD, Navsaria HA, Vadgama P (2009) Polypyrrole-based conducting polymers and interactions with biological tissues. J R Soc Interface 3:741–752.

[14] Kim DH, Richardson-Burns SM, Hendricks JL et al. (2007) Effect of immobilised nerve growth factor on conductive polymers: electrical properties and cellular response. Adv Funct Mater 17:79–86.

[15] Asplund M, von Holst H, Inganäs O (2008) Composite biomolecule/PEDOT materials for neural electrodes. Biointerphases 3:83–93.

[16] Thompson BC, Richardson RT, Moulton SE et al. (2010) Conducting polymers, dual neurotrophins and pulsed electrical stimulation—dramatic effects on neurite outgrowth. J Control Release 141:161–167.

[17] Abidian MR, Martin DC (2009) Multifunctional nanobiomaterials for neural interfaces. Adv Funct Mater 19:573–585.

[18] Wadhwa R, Lagenaur CF, Cui XT (2006) Electrochemically controlled release of dexamethasone from conducting polymer polypyrrole coated electrode. J Control Release 110:531–541.

[19] Gomez N, Schmidt CE (2007) Nerve growth factor-immobilised polypyrrole: bioactive electrically conducting polymer for enhanced neurite extension. J Biomed Mater Res A 81: 135–149.

[20] Green RA, Lovell NL, Poole-Warren LA (2009) Cell attachment functionality of bioactive conducting polymers for neural interfaces. Biomaterials 30:3637–3644.

[21] Cogan SF (2008) Neural stimulation and recording electrodes. Annu Rev Biomed Eng 10:275–309.

[22] Xiao YH, Li CM, Toh ML et al. (2008) Adenosine 50-triphosphate incorporated poly(3,4-ethylenedioxythiophene) modified electrode: a bioactive platform with electroactivity, stability and biocompatibility. J Appl Electrochem 38:1735–1741.

[23] Svirskis D, Travas-Sejdic J, Rodgers A et al (2010) Electro-chemically controlled drug delivery based on intrinsically conducting polymers. J Control Release 146:6–15.

[24] Sassolas A, Blum LJ, Leca-Bouvier BD (2012) Immobilisation strategies to develop enzymatic biosensors. Biotechnol Adv 30:489–511.

[25] Vela MH, de Jesus DS, Couto CMCM et al. (2003) Electroimmobilisation of MAO into a polypyrrole film and its utilization for amperometric flow detection of antidepressant drugs. Electroanalysis 15:133–138.

[26] Paul N, Muller M, Paul A et al. (2013) Molecularly imprinted conductive polymers for controlled trafficking of neurotransmitters at solid–liquid interfaces. Soft Matter 9:1364–1371.

[27] Piro B, Pham MC, Ledoan T (1999) Electrochemical method for entrapment of oligonucleotides in polymer-coated electrodes. J Biomed Mater Res A 46:566–572.

[28] Esrafilzadeh D, Razal JM, Moulton SE et al (2013) Multifunctional conducting fibres with electrically controlled release of ciprofloxacin. J Control Release 169:313–320.

[29] Boretius T, Schuettler M, Stieglitz T (2011) On the stability of poly-ethylenedioxythiopene as coating material for active neural implants. J Artif Organs 35:245–248.

[30] Green RA, Lovell NH, Wallace GG et al. (2008) Conducting polymers for neural interfaces: challenges in developing an effective long-term implant. Biomaterials 29:3393–3399.

[31] Thaning EM, Asplund MLM, Nyberg TA et al (2010) Stability of poly(3,4-ethylene dioxythiophene) materials intended for implants. J Biomed Mater Res B 93:407–415.

[32] Che J, Xiao Y, Zhu X et al (2008) Electro-synthesised PEDOT/ glutamate chemically modified electrode: a combination of electrical and biocompatible features. Polym Int 57:750–755.

[33] Rozlosnik N (2009) New directions in medical biosensors employing poly(3,4-ethylenedioxy thiophene) derivative-based electrodes. Anal Bioanal Chem

395:637–645.

[34] Santhosh P, Manesh KM, Uthayakumar S et al (2009) Fabrication of enzymatic glucose biosensor based on palladium nanoparticles dispersed onto poly(3,4-ethylenedioxythiophene) nanofibers. Bioelectrochemistry 75:61–66.

[35] Ho KC, Yeh WM, Tung TS et al (2005) Amperometric detection of morphine based on poly(3,4-ethylenedioxythiophene) immobilised molecularly imprinted polymer particles prepared by precipitation polymerisation. Anal Chim Acta 542:90–96.

[36] Nien PC, Chen PY, Hsu CY et al. (2011) On-chip glucose biosensor based on enzyme entrapment with pre-reaction to lower interference in a flow injection system. Sens Actuators B 157:64–71.

[37] Kros A, van Hövell SWFM, Sommerdijk NAJM et al. (2001) Poly-(3,4-ethylenedioxythiophene)-based glucose biosensors. Adv Mater 13:1555–1557.

[38] Schreml S, Szeimies R, Prantl L et al. (2009) Wound healing in the 21st century. J Am Acad Dermatol 63:866–879.

[39] Shao F, Liu L, Fan K, Cai Y, Yao J (2012) Ibuprofen loaded porous calcium phosphate nanospheres for skeletal drug delivery system. J Mater Sci 47:1054–1058. doi:10.1007/s10853-011-5894-9.

[40] Tsatsanis C, Androulidaki A, Venihaki M et al (2006) Signalling networks regulating cyclooxygenase-2. Int J Biochem Cell Biol 38:1654–1661.

Chapter 6

Controlled-Release Formulation of Perindopril Erbumine Loaded PEG-Coated Magnetite Nanoparticles for Biomedical Applications

Dena Dorniani[1], Aminu Umar Kura[2],
Mohd Zobir Bin Hussein[1], Sharida Fakurazi[2],
Abdul Halim Shaari[3], Zalinah Ahmad[2,4]

[1]Materials Synthesis and Characterization Laboratory (MSCL), Institute of Advanced Technology (ITMA), Universiti Putra Malaysia, 43400 Serdang, Selangor, Malaysia
[2]Vaccines and Immunotherapeutics Laboratory (IBS), Universiti Putra Malaysia, 43400 Serdang, Selangor, Malaysia
[3]Department of Physics, Faculty of Science, Universiti Putra Malaysia, 43400 Serdang, Selangor, Malaysia
[4]Chemical Pathology Unit, Department of Pathology, Faculty of Medicine and Health Sciences, Universiti Putra Malaysia, 43400 Serdang, Selangor, Malaysia

Abstract: Iron oxide nanoparticles (FNPs) were synthesized due to low toxicity and their ability to immobilize biological materials on their surfaces by the coprecipitation of iron salts in ammonia hydroxide followed by coating it with polyethylene glycol (PEG) to minimize the aggregation of iron oxide nanoparticles and enhance the effect of nanoparticles for biological applications. Then, the FNPs-PEG was loaded with perindopril erbumine (PE), an antihypertensive compound to form a new nanocomposite (FPEGPE). Transmission electron microscopy results showed that there are no significant differences between the sizes of FNPs and FPEGPE nanocomposite. The existence of PEG-PE was supported by the FTIR and TGA analyses. The PE loading (10.3%) and the release profiles from FPEGPE nanocomposite were estimated using ultraviolet-visible spectroscopy which showed that up to 60.8% and 83.1% of the adsorbed drug was released in 4223 and 1231min at pH 7.4 and 4.8, respectively. However, the release of PE was completed very fast from a physical mixture (FNPs-PEG-PE) after 5 and 7min at pH 4.8 and 7.4, respectively, which reveals that the release of PE from the physical mixture is not in the sustained-release manner. Cytotoxicity study showed that free PE presented slightly higher toxicity than the FNPs and FPEGPE nanocomposite. Therefore, the decrease toxicity against mouse normal fibroblast (3T3) cell lines prospective of this nanocomposite together with controlled-release behavior provided evidence of the possible beneficial biological activities of this new nanocomposite for nanopharmaceutical applications for both oral and non-oral routes.

Introduction

Recently, different types of nanoparticles-based therapeutic and diagnostic agents have been extensively studied to prolong the half-life of drug systemic circulation by reducing immunogenicity, sustained release of drugs in an environmentally responsive manner, lower frequency of administration in order to minimize systemic side effects of drugs for treatment of diabetes[1][2], asthma[3]–[5], allergy[6], infections[7][8], cardiovascular diseases[9], neurological diseases [10], cancers[11][12], pain, and so on. Therefore, a few pioneering therapeutic nanoparticles have been introduced into the pharmaceutical market.

Polymeric nanoparticles[13] have been used in nano-medicine to provide more effective and/or more convenient routes of administration, high encapsulation efficiency[14], lower therapeutic toxicity, extend the product life cycle, and ultimately reduce health-care costs.

Beside polymeric nanoparticles, the most common nanoparticles attracted more attention nowadays are poly-saccharide-based nanoparticles and metallic nanoparticles (typically iron oxide nanoparticles)[15] which could improve the therapeutic index of drugs by reducing the drug toxicity or enhancing drug efficacy. Previous studies showed that polymeric nanoparticles could be used to prevent the hemoglobin oxidation after loading into VAM41-polyethylene glycol (PEG)[16]. In the early 1990s, PEG, a non-toxic and non-immunogenic polymer was introduced into clinical uses in order to enhance the pharmacokinetics of various nanoparticle formulations and prolongs drug circulation half-life[17][18].

Controlled release of drug suggests plenty advantages over free drugs such as improved efficacy, reduced side effects, and improved patient compliance[18]. Previous reports showed that a variety of active compounds such as kojic acid[19], doxorubicin[20][21], 5-aminosalicylic acid [22], 6-mercaptopurine[23], 10-hydroxycamptothecin[24], gallic acid[25][26], folic acid[27], arginine[28], and anthranilic acid[29] can be loaded onto the surface of magnetic nanoparticles.

Heart disease is the first leading cause of death for both men and women globally and the number of people who die due to high blood pressure (BP) (hypertension) will increase to reach 23.3 million by 2030[30]. Therefore, to find a new nanocomposite for the treatment of hypertension, iron oxide nanoparticles can be selected due to low toxicity, low cost of production, ability to immobilize biological materials on their surfaces, potential for direct biodegradable, and suitable for surface modifications. To minimize the aggregation of iron oxide nanoparticles (FNPs), which causes due to dipole-dipole attractions between particles[31][32], PEG was coated on the surface of FNPs to create FNPs-PEG. Then, perindopril erbumine (PE), a long-acting angiotensin-converting enzyme inhibitor, which prevents various medical conditions indicating reduced systolic and diastolic BP in patients with mild-to-moderate hypertension, congestive heart failure and diabetic nephropathy[33][34] was loaded on the surface of FNPs-PEG and formed a new FPEGPE nanocomposite. The resulting nanocomposites (FPEGPE) was characterized for the structural and sustained release properties and evaluate the cytotoxic effects against normal fibroblast (3T3) cell lines as compared to free active drug and uncoated FNPs in order to improve the treatment of hypertension. In addition, the effect of PEG coating will be also evaluated.

2. Materials and Methods

2.1. Materials

Ferrous chloride tetrahydrate (FeCl$_2$·4H$_2$O ≥ 99%) and ferric chloride hexahydrate (FeCl$_3$·6H$_2$O, 99%) were obtained from Merck KGaA, Darmstadt, Germany. In order to coat FNPs with polymer, PEG with average M.W. 6000 was used, as a raw material from Acros Organics. PE (C$_{23}$H$_{43}$N$_3$O$_5$, with molecular weight 441.6g mol^{-1}) was purchased from CCM Duopharma (Klang, Malaysia) at 99.79% purity. Distilled deionized water (18.2MΩ cm^{-1}) was used throughout the experiments. In addition, all reagents used in this study were of analytical grade and were used without further purification.

2.2. Preparation of Magnetite Nanoparticles and FPEGPE Nanocomposite

The magnetite nanoparticles were synthesized by coprecipitation method as previously reported by Lee et al.[35]. The magnetite nanoparticles were prepared by mixing of 2.43g ferrous chloride tetrahydrate, 0.99g ferric chloride hexahydrate, and 80mL of deionized water in the presence of 6mL of ammonia hydroxide (25 mass%). The pH of the solution was keep at 10. The solution was ultrasonicated for 1h at room temperature. To remove all impurities, the precipitates were centrifuged and washed with deionized water for three times. For surface modification of FNPs, the collected magnetite was mixed with 2% PEG. After 24h stirring, the black precipitates mixture of magnetite-PEG was collected by a permanent magnet, washed for three times in order to remove the unbound PEG during the coating process. The 2% of drug solution, PE which was dissolved in deionized water was added to the magnetite-PEG, and the mixture was magnetically stirred at room temperature for 24h to facilitate the uptake of PE. Finally, the FPEGPE products (PE-loaded PEG-coated magnetite nanoparticle) were washed for three times and dried in an oven.

2.3. Cell Viability Study

Cell Culture

Mouse normal fibroblast (3T3) cell line was obtained from American Type Culture Collection (Manassas, VA, USA). They were maintained in DMEM medium (Dulbecco's Modified Eagle Medium, Gibco) supplemented with 10% fetal bovine serum and 1% antibiotics (100 units mL^{-1} penicillin 100mg mL^{-1} streptomycin). Cell's media were changed after every 2 days, and they were grown in a humidified incubator at 37°C (95% room air, 5% CO_2) and used for seeding and treatment after reaching 90% confluence.

The 3T3 cells were seeded at 1×10^5 cells mL^{-1} into 96-well plates and left overnight in a CO_2 incubator to become attached. Cytotoxic activity of the coated FNPs with PEG-PE, pure PE, and the uncoated FNPs was done after 72h exposure. A stock solution of 10mg mL^{-1} from nanoparticles, the FPEGPE nanocomposite, and pure PE was prepared in media and subsequently diluted to obtain the desired concentration of 0.47–50.0µg mL^{-1}. Wells containing cells and media only were used as control.

2.4. Cytotoxicity Study

The cytotoxic effect of the FNPs, FPEGPE nanocomposite, and pure drug (PE) on the cells was measured by the conventional MTT reduction assay as described previously[36]. MTT solution (5mg mL^{-1} in phosphate buffered saline—PBS) was added to the treated and control wells at 20µl final volume and then left in an incubator at 37°C. Media were discarded about 2h post MTT solution addition, and the reaction was stopped by gentle replacement with dimethyl sulfoxide (DMSO) 100µl per well. This is to dissolve the blue crystals formed due to the reduction of tetrazolium by living cells. The amount of MTT formazan produced was determined by measuring the absorbance at 570 and 630nm (background) using a microplate enzyme-linked immunosorbent assay reader (ELx800, BioTek Instruments, Winooski, VT, USA). All experiments were carried out in triplicate, and the results are presented as the mean ± standard deviation. Cell viability was expressed as a percentage of the value in untreated control cells and calculated as

$$\text{Cell viability (\%)} = \frac{[\text{Average}]\ \text{test}}{[\text{Average}]\ \text{control}} \times 100$$

2.5. Loading and Release Amounts of PE from FPEGPE Nanocomposite

The loading percentage of PE in the FPEGPE nanocomposite was measured spectrophotometrically. A measure of 11 mg of FPEGPE nanocomposite was dissolved into the mixture of 1:3mL HCl HNO$_3$ and marked it up to 25mL by deionized water and stirred for around 1h. Then, the amount of the PE in the sample was measured using a UV-Vis spectroscopy.

To study the release profiles of PE from FPEGPE nanocomposite, a PBS solution at two pH levels (7.4 and 4.8) was used at room temperature[37][38]. The cumulative released amount of PE into the solution was measured at preset time intervals at λ_{max} = 215nm by the UV-Vis spectrum. The rate of the release can be changed due to existing different anions such as Cl^-, HPO_4^{2-}, and $H_2PO_4^-$ in the PBS.

2.6. Instrumentation

In order to determine the crystal structure of the magnetite and FPEGPE nanocomposite, powder X-ray diffraction (XRD) patterns were obtained in a range of 5°–70° on an XRD-6000 diffractometer (Shimadzu, Tokyo, Japan) using CuKa radiation (λ = 1.5406Å) at 40kV and 30mA. Fourier transform infrared (FTIR) spectra of the materials were recorded over the range of 400–4000cm^{-1} on a Thermo Nicolet (AEM, Madison WI, USA) with 4cm^{-1} resolution, using the potassium bromide disk method. Thermogravimetric and differential thermogravimetric analyses (TGA-DTG) were carried out using a Mettler-Toledo instrument (Greifensee, Switzerland) in 150μl alumina crucibles in the range of 20°C–1000°C at a heating rate of 10°C min^{-1}. Magnetic properties were obtained by a Lakeshore 7404 vibrating sample magnetometer (VSM; Westerville, OH, USA). Morphology and particle size of the samples were determined by a transmission electron microscopy (TEM) (Hitachi, H-7100 at an accelerating voltage of 100kV). The UV-Vis spectrophotometer (Shimadzu 1650 series, Tokyo, Japan) was used to determine the optical and controlled-release properties of PE from the FPEGPE nanocomposite.

3. Results and Discussion

3.1. Powder XRD

The XRD patterns of the prepared magnetite nanoparticles as well as FPEGPE nanocomposite are shown in **Figure 1**. The inset shows the XRD patterns of pure PE and the PEG. Two main diffraction peaks appeared at $2\theta = 10.5°$

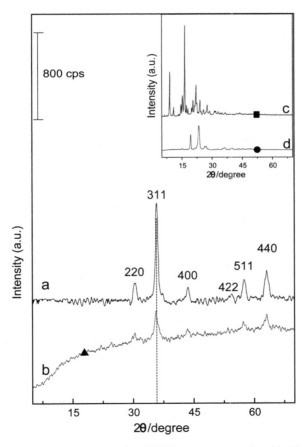

Figure 1. XRD patterns of FNPs (a), and FPEGPE nanocomposite (b). The inset shows the XRD patterns of pure PE (c) and pure PEG (d).

and 20.6° can be assigned to the pure PEG [**Figure 1(d)**][39]. The diffraction pattern of pure PE [**Figure 1(c)**] shows many intense sharp peaks in the fingerprint region, which was also previously reported[40]. **Figure 1(a)** shows six characteristic peaks of FNPs which are marked by their indices (220), (311), (400), (422), (511), and (440). These peaks confirm the formation of cubic inverse spinal structure of magnetite Fe_3O_4 nanoparticles[41].

Due to the same characteristic peaks that can be observed in FPEGPE nanocomposite, it can be proved that the modification of FNPs after coating with PEG-PE did not result in any phase change of magnetite FNPs[19][42]. Moreover, the absence of the characteristic diffraction peaks at (210), (213), and (300) indicates that the maghemite or the co-existence of maghemite (γ-Fe_2O_3) did not exist in both samples[43]. The mean grain size was measured using the Debye-Scherrer equation ($D = K\lambda/\beta\cos\theta$), where D is the mean grain size, K is the Scherrer constant (0.9), λ is the XRD wavelength (0.15418nm), β is the peak width of half maximum intensity, and θ is the Bragg diffraction angle. Therefore, the average crystal size of the FNPs was estimated to be 3nm.

3.2. Infrared Spectroscopy

Figure 2 provides the FTIR spectra of FNPs (a), pure PEG (b), free drug PE (c), and FPEGPE nanocomposite (d). The absorption peak at around 560cm^{-1} in magnetic FNPs spectrum relates to the stretching of Fe-O which was shifted to 577cm^{-1} after coating with PEG-PE [**Figure 2(d)**], confirming the presence of magnetite in FPEGPE nanocomposite. The characteristic band of pure PEG was appeared at 2889cm^{-1} [**Figure 2(b)**] and can be assigned to C-H stretching vibration which is shifted to 2872cm^{-1} in FPEGPE nanocomposite. Another two bands at 1468 and 1343cm^{-1} belong to the C-H bending vibration. The C-H bending vibration band at 1468cm^{-1} is shifted to 1445cm^{-1} after the coating process. In addition, two characteristic bands at 1281 and 1094cm^{-1} can be assigned to the O-H and C-O-H stretching vibration, respectively[44].

The FTIR spectra of PE [**Figure 2(c)**] show many intense, sharp absorption bands due to different functional groups present in molecules such as primary amine, secondary amine, ester, carboxylic acid, and methyl groups. The peak at 1247cm^{-1} is due to C_3C-N stretching[45], and it was shifted to 1251cm^{-1} after

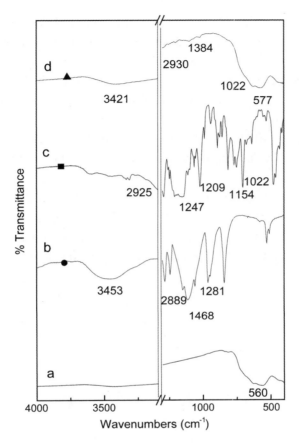

Figure 2. FTIR spectra of FNPs (a), pure PEG (b), pure PE(c), and FPEGPE nanocomposite (d).

coating process, demonstrating that PE was successfully loaded to FNPs-PEG. The band at 1022cm^{-1} is due to C–N stretching[46] and it is also present in FPEGPE nanocomposite. The band at 2925cm^{-1} indicates that CH in NH–CH–propyl is shifted to 2930cm^{-1} due to the coating procedure [**Figure 2(d)**]. The band appeared at 1154cm^{-1} is due to the symmetric stretching of C–N–C, which is also present in FPEGPE nanocomposite. This evidence strongly confirms the coating of PEG-PE on the surface of FNPs.

3.3. Thermogravimetric Analysis

In order to verify the coating and loading formation of FNPs with PEG-PE, the thermal behavior of the FNPs, free drug (PE), and FPEGPE nanocomposite was measured via thermogravimetric and differential thermogravimetric analyses (**Figure 3**). The FNPs show one stage weight loss which occurred at 286°C, which is corresponding to the loss of residual water with 4.1% weight loss [**Figure 3(a)**][32]. Thermogravimetric curve of free drug (PE) shows two main thermal stages. The first stage which is related to melting of PE was observed at 145°C with about 18% weight loss [**Figure 3(b)**]. The second mass reduction at 260°C with 78% weight loss is attributed to the decomposition and subtle combustion of PE[47].

For the FPEGPE nanocomposite, three major thermal events were observed [**Figure 3(c)**]. A slight mass reduction was observed up to 241°C (8.4% weight loss) which is likely due to the adsorbed water and decomposition of PE. The second mass reduction at 344°C might be due to the decomposition of PEG poly-

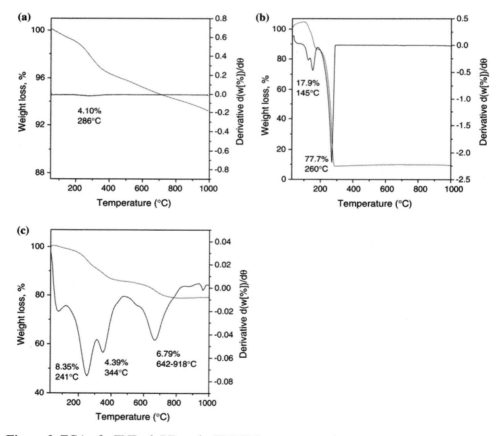

Figure 3. TGA of a FNPs, b PE, and c FPEGPE nanocomposite.

mer with weight loss of 4.4%. This was followed by the third stage in the region of 642°C up to 918°C with the mass reduction of 6.8%.

The temperature region after coating is clearly enhanced, due to the electrostatic attraction between the iron oxide surface and PEG-PE[40].

3.4. Magnetic Properties

In drug delivery system, superparamagnetism is having an important role in magnetic targeting carriers[48]. **Figure 4** shows the magnetic properties of FNPs [**Figure 4(a)**] and FNPs coated with PEG-PE (FPEGPE), by a VSM [**Figure 4(b)**]. Due to the lack of hysteresis loop in the magnetization curves and the saturation magnetization of 54.6 and 38.8emu g^{-1} for FNPs and FPEGPE nanocomposite, respectively, it can be confirmed that both samples have the superparamagnetic properties (i.e., no remanence effect) and were of soft magnets[24][42][48]. Method of synthesis and the particle size can be affected on the value of saturation magnetization; therefore, this value is usually lower than the theoretical value expected[49]. Saturation magnetization (Ms) of FNPs decreased after coating procedure, which is attributed to the effect of small-particle surface and also the exchange of electrons between the surface of Fe atoms and the PEG polymer coating[50]. In

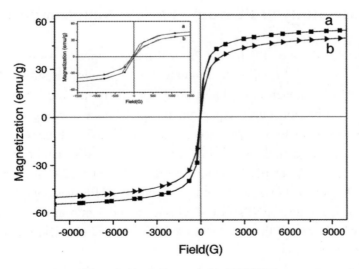

Figure 4. Magnetization plots of (a) FNPs and (b) FPEGPE nanocomposite. The inset shows the magnetic behavior under low magnetic fields.

addition, it can provide further supporting evidence of occurrence of the PEG on the surface of iron oxide magnetic nanoparticles.

3.5. Particle Size and Size Distribution

Magnetic nanoparticles with smaller size (< 30nm) show superparamagnetic properties[28] therefore, TEM together with a UTHSCSA ImageTool software were performed to determine the size, shape, and particle size distribution of the bare FNPs [**Figure 5(a)**, **Figure 5(c)**] and FPEGPE nanocomposite [**Figure 5(b)**, **Figure 5(d)**]. The particle size and size distribution of FNPs and FPEGPE nanocomposite were calculated by measuring the diameters of around 150 nanoparticles chosen randomly through the TEM images. FNPs are well-dispersed and have uniform size and shape, although some agglomeration exist due to the magnetization effect[24] and/or the dehydration process. **Figure 5** indicates that the pre-prepared FNPs and FPEGPE nanocomposite have spherical shape and were essentially monodisperse with the average size of 9 ± 2 and 13 ± 2 nm, for FNPs and FPEGPE nanocomposite, respectively. Due to increase of the particle size after coating, it is therefore proved that the PEG-PE was successfully coated on the surface of FNPs[51][52].

3.6. Release Study of PE from FPEGPE Nanocomposite

Release profiles of PE from FPEGPE nanocomposite were investigated in PBSs at pH 7.4 and 4.8 in order to simulate in vivo conditions similar to the body fluids and pH of stomach after digestion, respectively. By the UV spectrophotometer and a calibration curve equation, the percentage loading of PE into the FPEGPE nanocomposite was obtained to be around 10.3%. **Figure 6** provides the cumulative release quantities of PE from the FPEGPE nanocomposite were 60.883.1% and 83.1% at pH 7.4 and 4.8 after 4223 and 1231 min, respectively. The inset in **Figure 6** shows as expected the release of PE was completed very fast from a physical mixture (FNPs-PEG-PE) after 5 and 7 min at pH 4.8 and 7.4, respectively, which reveals that the release of PE from the physical mixture is not in the sustained-release manner.

The release rates of PE from the nanocomposite at pH 7.4 [**Figure 6(b)**] are

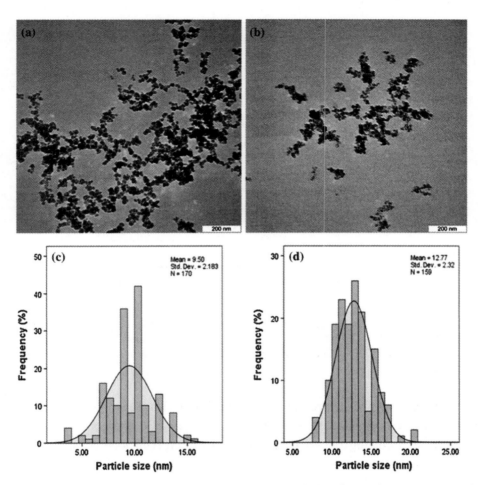

Figure 5. TEM micrographs for a FNPs with 200nm scale bar, b FPEGPE nanocomposite with 200nm scale bar, c particle size distribution of FNPs, and d particle size distribution of FPEGPE nanocomposite.

much slower than that at pH 4.8 [**Figure 6(a)**], and this behavior is attributed to the acidity of the media. At pH 7.4, the fast release of PE with a value of 44.2% at the initial 120min may be due to the release of PE anions adsorbed on the surface of the PEG polymer. However, it became much slower and more sustained after this initial time, which is attributed to the ion-exchange process between the PE anions and the anions present in the buffer solution. In our previous nanocomposite (FCPE) with chitosan was used as the coating polymer, the PE anion binding was found to be stronger compared to this nanocomposite (FPEGPE) due to the NH_3 active group that is present in chitosan in FCPE nanocomposite compared to

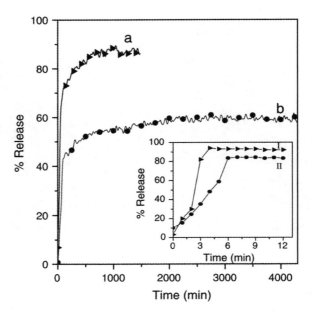

Figure 6. Release profiles of PE from the FPEGPE nanocomposite into (a) phosphate buffered solution at pH 4.8 and (b) pH 7.4. The inset shows the release profiles of PE from its physical mixture of FNPs-PEG-PE into phosphate buffered solution (I) at pH 4.8 and (II) at pH 7.4.

O–H active group that is present in PEG in FPEGPE nanocomposite. Therefore, the drug (PE) can be released more slowly from the FCPE carriers, when chitosan was used as a coating polymer compared to FPEGPE nanocomposite in which PEG was used as the coating polymer[40].

The release of drug molecules from FPEGPE nanocomposite can be described by three different kinetic models such as pseudo-first-order kinetic, pseudo-second-order kinetic, and parabolic diffusion model. Pseudo-first-order kinetic equation, $\ln(q_e - q_t) = \ln(q_e - k_1 t)$ [53][54] which is represented in the linear form, shows the release of PE from FPEGPE nanocomposite, and the decomposition rate depends on the amount of the drug in the nanocomposite. Therefore, if this kinetic model is followed, the plot of $\ln(q_e - q_t)$ versus t will be a linear from which the k_1 value can be obtained.

The pseudo-second-order kinetic model[54] that can be expressed in the linear form of $\frac{t}{q_t} = \frac{1}{k_2 q_e^2} + \frac{t}{q_e}$. The plot of $\frac{t}{q_t}$ versus t will be a linear and allows to measure of k_2. And the parabolic diffusion model[55] which can be represented as, $1-\dfrac{\left(\dfrac{M_t}{M_0}\right)}{t}=kt^{-0.5}$ equations. The q_e and q_t in the pseudo-first- and the pseudo-second-order kinetic models are the equilibrium release rate and the release rate at time t, respectively. In addition, k in all the three models is a constant and corresponding to the release amount, and M_0 and M_t in parabolic equation are the drug content remained in FPEGPE nanocomposite at release time 0 and t, respectively. The correlation coefficient (R^2) was used as criteria to select the best model for describing the release of drug from the nanocomposite.

With the use of these three kinetic models as mentioned earlier for the release kinetic data, it was found that the pseudo-second-order kinetic model was deemed more satisfactory for describing the release behavior of PE anions from FPEGPE nanocomposites compared to the other models used in this work [**Figure 7(a)**, **Figure 7(b)**, **Table 1**].

3.7. In Vitro Bioassay

The cell proliferation assay (MTT) is one of the several in vitro methods

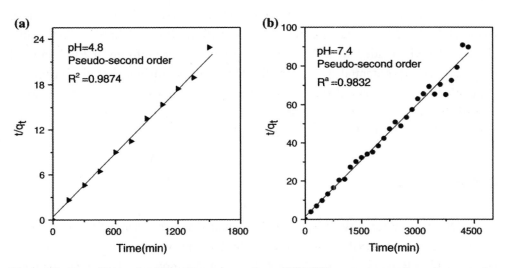

Figure 7. Data fitting for PE anion release from FPEGPE nanocomposite using pseudo-second-order kinetic model into PBS at pH 4.8 (a) and pH 7.4 (b).

Table 1. The correlation coefficient (R^2), rate constant (k), and half-time ($t_{1/2}$) obtained by fitting the PE anion release data from the FPEGPE nanocomposite into phosphate buffered solution at pH 4.8 and 7.4.

Aqueous solution	Saturated release (%)	R^2			Rate constant $(k)^a$ (mg min^{-1})	$t_{1/2}^a$ (min)
		Pseudo-first-order	Pseudo-second-order	Parabolic diffusion		
pH 4.8	60.8	0.3673	0.9874	0.4576	4.85×10^{-4}	29
pH 7.4	83.1	0.3294	0.9832	0.4164	2.42×10^{-4}	81

^{a} Estimated using pseudo-second-order kinetics

used to study toxicity potentials of active compounds[56][57]. **Figure 8** shows the toxicity effect of pure PE, FNPs, and the synthesized iron oxide coated with PEG-PE (FPEGPE) on normal fibroblast cell (3T3) at 72h post treatment using cell proliferation assay (MTT) between the dose ranges of 0–50μg mL^{-1}.

For the three compounds tested, dose-dependent properties were observed in which it decreases the cellular viability after 72h. Pure PE was found to have higher toxicity effect in a dose-dependent fashion compared to FPEGPE nanocomposite and FNPs. There was 20% decrease in cell viability at 50μg mL^{-1} exposure to PE, while FNPs and FPEGPE nanocomposite showed less than 10% decrease on fibroblast viability. Hence, pure drug PE causes a slightly higher toxicity to 3T3 cell in a dose-dependent manner than FPEGPE nanocomposite and FNPs.

We had reported earlier similar toxicity pattern, where PE-loaded iron oxide-chitosan (FCPE) nanoparticles tested on fibroblast cell line[40]. The results show that FNPs loaded with PE and coated with either chitosan or PEG showed decrease toxicity potential than pure PE. This may be attributed to the sustained, controlled release potential seen in both materials. The decrease toxicity prospective of this delivery system together with the controlled-release effects should be explored further for better drug delivery application.

4. Conclusion

Due to low toxicity effect of FNPs against mouse normal fibroblast (3T3) cell lines, magnetite FNPs which were synthesized by coprecipitation method were coated with PEG-PE to form the FPEGPE nanocomposite. The XRD patterns show pure phase magnetite FNPs with mean particle size of 9 ± 2nm. After

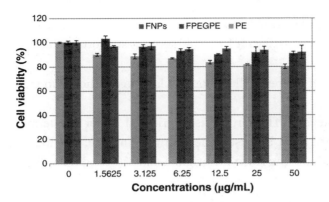

Figure 8. Cytotoxicity effects of FNPs, PE, and FPEGPE nanocomposite on 3T3 cells as determined by MTT assay after 72h of exposure.

coating, the FPEGPE nanocomposite composed of pure magnetite core with slightly bigger mean particle size of 13 ± 2nm. The FTIR shows the vibrational modes of the magnetite and the attachment of PEG-PE onto the surface of FNPs. VSM studies confirm the superparamagnetic properties of FNPs and the FPEGPE nanocomposite. Although, it was found that the release profiles of PE from FPEGPE nanocomposite into phosphate buffered solution are of controlled manner with the total release equilibrium of 60.8% and 83.1% when exposed to pH 7.4 and 4.8 at 4223 and 1231min, respectively, but due to the stronger binding of chitosan (NH_3 active group) compared to PEG (O–H active group), the drug (PE) can be released more slowly from the FCPE carriers, when chitosan was used as a coating polymer compared to FPEGPE nanocomposite in which PEG was used as the coating polymer. The cytotoxic effects of FNPs, free PE, and FPEGPE were determined against mouse normal fibroblast (3T3) cell lines and show slight decrease in the viability of 3T3 cells could be directly proportional to the concentrations used. However, pure PE presented slightly higher toxicity than the FNPs and FPEGPE nanocomposite. The decrease toxicity prospective of this new nanocomposite compared to pure drug (PE) together with controlled-release properties offers a great potential of this new nanocomposite to be used for the treatment of hypertension.

Acknowledgements

The authors would like to thank the Ministry of Science, Technology and

Innovation of Malaysia (MOSTI) under National Nanotechnology Initiative, Grant NND/NA/(1)/TD11-010 (Vot No. 5489100) for supporting this research.

Conflict of interest: All authors declare no conflict of interests in this work.

Open Access: This article is distributed under the terms of the Creative Commons Attribution License which permits any use, distribution, and reproduction in any medium, provided the original author(s) and the source are credited.

Source: Dorniani D, Kura A U, Hussein M Z B, et al. Controlled-release formulation of perindopril erbumine loaded PEG-coated magnetite nanoparticles for biomedical applications[J]. Journal of Materials Science, 2014, 49(24):8487−8497.

References

[1] Damgé C, Socha M, Ubrich N, Maincent P (2009) Poly(e-caprolactone)/eudragit nanoparticles for oral delivery of aspartinsulin in the treatment of diabetes. J Pharm Sci 99:879–889.

[2] Damgé C, Maincent P, Ubrich N (2007) Oral delivery of insulin associated to polymeric nanoparticles in diabetic rats. J Control Release 117:163–170.

[3] Oyarzun-Ampuero FA, Brea J, Loza MI, Torres D, Alonso MJ (2009) Chitosan-hyaluronic acid nanoparticles loaded with heparin for the treatment of asthma. Int J Pharm 381:122–129.

[4] Kumar M, Kong X, Behera AK, Hellermann GR, Lockey RF, Mohapatra SS (2003) Chitosan IFN-c-pDNA nanoparticle (CIN) therapy for allergic asthma. Genet Vaccines Ther 1:3.

[5] Wang W, Zhu R, Xie Q, Li A, Xiao Y, Li K, Liu H, Cui D, Chen Y, Wang S (2012) Enhanced bioavailability and efficiency of curcumin for the treatment of asthma by its formulation in solid lipid nanoparticles. Int J Nanomed 7:3667–3677.

[6] Kim M-H, Seo J-H, Kim H-M, Jeong H-J (2014) Zinc oxide nanoparticles, a novel candidate for the treatment of allergic inflammatory diseases. Eur J Pharmacol 738:31–39.

[7] Liu L, Xu K, Wang H, Tan PKJ, Fan W, Venkatraman SS, Li L, Yang Y-Y (2009) Self-assembled cationic peptide nanoparticles as an efficient antimicrobial agent. Nat Nanotech 4:457–463.

[8] Rai M, Yadav A, Gade A (2009) Silver nanoparticles as a new generation of antimi-

crobials. Biotechnol Adv 27:76–83.

[9] Anderson LJ, Holden S, Davis B, Prescott E, Charrier CC, Bunce NH, Firmin DN, Wonke B, Porter J, Walker JM (2001) Cardiovascular T2-star (T2*) magnetic resonance for the early diagnosis of myocardial iron overload. Eur Heart J 22:2171–2179.

[10] Jendelová P, Herynek V, Urdzikova L, Glogarová KI, Kroupová J, Andersson B, Bryja VTZ, Burian M, Hájek M, Syková E (2004) Magnetic resonance tracking of transplanted bone marrow and embryonic stem cells labeled by iron oxide nanoparticles in rat brain and spinal cord. J Neurosci Res 76:232–243.

[11] Leuschner C, Kumar CSSR, Hansel W, Soboyejo W, Zhou J, Hormes J (2006) LHRH-conjugated magnetic iron oxide nanoparticles for detection of breast cancer metastases. Breast Cancer Res Treat 99:163–176.

[12] Wang S, Chen KJ, Wu TH, Wang H, Lin WY, Ohashi M, Chiou PY, Tseng HR (2010) Photothermal effects of supramolecularly assembled gold nanoparticles for the targeted treatment of cancer cells. Angew Chem Int Ed 49:3777–3781.

[13] Yin Win K, Feng S-S (2005) Effects of particle size and surface coating on cellular uptake of polymeric nanoparticles for oral delivery of anticancer drugs. Biomaterials 26:2713–2722.

[14] He C, Hu Y, Yin L, Tang C, Yin C (2010) Effects of particle size and surface charge on cellular uptake and biodistribution of polymeric nanoparticles. Biomaterials 31:3657–3666.

[15] Jordan A, Scholz R, Maier-Hauff K, van Landeghem FKH, Waldoefner N, Teichgraeber U, Pinkernelle J, Bruhn H, Neumann F, Thiesen B (2006) The effect of thermotherapy using magnetic nanoparticles on rat malignant glioma. J Neuro-Oncol 78:7–14.

[16] Dessy A, Piras AM, Schirò G, Levantino M, Cupane A, Chiellini F (2011) Hemoglobin loaded polymeric nanoparticles: preparation and characterizations. Eur J Pharm Sci 43:57–64.

[17] Davis FF (2002) The origin of pegnology. Adv Drug Deliver Rev 54:457–458.

[18] Malam Y, Loizidou M, Seifalian AM (2009) Liposomes and nanoparticles: nanosized vehicles for drug delivery in cancer. Trends Pharmacol Sci 30:592–599.

[19] Hussein-Al-Ali SH, El Zowalaty ME, Hussein MZ, Ismail M, Dorniani D, Webster TJ (2014) Novel kojic acid-polymer-based magnetic nanocomposites for medical applications. Int J Nanomed 9:351–362.

[20] Unsoy G, Yalcin S, Khodadust R, Mutlu P, Gunduz U (2012) In situ synthesis and characterization of chitosan coated iron oxide nanoparticles and loading of doxorubicin. In: Nanocon. Brno, Czech Republic.

[21] Kaaki K, Hervé-Aubert K, Chiper M, Shkilnyy A, Soucé M, Benoit R, Paillard A, Dubois P, Saboungi M-L, Chourpa I (2011) Magnetic nanocarriers of doxorubicin coated with poly(ethylene glycol) and folic acid: relation between coating structure,

surface properties, colloidal stability, and cancer cell targeting. Langmuir 28: 1496–1505.

[22] Saboktakin MR, Tabatabaie R, Maharramov A, Ramazanov MA (2010) Synthesis and characterization of superparamagnetic chitosan-dextran sulfate hydrogels as nano carriers for colon-specific drug delivery. Carbohydr Polym 81:372–376.

[23] Dorniani D, Bin Hussein MZ, Kura AU, Fakurazi S, Shaari AH, Ahmad Z (2013) Preparation and characterization of 6-mercaptopurine-coated magnetite nanoparticles as a drug delivery system. Drug Des Devel Ther 7:1015–1026.

[24] Qu J-B, Shao H-H, Jing G-L, Huang F (2013) PEG-chitosancoated iron oxide nanoparticles with high saturated magnetization as carriers of 10-hydroxycamptothecin: preparation, characterization and cytotoxicity studies. Colloids Surf B 102:37–44.

[25] Dorniani D, Hussein MZ, Kura AU, Fakurazi S, Shaari AH, Ahmad Z (2011) Preparation of Fe3O4 magnetic nanoparticles coated with gallic acid for drug delivery. Int J Nanomed 7:5745–5756.

[26] Dorniani D, Kura AU, Hussein-Al-Ali SH, Hussein MZB, Fakurazi S, Shaari AH, Ahmad Z (2014) In vitro sustained release study of gallic acid coated with magnetite–PEG and magnetite– PVA for drug delivery system. Sci World J. doi: 10.1155/2014/ 416354.

[27] Sun C, Sze R, Zhang M (2006) Folic acid-PEG conjugated superparamagnetic nanoparticles for targeted cellular uptake and detection by MRI. J Biomed Mater Res A 78:550–557.

[28] Hussein-Al-Ali SH, Arulselvan P, Fakurazi S, Hussein MZ, Dorniani D (2014) Arginine-chitosan- and arginine-polyethylene glycol-conjugated superparamagnetic nanoparticles: preparation, cytotoxicity and controlled-release. J Biomater Appl. doi: 10.1177/0885328213519691.

[29] Hussein-Al-Ali SH, Arulselvan P, Hussein MZ, Fakurazi S, Ismail M, Dorniani D, El Zowalaty ME (2014) Synthesis, characterization, controlled release and cytotoxic effect studies of anthranilic acid loaded chitosan and polyethylene glycol-magnetic nanoparticles on murine macrophage RAW 264.7 cells. Nano. doi: 10.1142/S1793292014500167.

[30] Thom T, Haase N, Rosamond W, Howard VJ, Rumsfeld J, Manolio T, Zheng Z-J, Flegal K, O'donnell C, Kittner S (2006) Heart disease and stroke statistics—2006 update: a report from the American Heart Association Statistics Committee and Stroke Statistics Subcommittee. Circulation 113:e85–e151.

[31] Denkbas, EB, Kilic,ay E, Birlikseven C, Öztuürk E (2002) Magnetic chitosan microspheres: preparation and characterization. React Funct Polym 50:225–232.

[32] Li G-Y, Y-r Jiang, K-l Huang, Ding P, Chen J (2008) Preparation and properties of magnetic Fe$_3$O$_4$-chitosan nanoparticles. J Alloys Compd 466:451–456.

[33] Remková A, Kratochvil'ová H, Ďurina J (2008) Impact of the therapy by re-

nin-angiotensin system targeting antihypertensive agents perindopril versus telmisartan on prothrombotic state in essential hypertension. J Hum Hypertens 22: 338–345.

[34] Zannad F, Bernaud CM, Fay R (1999) Double-blind, randomized, multicentre comparison of the effects of amlodipine and perindopril on 24h therapeutic coverage and beyond in patients with mild to moderate hypertension. J Hypertens 17:137–146.

[35] Lee H, Shao H, Huang Y, Kwak B (2005) Synthesis of MRI contrast agent by coating superparamagnetic iron oxide with chitosan. IEEE Trans Magn 41:4102–4104.

[36] Kura AU, Hussein-Al-Ali SH, Hussein MZ, Fakurazi S, Arulselvan P (2012) Development of a controlled-release anti-parkinsonian nanodelivery system using levodopa as the active agent. Int J Nanomed 8:1103–1110.

[37] Wang C, Feng L, Yang H, Xin G, Li W, Zheng J, Tian W, Li X (2012) Graphene oxide stabilized polyethylene glycol for heat storage. Phys Chem Chem Phys 14:13233–13238.

[38] Chen T-J, Cheng T-H, Chen C-Y, Hsu SCN, Cheng T-L, Liu G-C, Wang Y-M (2009) Targeted Herceptin-dextran iron oxide nanoparticles for noninvasive imaging of HER2/neu receptors using MRI. J Biol Inorg Chem 14:253–260.

[39] Sangeetha J, Philip J (2012) The interaction, stability and response to an external stimulus of iron oxide nanoparticle-casein nanocomplexes. Colloids Surf A 406: 52–60.

[40] Dorniani D, Hussein MZB, Kura AU, Fakurazi S, Shaari AH, Ahmad Z (2013) Sustained release of prindopril erbumine from its chitosan-coated magnetic nanoparticles for biomedical applications. Int J Mol Sci 14:23639–23653.

[41] Calmon MF, de Souza AT, Candido NM, Raposo MIB, Taboga S, Rahal P, Nery JG (2012) A systematic study of transfection efficiency and cytotoxicity in HeLa cells using iron oxide nanoparticles prepared with organic and inorganic bases. Colloids Surf B 100:177–184.

[42] Dodi G, Hritcu D, Lisa G, Popa MI (2012) Core–shell magnetic chitosan particles functionalized by grafting: synthesis and characterization. Chem Eng J 203:130–141.

[43] Mürbe J, Rechtenbach A, Töpfer JR (2008) Synthesis and physical characterization of magnetite nanoparticles for biomedical applications. Mater Chem Phys 110:426–433.

[44] Shameli K, Bin Ahmad M, Jazayeri SD, Sedaghat S, Shabanzadeh P, Jahangirian H, Mahdavi M, Abdollahi Y (2012) Synthesis and characterization of polyethylene glycol mediated silver nanoparticles by the green method. Int J Mol Sci 13:6639–6650.

[45] Kipkemboi PK, Kiprono PC, Sanga JJ (2003) Vibrational spectra of t-butyl alcohol, t-butylamine and t-butyl alcohol, t-butyl-amine binary liquid mixtures. Bull Chem Soc Ethiopia 17:211–218.

[46] Smith BC (1999) Infrared spectral interpretation: a systematic approach. CRC Press,

Boca Raton, FL.

[47] Macêdo RO, Gomes do Nascimento T, Soares Aragâo CF, Barreto Gomes AP (2000) Application of thermal analysis in the characterization of anti-hypertensive drugs. J Therm Anal Calorim 59:657–661.

[48] Kayal S, Ramanujan RV (2010) Doxorubicin loaded PVA coated iron oxide nanoparticles for targeted drug delivery. Mater Sci Eng C 30:484–490.

[49] Debrassi A, Correâ AF, Baccarin T, Nedelko N, S'lawska-Wan-iewska A, Sobczak K, Dłuzewski P, Greneche J-M, Rodrigues CvA (2012) Removal of cationic dyes from aqueous solutions using N-benzyl-O-carboxymethylchitosan magnetic nanoparticles. Chem Eng J 183:284–293.

[50] Kumar R, Inbaraj BS, Chen BH (2010) Surface modification of superparamagnetic iron nanoparticles with calcium salt of poly(Y-glutamic acid) as coating material. Mater Res Bull 45:1603–1607.

[51] Qu J, Liu G, Wang Y, Hong R (2010) Preparation of Fe_3O_4-chitosan nanoparticles used for hyperthermia. Adv Powder Technol 21:461–467.

[52] L-y Zhang, X-j Zhu, H-w Sun, Chi G-r Xu, J-x Sun Y-l (2010) Control synthesis of magnetic Fe_3O_4–chitosan nanoparticles under UV irradiation in aqueous system. Curr Appl Phys 10:828–833.

[53] Hussein-Al-Ali SH, Al-Qubaisi M, Hussein MZ, Ismail M, Zainal Z, Hakim MN (2012) In vitro inhibition of histamine release behavior of cetirizine intercalated into Zn/Al-and Mg/Al-layered double hydroxides. Int J Mol Sci 13:5899–5916.

[54] Dong L, Yan L, Hou W-G, Liu S-J (2010) Synthesis and release behavior of composites of camptothecin and layered double hydroxide. J Solid State Chem 183: 1811–1816.

[55] Ho Y-S, Ofomaja AE (2006) Pseudo-second-order model for lead ion sorption from aqueous solutions onto palm kernel fiber. J Hazard Mater 129:137–142.

[56] Valiathan C, McFaline JL, Samson LD (2012) A rapid survival assay to measure drug-induced cytotoxicity and cell cycle effects. DNA Repair 11:92–98.

[57] Niles AL, Moravec RA, Riss TL (2008) Update on in vitro cytotoxicity assays for drug development. Expert Opin Drug Discov 3:655–669.

Chapter 7

Correlative EBSD and SKPFM Characterisation of Microstructure Development to Assist Determination of Corrosion Propensity in Grade 2205 Duplex Stainless Steel

C. Örnek1, D. L. Engelberg

Materials Performance Centre & Corrosion and Protection Centre,
School of Materials, The University of Manchester, Sackville Street Campus,
Manchester M13 9PL, UK

Abstract: Correlative electron backscatter diffraction (EBSD) and scanning Kelvin probe force microscopy (SKPFM) analysis has been carried out to characterise microstructure development and associated corrosion behaviour of as-received and 750°C heat-treated grade 2205 duplex stainless steel. High-resolution EBSD analysis revealed the presence of σ- and χ-phase, secondary austenite, Cr_2N, and CrN after ageing treatment. SKPFM Volta potential measurements confirmed the formation of discrete reactive sites, indicating local corrosion propensity in the microstructure. Cr_2N, σ-phase, and inter-granular χ-phase had the largest net cathodic activity, followed by CrN and intra-granular χ-phase showing medium electrochemical activity, with ferrite and austenite (including secondary austenite) showing net anodic activity. Corrosion screening confirmed selective corrosion of ferrite in the as-received and 75°C-aged conditions with the corrosion propensity of secondary phases staying in-line with SKPFM observations. Stress corrosion micro-cracks were also observed and are discussed in light of microstructure corrosion propensity.

1. Introduction

Duplex stainless steels (DSS) offer a synergistic combination of excellent mechanical properties with enhanced corrosion resistance, and these materials are now increasingly used, such as the manufacture of containers for the storage of intermediate-level radioactive waste (ILW)[1]. Heat treatment, welding, or prolonged exposure to elevated temperatures may lead to undesired phase reactions in these high-alloyed stainless steels[2]–[5]. For example, the ferrite can decompose into a series of meta-stable and thermodynamically stable phases, whilst the austenite has often been stated to be unaffected[2][4][6]. However, phase reactions can also occur in the austenite, increasing the volume fraction of secondary phase products in the microstructure[7]–[13]. Components made with large wall thicknesses, therefore, usually contain intermetallic phases due to variations in cooling rates after high temperature treatments, with the core of components typically containing large volume fractions[14][15].

Phase reactions are more favoured in the ferrite due to the enrichment of Cr and Mo, combined with far higher diffusion rates compared to the austenite [3][4][16]. For example, phase reactions in the temperature range between 250°C and 550°C have been known as '475°C embrittlement'[2]–[5][17][18] where in addition to the degradation of mechanical properties a significant reduction in corrosion performance has also been observed[17]–[21]. Phase reactions occurring in the temperature range between 600°C and 1000°C have become known as 'σ-phase embrittlement' where numerous secondary phases, such as Frank-Kasper phases (σ and χ) and, to some extent, R-phase can be formed[5][10][16][22]–[35]. These are often accompanied by the precipitation of nitrides (Cr_2N and CrN) and carbides[2][4][5][16][34][36][37].

Determination of the identity and the volume fraction of secondary phases have typically been carried out using image analysis of optical and/or electron microscopic micrographs, often supported by analytical semi-quantitative assessment of the chemical composition by energy-dispersive X-ray (EDX) analysis. However, uncertainty and lack of precision of these measurement methods did generally not allow to build-up a comprehensive mechanistic understanding of the behaviour of intermetallic phases during corrosion processes. Investigations to quantitatively describe microstructure development of aged duplex stainless steels at elevated temperatures, especially with respect to the formation of secondary phases, has often been associated with σ- and χ-phase formation based on eutectoid decomposition reaction of the ferrite. The effects of other precipitates such as Cr-nitrides were often not taken into account due to their small volume fractions and geometrical sizes, which makes them difficult to detect[3][4][16][26][34][38].

The electron backscatter diffraction (EBSD) technique has developed into a tool for precise mapping of the microstructure for texture, crystallographic phases, and identification of misorientation gradients with information about local plastic strain. Mapping of multiple phases over large areas with high spatial resolution of up to 10's of nm has become possible for quantitative identification of crystallographic information to allow in- and ex-situ observation of microstructure development[39]. EBSD combined with scanning Kelvin probe force microscopy (SKPFM) can provide crystallographic information about the microstructure with local Volta potential information at high spatial resolution (10's of nm's). The latter is a quantitative measure to describe electrochemical reactivity of a metal surface[40]–[44]. The knowledge about local Volta potential differences allows characterisation of corrosion processes, for example, to understand local micro-galvanic coupling which has importance in bi-phase alloys such as duplex stainless steels[13][45]–[50]. Selective dissolution of the ferrite phase in duplex stainless steels, for example, can be explained by the larger Volta potential difference with respect to the Pt reference (lower absolute Volta potential), compared to a smaller measured difference of the austenite[18][45][46]. With the introduction of cold work, grade 2205 duplex stainless steel tends to show preferential local corrosion sites associated with localised deformation in the austenite, with these regions related to local Volta potential extremes[45][51].

The purpose of the work reported in this paper was to link the precipitation of secondary phases after 750°C heat treatment using multi-scale EBSD analysis, to their Volta potential differences using SKPFM. Information from correlative EBSD and SKPFM analysis provides an insight into mechanistic understanding of the atmospheric corrosion and stress corrosion cracking (AISCC) behaviour of grade 2205 duplex stainless steel microstructure.

2. Experimental

A solution-annealed (as-received) grade 2205 duplex stainless steel plate of 10mm thickness was used in this work with a composition shown in **Table 1**. EDX spectroscopy measurements of the ferrite and the austenite phases were conducted to inform about alloy partitioning, with the results also provided in **Table 1**. Small miniature tensile samples were manufactured with an overall length of 50mm, a thickness of 1mm, with 25mm gauge length, and 3mm gauge width, as well as coupon samples with dimensions of 10mm × 10mm × 10mm (L × W × T). All samples were aged for 5h at 750°C, followed by air-cooling to room temperature. The surface of these samples was mechanically ground to 4000-grit using SiC paper, followed by a 3, 1, 1/4, and 0.1μm diamond paste polishing finish. A final fine-polishing treatment using a modified OP-S suspension (OP-S with a few drops of concentrated HNO_3 and H_2SO_4) was performed for one hour to achieve a smooth, strain-free surface finish for EBSD and SKPFM analysis.

2.1. Microstructure Analysis

The microstructures of as-received and heat-treated specimens were analysed by EBSD using an FEI Magellan high-resolution scanning electron micro-

Table 1. Chemical compositions (wt%) of grade 2205 duplex stainless steel used in this study.

Grade	C	Si	Mn	P	S	Cr	Ni	Mo	N	Fe
Plate	0.016	0.4	1.5	0.021	0.001	22.4	5.8	3.2	0.18	bal.
Ferrite	n.a.	0.5	1.7	n.a.	n.a.	25	4.3	4.3	n.a.	bal.
Austenite	n.a.	0.4	1.9	n.a.	n.a.	22.1	6.9	2.4	n.a.	bal.

Semi-quantitative EDX analysis of the ferrite and austenite composition is also provided to inform about element partitioning (n.a. = not analysed)

scope (SEM). Data acquisition was performed with a Nordlys EBSD detector from Oxford Instruments with AZtec 2.2 software. An accelerating voltage of 10–20kV was used with low current (spot size) to achieve effective spatial resolutions of 10's of nm[39][52]. High-resolution EBSD mapping was carried out with step-sizes between 56 and 75nm over an area of 100μm × 87μm. For the identification of crystallographic phases, the parameters listed in **Table 2** were used. The phase databases were included in AZtec software. Databases from HKL and Inorganic Crystal Structure Database (ICSD) were used for phase acquisition.

Indexing rates were typically close to 99% and at least six Kikuchi bands were selected for successful phase identification, with the minimum number of detected bands of the corresponding Kikuchi diffraction pattern for each phase given in **Figure 1**. Extra care was taken for indexing secondary phases, such as CrN, Cr_2N, and χ-phase. Data acquisition was first performed using all listed phases. When secondary precipitates were indexed, the dataset was re-assessed by de-selecting individual phases and re-mapping the same area without the corresponding phase. Phase identification proved to be reliable when the previously indexed region of a secondary phase was non-indexed, which confirms and validates the indexing procedure for CrN, Cr_2N, χ, and σ-phase. Indexing of chromium carbides ($Cr_{23}C_6$) for example, revealed to be problematic, since the ferrite or austenite were mis-indexed when the $Cr_{23}C_6$ phase was selected for acquisition. EBSD assessment of $Cr_{23}C_6$ was, therefore, excluded in our assessment. However, carbide formation in the 750°C temperature range is extremely retarded in modern duplex stainless steel, with nitrides usually formed in far larger quantities leading to a suppression of carbide formation[16][34].

Table 2. Database with crystallographic geometry parameters for EBSD phase identification.

Phase	a (Å)	b (Å)	c (Å)	α	β	γ	Space group	Database
Ferrite	2.87	2.87	2.87	90°	90°	90°	229	HKL
Austenite	3.66	3.66	3.66	90°	90°	90°	225	HKL
Cr_2N	4.75	4.75	4.43	90°	90°	120°	162	ICSD
CrN	2.97	4.12	2.88	90°	90°	90°	59	ICSD
Sigma	8.80	8.80	4.56	90°	90°	90°	136	HKL
Chi	8.92	8.92	8.92	90°	90°	90°	217	HKL
$Cr_{23}C_6$	10.6	10.6	10.6	90°	90°	90°	62	ICSD

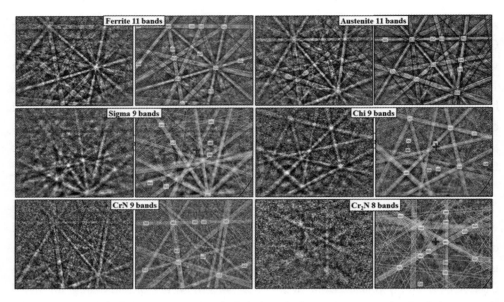

Figure 1. Kikuchi diffraction bands of all indexed phases in our study. Each image pair shows unsolved (left) and solved (right) Kikuchi pattern with the minimum number of bands detected for each phase.

All EBSD maps were processed using HKL Channel 5 software, and phase fractions and grain sizes were extracted. High-angle grain boundaries (HAGB's) were determined with misorientation in excess of 15, and low-angle grain boundaries (LAGB's) were defined with 1°–15°. Phase maps were generated with all phases detected, including phase and grain boundaries.

2.2. Scanning Kelvin Probe Force Microscopy

SKPFM and Magnetic Force Microscopy (MFM) measurements were carried out with a Dimension 3100 atomic force microscope (AFM) from *Veeco* interfaced with a *Nanoscope 3a* controller. Pt-coated *OSCM-PT* AFM probes from *Olympus* with 15nm nominal radius were used to map the surface topography and the corresponding Volta potential difference of the metal with respect to the tip. The scan size of AFM map was between 10 and 80µm, depending on the microstructural features to be characterised. The scan rate was adjusted to the tip velocity (10–30µm/s), which corresponds to a scan rate of 0.2–0.5 Hz. The images contained 512 × 512 pixels yielding effective spatial resolution between 20 and

156nm. All maps were processed using *Nanoscope* V1.5 software (*Bruker*). Topography/height and potential maps were flattened with 0th flattening order to achieve best contrast within the microstructure; therefore, all data are semi-quantitative only.

In this paper, and for the probes used, higher potentials indicate a larger potential difference and correspond to anodic sites due to a larger absolute work function difference between the bias-controlled AFM tip and the microstructure feature of the grounded sample. According to this definition, cathodic sites have, therefore, lower Volta potentials than their anodic counter-parts[53]. More comprehensive explanation about the meaning of the potential and the Kelvin probe technique can be found elsewhere[43]–[45][50][54][55].

2.3. Atmospheric-Induced Stress Corrosion Cracking Testing (AISCC)

AISCC susceptibility was investigated on as-received and 750°C heat-treated mini-tensile specimens, which were ground to 4000-grit using SiC sand papers. The heat-treated specimen, however, was afterwards electro-polished in an electrolyte of a mixture of 20% perchloric acid and 80% methanol at 20 V and at a temperature of −40°C. This was done to achieve smooth surface finish for phase identification after the corrosion test.

The as-received and heat-treated mini-tensile specimens were strained to 3% and 1%, respectively, in self-designed direct tension rigs[56]. A strain gauge was placed on the backside of each specimen and the sample extension monitored, in situ, using a LabVIEW programme. Water droplets containing $MgCl_2$ in different concentrations and volumes were applied onto the surface. The droplets were dispensed with an Eppendorf micropipette. Nominal deposition densities of magnesium chloride and chloride ions were calculated from area measurements as listed in **Table 3**. The deposited droplets changed their shape during exposure due to secondary spreading (see **Figure 2** for images after exposure); therefore, the determined 'initial' deposition densities most likely changed over time.

Table 3. Experimental conditions of the AISCC test.

Sample	Strain applied	No. of deposit	Volume (μl)	Droplet radius (mm)	Deposited MgCl$_2$ (μg/cm^2)	Deposited chloride (μg/cm^2)
As-received	3 %	1	0.5	1.78	1947	1450
Aged at 750 °C/5 h/air	1 %	1	0.5	1.78	20	14.5
		2	0.5	1.78	195	145
		3	0.5	1.78	1947	1450
		4	1.5	2.3	3319	2472
		5	2.5	2.8	3835	2856

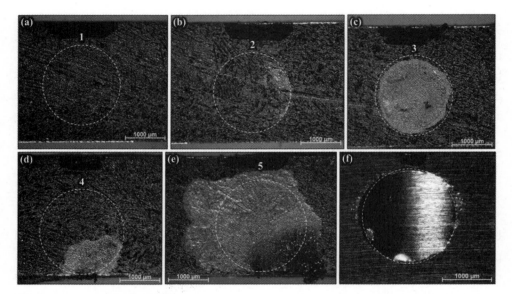

Figure 2. Stereo-microscopy images of exposed electrolyte droplets containing magnesium chloride a-e of specimen aged at 750°C (droplet 1−5) and f of as-received sample after exposure to 30% relative humidity at 50°C for 259 days.

The direct tension rigs were placed in a climatically controlled KBF Binder cabinet for 259 days at 50°C and 30% relative humidity (RH); however, fluctuations in RH up to 60% were observed during the last 2 weeks of exposure associated with equipment problems, therefore, the test was terminated. During the time of exposure, the direct tension rigs were periodically removed from the humidity chamber to assess whether corrosion had occurred. After terminating the test, the sample was rinsed in distilled water to dissolve salt and corrosion products, followed by an additional cleaning cycle in a 10wt% citric acid solution at 80°C for 2h to remove remaining corrosion products.

3. Results and Discussion

3.1. EBSD Microstructure Characterization

The microstructure of the as-received and heat-treated condition is shown as EBSD phase maps in **Figure 3(a, b)**. The as-received microstructure consisted of 44% ± 2% ferrite (δ) and 56% ± 2% austenite (γ). The ferrite formed the matrix with the austenite present in the form of island-like discrete grains or clusters of small grains. The shape of ferrite and austenite grains was elongated due to the hot rolling process during manufacture. The austenite contained large fractions of twin boundaries, shown by the straight lines (yellow) in **Figure 3(a)**, with interphase boundaries indicating more concave and convex shapes. The misorientation variation within ferrite and austenite grains was low. The average grain size of ferrite and austenite was 7μm ± 1μm and 6.5μm ± 1μm, respectively.

The specimen aged at 750°C in **Figure 3(b)** contained a large fraction of secondary phases, including σ-, χ-, chromium nitrides (Cr_2N, CrN), and secondary austenite ($γ_2$) which have formed during the ageing treatment[16][30]. The normalised phase fraction results are shown in **Table 4**. High-resolution EBSD maps allowed clear observations of the newly formed microstructure shown in **Figure 4(a)-(g)**. The austenite grain boundaries at the interphase had convoluted morphologies, with newly formed secondary austenite protruding from primary

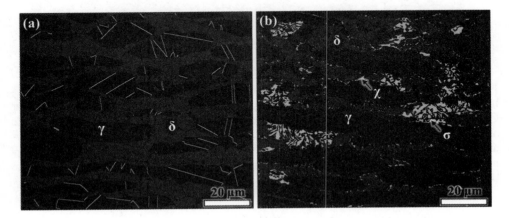

Figure 3. EBSD phase maps showing a as-received microstructure and b microstructure after heat treatment at 750°C for 5h. Black lines are phase and highangle grain boundaries. Note that Cr_2N and CrN are not visible at this scale (Color figure online).

Table 4. Normalised EBSD phase fractions.

Condition	δ	γ	σ	χ	CrN	Cr$_2$N
As-received	44	56	n/a	n/a	n/a	n/a
Aged 750 °C/5 h	26	65.5[a]	5.6	2.1	0.4	0.33

n/a not available

[a] Secondary austenite included

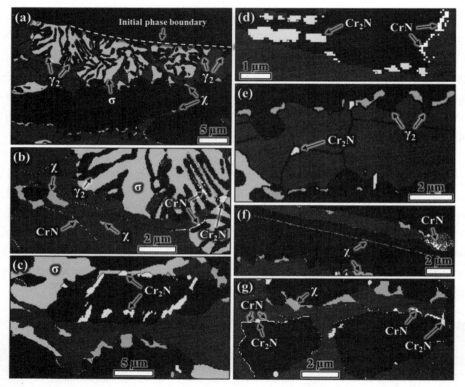

Figure 4. High-resolution EBSD phase maps of the specimen aged at 750°C for 5h showing a σ, χ, and γ$_2$ formation, b intra-granular χ phase formation in ferrite with inter-granular CrN formation in ferrite and at σ/γ$_2$ interphase boundaries, c, d Cr$_2$N formation in austenite and at existing δ/γ phase boundaries, with d CrN formation at δ/γ phase boundaries, e inter- and intra-granular Cr$_2$N formation in ferrite, f intra-granular CrN formation in δ and χ formation at boundaries and interphases, and g CrN, χ, and Cr$_2$N formation at interphase boundaries.

austenite islands as can be seen in **Figure 4(a)**.

The ferrite is thermodynamically unstable and has decomposed[3][4][16][34][38], with three mechanisms proposed for secondary austenite formation where the eutectoid equilibrium δ → σ + γ$_2$ he predominant reaction in the 700°C–900°C temperature range. The overall fraction of secondary austenite formed was approximately 12%, which was determined by the difference between pre- and post-ageing austenite contents. The eutectoid reaction mainly takes place at δ/γ-interphase boundaries due to increased diffusion rates, leading to segregation of Cr and Mo resulting in σ-phase formation[4][12][16][26][30][32][34]. The morphology of σ-phase often shows an allotriomorphic appearance due to its preferential growth along phase boundaries and towards ferrite grain interiors[57]. The surrounding regions are then depleted in Cr and Mo, resulting in the formation to secondary austenite[4][16][57]. The eutectoid secondary austenite can have a similar appearance to that of σ-phase, as can be seen in **Figures 3(b) and 4(a), (b)**. Secondary austenite in direct contact with primary austenite can also form by growth towards ferrite regions and assume coagulant-like shapes with similar grain orientation as the primary austenite, also shown in **Figure 4(e)**.

Austenite forming onto primary austenite has also been described as new austenite in the literature[58], but in our case all newly formed austenite induced by ageing is denoted as 'secondary austenite' (γ$_2$). Both types of secondary austenite have been found to contain reduced Cr contents, particularly if the formation of Cr-rich precipitates is involved, such as σ-phase[30][59]. The volumetric fraction of secondary austenite is typically twice that of formed σ-phase, clearly indicating faster reaction kinetics[30]. The corrosion performance of microstructures containing secondary austenite has been reported to cause reduced pitting corrosion resistance[37][60]. Therefore, the fraction of σ-phase alone is an insufficient parameter for microstructure assessment regarding the corrosion behaviour of duplex stainless steels.

The newly formed secondary austenite had similar grain orientation to the existing austenitic phase, with neither new high-angle or low-angle grain boundaries seen in most newly formed austenite grains. The secondary austenite seemed to have developed through grain growth from primary austenite grains, indicated by the white dashed line in **Figure 4(a)**, showing the approximate position of the initial δ/γ phase boundary. Secondary austenite formation was also observed through new austenite formation within ferrite regions associated with eutectoid σ-phase formation in the ferrite producing elongated lamellae-like shapes (allotriomorphic) also shown in **Figure 4(a), (b)**.

The decomposition of ferrite was accompanied by the formation of σ- and χ-phase. The σ-phase contained no high- and low-angle grain boundaries and showed allotriomorphic morphology. The σ-phase was located between pre-existing primary austenite islands, clearly evidencing that this phase formed by consuming ferrite. Ferrite-austenite phase boundaries and triple junctions of grain boundaries have been reported to act as nucleation sites for σ-phase[16][22]-[24][26]-[28][30][32]. The formation of σ-phase occurred at discrete regions in the microstructure, resulting in clusters of σ-phase and secondary austenite as can be seen in **Figure 3(b)**. Furthermore, σ-phase was also found in ferrite grains, highlighted in **Figure 4(c)**. The measured fraction of σ-phase was almost 6% (**Table 4**), which is in good agreement with the work of Michalska et al. who obtained 6%–7% σ-phase in grade 2205 after ageing at 750°C for 5h via conventional image analysis[26]. Elmer et al. also investigated σ-phase formation in grade 2205, and ageing at 750°C for 10h produced 22.7% of σ-phase determined via in situ synchrotron X-ray diffraction technique[24]. They reported 90% completion of phase reactions after 10h ageing, and predicted an overall fraction of 24% of σ-phase by Thermocalc calculations[24]. The size of σ-phase was estimated in **Figure 4** using horizontal and vertical ferret diameters of 1−7μm. However, the three-dimensional appearance may result in different 'structural' sizes due to the allotriomorphic shape of σ.

A total fraction of 2.1% χ-phase was measured after ageing at 750°C for 5h in the microstructure, shown in **Figure 4**. The χ-phase precipitated primarily at δ/γ phase boundaries, with expansion along the circumference of primary austenite grains but growth towards the ferrite. In addition, χ also consumed some austenite as can be seen in **Figure 4(a)-(g)**. However, χ-phase precipitation is not restricted to ferrite or δ/γ phase boundary regions. In austenitic stainless steels, χ, σ, and other secondary phases can also form by consumption of austenite grains[10]. Therefore, similar reactions are also expected to occur in duplex stainless steels. The formation of χ-phase in grade 2205 duplex stainless steel has been reported to occur at temperature approximately 75 Kelvin below that of σ-phase formation, with fastest transformation kinetics at 750°C[16][30]. Therefore, χ-phase formation has been reported to be more favoured at the initial stages of ageing[16][22][26][27][30]. Padilha et al., for example, observed qualitatively larger fractions of χ than σ-phase at early ageing stages, with the χ-phase consumed by σ-phase formation after prolonged ageing treatment at 750°C[30]. Several works have reported χ-phase as meta-stable[22][30], although computational isothermal sections of Fe-Cr-Ni-Mo-N systems containing 22% Cr and 5% Ni predicted χ as a stable phase, co-existing with σ-phase and Cr_2N[16].

The size of χ-phase was estimated in **Figure 4** using horizontal and vertical ferret diameters of 0.1–2μm. Transformation of v occurred at multiple sites, resulting in a more homogenous formation in the microstructure compared to the clustered σ-phase appearance. The χ-phase precipitates also seemed to pin interphase boundaries, causing bulging of the newly formed secondary austenite, as shown in **Figure 4(e)**. Moreover, it was noticed that χ-phase even nucleated within austenite grains, decorating part of a twin boundary, as shown in **Figure 4(f)**.

Chromium nitrides, both CrN and Cr_2N, were observed after ageing at 750°C. Both compounds can be seen in **Figure 4**, with the orthorhombic CrN and the hexagonal Cr_2N decorating δ/γ, δ/γ$_2$, and also σ/γ$_2$ interphase regions. The Cr_2N had more discrete dimensions in the form of larger areas than CrN (**Figure 4**). The morphology of Cr_2N was elongated and ellipsoid shaped. Their sizes varied between 130nm to 4μm with high aspect ratios. The CrN had smaller dimensions, typically between 80 and 620nm, but CrN was often found in clusters as highlighted in **Figure 4(f)**. The total fraction of Cr_2N was 0.33% and the fraction of CrN was 0.41%. There has been work published on Cr_2N and most is in agreement that Cr_2N is stable and co-exists with χ-phase and σ-phase precipitates[6][10][12]–[16][27][28][34][36][48][61][62], but far less is known about CrN [11][36].

The formation of Cr-nitrides is usually associated with increased ferrite contents in duplex stainless steels[11][36][61][62], and large fractions of nitrides have, therefore, been observed in welds and heat-affected zones[11][34]. The formation of Cr_2N has been reported to either occur during cooling, when ferrite is supersaturated with nitrogen leading to the formation of elongated and discretely-shaped intra-granular precipitates with $\langle 0001 \rangle_{Cr_2N} \square \langle 011\delta \rangle$ orientation relationship, or during isothermal ageing in the 700°C–900°C temperature range where inter-granular Cr_2N precipitates are formed on δδ and δγ-boundaries[16]. In our study, Cr_2N was observed not only at interphase boundaries, but also within austenite grains indicating that nitride precipitation may be possible in austenite as well, which can be clearly seen in **Figure 4(c), (d)**.

CrN formation occurred within the ferrite, mostly on δ/δ and δ/γ grain boundaries as shown in **Figure 4(b)-(g)**. No CrN was observed in the austenite, but only on γ$_2$/σ phase boundary regions as shown in **Figure 4(b)**. An orientation relationship between CrN and ferrite $\langle 110 \rangle_{CrN} \| \langle 011\delta \rangle$ and $\langle 001 \rangle_{CrN} \| \langle 110\delta \rangle$ has been reported[11]. In a grade 2505 duplex stainless steel with 0.14% N both types of Cr-nitrides were observed, when the microstructure was cooled with rapid cooling in the range of 40–150K/s[11]. The reported CrN precipitates showed a filmor platelet-like appearance, with Cr$_2$N having a rod-like shape, with the latter forming in larger sizes than CrN precipitates[56].

Both Cr-nitrides are typically enriched in Cr, N, Fe, and Mo, but Cr$_2$N contains more Cr than CrN, whilst the opposite holds true for the N content[36][62]. The extent of elemental depletion zones around these precipitates usually develops as a function of element enrichment within the precipitate and the volumetric size. Since Cr$_2$N is more enriched in Cr and has a larger size than CrN, it is, therefore, expected that Cr$_2$N yields larger Cr-depleted regions, resulting in an increased electrochemical activity associated with reduced corrosion resistance [13][36][47][48][61]. This assumption will be discussed in more detail in light of the SKPFM results.

3.2. SKPFM: Volta Potential Measurements

Maps showing the Volta potential differences measured over regions containing ferrite and austenite in the as-received microstructure with corresponding topography and magnetic frequency maps are shown in **Figure 5(a)-(c)**. Regions showing low Volta potential differences were determined to be non-ferromagnetic and hence austenitic as can be seen in the magnetic frequency map in **Figure 5(c)**, indicating net cathodic character. Vice versa, regions of high potential with net anodic character are ferritic, showing a magnetic frequency response. The Volta potential reflects the electronic activity of a metal, and the larger the potential difference with respect to Pt the lower the actual electronic activity leading to facilitated charge transfer during electrochemical reactions[40][41]. However, this describes thermo-dynamic equilibrium conditions, and kinetic information cannot be extracted.

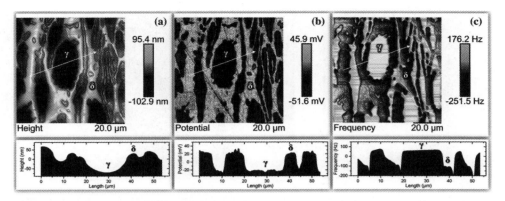

Figure 5. SKPFM analysis of as-received microstructure with a topography of a scanned area of 80μm² × 80μm² containing ferrite and austenite with a line profile measurement along the white arrow given below, b the corresponding Volta potential differences map with a line profile, and c correlated magnetic frequency measured over the same area (note a small off-set of the scanned area) (Color figure online).

The ferrite showed 50–70mV in average larger Volta potential values than the austenite indicating higher electrochemical activity. The Volta potential difference also indicated possible micro-galvanic activity between austenite and ferrite, at which the latter was expected to form the net anode whilst austenite the net cathode. The potential variation across ferrite and austenite interphases was smooth with only minor potential gradients. A smooth potential gradient is indicative for non-heterogeneous activity, such as selective attack. Earlier work on grade 2205 duplex stainless steel clearly demonstrated the susceptibility of ferrite when exposed to chloride-containing environment, with the introduction of plastic deformation leading to local Volta potential extremes (hot spots)[63].

A Volta potential differences map with topography to screen the 750°C-aged microstructure is given in **Figure 6(a), (b)**. The corresponding SEM image with EBSD phase map of the same region is given in **Figure 6(c), (d)**. Volta potential differences over ferrite and austenite in certain regions were nearly similar, which means that the driving force for galvanic coupling between those regions is reduced. However, local Volta potential extremes were developed indicating enhanced electronic activity and galvanic coupling between net anodic and net cathodic regions. The Volta potential variations within ferrite and austenite regions increased, indicating higher micro-galvanic activity within each phase. At this magnification, potential differences over secondary phases could not be discerned, and the mapped area in **Figure 6** was, therefore, rescanned with higher resolution

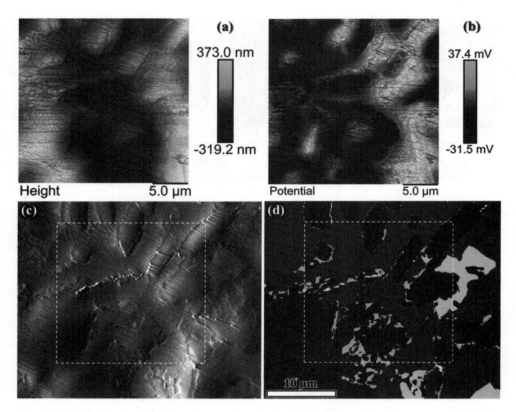

Figure 6. SKPFM analysis of aged microstructure with a topography map, b corresponding Volta potential map, c corresponding SEM micrograph in secondary electron imaging mode, and d corresponding EBSD phase map showing γ/γ_2 (blue), δ (red), χ-phase (green), σ-phase (aqua), Cr_2N (yellow), CrN (white), and grain/phase boundaries (black). Highlighted region in c, d indicate the scanned area (Color figure online).

as shown in **Figures 7** and **8**.

Figure 7 gives the Volta potential with corresponding EBSD map of an area containing intermetallic phases, showing numerous discrete regions with low and high potential variations and gradients. The line profile 1 in **Figure 7(a), (b), (e)** shows that Cr_2N precipitates have the lowest Volta potential followed by austenite regions, with the latter possibly secondary austenite. This indicates low electronic activity of these phases. However, large potential gradients were found surrounding Cr_2N precipitates which gives an indication of local micro-galvanic cells with enhanced electronic coupling activity. Such large gradients adjacent to second

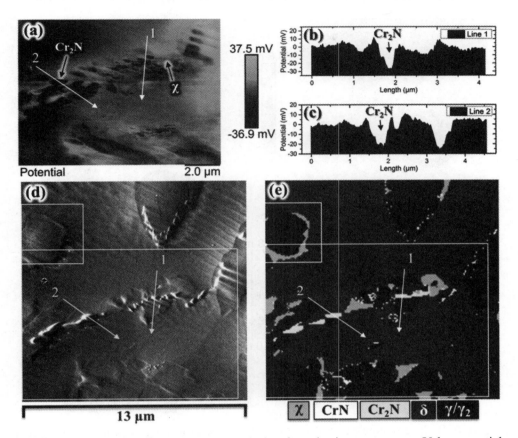

Figure 7. Medium-resolution SKPFM analysis of aged microstructure: a Volta potential map, b, c measured line profiles (1, 2) in a, d corresponding SEM image, and e corresponding EBSD phase map. The large highlighted area in e shows the SKPFM measured region in a (Color figure online).

phase precipitates are indicative of element depletion zones; hence, the adjacent regions often form local anodes. Cr_2N has been reported to contain higher Cr concentrations than CrN and χ, most likely resulting in concentration gradients during thermal treatments[11][36]. Volta potential gradients measured adjacent to CrN and χ precipitates were far lower than those measured adjacent to Cr_2N. Sathirachinda *et al.* investigated the effect of thermally treated microstructure of grades 2205 and 2507 duplex stainless steel with SKPFM[13], measuring potential difference of 10–15mV between γ and Cr_2N with respect to a Pt-Ir tip. The study concluded that Cr_2N has the highest practical nobility in the microstructure, which would mean net cathodic behaviour, in-line with our observations.

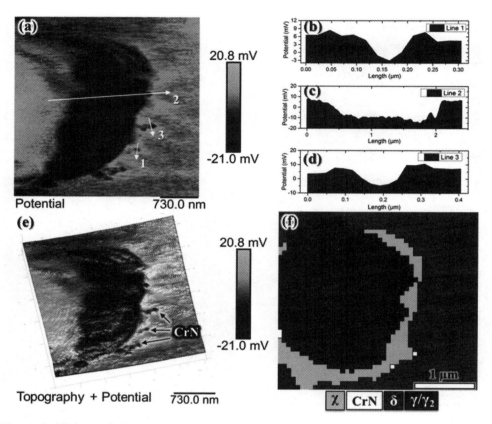

Figure 8. High-resolution SKPFM analysis of an area containing interphase χ (IP-χ) and CrN: a Volta potential map, b-d measured line profiles (1, 2, 3 in a), e 3D height map overlaid with Volta potential data, and f corresponding EBSD map (Color figure online).

The χ-phase showed variations of its electronic behaviour over different areas, with some regions indicating net cathodic activity, whereas others showed net anodic character as can be seen in **Figure 7(a), (e)**. The χ-phase at δ/γ boundaries was denoted as 'interphase/ inter-granular-χ' (IP-χ) and the χ-phase within ferrite labelled as 'in-ferrite/intragranular-χ' (IF-χ). The IP-χ had more net cathodic character, whilst IF-χ seemed to be net anodic, suggesting that they have different chemical compositions, with the IP-χ phase probably more enriched in Cr and Mo. However, the potential gradients adjacent to IP-χ were far steeper than those observed surrounding IF-χ, indicating the presence of depletion effects.

High-resolution Volta potential mapping was, therefore, performed on a region containing γ/γ$_2$, χ-phase, and CrN (**Figure 8**). Small finger-like globular features were apparent which were indexed by EBSD as CrN precipitates and shown in **Figure 8(e), (f)**. The CrN particles were next to χ-phase precipitates, embedded in the ferrite. Line profile measurements across the CrN precipitates showed a potential difference of 5–10mV higher than the surrounding ferrite phase, with the depleted areas around these precipitates indicating 5–6mV lower potentials than the ferrite. This clearly indicated that depleted regions would behave net anodic, whilst the CrN would behave net cathodic with respect to the ferrite phase. The IP-χ in **Figure 8** had 30–40mV larger Volta potentials than the IF-χ shown in **Figure 7** despite their comparable dimensions. This potential difference supported different chemical compositions of this phase. Seemingly, the cathodic character was not only limited to the χ-phase.

The potential differences measured over all crystallographic phases are shown in **Figure 9**. The Cr_2N phase showed the largest net cathodic behaviour, together with σ-and IP-χ. The intra-granular IF-χ phase showed net anodic properties whilst the IP-χ phase was more cathodic. Therefore, both χ-phase compositions are expected to behave differently. The nobility of CrN seemed to lie in-between the matrix phases and most other secondary phases. Volta potential values of ferrite and austenite had large scatters, with ferrite showing slightly more net anodic character than austenite. The scatter is partly due to large micro-galvanic activities observed within both matrix phases.

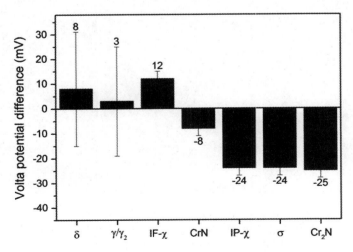

Figure 9. Comparison of relative Volta potential differences of all observed phases after ageing in descending order from left (ferrite) to right (Cr_2N).

3.3. Stress Corrosion Screening Tests

3.3.1. As-Received Microstructure

The as-received mini-tensile specimen dosed with 1450μg/cm² MgCl₂ was exposed to 50°C and 30% RH, and after removing and cleaning the sample substantial corrosion attack was observed, which can be seen in **Figure 10(a)**. Primarily the ferrite was corroded under the droplet deposit with some minor corrosion observed on the austenite. The same corrosion pattern has previously been reported under similar exposure conditions[45][63][64], with ferrite clearly the electrochemically more active phase leading to selective dissolution[17][18][44][45][56][64]–[72]. The ferrite phase is, usually, the electrochemically more active phase in 2205 duplex stainless steel due to its lower corrosion potential in contrast to the austenite in mild chloride-bearing environments leading to micro-galvanic coupling between ferrite and austenite[66][70][73]. Selective corrosion of the ferrite, therefore, is often manifested as the main corrosion mechanism whilst the

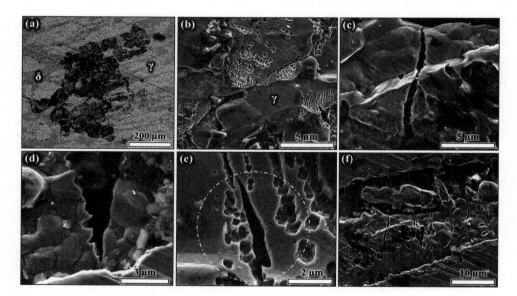

Figure 10. Corrosion morphology of as-received microstructure after exposure to 1450μg/cm² MgCl₂ showing a the entire corroded area, b localised intra-granular corrosion attack on austenite, c, d stress corrosion micro-cracks on austenite, e multiple corrosion sites on austenite with a crack inside a crack (within the highlighted area), f selectively dissolved ferrite regions with multiple cracks on austenite. The stress acted along the horizontal direction.

austenite is galvanically protected[17][63][72][74]–[77].

High-resolution SEM assessment confirmed the presence of localised attack in the form of superficial sub-micrometre sized intra-granular corrosion pits on the austenite, with typical images shown in **Figure 10(b), (e)**. The attack on the austenite seemed to be related to slip bands and strain. The evolution of strain localisation in correlation with the development of local Volta potential extremes in austenite leading to enhanced propensity to localised corrosion with the introduction of cold deformation was earlier demonstrated[45]. Pitting corrosion on the austenite tended to suppress the selective dissolution of the ferrite. Seemingly, heterogeneous nucleation of numerous discrete corrosion pits on the austenite indicated those sites as more susceptible, possibly associated with strain localisation.

Some stress corrosion micro-cracks were also found in the austenite, oriented perpendicular to the applied stress direction [stress acted horizontal in **Figure 10(c), (e), (f)**]. The longest crack observed was less than 15µm in length. Nucleation sites of cracks within existing cracks were also found, suggesting that some cracks may have stifled and stopped growing, before conditions for re-nucleating inside existing cracks were satisfied again. This is different from the classic stress corrosion cracking theory where cracks propagate with discrete steps. **Figure 10(e)** shows features related to crack nucleation inside an existing crack.

Retardation effects may have played an important part in crack development, for example, when the crack encounters γ/γ grain boundaries or δ/γ interphase boundaries. The stress corrosion micro-cracks observed were all transgranular in nature, and located in the austenite phase, as concluded from EBSD analysis seen in **Figure 11**. Slip planes are easily noticed from SEM images, which play an active role in crack nucleation and growth. The crack in **Figure 11** initiated on a grain with close 101 orientation and grew towards the grain interior before changing its direction slightly after encountering a boundary. A difference in crack opening is also observed with the crack wide open on one side of the grain boundary, and more tightly closed on the other.

3.3.2. Heat-Treated Microstructure

Figure 12(a) shows the specimen aged at 750°C and exposed to 14.5µg/cm²

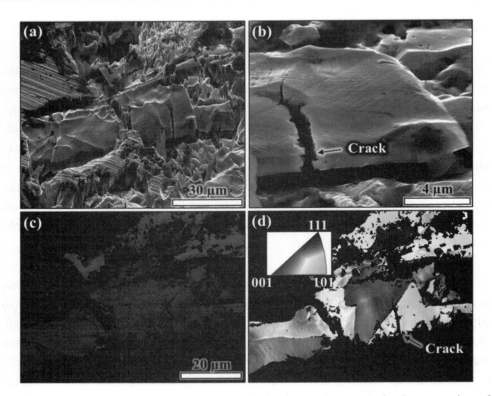

Figure 11. Corrosion morphology of as-received microstructure: a selective corrosion of ferrite (partially dissolved only) and b selective stress corrosion cracking of austenite with c corresponding EBSD phase map overlapped with band contrast map (austenite is blue and ferrite is red coloured), and d corresponding inverse pole figure map in x direction. Step size = 135nm (Color figure online).

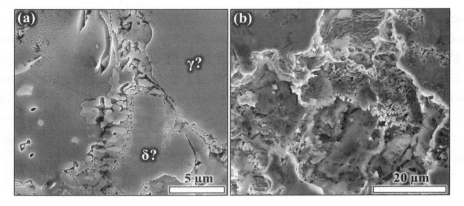

Figure 12. Corrosion morphology of the aged specimen with 1% elastic-plastic strain after 259 days exposure at 50°C and 30% RH: a minor corrosion attack under drop 1 (14.5μg/cm^2) and b corrosion area under drop 3 (1450μg/cm^2).

chloride for 259 days. Minor corrosion attack was only observed located at interphases and on the austenite, with some attack around secondary phases. Increasing the chloride concentration resulted in more severe attack as can be seen with a deposition density of 1450µg/cm^2 chloride (drop 3) in **Figure 12(b)**. The attacked area and corroded volume of the 750°C-aged sample was significantly larger than that observed on the as-received specimen exposed with equivalent deposition density of chloride in **Figure 10(a)**. However, no evidence for stress corrosion micro-cracking was found in these specimens. Stress corrosion cracking can only occur when the corrosion rate and crack velocity in that corrosion system have similar order of magnitude, but when corrosion reactions advance too fast, then crack nucleation and propagation are hindered.

The corrosion morphology of the aged sample after exposure to 145µg/cm^2 chloride (drop 2) indicated selective corrosion of the ferrite. However, high-resolution SEM analysis showed that selective corrosion of ferrite with some attack on the austenite occurred, shown in **Figure 13(a), (b)**. The σ-phase often remained in corroded regions, clearly indicating its net cathodic character, whereas ferrite and austenite were attacked, indicating their more net anodic character. In **Figure 13(b)**, however, secondary austenite remained unaffected after the corrosion test, indicated by the characteristic shape. A large number of smaller discrete precipitates can also be seen and these observations are in good agreement with the earlier Volta potential assessments, where such precipitates were measured to be net cathodic (**Figure 9**). Thus, these precipitates are assumed to be Cr$_2$N, χ, and possible CrN on basis of their Volta potential response and morphological appearance.

A large σ-phase precipitated with a transgranular crack was also found, shown in **Figure 13(c), (d)**. A number of smaller cracks were also found, all located in the σ-phase indicating its brittleness and susceptibility to cracking. The presence of these cracks indicates an increased microstructure crack nucleation propensity, which was clearly enhanced by the presence of the brittle σ-phase in the microstructure.

The corrosion morphology under a chloride deposition density of 2472µg/cm^2 (drop 4) is shown in **Figure 14(a)**. Selectively corroded ferrite regions can clearly be seen indicating the net anodic character during the corrosion process.

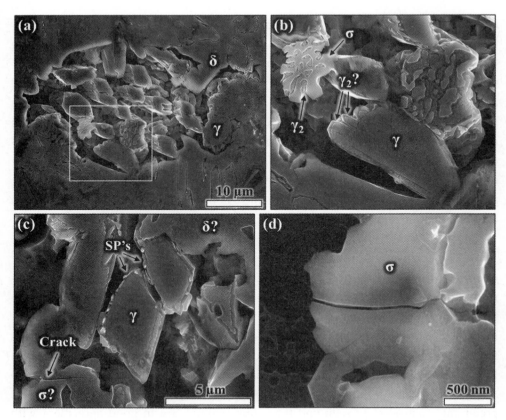

Figure 13. Corrosion morphology of the aged sample under drop 2 (145μg/cm^2) with a a corroded area, b higher-resolution image of the highlighted box in a, c selective dissolution of the ferrite with remainders of secondary phases, and cracked r, and d fracture in r phase.

Several corroded regions located along δ/γ interphases are also shown in **Figure 14(b)**, decorated with secondary phases (SP) that seemed almost unaffected by the corrosion process, supporting their net cathodic behaviour. Their discrete shape and morphology indicated that these secondary phases are possibly Cr$_2$N, χ, and/or CrN [**Figure 14(c), (d)**].

The same corrosion morphology as with exposure to lower chloride deposition densities was observed under the droplet containing 2856μg/cm^2 chloride (drop 5), showing selective attack on the ferrite with minor attack on the austenite, as can be seen in **Figure 15(a), (b)**. The most severe corrosion was observed under this droplet, due to largest deposition density, which in turn affects the corrosion potential. A large number of transgranular stress corrosion micro-cracks were

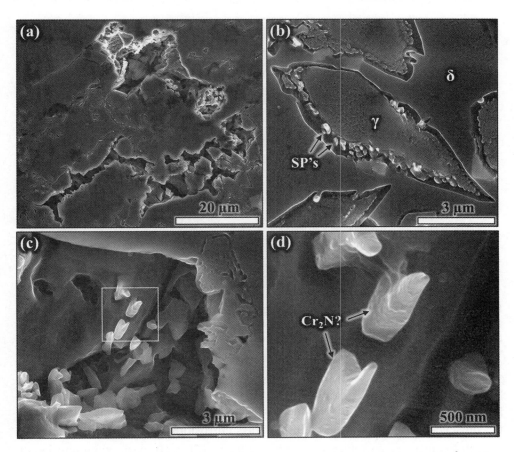

Figure 14. Corrosion morphology of the aged sample under drop 4 (2472μg/cm^2) with a selective corrosion of the ferrite, b selective attack along interphases with remaining secondary phases, c, d corrosion along interphase boundary regions with remaining secondary phases (possibly Cr$_2$N).

observed in areas containing σ-phase and austenite.

The austenite is susceptible to stress corrosion cracking due to its critical nickel content which was determined to be in the order of 7% (see **Table 1**)[78]. The Copson curve shows lowest time to failure, i.e. highest stress corrosion cracking susceptibility for chloride-induced stress corrosion cracking of stainless steels with nickel contents between ~7% and ~15%[78]. Stress corrosion microcracks were observed in regions containing secondary austenite and σ-phase as shown in **Figure 15(b)**, a region eutectoidically transformed from ferrite. Secondary austenite is depleted in Cr, Mo, and N, and therefore, more prone to corrosion.

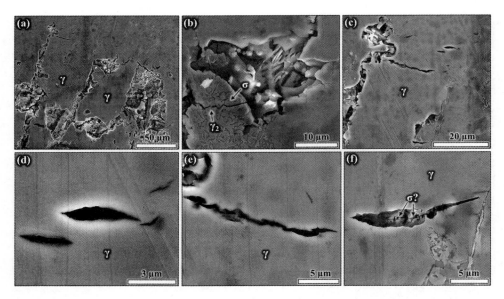

Figure 15. Corrosion morphology of the aged sample under drop 5 (2856μg/cm^2) with a selective corrosion of the ferrite and some attack on austenite, b transgranular cracks through an area containing σ-phase and austenite, c-f multiple micro-cracks nucleated within austenite.

The nickel content of secondary austenite is somewhat lower than that of primary austenite[30], and the cracking susceptibility may have been enhanced by the presence of sigma phase. Multiple stress corrosion cracking events were also seen in the primary austenite as shown in **Figure 15(c), (f)**, with maximum length of cracks in the aged microstructure of up to 30−35μm, spanning across several grains. The latter confirms stress corrosion cracking propensity of aged sample when exposed to environment with extremely high chloride concentrations.

3.3.3. Practical Relevance

Grade 2205 duplex stainless steel in the as-received condition is typically resistant to localised corrosion and stress corrosion cracking and the material only shows propensity towards corrosion and micro-cracking with exposure to harsh environmental conditions, such as the high chloride concentrations used in **Figure 10**. High deposition densities of chloride in high concentrations (close to deliquescence equilibrium of MgCl$_2$) led to a maximum crack length of 15μm, which showed minor stress corrosion cracking propensity of the as-received material.

Grade 2205 duplex stainless steel has superior stress corrosion cracking resistance to the most common austenitic counterpart alloys 304L and 316L which showed more severe crack developments in similar (or under less aggressive) conditions[74]-[76][79]-[84]. The 750°C-aged microstructure, however, was clearly more susceptible to localised corrosion and stress corrosion cracking, which may have implications for welding practice of duplex stainless steels. Grade 2205 duplex stainless steel can be rendered susceptible to AISCC at a temperature of 50°C, and the calculated threshold temperature (>50°C) for AISCC in grade 2205 should be re-visited under $MgCl_2$ exposure conditions[75].

4. Conclusions

The effect of 750°C ageing treatment on microstructure development using correlative EBSD and SKPFM techniques has been investigated. The crystallographic phases present were correlated to their local Volta potentials and corrosion behaviours.

1. CrN, Cr_2N, χ-, σ-, and secondary austenite ($γ_2$) phases precipitated in the microstructure with $γ_2$ and σ-forming the largest secondary phases in size and fraction. The formation of σ, $γ_2$, and χ was allotriomorphic, with the σ-phase showing no internal grain boundaries.

2. The σ-phase and $γ_2$ precipitated within ferrite and at δ/γ interfaces, and χ-phase at δ/γ interfaces only, whilst CrN/Cr_2N was found at δ/γ and σ/$γ_2$ interfaces and δ/δ boundaries.

3. The as-received microstructure showed 50–70mV Volta potential differences with δ acting as net anode and γ as net cathode. The Volta potential difference between δ and γ after ageing treatment decreased in certain regions, but Volta potential variations within each phase increased. Local Volta potential extremes were developed indicating enhanced corrosion susceptibility.

4. The σ-phase, Cr_2N, CrN, and inter-granular (IP) χ-phase indicated low electronic activities, i.e. expecting net cathodic behaviour, whilst the intra-granular (IF) χ-phase indicated only enhanced net anodic activity among secondary phases.

5. Selective corrosion of the ferrite was observed in the as-received and the 750°C-aged condition, with the overall corrosion propensity of secondary phases being in-line with SKPFM observations.

6. Stress corrosion micro-cracks were found in the austenitic phase in the as-received and 750°C heat-treated conditions after exposure to atmospheric $MgCl_2$ electrolyte.

Acknowledgements

The authors acknowledge Radioactive Waste Management (RWM) (NPO004411A-EPS02) and EPSRC (EP/I036397/1) for financial support. The authors are grateful for the kind provision of Grade 2205 Duplex Stainless Steel plate by Rolled Alloys. Special thanks also to Dr Christiano Padovani, Radioactive Waste Management Ltd. for valuable discussions.

Open Access: This article is distributed under the terms of the Creative Commons Attribution 4.0 International License (http://creativecommons.org/licenses/by/4.0/), which permits unrestricted use, distribution, and reproduction in any medium, provided you give appropriate credit to the original author(s) and the source, provide a link to the Creative Commons license, and indicate if changes were made.

Source: C. Örnek, D. L. Engelberg. Correlative EBSD and SKPFM characterisation of microstructure development to assist determination of corrosion propensity in grade 2205 duplex stainless steel[J]. Journal of Materials Science, 2016, 51(4):1–18.

References

[1] Padovani C (2014) Overview of UK research on the durability of container materials for radioactive wastes. Corros Eng, Sci Technol 49:402–409.

[2] Lo KH *et al* (2009) Recent developments in stainless steels. Mater Sci Eng R 65:39–104.

[3] Lula *et al* (1983) Duplex stainless steels. American Society for Metals, Mars.

[4] Charles J, Bernhardsson S (1991) Duplex stainless steels'91—volume 1. In: Duplex stainless steels'91. Beaune, Les editions de physique.

[5] Redjaïmia A *et al* (2002) Microstructural and analytical study of heavily faulted Frank-Kasper R-phase precipitates in the ferrite of a duplex stainless steel. J Mater Sci 37:4079–4091. doi:10.1023/ A:1020023500133.

[6] Byun S-H *et al* (2012) Kinetics of Cr/Mo-rich precipitates formation for 25Cr-6.9Ni-3.8Mo-0.3N super duplex stainless steel. Met Mater Int 18:201–207.

[7] Chung HM (1989) Spinodal decomposition of austenite in long-term-aged duplex stainless steel. Argonne National Lab, Lemont.

[8] Garner FA *et al* (1986) Ion-induced spinodal-like decomposition of Fe-Ni-Cr invar alloys. Nucl Instrum Methods Phys Res B 16:244–250.

[9] Horvath W *et al* (1998) Microhardness and microstructure of austenite and ferrite in nitrogen alloyed duplex steels between 20 and 500°C. Mater Sci Eng A 256:227–236

[10] Padilha AF, Rios PR (2002) Decomposition of austenite in austenitic stainless steels. ISIJ Int 42:325–327.

[11] Liao J (2001) Nitride precipitation in weld HAZs of a duplex stainless steel. ISIJ Int 41:460–467.

[12] Ramirez AJ *et al* (2004) Secondary austenite and chromium nitride precipitation in simulated heat affected zones of duplex stainless steels. Sci Technol Weld Join 9:301–313.

[13] Sathirachinda N *et al* (2010) Study of nobility of chromium nitrides in isothermally aged duplex stainless steels by using SKPFM and SEM/EDS. Corros Sci 52:179–186.

[14] Bruch D (2007) Investigations on microstructure, mechanical properties and corrosion resistance of large thickness duplex stainless steel forgings. Stainless steel world. KCI Word Publishing, Maastricht

[15] Bruch D *et al* (2008) Mechanical properties and corrosion resistance of duplex stainless steel forgings with large wall thicknesses, La Metallurgia Italiana, pp 1–7.

[16] Nilsson JO (1992) Super duplex stainless steels. Mater Sci Technol 8:685–700.

[17] Örnek C, Engelberg DL (2013) Effect of "475°C embrittlement" on the corrosion behaviour of grade 2205 duplex stainless steel investigated using local probing techniques. Corrosion management. The Institute of Corrosion, Northampton, pp 9–11.

[18] Örnek C *et al* (2015) Effect of 475°C embrittlement on microstructure development and mechanical properties of grade 2205 duplex stainless steel. Metall Mater Trans A

[19] Tavares SSM *et al* (2001) 475°C embrittlement in a duplex stainless steel UNS S31803. Mater Res 4:237–240.

[20] Park C-J, Kwon H-S (2002) Effects of aging at 475°C on corrosion properties of tungsten-containing duplex stainless steels. Corros Sci 44:2817–2830.

[21] Iacoviello F et al (2005) Effect of "475°C embrittlement" on duplex stainless steels localized corrosion resistance. Corros Sci 47:909–922.

[22] Calliari I et al (2009) Measuring secondary phases in duplex stainless steels. JOM 61:80–83.

[23] Cho H-S, Lee K (2013) Effect of cold working and isothermal aging on the precipitation of sigma phase in 2205 duplex stainless steel. Mater Charact 75:29–34.

[24] Elmer JW et al (2007) Direct observations of sigma phase formation in duplex stainless steels using in-situ synchrotron X-ray diffraction. Metall Mater Trans A 38:464–475.

[25] Hsieh C-C, Wu W (2012) Overview of intermetallic sigma (σ) phase precipitation in stainless steels. ISRN Metall 16.

[26] Michalska J, Sozańska M (2006) Qualitative and quantitative analysis of σ and χ phases in 2205 duplex stainless steel. Mater Charact 56:355–362.

[27] Nilsson JO et al (2000) Mechanical properties, microstructural stability and kinetics of σ-phase formation in 29Cr−6Ni−2Mo−0.38N superduplex stainless steel. Metall Mater Trans A 31:35–45.

[28] Nilsson J-O, Chai G (2007) The physical metallurgy of duplex stainless steels. International Conference & Expo DUPLEX 2007. Associazione Italiana di Metallurgia, Grado.

[29] Nilsson J-O, Liu P (1991) Aging at 400°C–600°C of submerged arc welds of 22Cr-3Mo-8Ni duplex stainless steel and its effect on toughness and microstructure. Mater Sci Technol 7:853–862.

[30] Padilha AF et al (2009) Chi-phase precipitation in a duplex stainless steel. Mater Charact 60:1214–1219.

[31] Pohl M et al (2007) Effect of intermetallic precipitations on the properties of duplex stainless steel. Mater Charact 58:65–71.

[32] Sieurin H, Sandström R (2007) Sigma phase precipitation in duplex stainless steel 2205. Mater Sci Eng 444:271–276.

[33] Cui J et al (2001) Degradation of impact toughness due to formation of R phase in high nitrogen 25Cr-7Ni-Mo duplex stainless steels. ISIJ Int 41:192–195.

[34] Karlsson L et al (1995) Precipitation of intermetallic phases in 22% Cr duplex stainless weld metals—the kinetics of inter-metallic phase formation in the temperature range 675°C–1000°C (1247°–1832°F) and effects on mechanical properties and corrosion resistance. Weld J-Incl Weld Res Suppl 74:28–38.

[35] Redjaïmia A et al (2008) Orientation relationships between the dferrite matrix in a duplex stainless steel and its decomposition products: the austenite and the χ and R Frank-Kasper phases. In: EMC 2008 14th European microscopy congress 1–5 September 2008, Aachen, Germany. Springer Berlin Heidelberg, pp. 479–480.

[36] Pettersson N et al (2015) Precipitation of chromium nitrides in the super duplex stainless steel 2507. Metall Mater Trans A 46:1062–1072.

[37] Ramirez AJ et al (2003) The relationship between chromium nitride and secondary austenite precipitation in duplex stainless steels. Metall Mater Trans A 34:1575–1597.

[38] Charles J, Bernhardsson S (1991) Duplex Stainless Steels '91—Volume 2. In: Duplex Stainless Steels '91. Les editions de physique, Beaune, Bourgogne, France.

[39] Humphreys FJ (2004) Characterisation of fine-scale microstructures by electron backscatter diffraction (EBSD). Scripta Mater 51:771–776.

[40] Li W et al (2006) Influences of tensile strain and strain rate on the electron work function of metals and alloys. Scripta Mater 54:921–924.

[41] Li W, Li DY (2006) Influence of surface morphology on corrosion and electronic behavior. Acta Mater 54:445–452.

[42] Bockris JOM et al (2002) Modern electrochemistry 2A. Fundamentals of electrodics, vol 2A. Kluwer Academic Publishers, New York, pp 1–817.

[43] Sadewasser S, Glatzel T (2012) Kelvin probe force microscopy measuring and compensating microscopy. Springer, Heidelberg.

[44] Örnek C et al (2015) A corrosion model for 475°C embrittlement in duplex stainless steel—a comprehensive study via scanning Kelvin probe force microscopy. Corros Sci.

[45] Örnek C, Engelberg DL (2015) SKPFM measured Volta potential correlated with strain localisation in icrostructure of cold-rolled grade 2205 duplex stainless steel. Corros Sci 99:164–171.

[46] Sathirachinda N et al (2008) Characterization of phases in duplex stainless steel by magnetic force microscopy/scanning Kelvin probe force microscopy. Electrochem Solid-State Lett 11:C41–C45.

[47] Sathirachinda N et al (2009) Depletion effects at phase boundaries in 2205 duplex stainless steel characterized with SKPFM and TEM/EDS. Corros Sci 51:1850–1860.

[48] Sathirachinda N et al (2011) Scanning Kelvin probe force microscopy study of chromium nitrides in 2507 super duplex stainless steel—implications and limitations. Electrochim Acta 56:1792–1798.

[49] Blücher DB et al (2004) Scanning Kelvin probe force microscopy: a useful tool for studying atmospheric corrosion of MgAl alloys in situ. J Electrochem Soc 151:B621–B626.

[50] Rohwerder M, Turcu F (2007) High-resolution Kelvin probe microscopy in corrosion science: scanning Kelvin probe force microscopy (SKPFM) versus classical scanning Kelvin probe (SKP). Electrochim Acta 53:290–299.

[51] Wang R et al (2014) Changes of work function in different deformation stage for

2205 duplex stainless steel by SKPFM. Procedia Mater Sci 3:1736–1741.

[52] Humphreys FJ et al (1999) Electron backscatter diffraction of grain and subgrain structures—resolution considerations. J Microsc 195:212–216.

[53] Frankel GS et al (2001) Characterization of corrosion interfaces by the scanning Kelvin probe force microscopy technique. J Electrochem Soc 148:B163–B173.

[54] Nonnenmacher M et al (1991) Kelvin probe force microscopy. Appl Phys Lett 58:2921–2923.

[55] Marcus P, Mansfeld F (2006) Analytical methods in corrosion science and engineering. Taylor & Francis Group, Boca Raton.

[56] Engelberg DL, Örnek C (2014) Probing propensity of grade 2205 duplex stainless steel towards atmospheric chloride-induced stress corrosion cracking. Corros Eng Sci Technol 49:535–539.

[57] Jackson EMLEM et al (1993) Distinguishing between Chi and Sigma phases in duplex stainless steels using potentiostatic etching. Mater Charact 31:185–190.

[58] Shiao JJ et al (1993) Phase transformations in ferrite phase of a duplex stainless steel aged at 500°C. Scr Metall Mater 29:1451–1456.

[59] Li H et al (2015) Austenite transformation behaviour of 2205 duplex stainless steels under hot tensile test. Steel Res Int 86:84–88.

[60] Liu H, Jin X (2011) Electrochemical corrosion behavior of the laser continuous heat treatment welded joints of 2205 duplex stainless steel. J Wuhan Univ Technol 26:1140–1147.

[61] Bettini E et al (2014) Study of corrosion behavior of a 2507 super duplex stainless steel: influence of quenched-in and isothermal nitrides. Int J Electrochem Sci 9:20.

[62] Kim J-S et al (2012) Mechanism of localized corrosion and phase transformation of tube-to-tube sheet welds of hyper duplex stainless steel in acidified chloride environments. Jpn Inst Met 53:9.

[63] Örnek C, Engelberg DL (2015) Effect of cold deformation mode on stress corrosion cracking susceptibility of 2205 duplex stainless steel, in preparation.

[64] Örnek C, Engelberg DL (2015) Environment-assisted stress corrosion cracking of grade 2205 duplex stainless steel under low-temperature atmospheric exposure, in preparation.

[65] Arnold N et al (1997) Chloridinduzierte Korrosion von Nichtrostenden Stä̈hlen in Schwimmhallen-Atmosphä̈ren Teil 1: elektrolyt Magnesium-Chlorid (30%). Mater Corros 48:679–686.

[66] Aoki S et al (2010) Dissolution behavior of α and γ phases of a duplex stainless steel in a simulated crevice solution. ECS Trans 25:17–22.

[67] Femenia M et al (2001) In situ study of selective dissolution of duplex stainless steel

2205 by electrochemical scanning tunnelling microscopy. Corros Sci 43:1939–1951.

[68] Pettersson RFA, Flyg J (2004) Electrochemical evaluation of pitting and crevice corrosion resistance of stainless steels in NaCl and NaBr. Outokumpu, Stockholm, Sweden.

[69] Deng B *et al* (2008) Critical pitting and repassivation temperatures for duplex stainless steel in chloride solutions. Electrochim Acta 53:5220–5225.

[70] Aoki S *et al* (2011) Potential dependence of preferential dissolution behavior of a duplex stainless steel in simulated solution inside crevice. Zairyo-to-Kankyo 60:363–367.

[71] Ebrahimi N *et al* (2011) Correlation between critical pitting temperature and degree of sensitisation on alloy 2205 duplex stainless steel. Corros Sci 53:637–644.

[72] Örnek C, Engelberg DL (2014) Kelvin probe force microscopy and atmospheric corrosion of cold-rolled grade 2205 duplex stainless steel. Eurocorr 2014. European Federation of Corrosion, Pisa, Italy, pp 1–10.

[73] Lee J-S *et al* (2014) Corrosion behaviour of ferrite and austenite phases on super duplex stainless steel in a modified green-death solution. Corros Sci 89:111–117.

[74] Prosek T *et al* (2008) Low temperature stress corrosion cracking of stainless steels in the atmosphere in presence of chloride deposits. NACE, vol. Paper No. 08484: NACE International, p 17.

[75] Prosek T *et al* (2014) Low-temperature stress corrosion cracking of austenitic and duplex stainless steels under chloride deposits. Corros Sci 70:1052–1063.

[76] Prosek T *et al* (2009) Low-temperature stress corrosion cracking of stainless steels in the atmosphere in the presence of chloride deposits. Corros Sci 65:13.

[77] Örnek C *et al* (2012) Effect of microstructure on atmospheric-induced corrosion of heat-treated grade 2205 and 2507 duplex stainless steels. Eurocorr 2012. Dechema, Istanbul, Turkey, pp 1–10.

[78] Cottis B *et al* (2010) Shreir's Corrosion, vol 2. Elsevier B.V., Manchester.

[79] Cook AB *et al* (2010) Preliminary evaluation of digital image correlation for in-situ observation of low temperature atmospheric-induced chloride stress corrosion cracking in austenitic stainless steels. ECS Trans 25:119–132.

[80] Cook AB *et al* (2014) Assessing the risk of under-deposit chloride-induced stress corrosion cracking in austenitic stainless steel nuclear waste containers. Corros Eng Sci Technol 49:529–534.

[81] Lyon SB *et al* (2010) Atmospheric corrosion of nuclear waste containers. In: DIAMOND'10 conference decommissioning, immobilisation and management of nuclear waste for disposal. Manchester, UK.

[82] Albores-Silva OE *et al* (2011) Effect of chloride deposition on stress corrosion

cracking of 316L stainless steel used for intermediate level radioactive waste containers. Corros Eng Sci Technol 46:124–128.

[83] Padovani C *et al* (2014) Corrosion control of stainless steels in indoor atmospheres—laboratory measurements under $MgCl_2$ deposits at constant relative humidity (Part 1). Corrosion 71:292–304.

[84] Padovani C *et al* (2014) Corrosion control of stainless steels in indoor atmospheres—practical experience (Part 2). Corrosion 71:246–266.

Chapter 8
Determination of Bone Porosity Based on Histograms of 3D μCT Images

M. Cieszko, Z. Szczepański, P. Gadzała

Institute of Mechanics and Applied Computer Science, Kazimierz Wielki University, Kopernika 1, 85-074 Bydgoszcz, Poland

Abstract: A new method is proposed for direct determination of bone porosity based on histograms of 3D μCT scans and for precise definition of the global image segmentation threshold, preserving assessed porosity in the reconstructed binary image of the bone sample. In this method, the normed histogram is considered to be a probability distribution of voxel density (CT number or gray level) in the scan. It is a linear combination of two distributions characterizing the frequency of occurrence of voxels of pore and matrix type with various densities. Volume porosity, in this model, defines the probability of pore voxel occurrence in the whole set of voxels in the scan of the sample. This parameter and the parameters of both probability distributions are determined by an optimization method. The new method was used to determine the porosity and segmentation thresholds for μCT images of two 3D samples of human cancellous bone. The results were compared with those determined by the standard method and Otsu's method. The new method allows the porosity and the image segmentation threshold to be determined even in cases where use of the other methods is questionable or impossible.

1. Introduction

Identification of the microscopic geometry of bone tissue and macroscopic parameters of its pore space structure is a very important issue in the study of the physical properties of such material. The internal bone structure determines its local mechanical properties and bone strength, as an element of the human skeleton, and also strongly influences processes that take place in the bone tissue.

There are many methods for identifying the microscopic pore geometry of porous materials and their macroscopic parameters, such as optical microscopy, ultrasonic microscopy and porosimetry, mercury porosimetry, electric spectroscopy, permeametry, and gas pycnometry. Microcomputed tomography (μCT)[1]–[3] is another of these methods. It is a very modern, nondestructive method used in various branches of science and engineering[4]–[7] for identification of the spatial structure of heterogeneous materials and small physical objects. In this method, as in the computed tomography applied in medical diagnostics, X-rays are used to achieve an image resolution of one micrometer.

Microtomographic images of samples of porous materials form a basis for the reconstruction of the microscopic pore space geometry or matrix architecture. This allows identification of the stochastic characteristics, microscopic and macroscopic parameters of the pore space and matrix structure, material constants, and their directional characteristics[8]–[18]. For this purpose, pure geometrical methods[9][14][19]–[21] and methods of simulation of physical processes at microscopic level[12][13][16][18] are used.

The accuracy of the parameters and coefficients obtained in this way is directly determined by the reconstruction quality of the microscopic pore space geometry. It depends not only on the image resolution of the sample, but also on the quality of the image segmentation, i.e., on the quality of transformation of the microtomographic image with various gray levels to a binary image. The crucial step for this process is image thresholding, which defines a limiting value of the gray level that separates all points (voxels) of the scan into two subsets constituting the matrix and the pore space. One can distinguish six groups of image thresholding methods[22]: local[23][24], global[25][26], based on the shape of the histogram[27], and using such tools as clustering[15], entropy[28], and fuzzy logic.

Thresholding is also crucial in the standard methods of determining the porosity parameter from microtomographic images of the material samples. For a binarized image of porous material, its porosity is a simple measure of the voxel volume fraction representing pores in the sample.

The most popular methods of global thresholding based on the shape of the histogram include what will be called here the standard method[25], and Otsu's method[29]. These are often used in microtomographic image analysis of human and animal bones[2][10][25][26][30]–[32].

In the standard method, the image segmentation threshold is defined as the voxel density (CT number or gray level) for which the frequency of voxel occurrence in the sample of bone scan reaches a minimum between the two extremes of the histogram corresponding to the pore and matrix types of voxels. In the graph of the cumulative histogram, this threshold value defines the location of the inflection point in the vicinity of which changes in the pore and matrix volume fractions in the sample are the smallest. It also corresponds to the ultimate changes in the voxel density. The standard method of determining the segmentation threshold has been implemented in some computer microtomographs.

In Otsu's method of thresholding, applied to µCT images of bone samples, the histogram of the scan is divided into two parts by the unknown value of the binarization threshold. After normalization, they are used as probability distributions of the density of voxels of two types (pore and skeleton) defined on two separate ranges. This makes it possible to define expressions for the mean voxel density of both classes as functions of the binarization threshold, and to define the so-called between-class variance of the mean densities of voxels in the scan. This is a measure of the deviation of the mean densities of both classes from the mean density of all voxels in the scan. Maximization of the value of this variance is the criterion for determining the optimum value for the scan binarization threshold. This method is often used for automatic threshold selection for image segmentation, and is implemented in the numerical computing environment MATLAB.

The aim of this paper is to present a new method for determining the porosity parameter and the binarization threshold for 3D µCT images of bone tissue in which the standard procedure for their assessment has been reversed. First, using the model-based approach, the bone porosity is determined directly from the histogram of the 3D μCT image. Next, the binarization threshold is calculated from a condition requiring the obtained porosity to be preserved in the reconstructed binary image of the bone sample.

In this paper, bone is considered as a macroscopically strongly inhomogeneous porous material with low porosity in regions of the cortical bone and with high porosity in regions of the cancellous bone. The spatial distribution of this parameter is a basic macroscopic characteristic of such a material, determining its mechanical properties, which are important in, for example, diagnostics of morbidities of the bone.

Due to the largely random nature of the origin of image blurring in μCT scans[3][4][33][34], a stochastic mixture model of a scan of the bone sample is proposed here. In this model, all voxels in the scan are considered to be of pore or matrix type, the density of which is a random variable, and the normalized histogram of the scan represents the probability distribution of this variable. This distribution is assumed to be a linear combination of two distributions describing the frequency of occurrence of vowels with various densities in the sets of voxels of pore and matrix type. The porosity in the proposed model defines the frequency (probability) of occurrence of voxels of pore type in the whole set of voxels in the scan of the bone sample.

The porosity parameter and parameters of the density distributions of voxels of pore and matrix type are determined by an optimization method implemented in the numerical computing environment MATLAB, i.e., by matching the mathematical model of the histogram to the histogram of the scan of the bone sample. In the applied method, the porosity parameter is calculated independently from an expression that minimizes a particular function of the approximation error. The obtained porosity allows the binarization threshold of the μCT image to be immediately determined, since the porosity should be preserved in the reconstructed binary image.

The proposed method was used to determine the porosities and the threshold values of two cubic samples of cancellous bone with various porosities taken from different places on the μCT scan of a human condyle. To make the samples statistically representative, a size limit was established above which their histograms do not change considerably. The results were compared with those obtained by the standard method[25][31] and Otsu's method[29]. It was shown that the porosity and threshold values of a sample with a small matrix fraction, as determined by the new method, are considerably smaller than those obtained using the standard method and Otsu's method. For porosity the differences are of a few percent, while for the segmentation threshold they are about 140% and 40%, respectively. This strongly influences the quality of reconstruction of the microscopic geometry of the bone sample.

The proposed method may be used for determining the porosity and the binarization threshold of representative samples of various porous materials. Its multiple use also enables identification of the spatial distribution of both parameters in μCT images of inhomogeneous porous materials, and consequently allows more precise reconstruction of their microscopic structure.

2. Characteristics of Bone Scan Samples

The new method of determining bone porosity based on a histogram of a 3D μCT image is presented using a scan of a human condyle performed on the microtomograph SkyScan 1172 with a voxel size of 17μm. The tomogram of one cross-sectional layer of the investigated bone is shown in **Figure 1(a)**.

The gray levels in this figure represent the CT numbers of particular voxels in the layer of values from 0 to 255, where 0 stands for black and represents voxels of pore type. For convenience, the CT number of voxel represented in the image by the voxel gray level will here be called the voxel density. The voxel density distribution in a scan corresponds to the mass density distribution in the scanned object, but does not represent this distribution directly.

We apply the methods of statistical analysis to investigate samples of the bone scan characterized locally by the density of voxels treated as a random variable. Due to the high macroscopic heterogeneity of bone, the voxel set in the whole

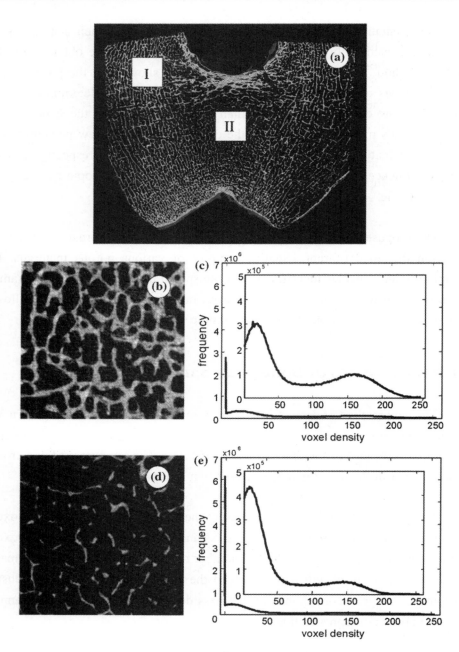

Figure 1. Microscopic representations of the investigated bone samples: a tomogram of one crosssectional layer of human condyle; b tomogram of one layer of sample I (400 × 400 × 1 voxels); d tomogram of one layer of sample II (400 × 400 × 1 voxels); c, e histograms of samples I and II respectively. Internal graphs contain enlarged plots of the continuous parts of the histograms.

bone scan cannot be considered as a study population, since all statistical characteristics of voxels have to be referenced to an area that can be recognized as homogeneous and representative in the statistical sense.

To make the analysis representative, two cubic samples of the scan of cancellous bone with sides of 400 voxels were taken from different places on the bone scan [**Figure 1(a)**]. One sample was taken from the lateral part of the bone scan (sample I) and the other from its central part (sample II). Enlarged images of one layer of both samples are shown in **Figure 1(b)** and **(d)**. Their histograms are presented in **Figure 1(c)** and **(e)**, respectively, and show the frequency of occurrence of voxels with the given density in the scans of the bone samples. After normalization, these curves can be interpreted as probability distributions of voxel density in the set of all voxels of the sample. Both histograms are discontinuous in the neighborhood of the point of zero density, and contain two visible extremes. In the range of lower values of the density, it corresponds to voxels of pore type, while in the range of higher values it corresponds to voxels of matrix type. The values of the histogram at the extreme points are different in both samples. This is caused by the larger volume fraction of pores in sample II in comparison with sample I, which is also visible in their tomograms.

The histograms of both samples of the bone scan also contain a considerable number of voxels with middle density values that cannot be uniquely attributed either to pores or to the matrix. This means that the choice of the threshold value of the density is very important for the segmentation process of samples of bone scans, and is crucial for the proper reconstruction of images of microscopic pore space geometry.

The dependence of the normalized histograms of both cubic bone samples on their size is shown in **Figure 2**. For the sake of clarity, the discontinuous part of the histograms occurring in the neighborhood of the point of zero density is omitted. This figure shows that the histograms in both cases depend on the size of the sample, and the differences between them decrease as the size increases. For samples with side length greater than 100 voxels, the histograms are almost the same. Samples of the limiting size can be considered statistically representative for calculations of macroscopic parameters and material characteristics, the definitions of which are based on the volumetric relations in the sample, e.g., for porosity.

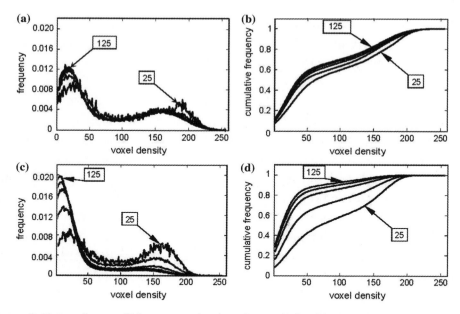

Figure 2. Dependence of histograms (a, c) and cumulative histograms (b, d) on the size of cubic samples of bone scans (side lengths: 25, 50 ...125 voxels): a, b sample I; c, d sample II.

Taking into account that the binarization threshold of the sample scan is uniquely related to the sample porosity (see "Determination of the binarization threshold" section), the binarization threshold should, therefore, also be determined for representative samples. Otherwise, both parameters will be functions of the sample size and hence it will not be possible to consider them as macroscopic quantities. This means that the whole bone, as a macroscopically strongly inhomogeneous material, should be characterized by a function defining the spatial distribution of the porosity, and the high accuracy of reconstruction of its microscopic pore space structure based on μCT images requires determination of the spatial distribution of the scan binarization threshold.

The analysis presented here is performed using samples with a side length of 125 voxels.

3. Model of the Bone Histogram

Taking into account that the origins of the blurring of microtomographic images are of a random nature[3][33][34], we derive a mathematical description of the histogram of the bone scan sample, taking a probabilistic mixture model of the histogram as a starting point. We assume that the quantized, three-dimensional sample of the scan of porous material constitutes a stochastic set of voxels with various densities (CT numbers) ρ represented in the scan by gray levels. The set of voxels in the scan sample form the overall population of the analyzed voxels, and their density is a random variable, the probability distribution function of which, denoted by $\psi(\rho)$, we identify with the normalized histogram of the sample of porous material's scan. This defines the frequency of occurrence of voxels with the given density in the whole set of voxels composing the sample of the scan. We assume, however, that the set of voxels in the sample consists of two separate subsets (subpopulations): voxels of pore type and of matrix type. The frequencies of voxel occurrence in these subsets are described by the probability distributions $\psi_p(\rho)$ and $\psi_m(\rho)$, respectively. Both functions are defined on the whole domain of real numbers. This means that the attribution of a voxel of given density to the pore type or matrix type subset is of a stochastic nature, being described by the probability distributions $\psi_p(\rho)$ and $\psi_m(\rho)$.

To derive the relationship between the voxel density distribution $\psi(\rho)$ in the scan sample and the distributions $\psi_p(\rho)$ and $\psi_m(\rho)$ in the subsets of voxels of pore and matrix type, we determine the probability of the event D^ρ of occurrence in the scan sample of voxels with density in the infinitesimal range $<\rho; \rho + d\rho>$. We introduce the following notation:

A—the event of occurrence of voxels of pore type in the scan sample.

B—the event of occurrence of voxels of matrix type in the scan sample.

Ω—the set of the elementary events.

Events A and B are disjoint and their union forms the certain event,

$$A \cap B = \Phi, A \cup B = \Omega \quad (3.1)$$

where Φ denotes the empty set.

Therefore, the probabilities of these events can be represented in the form

$$P(A) = f_v, \; P(B) = 1 - f_v. \tag{3.2}$$

The parameter f_v defines the frequency (probability) of occurrence of voxels of pore type in the set of all voxels in the sample of the scan. Therefore, this parameter can be interpreted as a measure of fraction of the pore voxels in the sample. We assume that value of parameter f_v is equal to the volume fraction of pores (porosity) in the bone sample, the μCT image of which is analyzed.

Taking into account that $D^p \subset \Omega$, and applying the total probability theorem, we have

$$P(D^p) = P(D^p|A)P(A) + P(D^p|B)P(B) \tag{3.3}$$

where $P(D^p|A)$ and $P(D^p|B)$ are conditional probabilities.

Since

$$P(D^p|A) = \psi_P(\rho)d\rho, \; P(D^p|B) = \psi_m(\rho)d\rho \\ P(D^p) = \psi(\rho)d\rho \tag{3.4}$$

from (3.3) we obtain the relation

$$\psi(\rho) = f_v \psi_p(\rho) + (1 - f_v)\psi_m(\rho) \tag{3.5}$$

in which the porosity f_v of the sample of porous material is present explicitly.

Such probabilistic models are commonly applied for the statistical analysis of data in various fields of scientific research[35]. This includes clustering, handing missing data, modeling heterogeneity, density estimation, pattern recognition, and machine learning.

4. Determination of Porosity and Other Model Parameters

The probabilistic mixture model (3.5) of the histogram of a bone scan sample allows determination of complete information about the statistical characteristics and internal structure of the voxel set in the sample. This includes the porosity parameter f_v and the parameters describing the density distributions $\psi_p(\rho)$ and $\psi_m(\rho)$ of voxels in the pore and matrix type subsets. These parameters will be estimated here by an optimization method, fitting the mathematical model of the histogram to the histogram of the bone scan sample. Due to the rather free choice of the quantization limits of the scan, they cannot be identified with the limit values of the voxel densities in their distributions. Therefore, estimators of these parameters are determined during the optimization process.

4.1. Probability Distribution Functions

We assume that the probability distribution functions $\psi_p(\rho)$ and $\psi_m(\rho)$ occurring in the formula (3.5) have the form

$$\psi_k(\rho) = C_k g_k(\rho) = C_k \frac{\left(\dfrac{\rho - a_k}{b_k - a_k}\right)^{\alpha_k - 1} \left(\dfrac{b_k - \rho}{b_k - a_k}\right)^{\beta_k - 1}}{\left(\dfrac{\rho - a_k}{b_k - a_k}\right)^{\alpha_k + \beta_k} \left(\dfrac{b_k - \rho}{b_k - a_k}\right)^{\alpha_k + \beta_k}} \qquad (4.1)$$

$(k = p, m)$

where

$$C_k = \frac{(\alpha_k + \beta_k)}{\pi(b_k - a_k)} \sin(\pi \frac{\alpha_k}{\alpha_k + \beta_k}) \qquad (4.2)$$

is the normalization coefficient.

This distribution can be obtained from the rational distribution of the form

$$\eta(x) = \frac{\alpha+\beta}{\pi x_0} \sin(\frac{\alpha}{\alpha+\beta}\pi) \frac{(x/x_0)^{\alpha-1}}{1+(x/x_0)^{\alpha+\beta}} \qquad (4.3)$$

by conversion of the random variable x, defined on the infinite domain of the positive real numbers, to a variable ρ given by the relation

$$\frac{x}{x_0} = \frac{\rho-a}{b-\rho} \qquad (4.4)$$

and defined on the interval <a, b>. Due to the form of function (4.1) we will call it a modified beta distribution.

Distributions (4.1) are defined on the finite domain of voxel densities and can take a skew form. Their parameters have to satisfy the following conditions:

$$b_k > a_k \quad \alpha_k + \beta_k > \alpha_k > 0 \quad (k = p,m)$$

In the case when

$$\alpha_k \geq 1 \quad \beta_k \geq 1$$

the distributions take finite values over the whole range of voxel densities.

Taking into account that the distributions (4.1) contain four parameters for each type of voxels, the theoretical model of the histogram given by the formula (3.5) is a function of nine parameters. This provides the model with high flexibility.

4.2. Optimization Rocedure

The parameters of the mixture model can be estimated by various methods[35], e.g., moment matching, spectral, direct optimization, minimum message length, and maximum likelihood (as with the expectation maximization (EM) algorithm). These methods estimate the structure parameters of the analyzed overall population as well the distribution parameters of its sub-populations based on sampling of the overall population.

The investigated set of voxels in the sample of the bone scan, unlike typical objects of statistical research, form an overall population with a known value for the frequency of occurrence of each element (voxel density) in the population, with the exception of voxels with zero density ascribed to them during the quantization process. The distribution of voxel density in the scan sample, represented by the normed cumulative histogram, is composed using information about the densities of all voxels in the sample of the scan which form the overall population of the investigated set of voxels, and not only a statistical sample of them. This allows the use of direct optimization methods instead of statistical methods to estimate the model parameters of the histogram.

The optimization procedure applied here is based on the multi-parameter nonlinear regression method, implemented in the numerical computing environment MATLAB. In this method, the best fit of the theoretical model to the experimental data is obtained by minimization of the sum of squared residuals of the model, i.e., the differences between the values of the data and the fitted model.

Denoting the normed histogram of the sample of the bone scan by $h(\rho_i)$, the residuals $r(\rho_i)$ of the model take the form

$$r(\rho_i) = h(\rho_i) - \psi(\rho_i) \qquad (4.5)$$

where ρ_i (i = 1, 2, ... N) is the voxel density in the bone scan sample.

Then, the objective function of the optimization problem can be defined in the form

$$E(f_v, p) = \sum_{i=1}^{N} r^2(\rho_i) = \sum_{i=1}^{N}(Y_i - f_v X_i)^2 \qquad (4.6)$$

where

$$\begin{aligned} X_i &= \psi_p(\rho_i) - \psi_m(\rho_i) = C_p g_p(\rho_i) - C_m g_m(\rho_i) \\ Y_i &= h(\rho_i) - \psi_m(\rho_i) = h(\rho_i) - C_m g_m(\rho_i) \end{aligned} \qquad (4.7)$$

and

$$P = [a_p, b_p, \alpha_p, \beta_p, a_m, b_m, \alpha_m, \beta_m]$$

represents the vector of parameters of distributions $\psi_p(\rho)$ and $\psi_m(\rho)$ given by relation (4.1).

Function (4.6) depends on nine parameters of the mixture model of the histogram: the porosity f_v and eight parameters of the distributions (4.1). Considering the normalization coefficients C_p and C_m as parameters, the distributions (4.1) become functions of five parameters that have to satisfy the normalization conditions (4.2). They play the role of constraints imposed on the model parameters. Such a change in the way of viewing the parameters of the objective function (4.6) is useful because it enables a reduction of the independent parameters that have to be determined directly in the optimization procedure.

Taking into account that the mixture model (3.5) of the histogram depends linearly on the porosity f_v and the normalization coefficients C_p and C_m, their optimum values for given values of the remaining parameters can be determined effectively from the condition for the minimum of the objective function (4.6). The first derivatives of function (4.6) with respect to the parameters f_v, C_p, and C_m give the conditions

$$\overline{X_i Y_i} - f_v \overline{X_i^2} = 0 \quad \overline{Y_i \psi_p^i} - f_v \overline{X_i \psi_p^i} = 0$$
$$\overline{Y_i \psi_m^i} - f_v \overline{X_i \psi_m^i} = 0 \tag{4.8}$$

where

$$\psi_k^i = \psi_k(\rho_i) = C_k g_k(\rho_i), \quad \overline{O_i} = \sum_{i=1}^{N} O_i / N$$

Due to relation (4.7)$_1$ only two of the conditions in (4.8) are independent. Condition (4.8)$_1$ can be used to define the optimum value of the porosity f_v; then the condition (4.8)$_2$, or equivalently (4.8)$_3$, defines the constraint imposed on the model parameters.

Applying condition (4.8)₁ the objective function (4.6) reduces to the form

$$E_p(p) \equiv E(f_v, p) = N(\overline{Y_i^2} - (\overline{X_i Y_i})^2 / \overline{X_i^2}) \tag{4.9}$$

The quantity $E_p(p)$ characterizes the error of the approximation, and its variance σ^2 is given by the formula

$$\sigma^2 = E_p / N = \overline{Y_i^2} - (\overline{X_i Y_i})^2 / \overline{X_i^2} \tag{4.10}$$

Since the number of data in the data set (in the sample of the bone scan) is very large, formally the law of large numbers and central limit theorem can be used to estimate the porosity distribution. Then expression (4.8)₁ defines the mean value of the porosity, and its variance $\sigma_{f_v}^2$ is given by

$$\sigma_{f_v}^2 = \sigma^2 \Big/ \sum_{i=1}^{N} x_i^2 = \left(\overline{Y_i^2} / \overline{X_i^2} - \left(\overline{X_i Y_i} / \overline{X_i^2} \right)^2 \right) \Big/ N \tag{4.11}$$

This allows the confidence interval $f_v \pm \Delta f_v$ for the porosity to be constructed for the assumed confidence level. In the optimization procedure applied here, the multiparameter nonlinear regression method is used repeatedly for each randomly chosen starting value of the model parameters. The procedure is stopped when the changes in the approximation error (4.9) become very small.

5. Determination of the Binarization Threshold

The purpose of the process of binarization of the scan of the sample of porous material is to produce a numerical representation of the sample's internal geometry which precisely reflects the geometry of the real object. This requires division of the whole set of voxels in the sample scan into two disjoint subsets representing the matrix and the pore space. Such a division is determined by the threshold value of voxel density, above which voxels are ascribed to the matrix, and below which they are ascribed to the pore space. In this case, the histogram of the sample scan with distributed voxel gray levels is transformed into a two-value histogram, and the sample image becomes binary. Therefore, the choice of the binarization threshold determines the quality of the reconstruction of the microscopic pore geometry and simultaneously defines the porosity of the sample.

The method proposed here for determining the porosity directly from the histogram of the sample scan, without prior reconstruction of its binary image, allows reversion of the order of determination of the porosity and the binarization threshold. This is possible because the binarization procedure should preserve the determined porosity of the sample. Therefore, the threshold of voxel density qt should reach the value for which the probability of occurrence of voxels in the sample scan with densities smaller than this threshold (voxels ascribed to pores) is equal to the porosity. We obtain the condition

$$\int_{a_p}^{\rho_t} \psi(\rho) d\rho = f_v \qquad (5.1)$$

This means that the threshold density ρ_t and its confidence interval $\rho_t \pm \Delta\rho_t$ can be directly determined from the cumulative histogram when the value of the porosity f_v and its confidence interval $f_v \pm \Delta f_v$ are known. From condition (5.1) we have

$$\Delta\rho_t = \Delta f_v / \psi(\rho_t) \qquad (5.2)$$

Considering (3.6), the condition (5.1) can be written in the form

$$\left(1 - f_v \int_{a_p}^{\rho_t} \psi_m(\rho) d\rho\right) = f_v \int_{\rho_t}^{b_m} \psi_p(\rho) d\rho \qquad (5.3)$$

which also allows another interpretation of this condition. The left side of equality (5.3) defines the probability of occurrence in the sample scan of voxels of matrix type with densities less than the threshold ρ_t ($\rho < \rho_t$), whereas the right side of this condition defines the probability of occurrence of voxels of pore type with densities greater than this threshold ($\rho > \rho_t$). The equality of these expressions means that the number of voxels of matrix type ascribed to pores in the process of binarization has to be equal to the number of voxels of pore type ascribed to the matrix. This provides the internal compatibility of voxel division in the sample scan by the threshold ρ_t with the stochastic division of this set determined in the optimization process based on the histogram model.

6. Results

The values of the estimators of the histogram model parameters for the bone sample scans I and II, as determined by the optimization method described in subsection 4.2, are presented in **Table 1**. This gives the results of five example optimizations for each sample with the best fit of the mathematical model, obtained during the optimization process performed for different randomly chosen starting values of the model parameters. The table also gives the standard deviation σ of the approximation error.

The graphs of the histograms of both samples and their example approximations described by the mixture model (3.5) and distributions (4.1) for the model parameter estimators I-1 and II-1 are shown in **Figure 3**, and distributions of the approximation error $r(\rho)$ are presented in **Figure 4**.

Table 2 gives the values of volume porosities and corresponding segmentation thresholds calculated from expressions $(4.8)_1$ and (5.1), respectively, for the first three parameter estimators presented in **Table 1**. The porosities and binarization thresholds are given together with their confidence intervals Δf_v and $\Delta \rho_t$, calculated from the variance (4.11) and from relation (5.2), respectively, for the confidence level 0.99.

For comparison, the estimators of the binarization threshold and the porosity parameter determined by the standard method and Otsu's method of image segmentation are also included in **Table 2**. Evaluation of the binarization threshold of sample II by the standard method is not clear, due to difficulties in establishing the position of the minimum of the histogram. In this work, the minima of both sample histograms are based on their local polynomial approximations.

Figure 5 illustrates how the segmentation threshold determined by the new and standard methods influences the quality of bone image reconstruction.

7. Discussion

The results of the optimization process described in the previous section show that the applied mixture model of the histogram and the modified beta distribution of the voxel density describe the histograms of both investigated samples of bone scans with high accuracy. This concerns both qualitative (**Figure 3**) and quantitative (**Figure 4**) fitting of the theoretical and experimental curves. From **Figure 4** it results that the oscillations of the approximation error of both sample histograms are greater in the range of lower densities; however, their relative values are comparable in the whole range. The mean values of the approximation error of the histogram are: $\sigma = 1.18 \times 10^{-4}$ and $\sigma = 1.25 \times 10^{-4}$ for samples I and II, respectively (**Table 1**), and the estimators of the porosity parameter determined in consecutive realizations of the optimization procedure take values lying within very narrow intervals. For 20 optimizations performed for each sample, with starting parameters generated randomly, the estimators obtained for the porosity lie in the intervals <0.6940, 0.7016> and <0.8849, 0.8853> for samples I and II, respectively. This is the case in spite of some instability of the determined parameter estimators for the pore voxel density distribution, observed for sample I (**Table 1**). This is caused mainly by the lack of information about the form of the histogram in the part cut-off during the process of quantization of the scan.

The small values of the approximation error of the histograms determine very narrow confidence intervals for the porosity parameter estimator, even for a confidence level of 0.99. They take the value $\Delta f_v = 0.002$ for both samples (**Table 2**). This interval cannot, however, define the confidence interval of the porosity

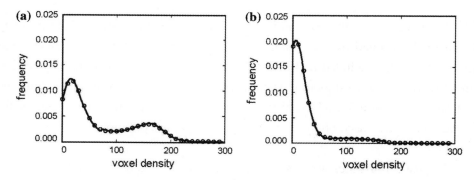

Figure 3. Histograms (circles) of sample I (a) and II (b) and their approximations (solid lines) described by the mixture model for parameter estimators I-1 and II-1.

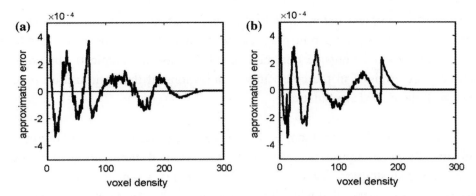

Figure 4. Distributions of the approximation error of the histograms of sample I (a) and II (b) from the mixture model for the parameter estimators I-1 and II-1.

Table 1. Estimators of the histogram model parameters for bone sample scans I and II of human condyle.

Estimation Number		$\sigma \times 10^4$	Estimators of the histogram model parameters								
			a_p	b_p	α_p	β_p	a_m	b_m	α_m	β_m	f_v
Sample I	I1	1.16	−157	178	6.45	3.49	72	275	1.22	3.20	0.699
	I2	1.17	−152	174	6.23	3.45	70	280	1.21	3.46	0.696
	I3	1.18	−155	177	6.37	3.48	72	278	1.20	3.39	0.698
	I4	1.18	−159	181	6.54	3.53	73	275	1.20	3.23	0.700
	I5	1.18	−155	177	6.37	3.48	72	278	1.20	3.39	0.698
Sample II	II1	1.25	−48	64	1.81	2.19	13	174	1.05	1.06	0.885
	II2	1.27	−49	65	1.85	2.25	12	179	1.07	1.11	0.885
	II3	1.27	−49	65	1.85	2.25	12	180	1.07	1.12	0.885
	II4	1.27	−49	65	1.86	2.25	12	182	1.07	1.15	0.885
	II5	1.27	−49	65	1.85	2.24	12	180	1.07	1.12	0.885

Table 2. Porosities and segmentation thresholds of samples I and II determined by the new and standard methods and by Otsu's method.

Estimation number		New method		Standard method		Otsu's method	
		$f_v \pm \Delta f_v$	$\rho_t \pm \Delta \rho_t$	ρ_t^s	f_v^s	ρ_t^o	f_v^o
Sample I	I1	0.699 ± 0.002	89 ± 1	91	0.699	89	0.695
	I2	0.696 ± 0.002	88 ± 1				
	I3	0.698 ± 0.002	89 ± 1				
Sample II	II1	0.885 ± 0.002	47 ± 1	112	0.956	65	0.914
	II2	0.885 ± 0.002	47 ± 1				
	II3	0.885 ± 0.002	47 ± 1				

parameter of sample I determined by the optimization procedure, since in each realization of the optimization process the estimator of the porosity parameter takes different values outside this interval. In the case of sample II, the porosity parameter determined in the consecutive realizations of the optimization process belongs to the confidence interval.

A solution to this problem is to consider the procedure of determining the porosity as a random process that provides the assumed minimum level of fitting to the model histogram. Then the porosity becomes a random variable of this process, the stochastic characteristics of which can be determined based on numerical experimental data. For 20 realizations of the optimization procedure performed for sample I, the obtained mean value of the porosity parameter is $f_v^I = 0.698$, and its standard deviation and confidence interval for the confidence level 0.99 take the values $\sigma_{f_v}^I = 0.003, \Delta f_v^I = 0.007$. Finally, for both samples the following results were obtained:

$$f_v^I = 0.698 \pm 0.007, f_v = 0.885 \pm 0.002$$

This allows the mean value of the scan binarization threshold ρ_t and its confidence interval $\Delta \rho_t$ to be determined. From relations (5.1) and (5.2) we have

$$\rho_t^I = 88 \pm 3, \rho_t = 47 \pm 1$$

The binarization threshold of sample II determined by the new method is much smaller than the thresholds determined by the standard method and Otsu's method (**Table 2**). As a consequence, the porosities determined by these methods are also considerably different. These differences take values of about 138% and 38%, respectively for the binarization threshold, and about 8% and 3% for the porosity. Nonetheless, the binarization threshold and the porosity of sample I determined by all three methods are almost the same. This substantial difference between the parameters of sample II determined by the new and standard methods is caused by the different levels of the models on which they are based. In the standard method, only information about one point of the histogram is used, whereas in the new method based on the stochastic model of the histogram, determination of image parameters is based on the information contained in the whole histogram. Therefore, the new method enables the porosity and the image segmentation threshold to be determined even in a case where application of the standard method is very difficult or impossible, i.e., when the histogram does not contain explicit extremes.

Detailed investigation of the relation between Otsu's method and the method proposed here is difficult on account of the different ways of using the histogram for determining the scan binarization threshold. Such a task would require detailed comparative analysis of both methods, which would be outside the scope of this work. Nevertheless, some qualitative evaluations can be formulated. The approach based on the mixture model of the histogram seems to be more fundamental. This method uses Eq. (5.1) for determining the scan binarization threshold, which can be interpreted as the primary definition of that threshold. Relation (5.1), written in the form (5.3), defines the statistical meaning of the thresholding process. Fulfillment of Eq. (5.3) provides the internal compatibility of the voxel division in the scan sample by the threshold density with the stochastic division of this set determined in the optimization process based on the histogram model. However, Otsu's method is formulated in a way that ensures that Eq. (5.1) is satisfied identically for any value of the scan binarization threshold. Instead of condition (5.1), an optimization procedure is proposed for determining the optimum value of the binarization threshold using the between-class variance of the voxel mean density in the scan sample as an objective function. The arbitrary choice of the objective function based on the mean densities of voxels of both classes ensures the simplicity of the optimization procedure, but it does not ensure its general nature.

The substantial difference between the binarization thresholds of sample II determined by the new method and the standard or Otsu's method is strongly influence the quality of reconstruction of its microscopic geometry. It is especially visible for the thresholds determined by the new and the standard methods, as shown in **Figure 5**.

8. Conclusion

A new method has been proposed here for determination of the bone porosity and image segmentation threshold based on the histogram of bone µCT scans. The novelty of this method consists in the use of a model-based approach that enables reversion of the procedure used in the other methods.

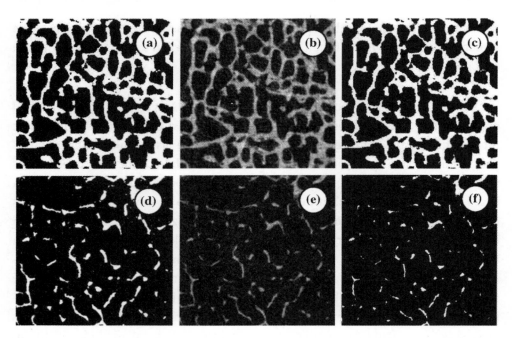

Figure 5. Influence of the binarization thresholds determined by the new method (images: a) for $\rho_t = 88$ and d for $\rho_t = 43$ and the standard method (images: c for $\rho_t = 91$ and f for $\rho_t = 112$) on the quality of image reconstruction. Tomogram b and its binary images a and c represent sample I, whereas tomogram e and its binary images d and f represent sample II.

Use of the model-based approach and the optimization method for identification of the porosity and the image segmentation threshold enables consideration of all of the information contained in the histogram, and not only information regarding the position of its separated points or the mean values of voxel densities, as in the standard method or Otsu's method. This ensures the high accuracy of both determined parameters, and improves the quality of the image reconstruction of the microscopic pore space geometry. As a consequence, other macroscopic parameters, such as tortuosity and permeability, determined by simulation of physical processes in the pore space, can also be identified with better precision.

Moreover, the new method allows the porosity and the image segmentation threshold to be determined even in cases where application of the other methods is questionable or impossible.

Open Access: This article is distributed under the terms of the Creative Commons Attribution License which permits any use, distribution, and reproduction in any medium, provided the original author(s) and the source are credited.

Source: Cieszko M, Szczepański Z, Gadzała P. Determination of bone porosity based on histograms of 3D µCT images[J]. Journal of Materials Science, 2015, 50(2):1–12.

References

[1] Feldkamp LA, Goldstein SA, Parfitt AM, Jesion G, Kleerekoper M (1989) The direct examination of three-dimensional bone architecture in vitro by computed tomography. J Bone Min Res 4:3–11.

[2] Rüegsegger P, Koller B, Muller R (1996) A microtomographic system for the nondestructive evaluation of bone architecture. Calcif Tissue Int 58:24–29.

[3] Davis GR, Wong F (1996) X-ray microtomography of bones and teeth. Physiol Meas 17:121–146.

[4] Stock SR (2008) Micro computed tomography: methodology and applications. CRC Press, Boca Raton.

[5] Mizutani R, Suzuki Y (2012) X-ray microtomography in biology. Review. Micron 43:104–115.

[6] Boller E, Cloetens P, Jl Baruchel *et al* (2006) Synchrotron X-ray microtomography: a high resolution, fast and quantitative tool for rock characterization. In: Desrues J, Viggiani G, Bésuelle P (eds) Advances in X-ray tomography for geomaterials, 7th edn. ISTE, London, pp 125–134. doi:10.1002/9780470612187.

[7] Tafforeau P, Boistel R, Boller E *et al* (2006) Applications of X-ray synchrotron microtomography for non-destructive 3D studies of paleontological specimens. Appl Phys A83:195–202. doi:10.1007/s00339-006-3507-2.

[8] Von Rietbergen B, Odgaard A, Kabel J, Huiskes R (1998) Relationships between bone morphology and bone elastic properties can be accurately quantified using high-resolution computer reconstructions. J Orthop Res 16:23–28.

[9] Biswal B, Manwart C, Hilfer R (1998) Three-dimensional local porosity analysis of porous media. Phys A 255:221–241.

[10] Barbier A, Martel C, Vernejoul MC, Triode F, Nys M, Mocaer G, Morieux C, Murakami H, Lacheretz F (1999) The visualization and evaluation of bone architecture in the rat using three-dimensional X-ray microcomputed tomography. J Bone Miner Metab 17:37–44.

[11] Lindquist WB, Venkatarangan A (1999) Investigating 3D geometry of porous media

from high resolution images. Phys Chem Earth A 25:593–599.

[12] Ams CH, Knackstedt MA, Pinczewski MV, Martys MS (2004) Virtual permeametry on microtomographic images. J Pet Sci Eng 45:41–46.

[13] Ams CH, Knackstedt MA, Pinczewski MV, Garboczi EJ (2002) Computation of linear elastic properties from microtomographic images: methodology and agreement between theory and experiment. Geophysics 67:1396–1405.

[14] Torquato S (2002) Statistical description of microstructures. Ann Rev Mater 32: 77–91.

[15] Taud H, Martinez-Angeles R, Parrot JF, Hernandez-Escobedo L (2005) Porosity estimation method by X-ray computed tomography. J Pet Sci Eng 47:209–217.

[16] Fedrich JT, Di Giovanni AA, Noble DR (2006) Predicting macroscopic transport properties using microscopic image data. J Geophys Res 111:B03201. doi:10.1029/2005JB00 3774.

[17] Sierpowska J, Hakulinnen MA, Toyras J, Day JS, Weinans H, Kiviranta I, Jurvelin JS, Lappalainnen R (2006) Interrelationships between electrical properties and microstructure of human trabecular bone. Phys Med Biol 51:5289–5303.

[18] Krabbenhaft K, Hain M, Wiggors P (2008) Computation of effective cement paste diffusivities from microtomographic images. Comp Methods Appl Sci 9:281–297.

[19] Torquato S (2002) Random heterogeneous materials: microstructure and macroscopic properties. Springer, New York.

[20] Hilfer R (1992) Local porosity theory flow in porous media. Phys Rev B 45: 7115–7121.

[21] Hilfer R (1993) Local porosity theory for electrical and hydrodynamical transport through porous media. Phys A 194:406–414.

[22] Rajagopalan S, Yaszemski MJ, Robb R (2004) Evaluation of thresholding techniques for segmenting scaffold images in tissue engineering. Proc SPIE 5370:1456–1465.

[23] Waarsing JH, Day JS, Weinans H (2004) An improved segmentation method for in vivo μCT imaging. J Bone Miner Res 19:1640–1650.

[24] Müller R, Hildebrand T, Rüegsegger P (1994) Non-invasive bone biopsy: a new method to analyze and display the three-dimensional structure of trabecular bone. Phys Med Biol 39:145–164.

[25] Ding M, Odgaard A, Hvid I (1999) Accuracy of cancellous bone volume fraction measured by micro-CT scanning. J Biomech 32:323–326.

[26] Beaupied H, Chappard C, Basillais A, Lespessailles E, Benhamou CL (2006) Effect of specimen conditioning on the microarchitectural parameters of trabecular bone assessed by micro-computed tomography. Phys Med Biol 51:4621–4634.

[27] Rosenfeld A, Torre P (1983) Histogram concavity analysis as an aid in threshold se-

lection. IEEE Trans Syst Man Cybern SMC 2:231–235.

[28] Kapur JN, Sahoo AKC, Wong A (1985) A new method for gray-level picture thresholding using the entropy of the histogram. Graph Model Image Process 29: 273–285.

[29] Otsu N (1979) Threshold selection method from gray-level histograms. IEEE Trans Syst Men Cybern SMC 9:62–66.

[30] Lima ICB, Oliveira LF, Lopes RT (2006) Bone architecture analyses of rat femur with 3D microtomographics images. J Radioanal Nucl Chem 269:639–642.

[31] Scanco Medical AG (1997) MicroCT 20 User's Guide, Software Revision 2.1:54–55.

[32] Palacio-Mancheno PE, Larriera AI, Doty SB, Cardoso L, Fritton SP (2014) 3D assessment of cortical bone porosity and tissue mineral density using high-resolution μCT: effects of resolution and threshold method. J Bone Miner Res 29:142–150.

[33] Van Geet M, Swennen R, Wevers M (2000) Quantitative analysis of reservoir rocks by microfocus X-ray computerized tomography. Sediment Geol 132:25–36.

[34] Ketcham RA, Carlson WD (2001) Acquisition optimization and interpretation of X-ray computed tomographic imagery: applications to geosciences. Comput Geosci 27:381–400.

[35] McLachlan G, Peal D (2000) Finite mixture models. Wiley, New York.

Chapter 9

Direct Reduction of Synthetic Rutile Using the FFC Process to Produce Low-Cost Novel Titanium Alloys

L. L. Benson[1], I. Mellor[2], M. Jackson[1]

[1]Department of Materials Science and Engineering, The University of Sheffield, Sheffield S1 3JD, UK

[2]Metalysis Ltd., Unit 2-Farfield Park, Manvers Way, Wath-Upon-Dearne, Rotherham S63 5DB, UK

Abstract: Typically, pure TiO_2 in pellet form has been utilised as the feedstock for the production of titanium metal via the solid state extraction FFC process. For the first time, this paper reports the use of loose synthetic rutile powder as the feedstock, along with its full characterisation at each stage of the reduction. The kinetics and mechanism of the reduction of synthetic rutile to a low oxygen titanium alloy have been studied in detail using a combination of X-ray diffraction, scanning electron microscopy, oxygen analysis, and X-ray fluorescence techniques. Partial reductions of synthetic rutile enabled a reaction pathway to be determined, with full reduction to a low oxygen titanium alloy occurring at 16h. Major remnant elements from the Becher process within the feedstock were followed throughout the process, with a particular emphasis placed on the reduction behaviour of iron within the alloy. Although impurities such as Fe, Al, and Mn are found in the feedstock and alloy, no major deviations from previously reported reaction mechanisms and phase transformations utilising a pure porous (25%–30% porosity)

TiO$_2$ precursor were found. Following reduction, the titanium alloy powder produced from synthetic rutile (approx. 3500ppm oxygen) has been consolidated via an emerging rapid sintering technique, and its microstructure analysed. This work will act as the baseline for future alloy development projects aimed at producing low-cost titanium alloys directly from synthetic rutile. Producing titanium alloys directly from synthetic rutile may negate the use of master alloy additions to Ti in the future.

1. Introduction

Since its publication in 2000[1], the FFC process, named after its inventors, Fray, Farthing and Chen, has generated a substantial amount of interest. Alongside significant academic research[2]–[13], a spin-off technology company, Metalysis was established to capitalise upon the new technology. Facilitating the extraction of metals from their ores and/or oxides, this process has the potential to drastically reduce the cost of the production of metals, particularly titanium. Titanium alloys are renowned or their superior properties—excellent biocompatibility, superb corrosion resistance, and a high strength to weight ratio. However, due to the expense of both extraction and fabrication, titanium alloys are also well known for their relatively high cost, achieving a status as a somewhat premium metal, despite the abundance of rutile and ilmenite ores. Hence, application of the FFC Metalysis process to produce titanium alloy powders, in conjunction with more economical downstream consolidation techniques, could lead to a step-change in the economics of titanium, allowing an affordable, streamlined production route of titanium metal components. Sectors outside of the aerospace industry are set to reap the largest reward, with the production of low-cost titanium being hailed as a long-term Holy Grail, particularly in the automotive industry[14]. Due to its commercially rewarding nature, a considerable amount of research has been undertaken to further develop the process. Much research has focused on taking advantage of its solid state extraction; by avoiding melting procedures, alloys previously prone to segregation effects can be produced in a much more cost-efficient manner[8][10]. Other research has focused on understanding the reaction pathway and kinetics of the process, facilitating the optimisation of the procedure[2]–[5][15].

A typical Metalysis FFC cell consists of a graphite anode and a cathode—

usually the metal oxide, immersed in a bath of molten $CaCl_2$[1]. Once a current is applied in an inert atmosphere at elevated temperatures, oxygen is ionised and diffuses through the electrolyte to the anode, releasing carbon oxides, CO and CO_2[2]. A detailed reaction mechanism published in 2005 by Schwandt et al.[2] has led to both a deeper understanding of reaction pathway, as well as improvements in the efficiency of the process in further work[4]. Equations (1)–(7) outline the overall reaction mechanism.

$$5TiO_2 + Ca^{2+} 2e^- \rightarrow Ti_4O_7 + CaTiO_3 \tag{1}$$

$$4Ti_4O_7 + Ca^{2+} + 2e^- \rightarrow 5Ti_3O_5 + CaTiO_3 \tag{2}$$

$$3Ti_3O_5 + Ca^{2+} + 2e^- \rightarrow 4Ti_2O_3 + CaTiO_3 \tag{3}$$

$$2Ti_2O_3 + Ca^{2+} + 2e^- \rightarrow 3TiO + CaTiO_3 \tag{4}$$

$$CaTiO_3 + TiO \rightarrow CaTi_2O_4 \tag{5}$$

$$CaTi_2O_4 + 2e^- \rightarrow 2TiO + Ca^{2+} + 2O^{2-} \tag{6}$$

$$TiO + 2(1-\delta)e^- \rightarrow Ti[O]\delta + (1-\delta)O^{2-} \tag{7}$$

On the initial application of the current, it is understood that Ca^{2+} ions are first incorporated into the cathode from the electrolyte, producing calcium titanate, $CaTiO_3$. Suboxides of titanium are sequentially formed, in order of decreasing oxygen content to reach Ti_2O_3. It is accepted in the field that the higher ordered Magnéli phases (Ti_xO_{2x-1}, where $4 \leq x \leq 9$) are likely to have formed prior to the formation of Ti_4O_7[3]. Suboxide TiO is eventually formed which combines with $CaTiO_3$ to produce dititanate Ca_2TiO_4. Destruction of this dititanate phase leads to the formation of equiaxed TiO and the release of Ca^{2+} and O^{2-} ions into the melt. Finally, TiO is reduced to form a solid solution of oxygen in titanium, Ti–O, concluding the reaction sequence. Metallic titanium is retrieved from the cathode, with oxygen levels as low as 1500ppm having been reported by technology company Metalysis[16].

Previously, the FFC process routinely made use of high purity TiO_2 pellet precursors, typically with around 25%–30% porosity[2]. Recent research, however, has now enabled the use of free flowing powder as the feedstock, relinquishing the requirement for a pressed preform[16]. Titanium's ore, rutile (TiO_2), can be obtained from naturally occurring deposits, (hence, known as natural rutile) which is purified via electrostatic and magnetic separation procedures to produce a high purity product[17]. Alternatively, TiO_2 is obtained from the more abundant iron-containing ore, ilmenite ($FeTiO_3$). Rutile created from ilmenite is known as synthetic rutile. A range of methods exist to produce synthetic rutile from ilmenite, such as the Becher, Benelite, Laporte, and Murso processes[17].

It is possible to produce synthetic rutile with varying degrees of purity. Variation in both the type and quantity of impurity present depends on both the source of ilmenite, as well as the method of extraction. Hence, alterations to the extraction process and/or ilmenite source can allow a degree of control over impurity character and content. Typically, synthetic rutile contains both Fe and Al impurities, which are common alloying additions for many commercial titanium alloys, reducing the dependence on costly master alloys further downstream.

Utilisation of synthetic rutile as a feedstock for the FFC process could further reduce the cost of titanium alloy powder production. Taking advantage of a synthetic rutile feedstock firstly minimises costs, as less processing steps are required compared to producing a pure precursor. In 2013, the estimated cost of synthetic rutile was $1400/t compared with $3000/t for that of pigment grade TiO_2[18]. Secondly, as synthetic rutile naturally contains both alpha and beta stabilisers, titanium alloys can be directly produced. Hence, with further research, a direct route from synthetic rutile to useable, low-cost titanium alloy powder for direct consolidation or near net shaping is a reality, particularly for non-aerospace applications. It is predicted that FFC synthetic rutile derived powder will be competitively priced compared with currently available titanium powders, the cost of which have been recently reported[19].

To take advantage of synthetic rutile feedstock, full characterisation of both the feedstock and the baseline titanium alloy product is necessary. Furthermore, to fully control the process, a thorough understanding of the reduction behaviour throughout electrolysis is required. Hence, this paper presents for the first time, the full characterisation of synthetic rutile at each stage of the reduction process. Partial experiments have allowed an insight into the reaction pathway and behaviour of the synthetic rutile throughout reduction. Previous partial reductions were undertaken using a 99.5% pure TiO_2 pressed preform pellets of 25%–30% porosity with an average particle size of 1–2μm. This set of experiments utilises loose synthetic rutile of particle size 150–212μm feedstock, which has been reduced and the mechanism compared with the original pure TiO_2 work[2]. Successful reductions allowed the consolidation of titanium powder derived from synthetic rutile via the spark plasma sintering (SPS) technique.

2. Methods and Materials

2.1. FFC Process Reductions

Reduction of synthetic rutile samples (150–212μm) supplied by Iluka Resources was achieved at both R&D scale (20g) and development scale (5kg). Partial reductions were completed at an R&D scale with powder for consolidation obtained from developmental scale reductions.

The R&D cells consist of a stainless steel retort and lid. Cathode and anode rods are inserted into the lid in addition to a thermocouple. 20g of each synthetic rutile sample was placed into a steel basket, lined with a stainless steel mesh. A connecting rod was attached to the basket, which screwed into the cathode rod. A carbon anode was screwed into the anode rods. 1.6kg of dried $CaCl_2$ salt contained in a ceramic crucible was placed in the retort before the lid was attached and the retort sealed. Exhaust and water pipes were attached to the lid to allow removal of heat and gases produced. Finally, an argon line is attached to ensure that the reaction occurs under an inert atmosphere. **Figure 1** displays the experimental setup utilised for R&D experiments.

After the cell had been sealed, the temperature was ramped to above 900°C. A salt sample was taken by dipping a rod into the molten salt via a viewing port. CaO content of the salt was then determined using an acid-base titration, facilitated by an autotitrator (Mettler Toledo autotitrator T50). Cathode and anodes were then lowered into the molten calcium chloride electrolyte. The depth of the anode

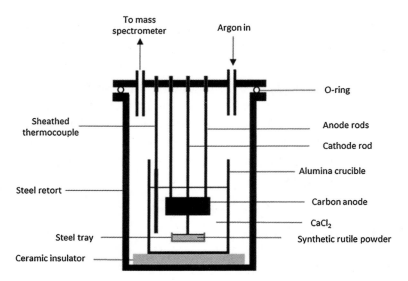

Figure 1. Experimental setup of an R&D reduction cell.

and cathode into the salt can be controlled by measuring the protrusion of the rods in the lid and was kept constant throughout the experiments. Experiments were run at constant current (5A), with an initial current ramp in place to avoid excessive voltages and chlorine formation. The average voltage for completed runs was 3.5V which was measured at the power supply. The anodic current density for these runs was typically around 550Am^{-2}. Each experiment was timed from the end of the ramp period. Off gases produced by the reaction (CO and CO_2) were monitored by mass spectroscopy. Full reductions were ground and washed until the water ran clear to remove any remaining salt. After washing the samples were dried, sieved and weighed.

Partial reductions were interrupted before completion by terminating the applied current and lifting the cathode tray out from the salt into the cooler upper section of the retort. After electrolysis, the partially reduced samples were washed in water and lightly ground before being soaked in hydrochloric acid (1 M). Samples were again washed with water and dried in air at around 80°C. Due to the powdered nature of the product and mechanical fragility of some phases, great care was taken to minimise material loss.

Once clean and dry, samples were analysed using X-ray diffraction (XRD) techniques outlined below, followed by microstructural studies, with the exception of the 2h partial, in which microstructure studies took place from a repeated non ground sample due to the mechanical fragility of the Ca_2TiO_4 needles. Each of the time intervals was completed at least twice.

2.2. X-Ray Diffraction (XRD) Analysis

Samples were ground before XRD analysis using a Bruker D2 phaser and Cu–Ka radiation. Samples were scanned throughout the range of 10°–80°. Phase analysis was completed using the International Centre for Diffraction Data (ICDD) database, with the following powder diffraction file (PDF) cards being utilised: Ti_2O_3 (01-074-0324), Ti_3O_5 (04-008-8183), $CaTiO_3$ (04-012-0563), TiO (04-004-9041), $CaTi_2O_4$ (04-010-0703), Ti_2O (04-005-4357), a-Ti (04-008-4973).

2.3. Morphology

Following mounting in conductive Bakelite, samples were subjected to grinding and polishing using silicon carbide papers, diamond paste and Struers OP-S polishing suspension. Scanning electron microscopy (SEM) images were taken using a JEOL-JSM6490LV using an acceleration voltage of 20keV.

2.4. Chemical Analysis

Oxygen content was measured using an Eltra ON-900. X-ray fluorescence (XRF) data were collected using a Phillips PW 2404. X-ray energy-dispersive spectroscopy (X-EDS) was performed using INCA software on the aforementioned scanning electron microscope. Finally, inductively coupled plasma-mass spectroscopy (ICP-MS) analysis was performed by Metalysis using a Thermo X series 2.

3. Consolidation

Consolidation was achieved via a spark plasma sintering technique using a FCT system e GmbH spark plasma sintering furnace-type HP D 25. 17g of powder was placed into a 20-mm-diameter graphite ring mould with two pistons, lined with graphite paper. A carbon felt jacket was added to prevent heat loss[20]. Once loaded into the SPS machine, the sample underwent a predetermined sintering cycle, involving a ramp rate of 100°C/min, an applied force of 15kN and a dwell time of 30min at 1200°C. Any remaining graphite paper was removed by grit blasting to yield a 10mm × 20mm consolidated synthetic rutile pellet.

4. Results and Discussion

Due to the nature of the experimentation, microstructural studies could not be completed in situ. Hence, further crystallographic transformations are possible during the cooling of the cell[15]. For example, the observed Ti_2O phase occurring at 4h of reduction, would not form in situ as this phase is a product of the Ti–O phase cooling to room temperature[4]. Other changes are possible such as the growth of $CaTi_2O_4$ needles during the cooling down of the sample.

4.1. Characterisation of Synthetic Rutile

The angular and porous morphology of synthetic rutile feedstock is shown in **Figure 2**. This porosity is inherent from its extraction route, the Becher process, which leaches out iron from ilmenite in an 'accelerated rusting' procedure[17][21]. Due to this extraction process, the synthetic rutile produced tends to be porous and shows a small level of chemical heterogeneity.

The synthetic rutile used for this particular study possesses a range of particle shapes and sizes. Likewise, the pores contained within the rutile are equally

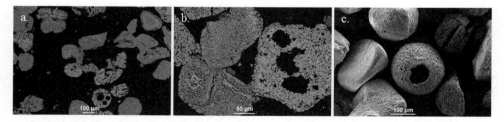

Figure 2. Synthetic rutile feedstock. a and b backscattered electron images of synthetic rutile. c secondary electron image of synthetic rutile feedstock.

as heterogeneous with a hierarchical range of pore sizes, shapes, and distributions. In some extremes, hollow particles can be observed as shown in **Figure 2(c)**. Remnant iron not removed from the Becher process can be identified as small white dots under Z contrast backscatter imaging. Chemistry of the synthetic rutile (150–212μm) was also analysed via XRF spectroscopy to reveal a composition of around 4% beta stabilising transition elements and less than 1% aluminium.

4.2. Partial Reductions of Synthetic Rutile

As discussed earlier, due to the high temperature, inert atmosphere and current required, in situ spectroscopy techniques were deemed impractical for microstructural studies. Instead, the reaction was stopped at timed intervals to gain an insight into the behaviour of synthetic rutile at various points throughout the reaction pathway. Samples were taken at 0.5, 1, 2, 4 and 8h, with full reduction occurring at 16h. Oxygen analysis, XRD, SEM and XRF techniques were used to characterise the partially reduced samples.

Samples reduced for less than 8h were generally dark in colour, with the exception of 30min reductions exhibiting a shade of brown-orange. However, the sample reverted back to black after acid washing overnight. It is thought this colour arises from iron precipitating out of solid solution. Both 4 and 8h samples required two overnight acid washes to remove the high level of CaO and $CaCO_3$ remaining within the sample.

Following the addition of acid to the sample during the washing procedure, some partially reduced samples exhibited a layer of floating scum on top of the aqueous phase. Only small samples of the scum were obtained and were analysed via X-EDS. After 1h of reduction, a gold scum formed, containing Ca, Cl, O, Ti and Fe, which gave rise to its colour. After 4 and 8h, a silver scum formed which consisted mainly of C, O and Ca and occasionally small amounts of titanium. It is likely that the scum formed contains a combination of free C and $CaCO_3$. $CaCO_3$ has been observed previously, and more detailed explanations can be found[8].

Each of the partial samples were subjected to XRD analysis, with a good reproducibility of the phases being found. Detection of phases was restricted by the limited sensitivity of the XRD technique, with only major phases being detected (>5%). Furthermore, the complex Ti–O system contains many suboxides, producing similar crystallographic peaks, (e.g. Ti_3O and Ti_6O) which can make identification of some phases strenuous[2]. **Figure 3** shows the XRD patterns produced from the reductions.

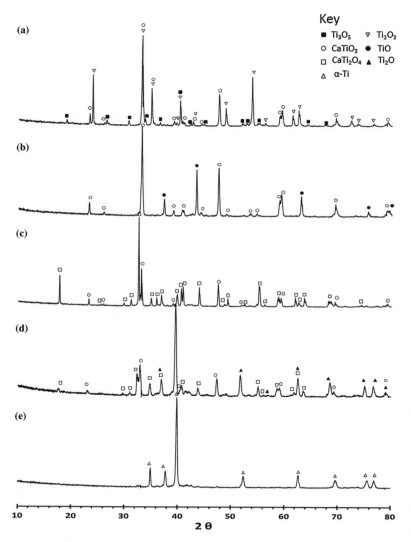

Figure 3. X-ray diffraction spectrum of partially reduced samples of synthetic rutile at various timed intervals a 0.5h showing Ti_3O_5, Ti_2O_3 and $CaTiO_3$ b 1h showing TiO and $CaTiO_3$ c 2h showing $CaTiO_3$ and $CaTi_2O_4$ d 4h showing Ti_2O, $CaTiO_3$ and $CaTi_2O_4$. e 8h showing α-Ti.

After 30min of electrodeoxidation, XRD reveals major phases of Ti_3O_5, Ti_2O_3 and $CaTiO_3$ [**Figure 3(a)**]. Requiring minimal atomic reconstruction, these phases have previously been observed by Schwandt *et al.*, following the reaction pathway via Equations (2) and (3)[2][3]. This work[2][3] has also deduced that the initial step of the reduction is the incorporation of Ca into the cathode, which is highlighted by the increase in Ca content and decrease in Ti content at 30min of polarisation, as illustrated by **Figure 4**. **Figure 4** also displays the oxygen content throughout the reduction process. Note, at high levels of oxygen these numbers are indicative values only as the instrument is calibrated for low levels of oxygen content.

The brown orange colour residue observed after 30min of reduction following removal of the salt is postulated to be due to the presence of iron, which had precipitated out of solid solution. Once washed in HCl, the iron was removed, reverting to a black powder with a substantially reduced iron content. Iron is a beta stabiliser and rapidly diffuses within titanium[22]. Furthermore, from a thermodynamic consideration, iron oxide reduces more readily than titania via the FFC process. Hence, as iron has precipitated out after 30min of polarisation, evidence from this work suggests that iron has reduced within this time frame and has migrated out of solid solution.

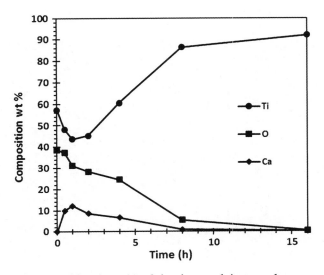

Figure 4. Chemical composition in wt% of titanium, calcium, and oxygen throughout the reduction of synthetic rutile as measured by XRF analysis. Calcium levels at 0 and 16h measured by ICP-MS due to low contents.

Morphology of the particles after 30min of reduction shows two distinct phases. Larger solid appearing areas have been analysed to reveal calcium, titanium and oxygen content. Therefore, these areas have been identified via X-EDS analysis, as the $CaTiO_3$ phase, shown in **Figure 5(b)**. Smaller pieces with a more fragmented appearance show a decreased presence of calcium and consist mainly of titanium and oxygen. Therefore, these areas have been identified as a combination of the titanium suboxide phases.

After reduction of synthetic rutile for 1h, the iron has reduced, shown in **Figure 5(e)** as the heavy, bright white phase. It is believed this iron-rich phase has diffusion bonded to the surrounding suboxide phase. Identification of the phase using X-EDS analysis was limited due to the large interaction volume with respect to the low volume fraction of the iron-containing phase. The stoichiometry of this intermetallic phase shows a wt% ratio of roughly 30:70 (Ti:Fe). Thus, it can be deduced that the intermetallic $TiFe_2$ phase has formed. $TiFe_2$ was not detected by XRD due to its low concentrations. Only two major phases consisting of cubic TiO and $CaTiO_3$ were detected by XRD at this point in the reduction [**Figure 3(b)**]. Again, the formation of the TiO phase is in agreement with the reaction mechanism outlined earlier by Schwandt et al., as shown in Equation (4)[2]. At this stage in the reduction, a kinetic barrier is expected as the lattice shape shifts from edge sharing Ti_2O_3 octahedra, to cubic titanium oxide[3]. **Figures 5(d)-(f)** depicts the

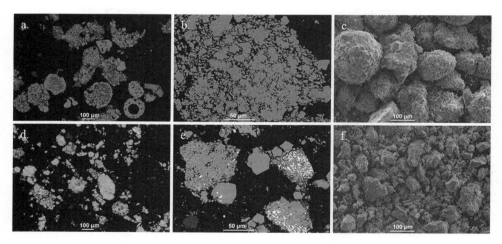

Figure 5. Partially reduced synthetic rutile. a and b show backscattered electron images of partially reduced synthetic rutile morphology after 0.5h and d and e after 1h. Figures c and f represent secondary electron images after 0.5 and 1h of reduction, respectively.

morphology observed at this stage in the process, illustrating the formation of agglomerate particles consisting of TiO and CaTiO$_3$.

Following the formation of TiO, the reaction proceeds to the formation of calcium dititanate at around 2h of polarisation. This stage features the largest reconstructive transformation throughout the reduction, with the formation of the lath-shaped acicular CaTi$_2$O$_4$ [Equation (5)] Due to the low mechanical stability of the dititanate phase, a further repeated partial reduction halted at 2h was performed in order to observe the needles before the sample was ground for XRD analysis. Backscattered electron images show clearly the presence of both titanates; the needle-like CaTi$_2$O$_4$ and the more faceted CaTiO$_3$ are shown in **Figure 6(b)** and **Figure 6(c)**. Formation of CaTi$_2$O$_4$ has previously been reported to grow in a "star burst" manner, which was also observed and shown in **Figure 6(c)**. This star burst pattern has previously been explained as a result of limited nucleation and fast anisotropic growth at low angles of misorientation[3]. **Figure 6(c)** also highlights an enlarged image of a faceted CaTiO$_3$ particle.

The presence of the iron as discrete TiFe$_2$ particles is still clear at this point in the reduction. However, occasionally, a thin layer of a less bright phase coats the TiFe$_2$ phase. It is thought that as the reaction proceeds, the iron within the TiFe$_2$ particles begins to diffuse into surrounding suboxides to form a thin layer of

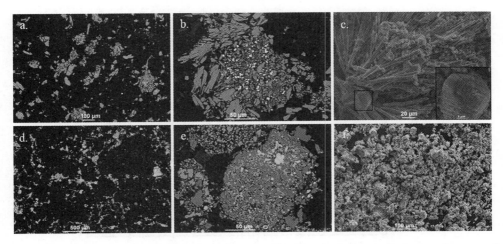

Figure 6. Partially reduced synthetic rutile. a and b show backscattered electron images of partially reduced synthetic rutile after 2h and d and e after 4h. Figures c and f represent secondary electron images after 2 and 4h of reduction, respectively.

TiFe around the TiFe$_2$ core. Formation of the TiFe phase is discussed in more detail at 4h of reduction.

At 4h, both titanates are still present, alongside newly formed Ti$_2$O [**Figure 3(d)**]. Note, an unidentified peak is observed in the XRD pattern at 42°, which persists at 8h of reduction. It has previously been reported that the Ti$_2$O phase is a product of the cooling of the Ti–O phase[4], which can be observed in the Ti–O phase diagram shown in **Figure 7**. During this stage of the reduction, destruction of the dititanate produces oxide ions alongside Ca^{2+}, resulting in the high content of CaO throughout the sample [Equation (6)]. Hence, extra acid washing was required at this stage to remove the calcium oxide. Morphology of the particles at this point is somewhat mixed, with some particles showing little resemblance to the initial spherical style particle they are evolved from. Secondary electron images reveal nodular growth of a globular nature.

Additionally, at 4h of reduction, bright, highly concentrated areas of iron begin to noticeably reduce in size and intensity. Areas of high iron concentration now display two clear phases, easily identifiable due to the difference in contrast. Porosity around the phase edges also affirm the formation of a second iron-containing phase. Again, identification of this second iron-containing phase (TiFe)

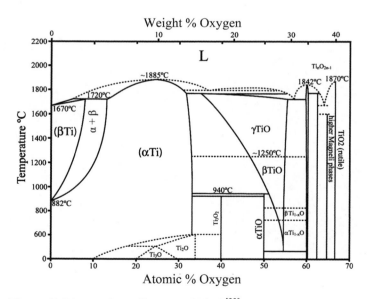

Figure 7. Binary phase diagram of Ti–O[23].

using X-EDS analysis was limited due to the large interaction volume with respect to the low volume fraction of the phase.

However, it appears the core TiFe$_2$, remains, with the second coating phase containing a lower concentration of iron. From the Ti–Fe–O phase diagram (**Figure 8**), the limited X-EDS data available and SEM images, it is plausible that the TiFe phase has formed a layer surrounding the TiFe$_2$ core.

After just 8h, a-titanium is identified as the major phase by XRD analysis [**Figure 3(e)**]. Note, a-Ti can contain approximately up to 15wt% oxygen[26] (**Figure 7**) and as shown by **Figure 4**, at 8h this sample contains around 5wt% oxygen. Other phases may be present, as shown by the presence of peaks at 32.5°, 36°, and 42° mentioned earlier, which were unable to be identified. An interconnected porous system is beginning to take shape, shown in **Figure 9(b)** which exhibits two separate phases, identified as alpha and beta titanium. At this point in the reduction, the final morphology begins to evolve, which is comparable to that of the final product shown in **Figure 9(d)** and **Figure 9(f)**. It is imperative to note at this stage in the reduction, no intermetallics or areas of excessive iron content can be

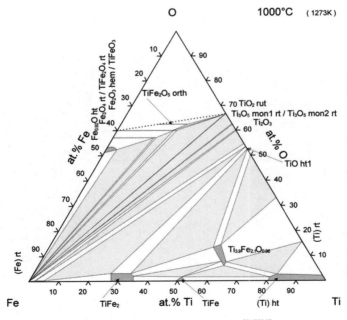

Figure 8. Tertiary Phase diagram of Ti–Fe–O[24][25].

found; the iron has fully dispersed throughout each particle.

Completion of the reduction occurs at 16h, giving a final oxygen content of approximately 4000ppm. The particles now have a much more sintered morphology. Again alpha and beta phases can be identified in the backscattered images [**Figure 9(e)**]. The predominant morphology, a fine scale alpha beta structure [**Figure 9(e)**], was found alongside a more lamellar structure which could be a product of slight local particle to particle chemical variation observed in the feedstock. Chemical composition of the reduced synthetic rutile was found to contain alpha stabilising aluminium, as well as 4.4% of beta stabilising transition elements.

Overall it appears as though the reaction pathway is in agreement with that previously reported by Schwandt et al.[2]; however, there are some significant differences in this system warranting further discussion. A new major feature is apparent in the formation and subsequent reduction of iron intermetallic phases. Initially iron present within the synthetic rutile is reduced to metallic Fe after 30min of reduction. Following the reduction of iron oxide is the formation of the intermetallic $TiFe_2$ as compact discrete particles, easily identifiable as the bright phase in backscattered images [**Figure 10(a)**]. As the reduction progresses the iron

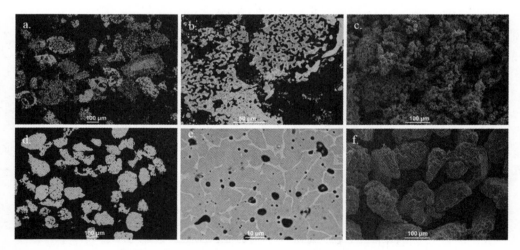

Figure 9. Partially and fully reduced synthetic rutile. a and b are backscattered electron images of partially reduced synthetic rutile after 8h and d and e after 16h (completion). Figures c and f represent secondary electron images after 8 and 16h of reduction, respectively.

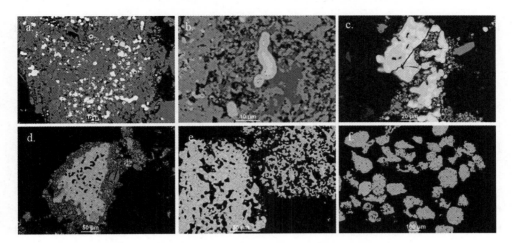

Figure 10. Backscattered electron images of a partial reduced synthetic rutile samples. a After 1h of reduction, highlighting the bright TiFe$_2$ phase. b, c and d After 4h of reduction with b Showing a TiFe$_2$ core surrounded by a TiFe layer. c Illustrates a bright TiFe$_2$ surrounded by titanium with iron content ranging from 40% to 50%. d Rare particle showing the first formation of titanium metal consisting mostly of beta titanium, surrounded by titanium suboxides and CaTiO$_3$. e Particles after 8h of reduction, left particle is beta rich and slightly more sintered whereas the right particle is alpha dominant and is less sintered. f Fully reduced material showing beta phase is well dispersed within final particles.

diffuses into the surrounding titanium suboxides, giving rise to TiFe. A coexistence of these phases can be seen in the titanium-iron phase diagram (**Figure 8**). Both intermetallic phases are observed in **Figure 10(b)** and **Figure 10(c)**, with the bright core of the iron-based particle consisting of TiFe$_2$ and the outside exhibiting the less bright TiFe phase. Further dispersion of the concentrated iron occurs with further diffusion of iron into surrounding suboxides. A solid solution of iron in titanium begins to appear with high contents of iron (up to 60Ti:40Fe) confirming its diffusion into the surrounding titanium [**Figure 10(c)**].

High local concentrations of β stabilising iron leads to a widening of the α + β and β phase fields at the expense of the a phase field (**Figure 7**). Widening of these phase fields allows a faster reduction within these areas due to the presence of beta titanium, in which oxygen diffuses at least two orders of magnitude faster than within alpha titanium. Occasionally, rare particles of high oxygen titanium metal are observed at 4h [**Figure 10(d)**]. The formation of titanium this early on in the reduction is likely due to the particles containing high quantities of iron, allowing the formation of the high oxygen beta phase and faster diffusion process.

Particles containing reduced titanium at 4h of reduction often contained significant quantities of the beta phase.

Iron continues to disperse throughout the reducing titanium metal and after 8h of reduction, no observable intermetallic compounds are found. Iron is found to be evenly dispersed throughout the beta phase of the resulting reduced particle. However, an observation was made that particles containing higher proportions of the beta phase (appearing lighter under backscatter imaging) were clearly more sintered than particles that were alpha dominant as shown in **Figure 10(e)**. This enhanced sintering is due to the faster diffusion rate of oxygen within beta titanium and as such, slightly faster reduction. However, at the end of reduction, most particles are sintered to a similar extent with iron fully dispersed throughout the beta phase of each particle [**Figure 10(f)**].

4.3. Consolidation and Forging

Due to the success of the R&D scale reductions (20g), the reduction of synthetic rutile was scaled up to developmental scale (5kg). A slightly different feedstock was chosen to reduce at development scale with a transition element concentration of 3.7% and aluminium content of less than 1%. At developmental scale reduced synthetic rutile samples have produced oxygen contents as low as around 2000ppm. A sample of 3500ppm oxygen containing around 3.4% transitional beta stabilising elements and less than 1% aluminium reduced synthetic rutile was consolidated using a spark plasma sintering technique. Spark plasma sintering is a consolidation technique in which powder is pressed together under a uni-axial pressure combined with elevated temperatures and an applied current[20][27]. **Figure 11** displays the microstructure achieved after consolidation as analysed by SEM.

Despite originating from a porous synthetic rutile feed-stock, containing slight chemical inhomogeneity between particles, following reduction and consolidation via spark plasma sintering, a fine α + β homogenous microstructure has emerged. Coarse pro-eutectoid acicular alpha regions dominate the microstructure, with a much finer eutectic α + β phase featured between the larger laths. Porosity

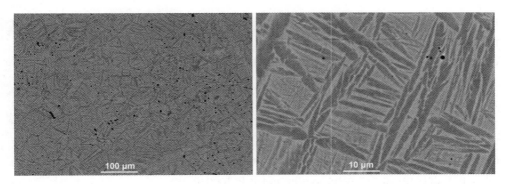

Figure 11. Backscattered image of consolidated reduced synthetic rutile of oxygen content ca. 3500ppm.

can be observed within the sample, and has been measured to give a density value of approx. 98.6%. Although this value is lower than the [99% commonly achieved by commercial titanium alloys[20], this porosity was obtained using standardised, non-optimised conditions. Furthermore, SPS or field-assisted sintering technology (FAST) is envisaged to be an intermediate step and a precursor to hot forging processes where further microstructure refinement and removal of residual porosity will be achieved.

5. Conclusion and Future Work

Partial reductions of synthetic rutile exploiting XRD, SEM and oxygen analysis revealed no major deviations from the reaction pathway previously outlined by Schwandt et al.[2], despite the presence of remnant elements from the Becher process. However, the monoclinic TiO phase was not found. Additionally, due to the presence of iron, new features of the reduction were noted, with particular emphasis on the reduction pathway of iron and the formation of intermetallics $TiFe_2$ and TiFe. These inter-metallics are deconstructed and the iron is fully dispersed throughout the reduced alloy product. However, slight inhomogeneity remains on a local level with some particles displaying a richer beta microstructure.

As previous work reported by Schwandt et al.[4][5] has shown that the reduction mechanism may be altered if conditions are changed such as CaO content, ramp rate and density of the preform, further work may be required to detect if the reduction mechanism of synthetic rutile responds in the same way. Further, due to the powdered nature of the feedstock, it was more difficult to establish if direct reduction as mentioned in previous work is occurring, hence future work may require the use of preform pellet to determine this.

Consolidation of synthetic rutile reduced titanium alloy powder has been successfully achieved, resulting in a fine α + β microstructure of 98.6% porosity using non-optimised sintering conditions. Further work will consist of optimising downstream processing as well as improving the mechanical properties of titanium alloy powders produced from synthetic rutile feedstock.

Overall the reduction of synthetic rutile has produced promising results. These results show that with further development, the FFC process is capable of producing low-cost titanium alloys produced directly from synthetic rutile. Further, the presence of beta stabilising elements appears to facilitate a faster and more efficient solid-state reduction[8]. Many of these beta stabilisers are prone to chemical segregation when produced using conventional melt processing techniques (such as beta flecks). Using a combination of Metalysis FFC powder and low-cost powder consolidation methods, a pathway is established that is entirely performed in the solid-state eliminating such segregation issues.

Current metastable β alloys such as Ti−10V−2Fe−3Al and Ti−5Al−5V−5Mo−3Cr are currently still hindered by their tendency to segregate during vacuum arc remelting, as well as high costs of downstream thermomechanical processing to homogenise chemistry and refine microstructure. These alloys have been notorious to displace, despite the production of alloys with improved properties, due to the high cost of qualification[28]. However, with further research, theses alloys may one day be overhauled by beta stabilised alloys produced directly from synthetic rutile, allowing the modernisation of current metastable beta alloys targeted towards non-aerospace sectors.

Acknowledgements

The authors gratefully acknowledge the financial support of the Engineering and Physical Sciences Research Council UK (EPSRC) through the Centre for Doctoral Training in Advanced Metallic Systems and technology company, Metalysis. The authors would like to thank Prof. Panos Tsakiropoulos for useful discussions, Nick Weston for his SPS guidance and Dr. Nik Reeves-McLaren for his assistance with XRD experimentation. Thanks also goes to the Metalysis technology team (Nader Khan, Greg Doughty, Joseph Campion, Terri Ellis, Matt Piper & Ruth Graham), analytical team (Dan Kitson, Basharat Ali & Becky Micklethwaite) and operations team.

Funding Funding for this research was provided by the Engineering and Physical Sciences Research Council and Metalysis.

Compliance with Ethical Standards

Conflict of interest Author Ian Mellor has a vested interest in the Metalysis Company. No other conflicts of interest.

Open Access: This article is distributed under the terms of the Creative Commons Attribution 4.0 International License (http://creativecommons.org/licenses/by/4.0/), which permits unrestricted use, distribution, and reproduction in any medium, provided you give appropriate credit to the original author(s) and the source, provide a link to the Creative Commons license, and indicate if changes were made.

Source: Benson L L, Mellor I, Jackson M. Direct reduction of synthetic rutile using the FFC process to produce low-cost novel titanium alloys[J]. Journal of Materials Science, 2016, 51(9):4250–4261.

References

[1] Chen GZ, Fray DJ, Farthing TW (2000) Direct electrochemical reduction of titanium dioxide to titanium in molten calcium chloride. Nature 407:361–364.

[2] Schwandt C, Fray DJ (2005) Determination of the kinetic pathway in the electrochemical reduction of titanium dioxide in molten calcium chloride. Electrochim Acta 51(1):66–76.

[3] Alexander DTL, Schwandt C, Fray DJ (2006) Microstructural kinetics of phase transformations during electrochemical reduction of titanium dioxide in molten calcium chloride. Acta Mater 54:2933–2944.

[4] Schwandt C, Alexander DTL, Fray DJ (2009) The electro-deoxidation of porous titanium dioxide precursors in molten calcium chloride under cathodic potential control. Electrochim Acta 54:3819–3829.

[5] Alexander DTL, Schwandt C, Fray DJ (2011) The electro-deoxidation of dense titanium dioxide precursors in molten calcium chloride giving a new reaction pathway. Electrochim Acta 56:3286–3295.

[6] Barnett R, Kilby KT, Fray DJ (2009) Reduction of tantalum pentoxide using graphite and tin-oxide-based anodes via the FFC-Cambridge process. Metal Mater Trans B 40B:150–157.

[7] Bhagat R (2008) The electrochemical formation of titanium alloys via the FFC Cambridge process, Imperial College London, PhD thesis.

[8] Bhagat R, Jackson M, Inman D, Dashwood R (2008) The production of Ti–Mo alloys from mixed oxide precursors via the FFC Cambridge process. J Electrochem Soc 155:E63–E69.

[9] Chen G, Fray DJ (2006) A morphological study of the FFC chromium. Trans Inst Min Metall C 115:49–54.

[10] Dring K, Bhagat R, Jackson M, Dashwood RJ, Inman D (2006) Direct electrochemical production of Ti-10W alloys from mixed alloys precursors. J Alloy Compd 419:103–109.

[11] Jackson BK, Jackson M, Dye D, Inman D, Dashwood RJ (2008) Production of NiTi via the FFC Cambridge process. J Electrochem Soc 155:E171–E177.

[12] Dring K, Dashwood R, Inman D (2005) Voltammetry of titanium dioxide in molten calcium chloride at 900 C. J Electrochem Soc 152(3):E104–E113

[13] Jiao S, Fray DJ (2010) Development of an inert anode for electro-winning in calcium chloride-calcium oxide melts. Metall Mater Trans B 41B:74–79.

[14] Jackson M, Dring K (2006) A review of advances in processing and metallurgy of titanium alloys. Mater Sci Technol 22:881–887.

[15] Bhagat R, Dye D, Raghunathan SL, Talling RJ, Inman D, Jackson BK, Rao KK, Dashwood RJ (2010) In situ synchrotron diffraction of the electrochemical reduction pathway of TiO_2. Acta Mater 58(15):5057–5062.

[16] Mellor I, Grainger L, Rao K, Deane J, Conti M, Doughty G, Vaughan D (2015) Titanium powder production via the Metalysis process, in Titanium powder metallurgy. Butterworth-Heinemann, Oxford, pp 51–67.

[17] Zhang W, Zhu Z, Cheng CY (2011) A literature review of titanium metallurgical processes. Hydrometallurgy 108:177–188.

[18] van Vuuren DS (2015) Direct titanium powder production by metallothermic processes, in titanium powder metallurgy. Butterworth-Heinemann, Oxford, pp 69–93.

[19] Froes FH (2013) Titanium powder metallurgy: developments and opportunities in a sector poised for growth. Powder Metall Rev 2(4):29–43.

[20] Weston NS, Derguti F, Tudball A, Jackson M (2015) Spark plasma sintering of commerical and development titanium alloy powders. J Mater Sci 50(14):4860–4878. doi:10.1007/s10853-015-9029-6.

[21] Iluka Resources (2012) Iluka's synthetic rutile production. http://www.iluka.com/docs/mineral-sands-briefing-papers/iluka's-synthetic-rutile-production-june-2012. Accessed 1 Oct 2015.

[22] Nakajim H, Koiwa M (1991) Diffusion in titanium. ISIJ Int 31(8):757–766.

[23] Murray JA, Wriedt HA (1994) In: Boyer RR, Welsch G, Collings EW (eds), Materials properties handbook: titanium alloys. Ohio: ASM International.

[24] Raghavan V (1989) The Fe–O–Ti (iron-oxygen-titanium) system, phase diagrams ternary iron alloys. Indian Inst Met 5:300–325.

[25] Raghavan V (2006) Ti–O–Fe phase diagram. ASM alloy phase diagrams database, Villars P, editor-in-chief, Okamoto H, Cenzual K (eds). http://www1.asminternational.org/AsmEnterprise/APD. Materials Park.

[26] Molchanova EL (1965) Phase diagrams of titanium alloys. Isreal Program for Scientific Translations, Jerusalem.

[27] Munir ZA, Anselmi-Tamburini U, Ohyanagi M (2006) The effect of electric field and pressure on the synthesis and consolidation of materials: a review of the spark plasma sintering method. J Mater Sci 41(3):763–777. doi:10.1007/s10853-006-6555-2.

[28] Cotton JD, Briggs RD, Boyer RR, Tamirisakandala S, Russo P, Shchetnikov N, Fanning JC (2015) State of the art in beta titanium alloys for airframe applications. JOM 67(6):1281–1303.

Chapter 10
Effect of Interfacial Area on Densification and Microstructural Evolution in Silicon Carbide-Boron Carbide Particulate Composites

Tom Williams[1], Julie Yeomans[1], Paul Smith[1], Andrew Heaton[2], Chris Hampson[3]

[1]University of Surrey, Guildford, UK
[2]Dstl, Salisbury, UK
[3]Morgan Advanced Materials, Stourport on Severn, UK

Abstract: A range of $SiC-B_4C$ composites have been prepared by pressureless sintering, using different proportions of two sizes of B_4C; 7 and 70μm. The interfacial area between the B_4C and SiC has been quantified and is shown to have a significant effect on both densification and the resultant microstructure of the composites. SiC/B_4C interfaces typically hinder densification. SiC/B_4C interfacial area is also shown to be related to grain growth and polytype distribution in the SiC. With more SiC/B_4C interfacial area, grain growth in the SiC is restricted and less of the SiC transforms from the starting 6H polytype to the 4H one. It is therefore suggested that it may be possible to use SiC/B_4C interfacial area as a means by which to engineering the microstructure.

1. Introduction

SiC and B_4C are of interest owing to their combination of high hardness, low density and high thermal tolerance, even when compared with other common engineering ceramics. However, B_4C is known to be difficult, and thus costly, to form into a dense body, particularly by pressureless sintering. This is because of its highly covalent structure and corresponding low self-diffusivity[1]. Despite these drawbacks, because it has a lower density than SiC and can display higher hardness, significant interest in its use remains. Given that SiC and B_4C have some capacity to act as sintering aids for each other, composite materials have been suggested as a pragmatic approach to producing carbide ceramic materials. Further, in spite of the associated difficulties, pressureless sintering of these materials is preferred to hot pressing techniques, if the resulting composites are to be commercially viable. Reaction bonding has also been used to produce these composites[2], but results in reduced hardness and strength compared with sintered materials[3] and so is of less interest for high performance applications. Use of coarser B_4C also has the potential to reduce the cost of the material. However, this may impair the ability of a composite to densify, given the low self-diffusivity of B_4C. This is significant since a primary consideration in components where hardness is important, such as those for wear applications, is that materials must reach a high percentage of their theoretical density (%TD). When using fine B_4C, composites with compositions ranging from 10 to 90wt% B_4C have been pressureless sintered to 98%TD, with only C additions[4]; this suggests that almost full densification is possible under the correct conditions. However, it has also been reported that B4C will not demonstrate self-densification with a particle size (as indicated by a median diameter, d_{50}, value) above ~8μm[5]. One potential solution to this is to surround a coarse B_4C material with a SiC matrix, which can be readily prepared with B and C additives to achieve good densification more readily than B_4C.

As well as %TD, the grain structure is important in determining the mechanical performance of the ceramic material. SiC has long been known to display discontinuous grain growth, both under liquid phase sintering (LPS)[6], and also under solid state sintering conditions[7][8]. Discontinuous grain growth is more common under pressureless sintering conditions than under pressure-assisted densification, since higher temperatures are typically required to produce adequate density in the sintered body. The use of a second inert phase, such as graphite, has been demonstrated to limit this[9]. Similar effects have been achieved in Al_2O_3, using B_4C as an inert phase[10][11]. However, the degree of control of grain growth that may be achieved using a second phase is not well understood. A distinct trend has been observed for B, (added as B_4C) in β-SiC[12]. It was observed that the addition of a small amount of B increased grain growth; this is in agreement with previous work suggesting that B typically increases diffusivity and mass transport in SiC[13]. However, further addition of B_4C inhibited grain growth. Since B has a low solid solubility in SiC[14], the remaining B_4C was likely to be inert. This suggests that the interfaces present between SiC and B_4C were inhibiting grain growth, as was observed with graphite addition[9] and when 5wt% B4C prevented discontinuous grain growth in a pressureless sintered SiC[15].

Sintering of SiC is also linked to changes in the polytype composition of the material, which may be important since it has been suggested that polytype transformations could act as a micro-plasticity mechanism under certain conditions[16]. Polytype transformation has been observed by Raman spectroscopy in SiC subject to machining[17] but is more commonly associated with certain grain growth mechanisms, including b to a transformation by dissolution-precipitation during LPS[18]. In α-SiC, it has also been observed that LPS 6H SiC tends to transform to 4H when abnormal grain growth occurs when an aluminium-boron-carbon (ABC) phase is used[19]-[21]. Further, in a related study, increased transformation from 6H to 4H was observed with increased sintering temperature for LPS α-SiC[22]. However, LPS seems to produce a smaller effect on polytype transformation in α-SiC than in β-SiC[23]. This may result from the greater thermal stability of 6H SiC compared with 3C. For both 3C and 6H SiC, however, the transformation to 4H is typically associated with the formation of a liquid phase and also therefore with grain growth.

Given that it appears that a second phase in these materials can influence the microstructural evolution in SiC, it is necessary to consider means of quantifying the degree of interaction between two phases. This can be accomplished using image segmentation methods, such as phase separation[24]. However, in the present work, the total SiC/B_4C interfacial area/unit volume in different material has been quantified directly.

In the present work, particulate SiC–B_4C composites have been produced by adding two relatively coarse grades of B_4C to a SiC matrix. The effects of the SiC/B_4C interfacial area in these composites are discussed. It is suggested that the interfacial area affects densification, together with material diffusivity and powder size effects. Further, the amount of SiC/B_4C interfacial area has a significant effect on the microstructural evolution, changing the grain size and type within the SiC matrix.

2. Experimental Procedure

All samples were prepared with an α-SiC starting powder, (SIKA Sintex 15C) with d_{50} measured as 0.8μm on a Malvern Mastersizer 3000, surface area $15m^2 g^{-1} \pm 1m^2 g^{-1}$ and 0.83wt% total O. Two sizes of B_4C were used with d50 values of approximately 7- and 70-μm (Sigma Aldrich research purity B_4C). According to LECO analysis provided by AMG analytical services, Rotherham, UK, the 7μm Sigma grade has 0.160wt% O and 70-μm Sigma grade has 0.041wt% O.

Powder blends were prepared by dispersion of both SiC and the B_4C powders in a slip. To facilitate densification in the SiC, 1wt% of 1-μm B_4C, H. C. Starck HS grade, and an organic C source (yielding approximately 4wt% C) were added to all powder blends. A fugitive binder system was also added to all blends. Materials with B_4C additions of 10wt%, 20wt% and 30wt% were prepared with both the 7- and 70-μm grades. All powder blends contained SiC as the remaining wt%, forming the matrix phase. A standard SiC, containing 1wt% of the 1-μm B_4C and the C source, but no coarse B_4C was also prepared as a reference material. The resulting slips were freeze dried and sieved through a 355-μm mesh. Discs of 20mm diameter, approximately 5mm thick were prepared, by uniaxial pressing at 345MPa ± 2MPa. Sintering of all parts was carried out in graphite trays in a graphite resistance furnace under Ar. A top temperature of 2125°C with a hold of 1h was used for all samples.

Sample mass was measured using a Mettler-Toledo balance to ±0.001 g and density was calculated using the Archimedes method in reverse osmosis (RO) water. The %TD was calculated from the density using the rule of mixtures and assuming densities of 3.20 and 2.52g cm^{-3} for SiC and B_4C, respectively. Five sample discs of each material were measured in this way. Samples were polished metallographically; the final step used a 1μm diamond slurry. Scanning electron microscopy (SEM) characterisation was carried out using a JEOL 6000 desktop SEM with a back scattered electron (BS) detector. Accelerating voltage was 10kV, working distance 19mm. X-ray diffraction (XRD) was carried out on a Hitachi Gen 3 model with a monochromater using Cu Kα radiation. Generator settings were 35kV and 40mA. Scan range was 5°–120° 2θ, with a 0.017° step and a 4s dwell. Samples were then etched by heating to 850°C in air for 1h and boiling in a Murakami's reagent. Reflected light microscopy was carried out on a Buhler instrument. Image analysis was carried out using ImageJ software and Rietveld refinement were carried out using General Structure Analysis Software (GSAS). Data for this refinement were obtained from the international crystal structures database[25][26][27]. For grain size analysis, direct measurement was used, by drawing lines along the long axes of grains on representative micrographs of each material and recording the lengths of these. At least 150 grains/material were measured.

To assess the amount of SiC/B_4C interfacial area in a given composite volume, an 'edge area' method was adopted, using BS SEM images of polished samples. All required image analysis was carried out using ImageJ software. The 'edge area' method used thresholding to select B_4C features in a given BS SEM image to produce a simple binary image and then the 'find edges' tool, a 3 by 3 sobel edge filter, to produce an image consisting solely of the lines representing the SiC/B_4C interfaces. B_4C is appreciably darker than SiC in these images and pores are darker again than the B_4C, producing three distinct shades. Thus, it is possible to select only B_4C. The stages to produce this image are shown in **Figure 1**. From examination of different material orientations, including through thickness, it appears that the B_4C distribution is generally isotropic. Hence, the perimeter length of the B_4C feature edges/unit area is related to the actual interfacial area/unit volume. The edge lines have finite thickness, and so the total edge area on the image was recorded as a fraction of the total image area. This value was converted to the interface perimeter/unit area and hence interface area/unit volume in 3D, by dividing it by the mean interface line thickness. Image smoothing was used to reduce the error from edge detection around pores. This was necessary because pores showed an appreciable contrast difference with the surrounding material, and it was otherwise impossible to use the 'find edges' tool without detecting pore edges. Five representative micrographs for each sample were analysed in this way.

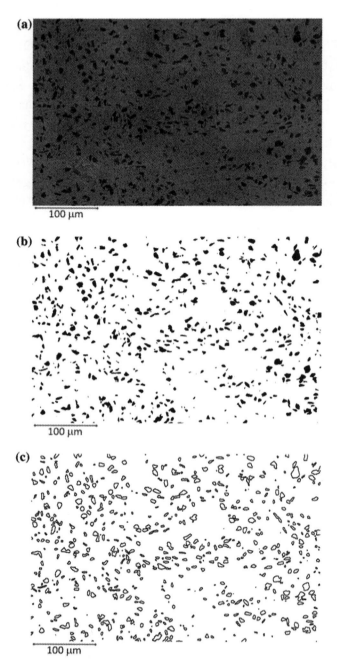

Figure 1. Images showing determination of SiC/B_4C interfacial area from a BS SEM micrograph—a original micrograph (in which B_4C is the darker phase), b B_4C features selected by thresholding shown in black, c perimeters of B_4C features with lines of finite thickness.

3. Results and Discussion

3.1. Interfacial Area

Example micrographs of the type used to measure the interfacial area for the composition ranges tested are shown in **Figure 2**. The interfacial area/unit volume measured in the composites is shown in **Figure 3** as a function of B_4C content. The graph shows that the interfacial area/unit volume between the phases is increased both by increasing the amount of B_4C and by reducing the B_4C size, thereby increasing the surface area to volume ratio. Critically, the SiC/B_4C interfacial area is comparable between materials with 10wt% 7-μm B_4C and 20wt% 70-μm B_4C and also to a lesser extent between materials with 20wt% 7-μm B_4C and 30wt% 70-μm B_4C. Hence, if the SiC/B_4C interfacial area is important to a given property, these two pairs of materials should demonstrate similar values.

3.2. Densification

The effect of B_4C content on %TD achieved in these composites after

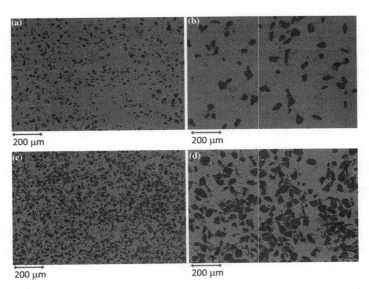

Figure 2. BS SEM micrographs showing a 10wt% 7-μmB_4C, b 10wt% 70-μmB_4C, c 30wt% 7-μm B_4C, d 30wt% 70-μmB_4C dispersed in SiC. B_4C is the darker phase in each micrograph.

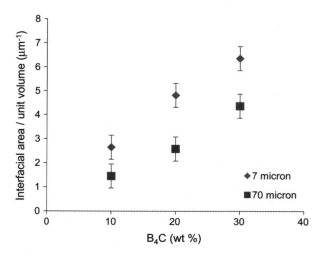

Figure 3. Effect of wt% B_4C content on interfacial area between the SiC and B_4C for samples containing two different B_4C sizes.

sintering for each B_4C size used is shown in **Figure 4**. **Figure 5** shows the same data plotted as a function of the SiC/B_4C interfacial area for these materials. This can be used to understand the observed relationships between addition level, size and %TD. In **Figure 5**, the data are compared with the standard SiC material, which is assumed to have no interfacial area/unit volume. Though the fine B_4C added as a sintering aid will generate some SiC/B_4C interfaces, the method employed to quantify the interfacial area for the coarse B_4C will not have selected the fine particles, due to the scale of the image used. Hence, in all materials measured, the fine B_4C contributes effectively nothing to the interfacial area/unit volume.

Figure 4 shows that increasing the wt% B_4C decreases the %TD achieved. Although final %TD is higher when using 70-μm B_4C, it might have been expected that the finer, 7-μm B_4C would have produced a higher degree of densification, as it had the greater curvature.

It is possible that the differing O content of the B_4C powders is affecting densification behaviour; O contamination may affect the diffusivity at the surfaces and grain boundaries of the SiC particles during sintering. However, while the 7-μm B_4C has approximately four times the O content of the 70-μm B_4C, the SiC has a far higher O content than either; 0.83wt% compared with 0.16wt% for the 7-μm B_4C. As a result of this, the total O content from the powders results mostly

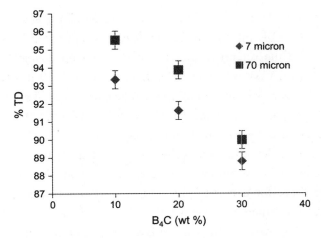

Figure 4. Effect of B$_4$C content on %TD for samples containing two different B$_4$C sizes.

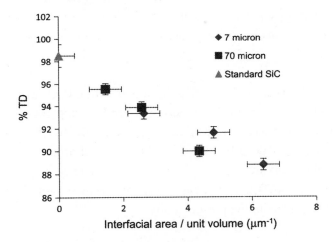

Figure 5. Effect of SiC/B$_4$C interfacial area/unit volume on %TD for samples containing two different B$_4$C sizes, compared with a standard SiC.

from SiC and does not vary much between materials. It is likely this total O content will affect the densification behaviour, since the B$_4$C is inert, with shrinkage occurring in the SiC. Finally, the four-fold increase in O content corresponds with a ten fold increase in specific surface area. Therefore, assuming that the O is found predominantly at the surface of the B$_4$C particles, the thickness of the O coating does not increase with decreasing particle size. Hence, the presence and amount of SiC/B$_4$C interfaces themselves may be a more important factor than the O content of each B_4C powder. This is supported by **Figure 5**, which shows that the %TD decreases as the SiC/B_4C interfacial area increases and this trend is constant across both B_4C sizes and the reference material containing no coarse B_4C. Therefore, samples with 7-µm B_4C show lower %TD than those with 70-µm B_4C at equal wt% addition because finer B_4C generates more interfacial area. This will reduce the number of SiC/SiC particle contacts in the material. As a result of this, less mass transport which contributes to densification occurs where SiC/B_4C interfaces exist. The driving force for densification is lowest when there is the greatest amount of SiC/B_4C interfacial area. Additionally, it seems likely that SiC/B_4C interfaces give a lower driving force for densification than SiC/SiC interfaces. This may be because of the lattice mismatch between SiC and B_4C raising the energy of these interfaces and also the low curvature of the B_4C particles reducing the driving force for densification of these particles. It should also be noted that the material with 30wt% 70-µm B_4C shows somewhat lower %TD than may be predicted from the trend observed in other materials. This may be because the numerous large B_4C grains, which do not shrink during densification, exert a more significant pressure during sintering[28].

Given the observation that with a d_{50} above ~8µm, B_4C cannot self densify[5], in these materials any B_4C/B_4C contacts will tend to trap porosity. However, it may be speculated that with a finer B_4C, which can effectively self-densify, additional driving force for densification would be created at any B_4C/B_4C contacts. Further, the higher curvature of the particles may increase the driving force for densification at the SiC/B_4C interfaces. These materials would therefore occupy a different regime to those in the present study.

3.3. Microstructure

The SiC/B_4C interfacial area can also influence grain growth in the SiC. Example SiC grain structures from different materials are shown in **Figure 6**. The large pore-like features in **Figure 6** are of a similar size to the B_4C features as shown in **Figure 2** and are therefore probably caused by the removal of B_4C. From comparison of micrographs before and after the etching process used, it appears that the etching tends to remove B_4C features, possibly because the interfaces between the coarse B_4C and the SiC matrix are weaker and so are more strongly affected by the etchant. The relationship between SiC/B_4C interfacial area and mean grain length

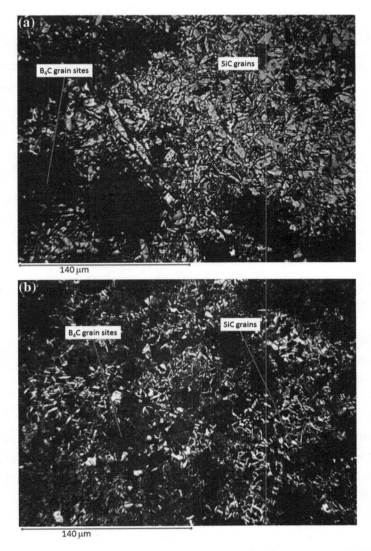

Figure 6. Reflected light micrograph of polished and etched composites with 20wt% B_4C, a 70-µm B_4C, b 7-µm B_4C.

for the materials produced for the present study is shown in **Figure 7**.

Figure 7 shows that, across both B_4C sizes, increasing the interfacial area between the phases seems to reduce SiC grain growth. With a greater SiC/B_4C interfacial area, there is a greater probability that a growing SiC grain will meet a SiC/B_4C interface and be physically stopped from growing. As with densification,

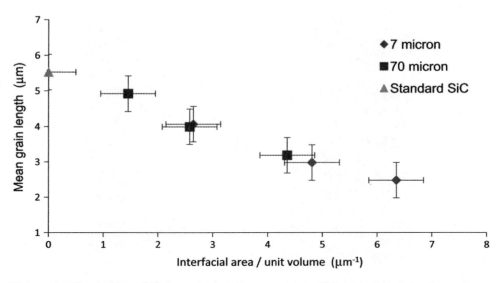

Figure 7. Effect of interfacial area/unit volume on mean SiC grain length in sintered samples with two sizes of B$_4$C, compared with standard SiC.

the O content could also affect the grain growth by altering the relative rates of different mass transport mechanisms. However, as previously noted, the total O content is likely to be important and this is dominated by the SiC, so it does not vary much.

Given that the addition of B$_4$C typically reduced the observed grain growth, it was hypothesised that it may also affect the SiC polytype composition in the sintered body. Example XRD data used to analyse the polytype composition in these materials are shown in **Figure 8**.

The effect of increasing interfacial area between the two phases on the vol% 4H formed, as determined by XRD and Rietveld refinement, for materials sintered at 2125°C, is shown in **Figure 9**. This graph shows that less 4H is observed as the interfacial area between the phases increases, which is also when there is less grain growth. As with other properties, this trend also continues to the standard SiC with no coarse B$_4$C. The error in this measurement was determined from the estimated error produced in the model. The vol% B$_4$C in the material has also been calculated in the refinement for all materials tested and compared with the known value from the starting mass ratio of the powders. This indicates that there may be a systematic error present in the refinement data, since the vol% B$_4$C in the material

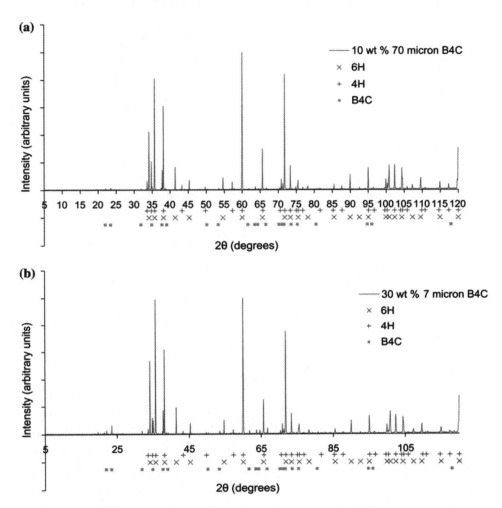

Figure 8. XRD data used in Rietveld refinement of sample materials analysed, a 10wt% 70-μm B_4C and b 30wt% 7-μm B_4C, with key peaks generated from the structure files of the two major SiC polytypes and B_4C phase used.

estimated by the refinement model is consistently lower than the known true value. This may result from the disorder within the B_4C structure causing uncertainty when fitting of the crystallographic data using existing structure models. Analysis of micrographs used to determine interfacial area in the sintered materials shows area % B_4C consistent with the theoretical volume % that should be present based on the wt% added to the powder blend, assuming the materials are isotropic. Therefore, the refinement model is definitely inaccurate for the B_4C. However, the refinements of all six materials tested are self consistent, having been analysed

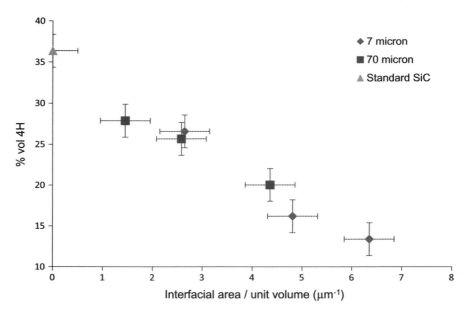

Figure 9. Effect of interfacial area as determined by the 'edge area' method on vol% 4H in sintered samples.

using the same B_4C structure file.

An increase in transformation to 4H with increased grain growth has been previously observed for LPS α-SiC[19]. From comparison with **Figure 7**, increase in vol% 4H corresponds to increasing grain length. This may be because the growth mechanism producing the larger SiC grains involves a transformation from 6H to 4H. However, the present system would normally be expected to give solid state sintering with no liquid phase present. It is possible that small volumes of a liquid phase may have formed, possibly a Si-B-C eutectic, which is believed to cause abnormal grain growth[13], or alternatively that the solid state grain growth mechanism also involves a polytype transformation. While further work would be needed to determine whether this is the operative mechanism in the present study, there does seem to be a correlation between grain growth, amount of 4H formed and the interfacial area in the material.

Concluding Remarks

This study has examined the effect of interfacial area/unit volume, a quantified microstructural parameter, on the sintering behaviour of SiC/B_4C particulate composites. The following conclusions can be reached:

- The amount of SiC/B_4C interfacial area in a composite has a controlling effect on the microstructure evolution of these materials.

- Densification is inhibited by SiC/B_4C interfaces at the sintering temperature tested.

- Increased SiC/B_4C interfacial area reduces the grain growth observed in the SiC matrix, possibly by physically impeding the growing SiC grains.

- Increased interfacial area correlates with decreased transformation of SiC from 6H to 4H during sintering, in parallel with reduced grain growth.

Acknowledgements

The Authors would like to acknowledge EPSRC and Dstl for funding this work, Morgan Advanced Materials for provision of equipment and expertise and Dr. Julia Percival for assistance with the GSAS software. This work was carried out as part of an Engineering Doctorate Programme in Micro- and Nano-Materials and Technologies at the University of Surrey, grant number EP/G037388/1.

Compliance with Ethical Standards

Conflict of interest: The authors declare that they have no conflict of interest.

Open Access: This article is distributed under the terms of the Creative Commons Attribution 4.0 International License (http://creativecommons.org/licenses/by/4.0/), which permits unrestricted use, distribution, and reproduction in any medium, provided you give appropriate credit to the original author(s) and the source, provide a link to the Creative Commons license, and indicate if changes were made.

Source: Williams T, Yeomans J, Smith P, *et al*. Effect of interfacial area on densification and microstructural evolution in silicon Carbide-Boron carbide particulate composites[J]. Journal of Materials Science, 2016, 51(1):353–361.

References

[1] Kuzenkova MA, Kislyi PS, Grabchuk BL *et al* (1979) The structure and properties of sintered boron carbide. J Less Common Met 67:217–223.

[2] Karandikar PG, Wong S, Evans G *et al* (2010) Microstructural development and phase changes in reaction bonded boron carbide. Ceram Eng Sci Proc 31:251–259.

[3] Pittari J III, Subhash G, Trachet A *et al* (2015) The rate-dependent response of pressureless-sintered and reaction-bonded silicon carbide-based ceramics. Int J Appl Ceram Technol 12:207–216.

[4] Vandeperre LJ, Teo JH (2014) Pressureless sintering of SiC-B_4C composites. Ceram Eng Sci Proc 34:101.

[5] Schwetz KA, Grellner W (1981) The influence of carbon on the microstructure and mechanical properties of sintered boron carbide. J Less Common Met 82:37–47.

[6] Padture NP (1994) In situ-toughened silicon carbide. J Am Ceram Soc 77:519–523.

[7] Stobierski L, Gubernat A (2003) Sintering of silicon carbide I. Effect of carbon. Ceram Int 29:287–292.

[8] Wereszczak AA, Lin H-, Gilde GA (2006) The effect of grain growth on hardness in hot-pressed silicon carbides. J Mater Sci 41:4996–5000. doi:10.1007/s10853-006-0110-z.

[9] Hamminger R (1989) Carbon inclusions in sintered silicon carbide. J Am Ceram Soc 72:1741–1744.

[10] Jung CH, Kim CH (1991) Sintering and characterization of Al_2O_3–B_4C composites. J Mater Sci 26:5037–5040. doi:10.1007/BF00549888.

[11] Lin X, Ownby PD (2000) Pressureless sintering of B4C whisker reinforced Al_2O_3 matrix composites. J Mater Sci 35:411–418. doi:10.1023/A:1004715300441.

[12] Górny G, Rączka M, Stobierski L *et al* (1997) Microstructure-property relationship in B4C–bSiC materials. Solid State Ion 101–103:953–958.

[13] Stobierski L, Gubernat A (2003) Sintering of silicon carbide II. Effect of boron. Ceram Int 29:355–361.

[14] Shaffer PTB (1970) Solubility of boron in alpha silicon carbide. Mater Res Bull 5:519–521.

[15] Magnani G, Beltrami G, Minoccari GL et al (2001) Pressureless sintering and properties of alphaSiC-B_4C composite. J Eur Ceram Soc 21:633–638.

[16] Shih CJ, Meyers MA, Nesterenko VF et al (2000) Damage evolution in dynamic deformation of silicon carbide. Acta Mater 48:2399–2420.

[17] Groth B, Haber R, Mann A (2014) Raman micro-spectroscopy of polytype and structural changes in 6H-silicon carbide due to machining. Int J Appl Ceram Technol 12:795–804.

[18] Sigl LS, Kleebe HJ (1993) Core/rim structure of liquid-phase-sintered silicon carbide. J Am Ceram Soc 76:773–776.

[19] Zhou Y, Tanaka H, Otani S et al (1999) Low-temperature pressureless sintering of alpha-SiC with Al_4C_3–B_4C–C additions. J Am Ceram Soc 82:1959–1964.

[20] Tanaka H, Zhou Y (1999) Low temperature sintering and elongated grain growth of 6H-SiC powder with AlB_2 and C additives. J Mater Res 14:518–522.

[21] Tanaka H, Yoshimura HN, Otani S et al (2000) Influence of silica and aluminum contents on sintering of and grain growth in 6H-SiC powders. J Am Ceram Soc 83:226–228.

[22] Tanaka H, Hirosaki N, Nishimura T et al (2003) Nonequiaxial grain growth and polytype transformation of sintered a-silicon carbide and b-silicon carbide. J Am Ceram Soc 86:2222–2224.

[23] Lee SK, Kim CH (1994) Effects of a-SiC versus b-SiC starting powders on microstructure and fracture toughness of SiC sintered with Al_2O_3–Y_2O_3 additives. J Am Ceram Soc 77:1655–1658.

[24] Gurland J (1967) A study of contact and contiguity of dispersions in opaque samples. In: Elias H (ed) Stereology. Springer, Berlin, pp 250–251.

[25] Morosin B, Mullendore AW, Emin D et al (1986) Rhombohedral crystal structure of compounds containing boron-rich icosahedra. AIP Conf Proc 86:70–86.

[26] Bind JM (1978) Phase transformation during hot-pressing of cubic SiC. Mater Res Bull 13:91–96.

[27] Capitani GC, Di Pierro S, Tempesta G (2007) The 6H-(SiC) structure model: further refinement from SCXRD data from a terrestrial moissanite. Am Miner 92:403–407.

[28] Bordia RK, Scherer GW (1988) On constrained sintering-III. Rigid inclusions. Acta Metall 36:2411–2416.

Chapter 11

Electrospun Green Fibres from Lignin and Chitosan: A Novel Polycomplexation Process for the Production of Lignin-Based Fibres

Makoto Schreiber[1,2], **Singaravelu Vivekanandhan**[1,3,4], **Peter Cooke**[5], **Amar Kumar Mohanty**[1,3], **Manjusri Misra**[1,3]

[1]Bioproducts Discovery and Development Centre, Department of Plant Agriculture, Crop Science Building, University of Guelph, Guelph, ON N1G 2W1, Canada
[2]Department of Physics, University of Guelph, Guelph, ON N1G 2W1, Canada
[3]School of Engineering, Thornborough Building, University of Guelph, Guelph, ON N1G 2W1, Canada
[4]Department of Physics, VHNSN College, Virudhunagar 626001, Tamil Nadu, India
[5]Core University Research Resources Laboratory, New Mexico State University, Las Cruces, NM 88003, USA

Abstract: Novel lignin-chitosan polyelectrolyte fibres were produced through a reactive electrospinning process. Poly-electrolyte formation between the anionic lignin and cationic chitosan was controlled through the pH of the solution. Through manipulating the polyelectrolyte complex formation, fibres could be effectively produced from two biopolymers, which are normally very difficult to electrospin on their own. Though minimal amounts of the petroleum-derived polyethylene oxide were introduced into the solution to enhance the spinnability of the polyelectrolyte solution, it could be easily removed from the fibres post spinning by washing with water. Thus, pure biopolymer fibres could be produced. The optimum composition of lignin to chitosan was identified through SEM, FTIR and TGA analysis of the electrospun fibres. Fluorescence spectra of the electrospun fibres reveal the homogeneous distribution of lignin and chitosan components throughout the fibre network.

1. Introduction

Carbon fibres have been studied extensively for their scientific and technological importance and are finding application in many fields including catalysis[1], composites[2]–[4], filtration[5]–[7], and alternative energy technologies[8]. Carbon fibres are advantageous due to being a light weight material with a high specific modulus and strength, good thermal/electrical conductivity and being structurally stable against various environmental changes[2][9]. In recent years, the demand for carbon fibres has steadily increased and it is expected to continue to rise in the following years[10]. The highest demands for carbon fibres are in the automobile and aerospace industries as their light weight allows for improved fuel efficiencies[11]. However, carbon fibres also have potential applications in other areas such as electronics, which do not require as high of a quality of the fibres as in the aerospace industry. A recent advancement in carbon fibre technology is their nanofabrication. Various functional properties of the carbon fibres can be improved by reducing their diameter down to nanometer scale dimensions. The main challenge with establishing wider carbon fibre technologies is their higher cost; traditional carbon fibres are sourced from petroleum resource-based feedstocks, which tend to continually increase in price. Currently, the main precursor material used in carbon fibre production is the synthetic/petroleum-based polymer polyacrylonitrile (PAN)[9]. Some carbon fibres are also produced from petroleum pitch and rayon[9]. Depending on the precursor material used, the properties of the carbon fibre can vary. For example, pitch-based carbon fibres have greater thermal conductivity than rayon-based carbon fibres, and PAN-based carbon fibres possess higher compressive and tensile strengths compared to pitch-based carbon fibres[5]. In addition to the precursor material, the processing method is also very important in fabricating carbon fibres with desired properties.

One method of producing carbon fibre precursors, with the potential of commercial applicability, is electrospinning. It has previously been demonstrated that electrospinning can successfully produce precursor fibres that can be converted into high quality carbon fibres with controlled fibre diameters and morphologies[12]–[19]. The majority of electrospun carbon fibre precursors reported in the literature are PAN-based. The high cost of PAN[18], depleting petroleum resources and the toxicity of its solvent, dimethylformamide (DMF)[20], has motivated research to look into alternative electrospinnable materials to produce cheaper and more environmentally friendly carbon fibres. The fact that petroleum-based carbon resources exhibit negative environmental impacts and are of limited availability further motivates research towards green carbon fibres[21].

Recently, a wide range of renewable resource-based materials have been investigated for the fabrication of carbon materials. Among them, lignin has been looked at as a very promising candidate as a precursor for carbon materials due to its (i) carbon-rich phenolic chemical structure, (ii) chemical compatibility and (iii) abundance in nature. Lignin is a constituent of most plant materials and along with cellulose and hemicellulose, forms their structural basis[22]. The structure of lignin is very complex and unlike most polymers, it is a three-dimensional branched structure[23]. Different species of plants will contain lignin composed of different ratios of its three basic monomeric units: p-coumaryl alcohol, coniferyl alchohol and sinapyl alcohol[24][25]. The greatest variation in the structure of lignin occurs as a result of the process of extracting lignin from the plant material, which may even add different functionalities to the lignin[24].

In the papermaking and cellulosic ethanol industries, lignin must be removed from the raw materials in a process known as delignification[26]. Due to this, lignin is massively produced as a co-product of these two industries. It is estimated that from the papermaking industry alone, 70 million tons are produced annually[26]. For the most part, this lignin is used as a low-energy density fuel by the factories[27] and 1%–2% of it is used for speciality products[24]. Thus, lignin is abundantly available and cheap. To establish lignin as an acceptable precursor material, it must be readily accessible from different sources. However, due to the variations in the properties and spinnability of the lignins based on their source and extraction methods, it is important to demonstrate the spinnability of a wide range of lignins, which may have different characteristics such as solubility and charge.

One of the main challenges associated with the production of lignin fibres through the electrospinning process are its low viscoelastic properties. A variety of work has demonstrated the benefits of blending lignin with small amounts of synthetic polymers such as polyethylene oxide (PEO) to improve the spinnability[28]–[31]. Recently, some work has emerged on the successful electrospinning of a variety of lignin-based fibres by utilising PEO[32]. Lignin has also been used to supplement PAN for carbon fibre precursor fibre production[19]. Pure lignin electrospinning has also been demonstrated through the use of co-axial electrospinning in which two concentric capillaries are used; the inner capillary containing the lignin solution while ethanol flowed from the outer capillary to stabilize the lignin jet[33][34]. Different types of lignin are soluble in different solvents such as DMF[19][32], ethanol[33][34] and water[32]. Water would be the most environmentally friendly solvent to use due to its abundance and non-toxicity. However, the majority of lignin is water insoluble; the most well known water-soluble lignins being sulphur lignin (SL) and lignosulfunate (LS)[32]. In our previous study[35], we have used a water-soluble and sulphur-free anionic sodium carbonate lignin (SCL), which is obtained through a soda pulping process. In this study, PEO was used in order to enhance the spinnability of sodium carbonate lignin. However, the strong anionic nature of the SCL hindered it from continuous electrospinning.

In this paper, we attempt to neutralise the inherent negative charge of the lignin by combining SCL with another cationic polymer. Chitosan (CS), an abundant biopolymer derived from the naturally occurring chitin, is a cationic polymer and thus may be able to neutralise the anionic charges on the SCL. In addition, CS is soluble in acidic aqueous solutions, has been found to produce small diameter fibres with low levels of beading defects[36][37] and has been demonstrated to be carbonizable[38]–[40]. Therefore, it would maintain the green nature and the desired morphology of a carbon fibre precursor. CS itself has been shown to be a difficult material to electrospin due to difficulties in creating solutions with a high enough polymer concentration and chain entanglement while still having a low enough viscosity to electrospin[41][42]. However, as with lignin, chitosan has been successfully electrospun through blending with other easily electrospun polymers such as PEO[36][37][43][4].

When a polyanion and polycation interact, it is well known that a structure known as a polyelectrolyte complex (PEC) can form. PECs are polymeric structures in which the constituent polyions are bound together by ionic link-ages[45]. In general, PECs formed between two strong polyions will have a 1:1 ionic stoichiometry ratio[46]. However, if the constituent polyions are mixed in a non 1:1 ionic stoichiometry, the resultant PECs can be strongly overcharged rather than neutrally charged[46]. Thus, in the present study, we take advantage of the PEC forming ability of SCL and CS in order to improve their electrospinnability while reducing the PEO content. SCL, CS and PEO are blended in aqueous solutions of acetic acid and subsequently electrospun.

2. Experimental

2.1. Materials

SCL was purchased from Northway Lignin Chemical as Polybind 300 (liquid). Before use, the SCL was dried in a 100°C laboratory oven until dry. The dry weight of the lignin was ~50% of the as-received wet lignin weight. Medium molecular weight CS (viscosity average Mw 190−310kDa, 75%−85% deacetylation) and 5,000,000 Mw PEO were purchased from Sigma-Aldrich and used as-received without further purification. Acetic acid (C99.5% pure) was purchased from Acros Organics. Deionized water was used as the primary solvent.

2.2. Method

2.2.1. Solution Preparation

Various electrospinning solutions were prepared from a blend of SCL, CS and PEO. All polymer concentrations are reported as w/v % (g/ml). The solutions were prepared with 0.6% PEO, 1.5% chitosan and 1.5%, 2.0%, 2.5%, or 3.0% SCL. Initially, CS and PEO were dissolved in acetic acid and deionized water, while SCL was dissolved in deionized water. The two solutions were then mixed under constant stirring conditions to achieve the desired polymer concentrations in a 40(v/v) % acetic acid solution. The solutions are labelled as 1.5, 2.0, 2.5 and 3.0L based on their respective lignin content.

2.2.2. Electrospinning

Electrospinning was performed in a NANON-01A electro-spinning setup, MECC Co., Ltd. Japan. A custom-made plate collector, which allowed for a variation in the height of the collector (the collector, which came with the machine had a fixed height), was employed to achieve a greater working distance. Solutions were electrospun using a 24 gauge needle, 14kV applied voltage, 0.1ml/h flow rate and a 22.5cm working distance in ambient conditions.

2.2.3. Characterization

Solution viscosities were measured using a Brookfield DV-II + Pro viscometer at room temperature with a #4 spindle at 100rpm. Each measurement had an inherent 1% error, and three independently prepared solutions were measured to calculate an average and standard deviation. Solution surface tensions were measured using the Du Nuoy ring method with the ring attached to a microbalance at room temperature where the ring was slowly pulled out of the solution, and the surface tension being recorded just before the ring broke contact with the solution surface. Each measurement had an inherent 5% error, and three independently prepared solutions were measured to calculate an average and standard deviation. The resulting fibre morphologies were analysed using an FEI-Inspect S50 scanning electron microscope (SEM) operated at 15kV. The fibres were sputter-coated with ~20nm of gold to make them conductive. The obtained SEM images were further analysed using the ImageJ software to measure the fibre diameters using over 100 measurements for each formulation to obtain an average and standard deviation. A model TCS SP5 II laser scanning confocal microscope (LCM) (Leica Microsystems, Exton, PA) was used in the 'xyλ' mode to obtain fluorescence emission spectra of the resulting fibres and the constituent materials as well as to resolve the spatial distribution of the constituent materials within the fibres. Thermal characteristics of the fibres were measured using a TA instruments thermogravimetric analyser (TGA) Q500. TGA thermograms were obtained from 20 to 800°C at a ramp rate of 10°C/min. Heating took place in a nitrogen atmosphere injected at a flow rate of 15ml/min. Structural characteristics of the fibres and constituent polymers were investigated using a Thermo Scientific Nicolet 6700 Fourier transform infrared (FTIR) spectrometer with a GladiATR single reflection ATR accessory operated between 400 and 4000cm^{-1}. The spectra were collected at a resolution of 4cm^{-1} with 32 scans per sample at room temperature. Elemental analysis was performed on the fibres using a model S3400-N SEM (Hitachi Hitachi-Technologies, Pleasanton, CA) coupled to a Noran System 6 energy dispersive x-ray spectrometer (Thermo Electron Corp., Madison, WI). To confirm that the resultant fibres were composed of PECs, the fibres were soaked in water for at least a day and analysed through SEM, LCM, EDS, FTIR and TGA.

3. Result and Discussion

3.1. Polyelectrolyte Complex of Sodium Carbonate Lignin and Chitosan

Based on the polyionic nature of the SCL and CS, it was expected that they would form PECs with each other (**Figure 1**). It was undesirable for them to form PECs in solution as it would cause clogging of the needle. It was observed that above an acetic acid concentration of 40% in water, no precipitates formed in solution. This indicates that the pKa of the anionic groups on the SCL is about 4.3, below, which the carboxyl groups on the SCL are protonated and thus unable to form PECs. It was expected that during the electrospinning process, as solvent is evaporated from the solution, the pH rises above the pKa of the carboxyl groups on the SCL, thus allowing for PEC formation to occur.

3.2. Solution Properties and Their Electrospinning Behaviour

The properties of the various prepared solutions are shown in **Figure 2**. It was found that the viscosity of the solutions dropped dramatically with increasing SCL content until a saturation point was reached. This is in agreement with our previous report[35] and is due to the low intrinsic viscosity of lignin[47]. Although the viscosity decreased dramatically, the high viscosity imparted by the chitosan still allowed for the solutions to be electrospinnable. The surface tension of the solutions was found to remain constant across the different formulations. This was

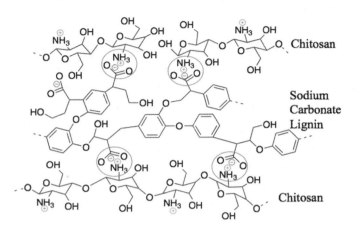

Figure 1. Schematic of polyelectrolyte complexation between sodium carbonate lignin and chitosan. (Circled regions indicate the ionic bonds.)

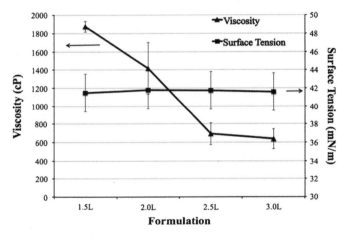

Figure 2. Viscosity and surface tension properties of the various prepared solutions.

expected based on the report by Geng et al.[48], which showed that in chitosan-based solutions, surface tension was greatly affected by the acetic acid concentration but not by polymer concentration.

All solutions were successfully electrospun using the applied electrospinning parameters. However, only the electrospinning jet of the 2.0L solution formed a stable region (region where the jet underwent no instabilities) whilst the 1.5, 2.5 and 3.0L solutions all had no stable region. In addition, of all the formulations, only the 2.0L solution deposited as a flat, dense fibre mat. The other formulations formed floating fibres (fibres that rise off from the plate collector towards the capillary caused by the deposited fibres having a high net charge density so that they repel each other and are attracted towards the applied voltage). Floating fibres are undesirable as they reduce the working distance available in which the electrospinning jet can elongate and solidify and also change the electric field. These effects cause the deposition of fibres with uneven fibre distributions, more beading defects and fibre bundling. The 1.5L and 2.5L fibres deposited as a loose, cotton-like mat. The morphology of the 1.5L mat was better than that of the 2.5L mat as the 2.5L mat formed some fibre bundles. The 3.0L mat was composed predominantly of fibre bundles. The 2.0L formulation (4:3 SCL:CS ratio) produced the best fibre mats and thus likely contained an almost stoichiometric charge ratio of SCL and CS functional groups. It may be expected that the use of CS with a higher degree of deacetylation would allow for optimised fibres with higher SCL:CS ratios to be produced due to the greater number of charged groups per CS molecule.

3.3. Characterization of Electrospun Fibres

In order to confirm the PEC nature of the electrospun fibres, the fibre mats were soaked in water for at least a day. This was to test if the fibre structure was retained after soaking as the predominant component, SCL, is water soluble. As PEO does not take place in PEC formation and is water soluble, it is also expected that PEO can be removed from the fibres through the water-soaking process. The fibres before and after being soaked in water were characterised for their morphological (SEM), chemical (EDS), thermal (TGA), spectroscopic (fluorescence) and structural (FTIR) properties.

3.3.1. SEM/EDS Characterization

SEM micrographs of the four different fibre mats produced are shown in **Figure 3**. The 2.0L fibres were found to possess the best fibre morphology out of the four samples. In the 2.0L fibres, beading was not observed while the 1.5L had some, the 2.5L more, and the 3.0L fibres were composed almost entirely of beads. The diameters of the 2.0L fibres were also much smaller and more uniform than the other fibres at 253nm ± 77nm. The 1.5L fibres possessed diameters of 1.92µm ± 0.87µm.

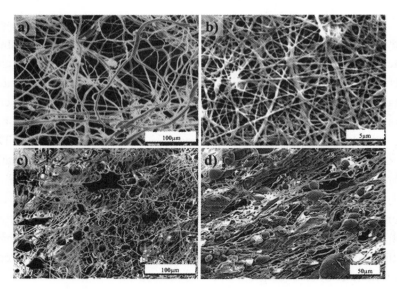

Figure 3. SEM images of a 1.5L, b 2.0L, c 2.5L and d 3.0L electrospun fibres.

The diameters of the 2.5L fibres were much too varied and the 3.0L fibres were too beaded to measure. The morphology of the 2.0L fibres was not perfect and showed some fusing between fibres indicating that the solvent did not completely evaporate upon deposition. However, for the desired application as carbon fibres, fused fibres may be beneficial as they provide greater electrical contact between the fibres and thus a greater electrical conductivity throughout the fibre mat. Thus, based on both the macroscale fibre mat morphologies and microscale fibre morphologies, the 2.0L formulation was optimal and was the focus of subsequent characterisations.

Figure 4(a) and **Figure 4(b)** shows the SEM micrographs of the 2.0L fibres after being soaked in water (2.0L-W). The fibres can be observed to have retained their structure and be very similar in morphology to the as-spun 2.0L fibres with diameters of 248nm ± 56nm. Thus, the fibres appear stable in water; supporting their PEC composition. **Figure 4(c)** shows the EDS spectra of the 2.0L and 2.0L-W fibres. The N peak of chitosan did not show up in the EDS spectra due to the N x-ray fluorescence being quenched by O. It is clearly observed that while the samples contained sodium before being soaked in water, after soaking, there was no visible peak due to sodium (originating from the SCL). As is well established of PECs, upon forming, the couterions originally associated with the constituent polyions are excluded. Thus, through soaking the fibres in water, the free sodium

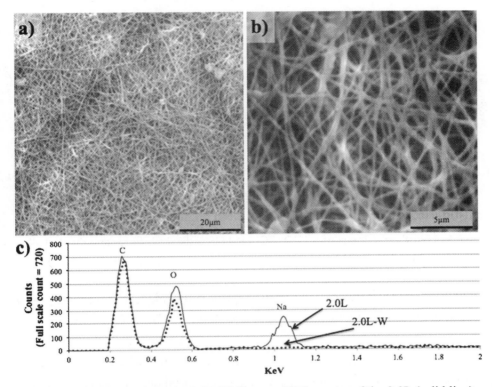

Figure 4. a,b SEM images of the 2.0L-W fibres. c EDS spectra of the 2.0L (solid line) and 2.0L-W (dotted line) fibres.

ions could easily be washed out of the fibres. Removal of these sodium ions would lead to a more pure carbon fibre.

3.3.2. Thermal Properties

The TGA thermograms of the PEO, SCL and CS are shown in **Figure 5(a)** and the 2.0L and 2.0L-W fibres are shown in **Figure 5(b)**. The PEO and CS exhibit one-step degradations with the degradation of PEO being very rapid upon onset and leaving virtually no residue while the degradation of CS is more gradual and leaves some residue. SCL exhibits a three-step degradation over a wide temperature range, which is common of lignin due to differences in the thermal stability of oxygen-containing functional groups[49]. The first degradation is likely due to cleavage of the functional groups, while the second is likely due to rearrangement of the phenolic backbone[49]. The third degradation is not usually observed in

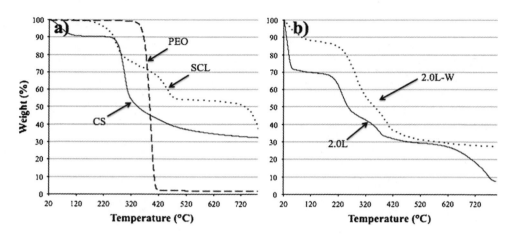

Figure 5. TGA thermograms of a PEO, SCL and CS b the 2.0L fibres and the 2.0L-W fibres.

lignins and may be due to volatilization of the sodium counterions present in SCL. The first degradation of SCL and CS are very close.

The 2.0L fibres, as well as the other fibres (not shown), exhibit a three-step degradation. The first degradation peak matches the first degradation peaks of SCL and CS. The second corresponds to that of PEO and the third to the third degradation of SCL. In comparing the degradation curve of the 2.0L-W fibres to the 2.0L fibres, it can be seen that the second degradation peak is greatly diminished while the third peak disappears. The diminishment of the second peak supports the removal of PEO from the fibres with water soaking. As the PEO used was of an extremely high molecular weight, it is expected that the smaller fractions diffused out of the fibres first and the larger fractions may require more time to diffuse out completely. The removal of PEO from the fibres is desirable as PEO has a low melting point (66°C–70°C), which makes it an undesirable component in the thermal stabilization and carbonization process. The disappearance of the third peak supports that the peak was caused by the sodium ions as the EDS data revealed large amounts of sodium to be removed from the fibres upon water soaking. Though EDS only characterised the surface of the fibres, the disappearance of this TG peak indicates that the sodium was also removed from the bulk as well. The first peak also appears to have been slightly shifted to a higher onset of degradation indicating some enhanced thermal stability.

3.3.3. Fluorescence Characterization

Lignin, due to its aromatic structure is well known to fluoresce; the fluorescence most likely arising from groups such as stilbene, biphenyl and phenylcoumarin[50]. In their pure forms, CS and PEO do not fluoresce. However, fluorescent impurities can exist in unpurified samples, which give them fluorescent peaks. The fluorescence emission spectra of PEO, SCL, CS, the 2.0L fibres and the 2.0L-W fibres are shown in **Figure 6(a)**. Both CS and PEO share a peak emission at the wavelength of 544nm; indicating the presence of a similar impurity. SCL has a broad peak emission between 583–598nm, a shoulder at 558nm and another slight shoulder at 647nm. The 2.0L fibres before and after being soaked in water possess nearly identical emission spectra. The first peak occurs at 549nm and is slightly

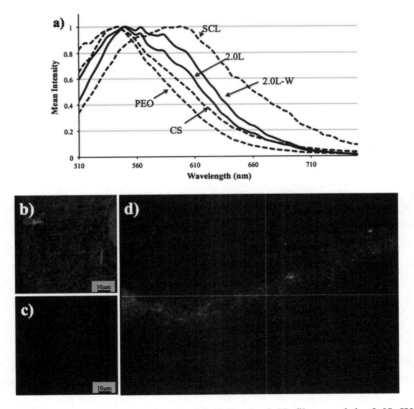

Figure 6. A Fluorescence spectra of PEO, CS, SCL, the 2.0L fibres and the 2.0L-W fibres. Topographical fluorescence microscopy images of the 2.0L fibre mat imaged between b 570–620nm and c 520–560nm. d Crosssectional fluorescence microscopy image of the 2.0L fibre mat.

shifted compared to the emission due to CS and PEO. The remaining emission peaks on the fibres are due to lignin. The peak at 563nm matches one shoulder from SCL and is quite prominent. A peak and a shoulder appear at 583 and 598nm, respectively, and arise from the extremes of the broad peak on SCL. The final shoulder occurs at 661nm and is greatly shifted from the slight peak on lignin at 647nm.

The slight and large red-shift in the peaks at 549 and 647nm, respectively, in the spectra of the 2.0L fibres compared to the spectra of the constituent materials is indicative of some conformational changes occurring in the polymers upon PEC formation and electrospinning. With a conformational change, the degree of p-p interactions between aromatic rings on the SCL can change[51]—a red-shift indicating closer packing of the polymers[52]. Thus, upon PEC formation between SCL and CS during the electrospinning process, the constituent polymers could be thought to pack much closer together than in their raw states. The linear structures of PEO and CS prevent much changes to their packing, and thus no shifts are visible in peaks corresponding to the fluorescence of either PEO or CS. SCL on the other hand is branched and through PEC formation, it would be expected to align somewhat with the linear CS chains, which is supported by the the large shift in one of the SCL peaks.

The fibres were imaged through confocal fluorescent microscopy in order to investigate the distribution of the component polymers throughout the fibres [**Figures 6(b)-(d)**]. Two fluorescence channels were used: one channel operated between 520−560nm (coloured red) and another channel operated between 570−620nm (coloured green) corresponding to the peak fluorescence emissions of CS and SCL, respectively. Topographical images of the fibre imaged using the SCL and CS peak emissions are shown in **Figure 6(b)** and **Figure 6(c)**, respectively. From these images, it can be observed that the two components are evenly distributed throughout the fibres. However, a cross-sectional image of a fibre mat [**Figure 6(d)**] reveals some inhomogeneity in the polymer distribution through the fibre mat. The bottom of the fibre mat contains more CS while the top more SCL. This inhomogeneity may arise from the charges applied during the electrospinning process. Initially, the collector is negatively charged with respect to the needle, causing the positively charged CS to be more attracted to it then SCL, resulting in the deposition of PECs with a higher CS to SCL ratio. As the fibres continue to be deposited, an insulating layer develops over the collector; reducing the voltage bias and allowing more SCL to compose the deposited PEC fibres. Thus, during the deposition of the fibres, the PEC composition may vary slightly.

3.3.4. FTIR Characterization

The FTIR spectra of PEO, CS, SCL and the 2.0L and 2.0L-W fibres are shown in **Figure 7**. It can be observed that the spectrum of the 2.0L fibres is quite close to that of the SCL. No noticeable variation was observed between the FTIR spectra of the electrospun fibre mats produced from the various solutions (data not shown); as would be expected from simple variations in polymer composition. The effects of the PEO signal are minimal in the fibre spectrum as PEO is the smallest constituent of the fibre. The CS curve is also similar to and not as intense as the SCL signals and so are not very distinct in the 2.0L spectrum. Comparing the SCL curve to the 2.0L curve, some slight shifting is observed from 1570 and 1410cm^{-1} in SCL to 1540 and 1400cm^{-1} in the fibres, respectively (C=O stretching of carbonyl groups, circled in **Figure 7**). Thus, there is some change in the structure of SCL during PEC formation as expected.

A surprising result was the dramatic change in the spectra between the 2.0L and 2.0L-W fibres. The two peaks due to the C=O stretching of carbonyl groups

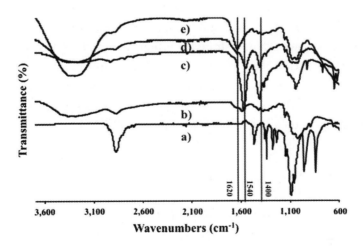

Figure 7. FTIR spectra of a PEO, b CS, c SCL, d 2.0L fibres and e 2.0L-W fibres.

disappear at 1540 and 1400cm^{-1} and a new peak, unobserved in any of the constituent materials, appears at 1620cm^{-1}. Further studies will be required to elucidate the exact mechanism of this change but it can be speculated that the removal of the associated counterions and PEO from the fibres upon water soaking allowed the ionic PEC bonds to become much stronger. Thus, the stretching of the carbonyl groups could be largely extinguished, while the stretching of the aromatic diene bonds become much more prominent.

Thus, based on the increased thermal stability as seen from the TGA studies and the significant structural changes as seen from FTIR studies, it was revealed that soaking the PEC fibres in water significantly enhanced the stability of the SCL-CS-PEC structures.

4. Conclusion

Polyelectrolyte complex (PEC) fibres of SCL and CS were successfully produced by manipulating the rapid solvent evaporation that takes place during the electrospinning process to change the solution pH and induce PEC formation as the fibres are being formed. SCL-CS-PEO blend solutions with varying SCL contents were electrospun. The solution containing a 4:3 SCL:CS ratio (2.0L solution) produced the best fibre morphologies due to the charges on SCL and CS being stoichiometrically balanced. Through soaking the fibres in water, the PEC composition of the fibres was confirmed and the ability to remove PEO and sodium ions present in the fibres was demonstrated. Further, it was revealed that soaking the fibres in water increased the strength of the PEC bonds leading to an enhanced thermal stability. Thus, a novel method of producing pure-biopolymer electrospun fibres with green solvents was demonstrated. Future works are required to explore the thermal stabilization and carbonization of these PEC fibres.

Acknowledgments

The financial support from the Natural Sciences and Engineering Research Council (NSERC), Canada for the Discovery grant individual to Manjusri Misra; NSERC NCE AUTO21; the Ontario Ministry of Agriculture, Food, and Rural Affairs (OMAFRA) New Directions and Alternative Renewable Fuels research program; and the Ontario Research Fund (ORF) Research Excellence (RE) Round-4 from the Ontario Ministry of Economic Development and Innovation (MEDI) to carry out this research is gratefully acknowledged. The Department of Physics is acknowledged for use of its SEM facilities. Some of the microscopic imaging was done with equipment supported by the NSF through an MRI Award # 0959817.

Open Access: This article is distributed under the terms of the Creative Commons Attribution License which permits any use, distribution, and reproduction in any medium, provided the original author(s) and the source are credited.

Source: Schreiber M, Vivekanandhan S, Cooke P, *et al*. Electrospun green fibres from lignin and chitosan: a novel polycomplexation process for the production of lignin-based fibres[J]. Journal of Materials Science, 2014, 49(23):7949−7958.

References

[1] Serp P (2003) Carbon nanotubes and nanofibers in catalysis. Appl Catal A 253:337.

[2] Chand S (2000) Carbon fibres for composites. J Mater Sci 35:1303.

[3] Park SB (1991) Experimental study on the engineering properties of carbon fiber reinforced cement composites. Cem Concr Res 21:589.

[4] Matsuo T (2008) Fibre materials for advanced technical textiles. Text Prog 40:87

[5] Thiruvenkatachari R, Su S, An H, Yu XX (2009) Post combustion CO_2 capture by carbon fibre monolithic adsorbents. Prog Energy Combust Sci 35:438.

[6] Heijman SGJ, Hopman R (1999) Activated carbon filtration in drinking water production: Model prediction and new concepts. Colloids Surf A 151:303.

[7] Martín-Gullón I, Font R (2001) Dynamic pesticide removal with activated carbon fibers. Water Res 35:516.

[8] Suarez-Garcia F, Vilaplana-Ortegoa E, Kunowskya M, Kimurab M, Oyac A, Linares-Solano A (2009) Activation of polymer blend carbon nanofibres by alkaline hydroxides and their hydrogen storage performances. Int J Hydrogen Energy 34:9141.

[9] Huang X (2009) Fabrication and properties of carbon fibers. Materials 2:2369.

[10] Roberts T (2006) The carbon fibre industry: Global strategic market evaluation 2006-2010. Materials Technology Publication.

[11] ORNL (2000) ORNL Review: Carbon-fiber composites for cars, vol 33. National

Laboratory Review, Oak Ridge.

[12] Liu CK, Lai K, Liu W, Yao M, Sun RJ (2009) Preparation of carbon nanofibres through electrospinning and thermal treatment. Polym Int 58:1341.

[13] Pashaloo F, Bazgir S, Tamizifar M, Faghihisani M, Zakerifar S (2009) Preparation and characterization of carbon nanofibers via electrospun PAN nanofibers. Textile Science and Technology Journal 3:1.

[14] Moon S, Farris RJ (2009) Strong electrospun nanometer-diameter polyacrylonitrile carbon fiber yarns. Carbon 47:2829.

[15] Zhou Z, Lai C, Zhang L, Qian Y, Hou H, Reneker DH, Fong H (2009) Development of carbon nanofibers from aligned electro-spun polyacrylonitrile nanofiber bundles and characterization of their microstructural, electrical, and mechanical properties. Polymer 50:2999.

[16] Zhang L, Hsieh YL (2009) Carbon nanofibers with nanoporosity and hollow channels from binary polyacrylonitrile systems. Eur Polym J 45:47.

[17] Kim C, Jeong YI, Ngoc BTN, Yang KS, Kojima M, Kim YA, Endo M, Lee JW (2007) Synthesis and characterization of porous carbon nanofibers with hollow cores through the thermal treatment of electrospun copolymeric nanofiber webs. Small 3:91.

[18] Inagaki M, Yang Y, Kang F (2012) Carbon nanofibers prepared via electrospinning. Adv Mater 24:2547.

[19] Seo DK, Jeun JP, Bin Kim H, Kang PH (2011) Preparation and characterization of the carbon nanofiber mat produced from electrospun pan/lignin precursors by electron beam irradiation. Rev Adv Mater Sci 28:31.

[20] Kennedy GJ, Sherman H (1986) Acute and subchronic toxicity of dimethylformamide and dimethylacetamide following various routes of administration. Drug Chem Toxicol 9:147.

[21] Hill J, Nelson E, Tilman D, Polasky S, Tiffany D (2006) Environmental, economic, and energetic costs and benefits of bio-diesel and ethanol biofuels. PNAS 103:11206.

[22] Pouteau C, Baumberger S, Cathala B, Dole P (2004) Lignin–polymer blends: evaluation of compatibility by image analysis. CR Biol 327:935.

[23] Faulon JL, Hatcher PG (1994) Is there any order in the structure of lignin? Energy Fuels 8:402.

[24] Lora JH, Glasser WG (2002) Recent industrial applications of lignin: a sustainable alternative to nonrenewable materials. J Polym Environ 10:39.

[25] Whetten RW, MacKay JJ, Sederoff RR (1998) Recent advances in understanding lignin biosynthesis. Annu Rev Plant Physiol Plant Mol Biol 49:585.

[26] Satheesh Kumar MN, Mohanty AK, Erickson L, Misra M (2009) Lignin and its ap-

plications with polymers. J Biobased Mat Bioenergy 3:1.

[27] Luo J (2010) Lignin-based carbon fiber. MSc thesis. The University of Maine.

[28] Kubo S, Kadla JF (2004) Poly(ethylene oxide)/organosolv lignin blends: Relationship between thermal properties, chemical structure, and blend behavior. Macromolecules 37:6904.

[29] Kubo S, Kadla JF (2005) Kraft lignin/poly(ethylene oxide) blends: effect of lignin structure on miscibility and hydrogen bonding. J Appl Polym Sci 98:1437.

[30] Kubo S, Kadla JF (2006) Effect of poly(ethylene oxide) molecular mass on miscibility and hydrogen bonding with lignin. Holzforschung 60:245.

[31] Kadla JF, Kubo S (2004) Lignin-based polymer blends: analysis of intermolecular interactions in lignin-synthetic polymer blends. Compos A 35:395.

[32] Dallmeyer I, Ko F, Kadla JF (2010) Electrospinning of technical lignins for the production of fibrous networks. Wood Chem Technol 30:315.

[33] Ruiz-Rosas R, Bedia J, Lallave M, Loscertalesc IG, Barrero A, Rodríguez-Mirasol J, Cordero T (2010) The production of submicron diameter carbon fibers by the electrospinning of lignin. Carbon 48:696.

[34] Lallave M, Bedia J, Ruiz-Rosas R, Rodríguez-Mirasol J, Cordero T, Otero JC, Marquez M, Barrero A, Loscertales IG (2007) Filled and hollow carbon nanofibers by coaxial electrospinning of alcell lignin without binder polymers. Adv Mater 19:4292.

[35] Schreiber M, Vivekanandhan S, Mohanty AK, Misra M (2012) A study on the electrospinning behaviour and nanofibre morphology of anionically charged lignin. Adv Mat Lett 3:476.

[36] Zhang YZ, Su B, Ramakrishna S, Lim CT (2008) Chitosan nanofibers from an easily electrospinnable UHMWPEO-doped chitosan solution system. Biomacromolecules 9:136.

[37] Chang W, Ma G, Yang D, Su D, Song G, Nie J (2010) Electro-spun ultrafine composite fibers from organic-soluble chitosan and poly(ethylene oxide). J Appl Polym Sci 117:2113.

[38] Bengisu M, Yilmaz E (2002) Oxidation and pyrolysis of chitosan as a route for carbon fiber derivation. Carbohydr Polym 50:165.

[39] Zawadzki J, Kaczmarek H (2010) Thermal treatment of chitosan in various conditions. Carbohydr Polym 80:394.

[40] Kaczmarek H, Zawadzki J (2010) Chitosan pyrolysis and adsorption properties of chitosan and its carbonizate. Carbohydr Res 345:941.

[41] Kriegel C, Arrechi A, Kit K, McClements DJ, Weiss J (2008) Fabrication, functionalization, and application of electrospun biopolymer nanofibers. Crit Rev Food Sci Nutr 48:775.

[42] Desai K, Kit K, Li J, Zivanovic S (2008) Morphological and surface properties of electrospun chitosan nanofibers. Biomac-romolecules 9:1000.

[43] Kriegel C, Kit KM, McClements DJ, Weiss J (2009) Electrospinning of chitosan-poly(ethylene oxide) blend nanofibers in the presence of micellar surfactant solutions. Polymer 50:189.

[44] Ojha SS, Stevens DR, Hoffman TJ, Stano K, Klossner R, Scott MC, Krause W, Clarke LI, Gorga RE (2008) Fabrication and Characterization of Electrospun Chitosan Nanofibers Formed via Templating with Polyethylene Oxide. Biomacromolecules 9:2523.

[45] Michaels AS (1965) Polyelectrolyte complexes. Ind Eng Chem 57:32.

[46] Thünemann AF, Müller M, Dautzenberg H, Joanny JF, Löwen H (2004) Polyelectrolyte complexes. Adv Polym Sci 166:113.

[47] Goring D (1962) The physical chemistry of lignin. Pure Appl Chem 5:233.

[48] Geng X, Kwon O, Jang J (2005) Electrospinning of chitosan dissolved in concentrated acetic acid solution. Biomaterials 26:5427.

[49] Brebu M, Vasile C (2009) Thermal degradation of lignin—A review. Cellul Chem Technol 9:353.

[50] Machado AEH, Nicodem DE, Ruggiero R, da Perez SD, Castellan A (2001) The use of fluorescent probes in the character-ization of lignin: the distribution, by energy, of fluorophores in Eucalyptus grandis lignin. J Photochem Photobiol A 138:253.

[51] Zhu Z, Zhang L, Smith S, Fong H, Sun Y, Gosztola D (2009) Fluorescence studies of electrospun MEH-PPV/PEO nanofibers. Synth Met 159:1454.

[52] Kong F, Wu XL, Yuan RK, Yang CZ, Siu GG, Chu PK (2006) Optical emission from the aggregated state in poly [2-methoxy-5-(20-ethyl-hexyloxy)-p-phenylene vinylene]. J Vac Sci Technol, A 24:202.

Chapter 12

Enhancement of Mechanical Properties of Biocompatible Ti-45Nb Alloy by Hydrostatic Extrusion

K. Ozaltin[1], W. Chrominski[1], M. Kulczyk[2], A. Panigrahi[3], J. Horky[3], M. Zehetbauer[3], M. Lewandowska[1]

[1]Faculty of Materials Science and Engineering, Warsaw University of Technology, Woloska 141, 02-507 Warsaw, Poland
[2]Institute of High Pressure Physics, Polish Academy of Sciences, Sokolowska 29/37, 01-142 Warsaw, Poland
[3]Physics of Nanostructured Materials, Faculty of Physics, University of Vienna, Boltzmanngasse 5, 1090 Vienna, Austria

Abstract: β-Type titanium alloys are promising materials for orthopaedic implants due to their relatively low Young's modulus and excellent biocompatibility. However, their strength is lower than those of α- or α + β-type titanium alloys. Grain refinement by severe plastic deformation (SPD) techniques provides a unique opportunity to enhance mechanical properties to prolong the lifetime of orthopaedic implants without changing their chemical composition. In this study, β-type Ti-45Nb (wt%) biomedical alloy in the form of 30mm rod was subjected to hydrostatic extrusion (HE) to refine the microstructure and improve its mechanical properties. HE processing was carried out at room temperature without intermediate annealing in a multi-step process, up to an accumulative true strain of 3.5. Significant microstructure refinement from a coarse-grained region to an ultrafine-grained one was observed by optical and transmission electron microscopy. Vickers hardness measurements ($HV_{0.2}$) demonstrated that the strength of the alloy increased from about 150 to 210 $HV_{0.2}$. Nevertheless, the measurements of Young's modulus by nanoindentation showed no significant changes. This finding is substantiated by X-ray diffraction analyses which did not exhibit any phase transformation out of the bcc phase being present still before processing by HE. These results thus indicate that HE is a promising SPD method to obtain significant grain refinement and enhance strength of b-type Ti-45Nb alloy without changing its low Young's modulus, being one prerequisite for biomedical application.

1. Introduction

Pure titanium and its alloys are frequently used in biomedical applications as implants due to their excellent mechanical properties, high corrosion resistance[1]–[4] and better strength/weight ratio compared to other metallic materials[5][6]. Apart from mechanical properties, they also exhibit good biocompatibility, which is an essential requirement to avoid inflammation and long-term diseases[7].

α-Type commercially pure titanium (CP-Ti) and α + β-type Ti-6Al-4 V are well-known biomaterials which are commonly used as orthopaedic implants so far[4][8]. However, their Young's modulus of about 100−110GPa[1] is well above the Young's modulus of bones (10−30GPa)[1][9][10]. This mismatch might cause absorption of bone and premature failure of the implant during utilization as a result of stress shielding effect[1][2][10][11]. In addition, Ti-6Al-4 V alloy contains toxic elements such as Al, which causes neurological diseases like Alzheimer, Osteomalacia and metabolic bone diseases[12], and V which is incompatible with tissue in animals[5]. On the other hand, β-type titanium alloys are promising biometallic materials since they usually contain non-toxic elements and exhibit relatively low Young's modulus compared to α- and α + β-type alloys[2][13].

Among β-type titanium alloys, Ti-Nb binary alloys consisting of 40wt%−45wt% of β-stabilizer non-toxic[13] and non-allergic[14] niobium are attractive orthopaedic materials due to their lowest Young's modulus of ~62GPa[2][8]. On the other hand, mechanical strength of b-type alloys is significantly lower than those of CP-Ti and Ti-6Al-4 V. Enhancement in mechanical properties to prolong the lifetime of the implant without alloying is possible by severe plastic deformation (SPD)[4][15][16]. In recent years, a number of SPD techniques have been developed, including equal-channel angular extrusion (ECAP)[17][18], high-pressure torsion (HPT)[10][19], accumulative roll bonding (ARB)[20] and hydrostatic extrusion (HE)[21][22]. They bring about grain refinement down to sub-micron scale[21]. The characteristic features of SPD materials include deformation-induced high-angle grain boundaries[23] and high dislocation density inside the grains[24]. Also, the reduction of grain size contributes to enhancement of both the strength and the fatigue limit[25], which is quantitatively described by the Hall-Petch relationship[26][27].

It has been demonstrated that HE is an efficient method to reduce the grain size in metallic materials[28][29]. The grain size depends on the material being processed and process condition, e.g. accumulated strain. The typical grain size ranges from 500nm for pure Al, 60nm for a 7475 Al alloy and 50nm for CP-Ti[21][28][30]–[32]. In this paper, we report recent results on microstructure and mechanical properties of β-type Ti-45Nb biomedical titanium alloy processed by hydrostatic extrusion.

2. Experimental

The material studied was a β-type Ti-45Nb (wt%) alloy with chemical composition given in **Table 1**.

Ti-45Nb alloy was supplied in the form of hot-extruded bar, 42mm in diameter, featuring an average grain size of 23µm, hereafter denoted as coarse-grained (CG) sample. HE was performed at room temperature (293K) in a multi-step process (six passes). The initial diameter was 30mm, whereas the final one was 5mm, which correspond to a total true strain of 3.5 (the true strain is calculated by $\varepsilon = 2\ln(d_i/d_f)$, where d_i is the initial and d_f is the final diameter, respectively). The direction of the HE was parallel to the longitudinal direction of the bar and the measured hydrostatic pressure was about 1GPa during each pass. The parameters

Table 1. Chemical composition of Ti45Nb alloy.

Element	wt%	wt% error	at.%	at.% error
Ti	53.92	±0.81	69.22	±1.04
V	0.51	±0.46	0.62	±0.55
Nb	45.57	±0.83	30.16	±0.55

of all stages of HE are summarized in **Table 2**. HE-processed samples are denoted HEx, where x stands for sample's diameter.

The microstructure was observed using optical and transmission electron microscope (TEM) Jeol JEM 1200 (operated at 120kV). All TEM samples were prepared from transverse sections of the specimens. The revealed micro-structures were quantitatively described in terms of grain size using equivalent diameter d_2, which is defined as the diameter of a circle which has the same surface area as the measured grains. Phase characterizations were carried out by X-ray diffraction (XRD) analysis (AXS Bruker D8 diffractometer) using a Cu K_α radiation (λ = 0.154nm) with 0.8mm spot size for transverse sections of all samples after mirror-like polishing.

Microhardness tests were performed using a Zwick Roell ZHU 2.5 testing facility by means of Vickers hardness. All indentations were carried out using a load of 200g ($HV_{0.2}$) for ten measurements on transverse section of each specimen. Then, the average value and the standard deviation were calculated (see **Figure 1**). Micro-tensile specimens were prepared in dumbbell-shaped form by spark erosion machine along the extrusion direction with a total length of 8mm, 0.75mm width, and 0.45mm thickness with the parallel gauge length of 2.5mm. Tensile tests were carried out for CG and HE5 samples at room temperature with constant crosshead speed and initial strain rate of 10^{-3} s^{-1} (see **Figure 2**).

Nanoindentation measurements were carried out with a load rate of 100mN/20s using a Vickers indenter (creep at 100mN for 30s and thermal drift at 10mN for 60s) to reveal changes in Young's modulus. The method of Oliver and Pharr[33] was used to determine the Young's modulus from load-displacement curves (the Poisson's ratio was assumed to be 0.41). Twenty-five indentations have been applied per specimen. The average values of Young's modulus as well as of standard deviation were calculated (see **Figure 3**).

Table 2. Parameters of different stages of hydrostatic extrusion.

Sample names	Staged	Initial dia (mm)	Final dia (mm)	Reduction factor (Area)	Accumulative reduction factor	True strain	Accumulative true strain	Pressure (MPa)
HE15	1st	30	15	4	4	1.38	1.38	565
HE10	2nd	14.5	10	2.1	8.4	0.74	2.12	704
HE8	3rd	10	8	1.56	13.1	0.44	2.57	843
HE7	4th	8	7	1.3	17	0.26	2.83	791
HE6	5th	7	6	1.36	23.2	0.31	3.14	750
HE5	6th	6	5	1.44	33.4	0.36	3.2	825

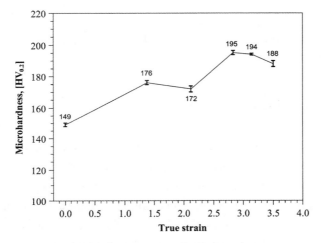

Figure 1. Microhardness as a function of true strain.

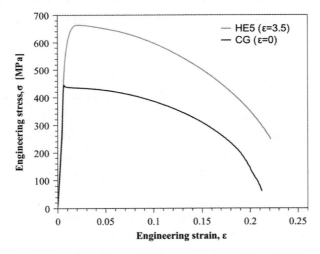

Figure 2. Stress-strain curves of materials CG and HE5.

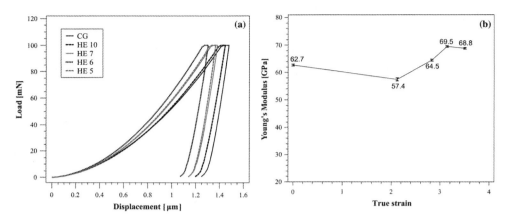

Figure 3. Nanoindentation results of CG and HE-processed samples: a load versus displacement curves and b Young's modulus values.

3. Results and Discussion

HE processing induces a significant increase in the microhardness, as illustrated in **Figure 1**. The most pronounced hardening is observed for the low values of applied strain; then the microhardness tends to stabilize and shows even a decrease for the highest applied strain. Especially the latter can be attributed to recovery and recrystallization processes, which are likely to take place during subsequent passes of HE[32], and also occur in other SPD techniques as cold rolling[34] and HPT[35][36], respectively. In general, the microhardness increased from 149 $HV_{0.2}$ for CG sample to 195 $HV_{0.2}$ obtained for HE7 sample (which corresponds to a true strain of 2.83).

Figure 2 shows the stress-strain curves of CG and HE5 specimens obtained by micro-tensile tests. The ultimate tensile strength increased by 50% from 445 to 663MPa and yield stress increased by 45% from 430 to 620MPa. It is worth noting that the fracture strain of HE5 sample has only slightly changed from 23% to 21.7%, despite the fact that SPD materials generally exhibit reduced ductility[37]. This is a favourable characteristic for metals intended for orthopaedic applications.

Figure 3 shows load versus displacement curves and the changes in Young's modulus during the consecutive HE processing steps. The values of Young's modulus measured with the highest HE deformation did not exceed 68.8GPa which is only slightly higher than 62.7GPa measured for the CG samples. This increase corresponds to about 9% only, and thus fulfils the requirement that any strengthening procedure in Ti-Nb biomedical alloy should not increase the Young's modulus too much in order to avoid the stress shielding effect and thus the gradual deterioration of bone material, as it may occur with conventional a-type titanium alloys having a Young's modulus of about 110GPa. It should be noted that Matsumoto et al.[38] reported for another β-type alloy (Ti-35Nb-4Sn) that Young's modulus decreases as a function of strain applied in cold rolling. This was attributed to the transformation of β-phase to orthorhombic α″ one. The alloy investigated in the present study has higher Nb content, which not only stabilizes β-phase but also facilitates β to ω transformation, as reported and discussed within this paper later on. Such a transformation may result in an increase of Young's modulus, since ω-phase is stiffer than the β one.

In order to maintain the Young's modulus at a relatively low level typical of β-Ti alloys, it is essential to avoid any phase transformation during processing. **Figure 4** shows XRD patterns of all samples before and after different strains of HE. The results show that all peaks correspond to the β-phase, which suggests that no phase transformation takes place during HE processing. The variations in the peak intensity can be attributed to the changes in crystallographic texture, which is a consequence of the HE process. One could also argue that these changes in texture cause the variations in microhardness and/or Young's modulus shown in **Figure 1** and **Figure 3(b)**, respectively. However, two facts speak against this suspicion: (i) Vickers microhardness is known to be not too sensitive to lattice orientation and thus to texture changes; (ii) calculations of changes in Young's modulus according to texture changes cannot reflect the experimental data, according to recent investigations which will be published elsewhere[39].

The microstructure of the CG sample as inspected by optical microscopy consists of equiaxed β-grains with an average grain size of 23μm, as illustrated in **Figure 5(a)**. HE processing brings about a significant microstructure refinement [**Figure 5(b)**]. However, the microstructure after HE is quite heterogeneous and, as illustrated in **Figure 6**, some typical regions can be distinguished.

The first typical region consists of relatively small elongated grains, as shown in **Figure 6(a)**. The diffraction pattern in the form of rings suggests different

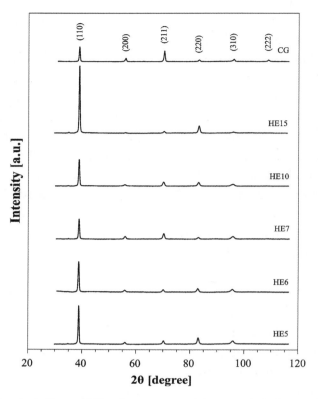

Figure 4. X-ray diffraction pattern of the samples after each stage of HE.

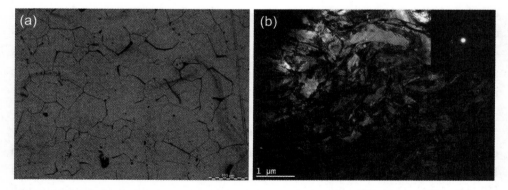

Figure 5. General view of microstructure of Ti-45Nb: a CG and b HE5 samples.

orientations of individual grains, and the presence of high-angle grain boundaries between them. The average grain size was found to be about 300nm. This region seems to be the most advanced stage of grain refinement reported in this

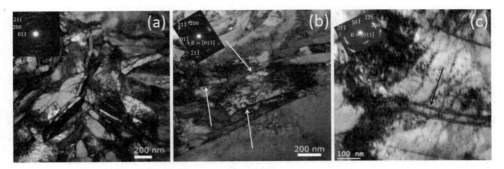

Figure 6. Images of different areas in HE5 specimens extruded with a true strain of 3.5: a small grains, b elongated grains with relatively large dimensions and c a typical grain representing dislocation substructure and twins. The images are orientated perpendicularly to the extrusion direction. White and black arrows indicate low-angle grain boundaries and twins, respectively.

study. Inside the small grains, the density of dislocations is relatively high, but they are not arranged in special sub-structures. The amplitude contrast changes within some grains which can be attributed to the existence of residual stresses.

The remaining microstructure shows a deformation substructure. It consists of relatively large grains with $\langle 011 \rangle$ direction parallel to the extrusion direction (the diffraction pattern was taken at zero tilt). Inside the grains, low-angle grain boundaries are present [**Figure 6(b)**] and some deformation twins with habit plane of {121} typical for bcc metals [**Figure 6(c)**] can be recognized.

As the microstructure exhibits different types of sub-structures, each of them will give different reactions to applied stress. High-angle grain boundaries and mechanical twins seem to act as obstacles to dislocation motion causing a strengthening effect[40][41]. However, the presence of relatively large grains with dislocations and low-angle grain boundaries indicate that the material can even accumulate more strain[42], so that it allows applying further HE stages based on the same requirements. The co-existence of small and large grains allows for both enhancement of strength and almost stable strain[43]. This can explain why the values of fracture strain observed for both CG and UFG specimens are almost identical.

Compared to CP-Ti (HE processed at ambient temperature with a total true strain of 3.77) examined by Topolski *et al.*[44], the improvement of mechanical strength by HE in Ti-45Nb is lower (**Figures 1, 2**). In the case of CP-Ti, the improvements were between 110% (ultimate tensile strength) and about 250% (yield strength).

The reason of the different efficiency of HE in terms of strengthening in those materials can be attributed to less advanced process of grain refinement in the Ti-45Nb alloy. In the case of CP-Ti, a homogeneous microstructure with an average grain size of 55nm was obtained, while for the current β-type titanium alloy, only a fraction of micro-structure exhibited grains with sizes below 1μm. The different potentials for grain refinement can be attributed to different mechanisms of plastic deformation taking place in hcp (α) and bcc (β) titanium. In α-Ti, grain refinement occurs via the combination of various twinning modes and dislocation glide[45]. Detailed studies were done for different ECAP routes and have shown that the final microstructure strongly depends on the number of active slip systems induced by stress conditions defined by deformation route. Such an analysis has not been done for β-Ti. However, deformation in bcc metals is expected to be significantly different from deformation in hcp metals since the close-packed direction ⟨111⟩ belongs to many planes in contrast to the hcp case. As a result, the formation of dislocation jogs in bcc lattice structure occurs rarely compared to the hcp lattice[42], and the grain refinement is less advanced.

Careful inspection of the microstructure revealed in some places a zig-zag structure (**Figure 7**), which can be described as a combination of ω-phase with deformation twins caused by shearing[46]. Diffraction pattern presented as an inset displays diffusive lines between spots of (200) and (110) planes in [011] zone axis pattern. Such lines may be caused by scattering by ω-phase as reported by Yano *et al.*[47]. The presence of small amount of ω-phase can explain the observed variations in Young's modulus in HE-processed samples as Young's modulus of ω-phase is higher than that of β-phase[48][49]. However, this will be a subject of more detailed investigation in near future.

4. Conclusions

In the present study, it has been shown that HE is an efficient technique in tailoring mechanical properties of β-type Ti-Nb alloy. The strength was

Figure 7. TEM image of a zig-zag structure in HE5 sample (extruded to a true strain of 3.5), diffused spots from ω-phase are indicated by an arrow.

significantly improved by 45% due to grain boundary and dislocation strengthening mechanisms. The Young's modulus was kept at a low level (below 70GPa), thanks to maintaining almost a single-phase β-structure during HE processing. Furthermore, HE processing did not deteriorate plasticity of high-strength material which can be attributed to the presence of large grains, which still have ability to accommodate plastic deformation during further straining.

Acknowledgements

This work was supported within the EU 7th framework programme FP7/2007-13 under Marie-Curie project Grant No. 264635 (BioTiNet-ITN).

Open Access: This article is distributed under the terms of the Creative Commons Attribution License which permits any use, distribution, and reproduction in any medium, provided the original author(s) and the source are credited.

Source: Ozaltin K, Chrominski W, Kulczyk M, *et al*. Enhancement of mechanical properties of biocompatible Ti-45Nb alloy by hydrostatic extrusion[J]. Journal of

Materials Science, 2014, 49(20):6930–6936.

References

[1] Calin M, Gebert A, Ghinea AC, Gostin PF, Abdi S, Mickel C, Eckert J (2013) Designing biocompatible Ti-based metallic glasses for implant applications. Mater Sci Eng C 33:875–883.

[2] Godley R, Starosvetsky D, Gotman I (2006) Corrosion behavior of a low modulus β−Ti−45% Nb alloy for use in medical implants. J Mater Sci Mater Med 17:63–67.

[3] Kuroda D, Niinomi M, Morigana M, Kato Y, Yashiro T (1998) Design and mechanical properties of new b type titanium alloys for implant materials. Mater Sci Eng A 243:244–249.

[4] Purcek G, Saray O, Karaman I, Yapici GG, Haouaoui M, Maier HJ (2009) Mechanical and wear properties of ultrafine-grained pure Ti produced by multi-pass equal-channel angular extrusion. Mater Sci Eng A 517:97–104.

[5] Hon YH, Wang JY, Pan YN (2003) Composition/phase structure and properties of titanium–niobium alloys. Mater Trans 41: 2384–2390.

[6] Martins GV, Silva CRM, Nunes CA, Trava-Airoldi VJ, Borges JLA, Machado JPB (2010) Beta Ti–45Nb and Ti–50Nb alloys produced by powder metallurgy for aerospace application. Mater Sci Forum 660–661:405–409.

[7] Bauer S, Schmuki P, Mark KVD, Park J (2013) Progress in materials science engineering biocompatible implant surfaces part I : materials and surfaces. Prog Mater Sci 58:261–326.

[8] Gostin PF, Helth A, Voss A, Sueptitz R, Calin M, Eckert J, Gebert A (2013) Surface treatment, corrosion behavior, and apatite-forming ability of Ti-45Nb implant alloy. J Biomed Mater Res Part B 101B:269–278.

[9] Nakai M, Niinomi M, Hieda J, Yilmazer H, Tokada Y (2013) Heterogeneous grain refinement of biomedical Ti-29Nb-13Ta-4.6Zr alloy through high-pressure torsion. Sci Iran F 20: 1067–1070.

[10] Williams JC, Lutjering G (2007) Titanium, 2nd edn. Springer, New York.

[11] Niinomi M (2008) Review: mechanical biocompatibilities of titanium alloys for biomedical applications. J Mech Behav Bio- med 1:30–42.

[12] Gepreel MAH, Niinomi M (2013) Biocompatibility of Ti-alloys for long-term implantation. J Mech Behav Biomed 20:407–415.

[13] Niinomi M (1998) Mechanical properties of biomedical titanium alloys. Mater Sci Eng A 243:231–236.

[14] Yilmazer H, Niinomi M, Nakai M, Cho K, Hieda J, Todaka Y, Miyazaki T (2013) Mechanical properties of a medical b-type titanium alloy with specific microstructural evolution through high-pressure torsion. Mater Sci Eng C 33:2499–2507.

[15] Valiev RZ, Gunderov DV, Lukyanov AV, Pushin VG (2012) Mechanical behaviour of nanocrystalline TiNi alloy produced by severe plastic deformation. J Mater Sci 47:7848–7853. doi:10. 1007/s10853-012-6579-8.

[16] Sordi VL, Ferrante M, Kawasaki M, Langdon TG (2012) Microstructure and tensile strength of grade 2 titanium processed by equal-channel angular pressing and by rolling. J Mater Sci 47:7870–7876. doi:10.1007/s10853-012-6593-x.

[17] Saitova LR, Hoppel HW, Goken M, Semenova IP, Raab GI, Valiev RZ (2009) Fatigue behavior of ultrafine-grained Ti-6Al-4 V 'ELI' alloy for medical applications. Mater Sci Eng A 503:145–147.

[18] Valiev RZ, Islamgaliev RK, Alexandrov IV (2000) Bulk nano-structured materials from severe plastic deformation. Prog Mater Sci 45:103–189.

[19] Pippan R (2009) High pressure torsion—features and applications. In: Zehetbauer MJ, Zhu YT (eds) Bulk nanostructured materials. Wiley, Weinheim, p 217.

[20] Estrin Y, Vinogradov A (2010) Fatigue behaviour of light alloys with ultrafine grain structure produced by severe plastic deformation: an overview. Int J Fatigue 32:898–907.

[21] Lewandowska M, Kurzydlowski KJ (2005) Thermal stability of a nanostructured aluminium alloy. Mater Charact 55:395–401.

[22] Kulczyk M, Pachla W, Mazur A, Sus-Ryszkowska M, Krasilni-kov N, Kurzydlowski KJ (2007) Producing bulk nanocrystalline materials by combined hydrostatic extrusion and equal-channel angular pressing. Mater Sci Pol 25:991–999.

[23] Chen Y, Li J, Tang B, Kou H, Zhang F, Chang H, Zhou L (2013) Grain boundary character distribution and texture evolution in cold-drawn Ti–45Nb wires. Mater Lett 98:254–257.

[24] Park CH, Park JW, Yeom JT, Chun YS, Lee CS (2010) Enhanced mechanical compatibility of submicrocrystalline Ti–13Nb–13Zr alloy. Mater Sci Eng A 527:4914–4919.

[25] Zherebtsov S, Lojkowski W, Mazur A, Salishchev G (2010) Structure and properties of hydrostatically extruded commercially pure titanium. Mater Sci Eng A 527:5596–5603.

[26] Zherebtsov S, Mazur A, Salishchev G, Lojkowski W (2008) Effect of hydrostatic extrusion at 600°C–700°C on the structure and properties of Ti-6Al-4V alloy. Mater Sci Eng A 485:39–45.

[27] Zherebtsov S, Salishchev G, Lojkowski W (2009) Strengthening of a Ti-6Al-4 V titanium alloy by means of hydrostatic extrusion and other methods. Mater Sci Eng A 515:43–48.

[28] Lewandowska M, Kurzydlowski KJ (2008) Recent development in grain refinement by

hydrostatic extrusion. J Mater Sci 43:7299–7306. doi:10.1007/s10853-008-2810-z.

[29] Pachla W, Kulczyk M, Swiderska-Sroda A, Lewandowska M, Garbacz H, Mazur A, Kurzydlowski KJ (2006) Nanostructuring of metals by hydrostatic extrusion. Proc Int Esaform Conf Mater Form 16:535–538.

[30] Lewandowska M, Wawer K (2007) Optimization of particle size and distribution by hydrostatic extrusion. Mater Sci Forum 561–565:869–872.

[31] Pakiela Z, Garbacz H, Lewandowska M, Druzycka-Wiencek A, Sus-Ryszkowska M, Zielinski W, Kurzydlowski KJ (2006) Structure and properties of nanomaterials produced by severe plastic deformation. Nukleonika 51:19–25.

[32] Pachla W, Kulczyk M, Sus-Ryszkowska M, Mazur A, Kurzyd-lowski KJ (2008) Nanocrystalline titanium produced by hydro-static extrusion. J Mater Process Technol 205:173–182.

[33] Oliver WC, Pharr GM (2004) Review: measurement of hardness and elastic modulus by instrumented indentation: advances in understanding and refinements to methodology. J Mater Res 19:3–20.

[34] Zehetbauer M, Trattner D (1987) Effects of stress aided static recovery in iteratively cold-worked aluminium and copper. Mater Sci Eng 89:93–101.

[35] Schafler E (2010) Effects of releasing the hydrostatic pressure on the nanostructure after severe plastic deformation of Cu. Scr Mater 62:423–426.

[36] Schafler E (2011) Strength response upon pressure release after high pressure torsion deformation. Scr Mater 64:130–132.

[37] Zehetbauer MJ, Valiev RZ (2004) Nanomaterials by severe plastic deformation. Wiley, Weinheim Matsumoto H, Watanabe S, Hanada S (2007) Microstructures and

[38] mechanical properties of metastable b TiNbSn alloys cold rolled and heat treated. J Alloys Compd 439:146–155.

[39] Panigrahi A, Sulkowski B, Ozaltin K, Horky J, Lewandowska M, Waitz T, Skrotzki W, Zehetbauer M (2014) Mechanical properties and texture evolution in biocompatible Ti–45Nb alloy processed by severe plastic deformation (unpublished).

[40] Hall EO (1951) The deformation and ageing of mild steel: III discussion of results. Proc Phys Soc B64:747–753.

[41] Petch NJ (1953) The cleavage strength of polycrystals. J Iron Steel Inst 174:25–28.

[42] Kuhlmann-Wilsdorf D (1989) Theory of plastic deformation: properties of low energy dislocation structures. Mater Sci Eng A 113:1–41.

[43] Estrin Y, Vinogradov A (2013) Extreme grain refinement by severe plastic deformation: a wealth of challenging science. Acta Mater 61:782–817.

[44] Topolski K, Garbacz H, Kurzydlowski KJ (2008) Nanocrystalline titanium rods processed by hydrostatic extrusion. Mater Sci Forum 584–586:777–782.

[45] Shin DH, Kim I, Kim J, Kim YS, Semiatin SL (2003) Micro-structure development during equal-channel angular pressing of titanium. Acta Mater 51:983–996.

[46] Xing H, Sun J (2008) Mechanical twinning and omega transition by ⟨111⟩ {112} shear in a metastable b titanium alloy. Appl Phys Lett 93:031908.

[47] Yano T, Murakami Y, Shindo D, Hayasaka Y, Kuramoto S (2010) Transmission electron microscopy studies on nanometer-sized ω phase produced in gum metal. Scr Mater 63:536–539.

[48] Tane M, Nakano T, Kuramoto S, Niinomi M, Takesue N, Nakajima H (2013) ω transformation in cold-worked Ti-Nb-Ta-Zr-O alloys with low body-centered cubic phase stability and its correlation with their elastic properties. Acta Mater 61:139–150.

[49] Banerjee D, Williams JC (2013) Perspectives on titanium science and technology. Acta Mater 61:844–879.

Chapter 13

Evolution of Microstructure in AZ91 Alloy Processed by High-Pressure Torsion

Ahmed S. J. Al-Zubaydi[1,2], Alexander P. Zhilyaev[3,4], Shun C. Wang[1], P. Kucita[1], Philippa A. S. Reed[1]

[1]Materials Research Group, Faculty of Engineering and the Environment, University of Southampton, Southampton SO17 1BJ, UK
[2]Branch of Materials Science, Department of Applied Sciences, University of Technology, Baghdad, Iraq
[3]Institute for Problems of Metals Superplasticity, Russian Academy of Sciences, Khalturina 39, Ufa, Russia 450001
[4]Research Laboratory for Mechanics of New Nanomaterials, St. Petersburg State Polytechnical University, St. Petersburg, Russia 195251

Abstract: An investigation has been conducted on AZ91 magnesium alloy processed in high-pressure torsion (HPT) at 296, 423 and 473K for different numbers of turns. The microstructure has altered significantly after processing at all processing temperatures. Extensive grain refinement has been observed in the alloy processed at 296K with apparent grain sizes reduced down to 35nm. Segmentation of coarse grains by twinning has been observed in the alloy processed at 423K and 473K with average apparent grain sizes of 180nm and 250nm. Substantial homogeneity in microhardness has been observed in the alloy processed at 296K compared to that found at 423K and 473K. The ultrafine-grained AZ91 alloy exhibited a significant dependence of the yield strength on grain size as shown by the microhardness measurements, and it obeys the expected Hall-Petch relationship. The alloying elements, fraction of nano-sized particles of b-phase, and the dominance of basal slip and pyramidal modes have additional effects on the strengthening of the alloy processed at 296K.

1. Introduction

Magnesium alloys are promising alternatives to replace denser materials, such as steel and aluminium alloys, with the objective of meeting requirements to save fuel by manufacturing light weight/high strength parts[1]. The mechanisms of deformation in magnesium alloys at room temperature are basal slip and twinning, which result in a limitation in their workability at room temperature[2]. The limited ductility and workability of these alloys can be improved at higher temperatures by the activation of additional slip systems[1]. Thermo-mechanical processing is used to improve the workability of these alloys, although such processing is associated with grain growth and a greater consumption of energy[3]. Several processing routes have been introduced to achieve optimization of the microstructure, and these routes include dynamic recrystallization under high-temperatures in ECAP processing[4], HPT processing[5][6], ECAP processing at relatively low temperatures assisted by a back-pressure[7], or through the use of a higher channel angle of pressing die in ECAP processing[8]. The majority of the earlier work in SPD processing of magnesium alloys, especially for AZ91 alloy, has been conducted using ECAP at elevated temperatures (≥473K)[2][4][9] with resultant grain refinement being achieved in the micrometre range. The AZ91 alloy (Mg-9wt%Al-1wt%Zn-0.3wt%Mn) is a common alloy in the Mg-Al-Zn family. This alloy has a good strength-to-density ratio, good corrosion resistance and ease of production and machining[3]. To date, only one investigation has been conducted on Mg-9wt%Al alloy[6] using HPT at room temperature. The development of microstructure and microhardness across horizontal and vertical cross-sections of AZ91 samples processed by HPT has not been reported to date. This research describes the microstructural homogeneity and development of microhardness in AZ91 alloy after processing by HPT at different processing temperatures. The dislocation density, distribution of β-phase and Hall-Petch relationship have also been investigated.

2. Experimental Materials and Procedures

AZ91 alloy (Mg-9%Al-1%Zn) in the form of an extruded rod was used in this work, the alloy was supplied by Magnesium Elektron Co. (Manchester, UK). Thin discs were made of the extruded rod with thicknesses of 1.5mm and final thicknesses of 0.85mm. The HPT processing was conducted at 296, 423 and 473K using a HPT facility that has been previously discussed in detail elsewhere[10]. The HPT processing was conducted under a quasi-constrained condition at a speed of 1 rpm using an applied pressure of 3.0GPa for differing numbers of turns: N = 1/2, 1, 5 and 10 turns. The as-received and processed microstructures were observed using optical microscopy (OM, OLYMPUS-BX51, Japan) and scanning electron microscopy (SEM, JEOL JSM-6500F, Japan). Subsequently, a transmission electron microscope (TEM, JEOL JEM-3010) was used for microstructural observation of the alloy after HPT processing. The chemical compositions of the as-received and processed alloy were analysed using energy-dispersive spectroscopy (EDS). The area fraction and average size of the β-phase particles in the as-received alloy and processed alloy were determined by ImageJ software using a point count technique[11]. X-ray diffraction was used to determine the crystallite size and dislocation density in the processed alloy using an XRD facility (D2 Phaser, Germany). The diffraction data were analysed using Rietveld refinement based software program (MAUD). Microstructural observations and microhardness testing were conducted over the horizontal and vertical cross-sections that are illustrated schematically in **Figure 1(a), (b)**. The microhardness measurements of the processed disc were conducted using a Vickers microhardness tester (FM-300, Japan) and using an applied load of 100gf and a dwell time of 15s. The microhardness data were recorded at separation distances of 0.3 and 0.1mm throughout the entire horizontal and vertical cross-sections, as reported earlier[5][12].

3. Experimental Results

The microstructure of the AZ91 magnesium alloy prior to and after HPT processing is shown in **Figure 2**. The as-received AZ91 alloy has an average grain size of 30μm and an average value of Vickers microhardness of 70 ± 5. The initial and processed microstructures consist of two main phases: α-Mg matrix, β-phase and Al_8Mn_5 particles as shown in **Figure 2(a), (b)**. The chemical analysis obtained

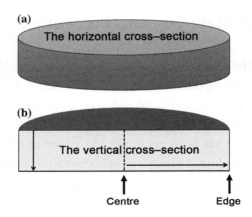

Figure 1. An illustration of a HPT disc shows a the horizontal crosssection, and b The vertical cross-section. These cross-sections were used in the microstructural and microhardness observations. The arrow from the centre to the edge refers to the longitudinal (radial) direction, whereas the arrow from the upper surface to lower surface refers to the through-thickness (vertical) direction.

by EDS of alloying elements in the alloy processed at 296K for N = 5 turns is shown in **Figure 3**. The alloy constituents were identical before and after HPT as shown earlier[13]. The processed microstructure at 296K showed extensive grain refinement, and the original decoration of the grain boundaries by b-phase disappeared with increasing number of turns as shown in **Figure 2(b), (c)**. The β-phase fragmented into nano-sized particles as observed in **Figures 2(b)-(d)** and appears aligned along the direction of torsional straining. A strong degree of grain refinement after processing at 296K was observed with an apparent grain size down to 500 and 50nm observed after N = 1/2 and 1 turn, respectively, as shown in **Figure 2(e), (f)**. A reduction in the crystallite size from 60 to 35nm was found with increasing number of turns up to N = 10 turns. The processed microstructures at 296K across the vertical cross-sections are shown in **Figure 4**. The microstructure seems slightly deformed with the presence of twinning as shown in **Figure 4(a)**. Shear bands decorated by the β-phase were observed aligned parallel to the radial direction across the vertical cross-section as observed in **Figure 4(b)**. Recorded peaks by XRD as shown in **Figure 5** are prismatic planes basal plane and pyramidal planes $(10\bar{1}0)$, $(11\bar{2}0)$, $(20\bar{2}0)$, basal plane (0002) and pyramidal planes $(10\bar{1}1)$, $(10\bar{1}2)$, $(10\bar{1}3)$. The microstructures of the alloy processed at 423 and 473K are shown in **Figure 6**. The samples showed twinning, and the distribution of twinning increased and spread gradually with increasing number of turns. The microstructures were effectively refined by the segmentation of the coarse grains

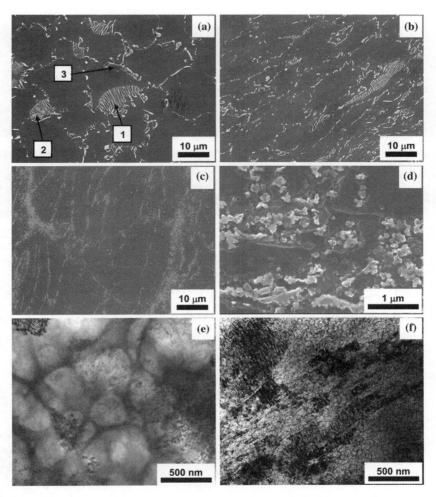

Figure 2. Microstructural observations using SEM for a the as-received alloy, b the alloy processed for N = 1 turn (296K), c the alloy processed for N = 10 turns (296K) and d the nano-sized particles of β-phase in the alloy processed for N = 10 turns (296K), and TEM observation of the alloy processed for e N = 1/2 turn (296K) and f N = 1 turn (296K). The corresponding numbers (1, 2, 3) in the micrograph a represent the lamellar, agglomerate forms of the β-phase ($Mg_{17}Al_{12}$) and Al_8Mn_5 particle, respectively.

by twinning as observed in **Figure 6(a), (b)**. However, grain growth has been observed at 473K with increasing number of turns up to N = 5 turns as shown in **Figure 6(b)**. The apparent area fraction has increased (which may reflect a sampling effect once the second phase is more homogeneously distributed), and the average size of the β-phase particles has been refined down to 200nm in the processed alloy compared to the as-received alloy as shown in **Figure 7**. A gradual

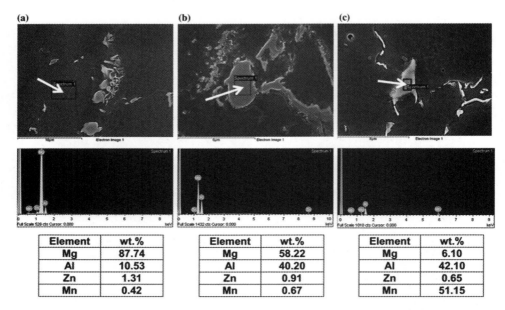

Figure 3. The chemical analysis with weight fractions of the alloy processed at 296K for N = 5 turns showing a α-Mg matrix, b β-phase ($Mg_{17}Al_{12}$), c Al_8Mn_5 particle.

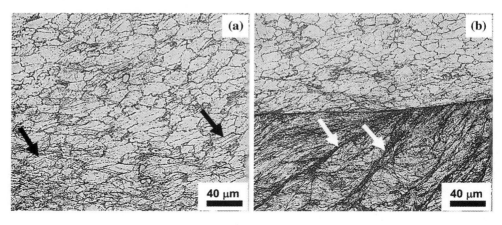

Figure 4. The microstructures of the alloy processed at 296K as observed along the vertical cross-sections for a N = 1 turn and b N = 5 turns. The black and white arrows refer to the twinning and shear bands decorated by the β-phase, respectively.

development in the microhardness over the horizontal and vertical cross-sections has been achieved with increasing number of turns up to N = 10 turns as shown in **Figure 8** and **Figure 9**. The distributions of microhardness were relatively lower for the alloy processed at 423 and 473K than at 296K. A significant increase in

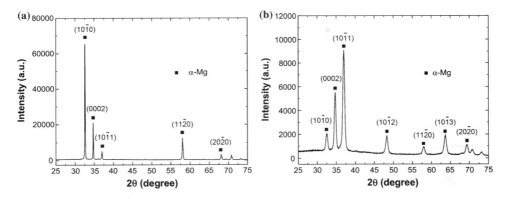

Figure 5. XRD diffraction patterns for a the as-received alloy and b the alloy processed at 296K for N = 10 turns.

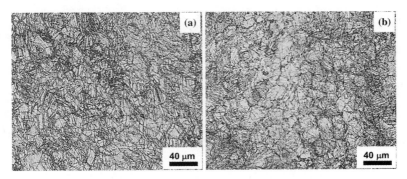

Figure 6. The microstructures of the alloy as observed across the horizontal cross-sections after HPT processing at a 423K (N = 5 turns) and b 473K (N = 5 turns).

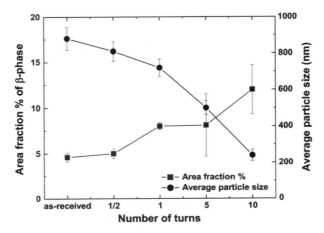

Figure 7. The area fraction and average size of the β-phase particles in the as-received alloy and processed alloy at 296K for different number of turns.

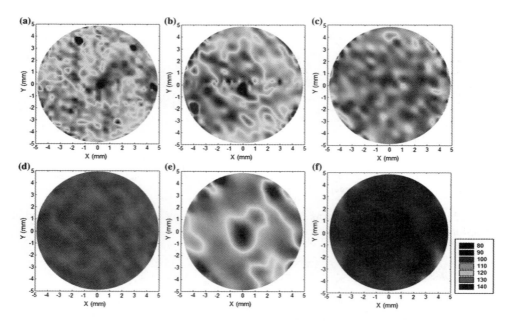

Figure 8. The colour-coded maps of the microhardness over the horizontal cross-sections of the AZ91 discs processed forf a N = 1/2 turn (296K), b N = 1 turn (296K), c N = 5 turns (296K), d N = 10 turns (296K), e N = 10 turns (423K) and f N = 10 turns (473K). The small inset in the figure shows the scale of the microhardness with regard to each colour (Color figure online).

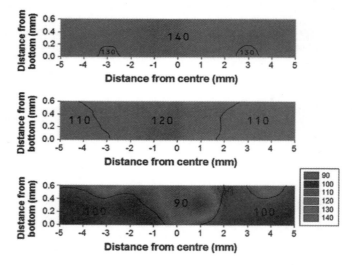

Figure 9. The colour-coded maps of the microhardness distributions over the vertical cross-sections of the AZ91 discs processed for N = 10 turns at 296K (upper), 423K (centre) and 473K (lower). The small inset in the figure shows the scale of the microhardness with regard to each colour (Color figure online).

the microhardness has been observed as shown in **Figure 10**, with increasing equivalent strain imposed during HPT for the alloy processed at 296K. A significant dependency of the microhardness on the crystallite size of the AZ91 alloy processed at 296K is shown in **Figure 11**. The lower processing temperature leads to finer crystallite size, higher microhardness and dislocation density, and at elevated temperatures, these outcomes decreased significantly as the number of turns increased as shown in **Figure 12**.

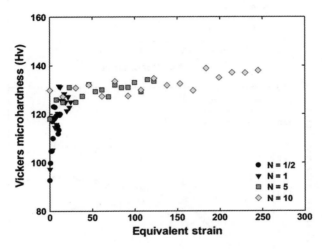

Figure 10. Correlation of the measured microhardness with the equivalent strain imposed by HPT processing for the alloy processed at 296K for different number of turns.

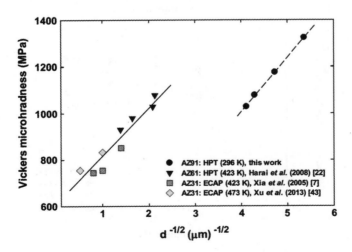

Figure 11. The Hall-Petch relationship for the ultrafine-grained AZ91 alloy in the current work and for AZ31 and AZ61 alloys processed by HPT and ECAP.

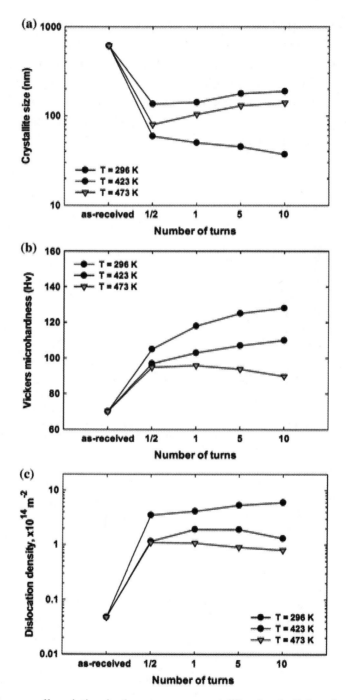

Figure 12. The overall variation in the average a crystallite size, b dislocation density and c microhardness for the AZ91 alloy after HPT at different processing temperatures.

4. Discussion

4.1. Feasibility of HPT Processing of AZ91 Magnesium Alloy

The TEM and XRD revealed the occurrence of extensive grain refinement in the AZ91 alloy due to the imposition of a very high plastic strain by HPT at 296K. However, for the sample processed for N = 1/2 turn, it is noteworthy that the value of the crystallite size obtained by XRD was significantly lower than the apparent grain size measured by TEM. This difference between the measurements via XRD and TEM is expected in SPD-processed materials, because the grains in these materials are made of subgrains and/or dislocation cells. Thus, coherent scattering of the X-ray from these substructures represents the (smaller) mean crystallite size rather than grains which can be more easily observed in TEM[14]. The feasibility of HPT processing at 296K for the AZ91 magnesium alloy can be attributed to the presence of hydrostatic pressure, which prevents propagation of fracture during processing[6]–[8]. Furthermore, the geometry of the processing zone constrains the alloy within a specific volume as illustrated earlier and thus activation of twinning[8][15][16]. The XRD observations indicate the orientation of the processed microstructure towards twinning and basal deformation modes under HPT conditions that facilitate processing at room temperature[16]. The unidirectional nature of straining during HPT processing may have contributed to re-orientation of the microstructure towards easy slip[17]. The twinning activity has persisted in the processed alloy at 296K with increasing number of turns, which confirms its accommodation for the higher imposed strain produced by HPT[18].

4.2. Grain Refinement in AZ91 Alloy

The relatively high content of aluminium in the AZ91 magnesium alloy leads to a significant reduction in the stacking fault energy through the solute– dislocation inter-action and results in smaller grain sizes under SPD processing[19]. The effect of dynamic recovery was absent as the alloy has been processed at room temperature. It is anticipated that the homogeneity developed gradually with further straining at room temperature as mentioned by several investigators[20]–[23]. The grain refinement in the processed alloy at 423K has developed efficiently by twinning intersections and the grain subdivision mechanism. At this temperature, dynamic recrystallization was absent or had a minor effect on the refinement process compared to the twinning activity. It is likely that dynamic recrystallization may have contributed to grain refinement in the processed alloy at 473K. However, the formation and fragmentation of twinning appears to be the dominant mechanism for refinement at 473K. The HPT-processed alloy at 423 and 473K has significantly refined apparent grain sizes of 180 and 250nm, respectively, which are finer than in the previously reported ECAP[20][24]–[26], FSP[27] and ARB-processed alloys[28]. In the aforementioned SPD techniques, grain refinement occurs mainly by dynamic recrystallization with resultant microstructures of micrometre size grains. The severe levels of deformation in the alloy and the deformation incompatibility between α-Mg matrix and β-phase have resulted in fragmentation of the β-phase[29]. The significant dispersion of nano-sized particles of the β-phase during processing had a pinning effect on grain growth at a higher number of turns and elevated temperatures[23]. The alloy processed at 296K showed microstructural homogeneity at the initial stage of HPT processing rather than the heterogeneity observed in the alloy processed at 423 and 473K, which required further processing turns and/or higher processing temperature to achieve a reasonable homogeneity[10]. The temperature rise expected during HPT processing at room temperature does not exceed 293K for samples processed at 296K for N = 10 turns. This value of temperature rise has been calculated using the equation stated in[30], and is similar to the experimental value (290K) measured directly from the thermocouple located in the upper anvil. The low value of temperature rise can be attributed to (1) the heat loss from relatively small samples in contact with the much larger HPT anvils and (2) due to the low strain rates of deformation in HPT processing[31]. A further factor, making the heat generated low, is the lower friction expected between the relatively lower strength magnesium alloy and the high strength (high speed tool steel) anvils[15][31]. As a result, any temperature rise due to processing is considered negligible and unlikely to produce any occurrence of recrystallization or grain growth during processing at room temperature[32].

4.3. Development of Microhardness

The initial heterogeneity of microstructures leads to an initial heterogeneity in the distribution of microhardness[10][33][34]. The difference in grain sizes at the centre and edge regions was diminished by further straining, where a gradual evolution towards homogeneity was found in the observed microstructure and microhardness at both centre and edge regions at a higher number of turns[10][35]. The existence of misalignment between the anvils at a high number of turns causes an additional deformation at the centre region of the processed disc, which appears as an increase in the measured microhardness[36][37]. The development of microhardness after HPT processing depends on the stacking fault energy of the alloy[10][19]. The AZ91 alloy with a low stacking fault energy[38] shows a slow rate of dynamic recovery during processing at room temperature, thus strain hardening occurs at a fast rate during processing[20][32]. The AZ91 alloy processed in HPT showed an earlier saturation in the microhardness distribution than for the AZ31 alloy[21] processed in HPT at room temperature. The stacking fault energy is lower, and the fraction of particles of β-phase is higher in the AZ91 alloy than for the AZ31 alloy[38][39]. Therefore, the evolution of grain refinement and strain hardening occurred at faster rates in the AZ91 alloy than for the AZ31 alloy. The overall microhardness values for the alloy processed at 473K were significantly lower than for their counterparts processed at 296 and 423K, due to the variation in dislocation density with processing temperature[14]. However, the level and homogeneity of strengthening are still higher when processing by HPT at elevated temperatures than observed in ECAP[26], and FSP[27], where strengthening has been lowered by dynamic recrystallization, over-ageing and precipitate coarsening[26][27]. The microhardness distributions in the AZ91 alloy are heterogeneous along the through-thickness directions in the initial stage of deformation. This is supported by the differences in microstructural observations along this direction. A sufficient high number of turns may reduce heterogeneity, by filling the alloy in-between the anvils and achieving a significant sticking condition which then increases the deformation and microstructural homogeneity[5][12][15][40][41]. The distribution of microhardness along the vertical and horizontal cross-sections showed considerable consistency for the current alloy processed at each specific processing temperature. This indicates the development of microstructural homogeneity with increasing imposed strain at each condition[8]. This consistency in the AZ91 alloy has not been observed in the AZ31 alloy or AZ91 alloy processed by ECAP[26] and FSP[27]. This is attributed to the difference in the aluminium content and stacking fault energy in both alloys, which control the extent of grain refinement, dislocation density, achieved homogeneity and resultant mechanical properties[14][21][38]. The behaviour of strain hardening and homogeneity of microhardness in the AZ91 alloy follows a standard model of hardness evolution with increasing equivalent strain reported in earlier work[20].

4.4. The Effect of the Equivalent Strain on the Hall-Petch Relation and Dislocation Density

The increase in the equivalent strain resulted in an evolution in microstructure and a gradual development in the microhardness[10]. The strength of the alloy in terms of its microhardness improved significantly with grain refinement at room temperature. This proportionality has been expressed by the Hall-Petch relationship for hardness measurements: $H_v = H_0 + k_H d^{-1/2}$ [42]. The effect of grain refinement on the strength of the ultrafine-grained alloy AZ91 alloy showed a significant consistency with this Hall-Petch relationship. The material constants are H_0 = 76MPa and k_H = 233MPa μm$^{1/2}$, which is relatively higher than those found for AZ31 and AZ61 alloys (H_0 = 647–697MPa and k_H = 118–170MPa μm$^{1/2}$)[7][22][43]. Thus, the ultrafine-grained AZ91 alloy shows a relatively higher level of hardness than for AZ31 and AZ61 alloys processed by HPT and ECAP processing at room temperature and elevated temperatures[21][22][44]. The difference in k_H can be attributed to the difference in alloying constituents in the mentioned alloys, where the high content of alloying element in the AZ91 alloy resulted in a lowering of its stacking fault and thus a finer microstructure and a higher dislocation density in the AZ91 alloy after processing than in the AZ61 and AZ31 alloy[19][21][22]. The evolution in dislocation density with increase of imposed strain in HPT has a major effect on the achieved strengthening in the AZ91 alloy. The evolution of dislocation density is affected by the fraction of nano-sized particles of β-phase, value of applied pressure in HPT, and value of stacking fault energy. The widely distributed b-phase fine particles are reported as acting as barriers for mobile dislocations during deformation[39]. The high value of applied pressure has also been reported to enhance the obstruction of defect migration in the processed material and then promotes the suppression of dislocation annihilation[45][46]. The low stacking fault energy in the AZ91 alloy leads to a significant inhibition of dislocation cross-slip, and formation of a high density of planar arrays of dislocations has also been reported[10][38].

5. Conclusions

1. AZ91 magnesium alloy has been effectively processed in HPT processing at room temperature with an ultrafine-grained microstructure down to 35nm. The

alloy processed at 423 and 473K has been significantly refined by twinning segmentation of the original grains into fine grains with average apparent grain sizes of 180 and 250nm, respectively.

2. Fragmentation and alignment of the β-phase in the direction of torsional strain have been observed during processing. This phase has been refined down to nanometre sizes with a higher fraction as the number of turns increased, indicating the very high level of plastic deformation that is imparted to the alloy during HPT.

3. Existence of twins at all processing temperatures and their distribution was proportional to processing temperature and the number of turns. The occurrence of twinning has been induced by the need for re-orientation of the microstructure towards the slip direction and to accommodate severe plastic deformation.

4. Lower processing temperature has resulted in homogenous microstructure and significant development of strength. Higher processing temperatures have resulted in heterogeneous microstructures especially in the initial stages of HPT and this heterogeneity decreased gradually at higher numbers of turns.

5. A considerable dislocation density has developed with increasing the number of turns at lower processing temperature rather than at higher processing temperatures. The values of dislocation density after HPT were higher than earlier reported data for the same alloy.

6. The ultrafine-grained AZ91 alloy follows the Hall-Petch relationship, and this emphasizes the significant dependence of strength on grain size. The higher alloying content, fraction of nano-sized particles of b-phase and the dominance of basal slip and pyramidal modes after processing also have a significant effect on the strengthening of the alloy processed at 296K.

Acknowledgements

One of the authors (Ahmed S. J. Al-Zubaydi) is grateful to The Higher Committee for Education Development (HCED) of the Government of Iraq for the provision of Ph.D. scholarship.

Funding this work was supported in part by the Russian Science Foundation under Grant No. 14-29-00199 (APZ).

Compliance with Ethical Standards

Conflict of Interest: The authors declare that they have no conflict of interest.

Open Access: This article is distributed under the terms of the Creative Commons Attribution 4.0 International License (http://creativecommons.org/licenses/by/4.0/), which permits unrestricted use, distribution, and reproduction in any medium, provided you give appropriate credit to the original author(s) and the source, provide a link to the Creative Commons license, and indicate if changes were made.

Source: Al-Zubaydi A S J, Zhilyaev A P, Wang S C. Evolution of microstructure in AZ91 alloy processed by high-pressure torsion[J]. Journal of Materials Science, 2016, 51(7):3380–3389.

References

[1] Ravi Kumar NV, Blandin JJ, Desrayaud C, Montheillet F, Suéry M (2003) Grain refinement in AZ91 magnesium alloy during thermomechanical processing. Mater Sci Eng A 359:150–157.

[2] Yamashita A, Horita Z, Langdon TG (2001) Improving the mechanical properties of magnesium and a magnesium alloy through severe plastic deformation. Mater Sci Eng A 300:142–147.

[3] Kubota K, Mabuchi M, Higashi K (1999) Processing and mechanical properties of fine-grained magnesium alloys. J Mater Sci 34:2255–2262. doi:10.1023/A:1004561205627.

[4] Mabuchi M, Iwasaki H, Yanase K, Higashi K (1997) Low temperature superplasticity in an AZ91 magnesium alloy processed by ECAE. Scr Mater 36:681–686.

[5] Figueiredo RB, Langdon TG (2011) Development of structural heterogeneities in a magnesium alloy processed by high-pressure torsion. Mater Sci Eng A 528:4500–4506.

[6] Kai M, Horita Z, Langdon TG (2008) Developing grain refinement and superplastic-

ity in a magnesium alloy processed by high-pressure torsion. Mater Sci Eng A 488:117–124.

[7] Xia K, Wang JT, Wu X, Chen G, Gurvan M (2005) Equal channel angular pressing of magnesium alloy AZ31. Mater Sci Eng A 410–411:324–327.

[8] Valiev RZ, Zhilyaev AP, Langdon TG (2013) Bulk nanostructured materials: fundamentals and applications. Wiley, New Jersey.

[9] Gubicza J, Máthis K, Hegedűs Z, Ribárik G, Tóth AL (2010) Inhomogeneous evolution of microstructure in AZ91 Mg-alloy during high temperature equal-channel angular pressing. J Alloy Compd 492:166–172.

[10] Zhilyaev AP, Langdon TG (2008) Using high-pressure torsion for metal processing: fundamentals and applications. Prog Mater Sci 53:893–979.

[11] Vander Voort GF (2004) Metals handbook: metallography and microstructures, vol 9. ASM International, Cleveland.

[12] Kawasaki M, Figueiredo RB, Langdon TG (2011) An investigation of hardness homogeneity throughout disks processed by high-pressure torsion. Acta Mater 59:308–316.

[13] Al-Zubaydi ASJ, Zhilyaev AP, Wang SC, Reed PAS (2015) Superplastic behaviour of AZ91 magnesium alloy processed by high-pressure torsion. Mater Sci Eng A 637:1–11.

[14] Máthis K, Gubicza J, Nam NH (2005) Microstructure and mechanical behavior of AZ91 Mg alloy processed by equal channel angular pressing. J Alloy Compd 394: 194–199

[15] Hohenwarter A, Bachmaier A, Gludovatz B, Scheriau S, Pippan R (2009) Technical parameters affecting grain refinement by high pressure torsion. Int J Mater Res 100:1653–1661.

[16] Myshlyaev MM, McQueen HJ, Mwembela A, Konopleva E (2002) Twinning, dynamic recovery and recrystallization in hot worked Mg–Al–Zn alloy. Mater Sci Eng A 337:121–133.

[17] del Valle JA, Pérez-Prado MT, Ruano OA (2003) Texture evolution during large-strain hot rolling of the Mg AZ61 alloy. Mater Sci Eng A 355:68–78.

[18] Al-Samman T, Gottstein G (2008) Room temperature formability of a magnesium AZ31 alloy: examining the role of texture on the deformation mechanisms. Mater Sci Eng A 488:406–414.

[19] Zhao YH, Liao XZ, Zhu YT, Horita Z, Langdon TG (2005) Influence of stacking fault energy on nanostructure formation under high pressure torsion. Mater Sci Eng A 410–411:188–193.

[20] Kawasaki M, Figueiredo RB, Huang Y, Langdon TG (2014) Interpretation of hardness evolution in metals processed by high-pressure torsion. J Mater Sci 49: 6586–6596. doi:10.1007/s10853-014-8262-8.

[21] Stráská J, Janeček M, Gubicza J, Krajňák T, Yoon EY, Kim HS (2015) Evolution of microstructure and hardness in AZ31 alloy processed by high pressure torsion. Mater Sci Eng A 625:98–106.

[22] Harai Y, Kai M, Kaneko K, Horita Z, Langdon TG (2008) Microstructural and mechanical characteristics of AZ61 magnesium alloy processed by high-pressure torsion. Mater Trans 49:76–83.

[23] Matsubara K, Miyahara Y, Horita Z, Langdon TG (2003) Developing superplasticity in a magnesium alloy through a combination of extrusion and ECAP. Acta Mater 51:3073–3084.

[24] Braszczyńska-Malik KN, Froyen L (2005) Microstructure of AZ91 alloy deformed by equal channel angular pressing. Int J Mater Res (former: Zeitschrift für Metallkunde) 96:913–917.

[25] Braszczyńska-Malik KN (2009) Spherical shape of c-Mg17Al12 precipitates in AZ91 magnesium alloy processed by equal-channel angular pressing. J Alloy Compd 487:263–268.

[26] Zhao Z, Chen Q, Hu C, Shu D (2009) Microstructure and mechanical properties of SPD-processed an as-cast AZ91D + Y magnesium alloy by equal channel angular extrusion and multi-axial forging. Mater Des 30:4557–4561.

[27] Cavaliere P, De Marco PP (2007) Superplastic behaviour of friction stir processed AZ91 magnesium alloy produced by high pressure die cast. J Mater Process Technol 184:77–83.

[28] Pérez-Prado MT, Valle D, Ruano OA (2004) Grain refinement of Mg–Al–Zn alloys via accumulative roll bonding. Scr Mater 51:1093–1097.

[29] Kim WJ, Hong SI, Kim YH (2012) Enhancement of the strain hardening ability in ultrafine grained Mg alloys with high strength. Scr Mater 67:689–692.

[30] Figueiredo RB, Pereira PHR, Aguilar MTP, Cetlin PR, Langdon TG (2012) Using finite element modeling to examine the temperature distribution in quasi-constrained high-pressure torsion. Acta Mater 60:3190–3198.

[31] Edalati K, Miresmaeili R, Horita Z, Kanayama H, Pippan R (2011) Significance of temperature increase in processing by high-pressure torsion. Mater Sci Eng A 528:7301–7305.

[32] Al-Zubaydi A, Figueiredo RB, Huang Y, Langdon TG (2013) Structural and hardness inhomogeneities in Mg-Al-Zn alloys processed by high-pressure torsion. J Mater Sci 48:4661–4670. doi:10.1007/s10853-013-7176-1.

[33] Zhilyaev AP, Nurislamova GV, Kim B-K, Baro' MD, Szpunar JA, Langdon TG (2003) Experimental parameters influencing grain refinement and microstructural evolution during high-pressure torsion. Acta Mater 51:753–765.

[34] Zhilyaev AP, Oh-ishi K, Langdon TG, McNelley TR (2005) Microstructural evolution in commercial purity aluminum during high-pressure torsion. Mater Sci Eng A

410–411:277–280.

[35] Zhilyaev AP, Lee S, Nurislamova GV, Valiev RZ, Langdon TG (2001) Microhardness and microstructural evolution in pure nickel during high-pressure torsion. Scr Mater 44:2753–2758.

[36] Vorhauer A, Pippan R (2004) On the homogeneity of deformation by high pressure torsion. Scr Mater 51:921–925.

[37] Huang Y, Kawasaki M, Langdon TG (2013) Influence of anvil alignment on shearing patterns in high-pressure torsion. Adv Eng Mater 15:1–755.

[38] Somekawa H, Hirai K, Watanabe H, Takigawa Y, Higashi K (2005) Dislocation creep behavior in Mg-Al-Zn alloys. Mater Sci Eng A 407:53–61.

[39] Tahreen N, Chen DL, Nouri M, Li DY (2014) Effects of aluminum content and strain rate on strain hardening behavior of cast magnesium alloys during compression. Mater Sci Eng A 594:235–245.

[40] Figueiredo RB, Aguilar MTP, Cetlin PR, Langdon TG (2012) Analysis of plastic flow during high-pressure torsion. J Mater Sci 47:7807–7814. doi:10.1007/s10853-012-6506-z.

[41] Kawasaki M, Figueiredo RB, Langdon TG (2012) Twenty-five years of severe plastic deformation: recent developments in evaluating the degree of homogeneity through the thickness of disks processed by high-pressure torsion. J Mater Sci 47: 7719–7725. doi:10.1007/s10853-012-6507-y.

[42] Furukawa M, Horita Z, Nemoto M, Valiev RZ, Langdon TG (1996) Microhardness measurements and the Hall-Petch relationship in an Al-Mg alloy with submicrometer grain size. Acta Mater 44:4619–4629

[43] Xu J, Shirooyeh M, Wongsa-Ngam J, Shan D, Guo B, Langdon TG (2013) Hardness homogeneity and micro-tensile behavior in a magnesium AZ31 alloy processed by equal-channel angular pressing. Mater Sci Eng A 586:108–114.

[44] Chang S-Y, Lee S-W, Kang KM, Kamado S, Kojima Y (2004) Improvement of mechanical characteristics in severely plastic-deformed Mg alloys. Mater Trans 45:488–492.

[45] Setman D, Schafler E, Korznikova E, Zehetbauer MJ (2008) The presence and nature of vacancy type defects in nanometals detained by severe plastic deformation. Mater Sci Eng A 493:116–122.

[46] Gubicza J, Dobatkin SV, Khosravi E, Kuznetsov AA, Lábár JL (2011) Microstructural stability of Cu processed by different routes of severe plastic deformation. Mater Sci Eng A 528:1828–1832.

Chapter 14

Fabrication of Bioactive Polycaprolactone/Hydroxyapatite Scaffolds with Final Bilayer Nano-/Micro-Fibrous Structures for Tissue Engineering Application

Izabella Rajzer

Division of Materials Engineering, Department of Mechanical Engineering Fundamentals, ATH University of Bielsko-Biala, Willowa 2 Street, 43-309 Bielsko-Biała, Poland

Abstract: In this study, two techniques, namely electrospinning and needle-punching processes, were used to fabricate bioactive polycaprolactone/hydroxyapatite scaf-folds with a final bilayer nano-/micro-fibrous porous structure. A hybrid scaffold was fabricated to combine the beneficial properties of nanofibers and microfibers and to create a three-dimensional porous structure (which is usually very difficult to produce using electrospinning technology only). The first part of this work focused on determining the conditions necessary to fabricate nano- and micro-fibrous components of scaffold layers. A characterization of scaffold components, with respect to their morphology, fiber diameter, pore size, wettability, chemical composition and mechanical properties, was performed. Then, the same process parameters were applied to produce a hybrid bilayer scaffold by electrospinning the nanofibers directly onto the micro-fibrous nonwovens obtained in a traditional mechanical needle-punching process. In the second part, the bioactive character of a hybrid nano-/micro-fibrous scaffold in simulated body fluid (SBF) was assessed. Spherical calcium phosphate was precipitated onto the nano-/microfibrous scaffold surface proving its bioactivity.

1. Introduction

A number of fabrication technologies have been applied to produce an ideal scaffold for bone tissue engineering[1]. Fiber-based structures represent a wide range of morphological and geometric possibilities that can be tailored for each specific tissue engineering application[2][3]. Recently, electrospinning of nanofibers has been widely researched and numerous composite nanofibers, containing bioactive molecules, have been prepared to match the requirements for bone tissue scaffolds[4]–[6]. Current developments include the fabrication of nanofibrous scaf-folds which, due to their similarity to the extracellular matrix (ECM), can provide chemical, mechanical and biological signals to respond to the environmental stimuli[7]. Several investigators have developed electrospun scaffolds that combine degradable polymers such as polycaprolactone (PCL) with nano-hydroxyapatite (n-HAp)[8][9]. Many studies have discovered that a composite of PCL/n-HAp biomaterials generally favors calcium phosphate mineralization followed by an osteogenic differentiation process[10]–[12]. Several studies have showed that, by introducing a bone-bioactive inorganic component (n-HAp) into the PCL matrix, better interaction and improved cell adhesion with the biological environment can be achieved[13]. It is well known that inter-connected pores in tissue engineering scaffolds are essential for cell growth, migration, vascularization, and tissue formation[14]. Electrospinning is a process that can generate a fibrous scaffold with high porosity, interconnected pores, a large surface-area-to-volume ratio and a variable fiber diameter[15]. However, the small pore size of electrospun nanofibrous scaffolds may limit cellular infiltration[16]. Pore size below cellular diameter cannot allow cell migration within the structure[17]. Porosity and thereby cell penetration could be enhanced in scaffolds by using the strategy of creating a 3D nano-/micro-architecture, where nanofibers are combined with microfibers[18]–[22]. Nano-/micro-fibrous scaffolds have an innovative structure, inspired by an ECM that combines a nano-network, aimed to promote cell adhesion, with a micro-fiber mesh to generate the 3-D structure[23][24]. Although micro-fibrous scaffolds are not on the same size scale as ECM components, they could potentially be advantageous because they are composed of larger pores as compared to nano-fibrous scaffolds[25]. Conventional textile technologies, which are simple and nontoxic (such as the needle-punching technique), can generate highly porous, micro-fibrous nonwovens with interconnected pores suitable for tissue engineering scaffolds[26]. Moreover, modification of fibers using bioactive ceramic particles can lead to the development of new materials for the manufacturing of implants which can establish direct chemical bonds with bone tissue after implantation[27][28]. A nano-fibrous membrane in a bilayer scaffold can also act as a barrier membrane which serves two functions: it is permeable to nutrients yet, when coupled with a micro-fibrous scaffold, allows for the proliferation of cells on both sides of the barrier while at the same time preventing the unwanted migration of cells across the barrier[29].

This study focused on producing bilayer hybrid nano-/micro-fibrous composite scaffolds of poly(ε-caprolactone) modified with hydroxyapatite nanoparticles for tissue engineering application. The first part of this work concentrated on determining the conditions necessary to produce nano- and micro-fibrous components of scaffold layers. A characterization of scaffold components with respect to their morphology, fiber diameter, pore size, wettability, chemical composition, and mechanical properties was performed. Then, the same process parameters were applied to produce a hybrid bilayer scaffold by electrospinning nanofibers directly onto the micro-fibrous nonwovens obtained in a traditional mechanical needle-punching process. In the second part, the bioactive character of the hybrid nano-/micro-fibrous scaffold was assessed.

2. Materials and Methods

2.1. Scaffold Fabrication

2.1.1. Nano-Fibrous Scaffolds

In this study, nano-fibrous scaffolds were produced by electrospinning. Poly (ε-caprolactone) from Sigma-Aldrich, with an average molecular weight of 80kDa, was dissolved into equal parts of chloroform and methanol (POCH, Poland) at a concentration of 6% (wt/vol). For the composite scaffold, the PCL solution was mixed with 20wt% of n-HAp (AGH, Cracow, Poland)[30]. The solutions were loaded into a 10mL plastic syringe with a stainless-steel blunt needle (diameter 0.7mm). The needle was connected to a high-voltage power supply which generated a voltage of 30kV. The flow rate of the solution was 1.5mL/h. The fibers were collected on a rotary drum (diameter: 60mm; length: 300mm; rotation speed: 230rpm; linear velocity: 0.72m/s) wrapped with a piece of silica-coated paper which was placed at a distance of 15cm from the needle tip. The deposition was perform for a duration of 3h. Two types of electrospun scaffolds were formed: (1) pure PCL (nPCL) and (2) PCL modified with n-HAp (nPCL/n-HAp).

2.1.2. Micro-Fibrous Scaffolds

Micro-fibrous scaffolds were produced using a traditional needle-punching process. Before scaffold fabrication, pure PCL and composite PCL/n-HAp fibers were produced using the melt spinning method, with a prototype laboratory spinning machine PROMA (Torun, Poland) as previously described[31]. In order to obtain composite fibers, n-HAp powder (5wt%) was premixed into the PCL polymer before melting. This amount of n-HAp powder was chosen based on previous studies in which the higher concentration of nanoparticles resulted in a significant decrease in the mechanical properties of the fibers[32]. The premixing process was applied in order to obtain homogeneously mixed components. In the premixing process, hydroxyapatite powder and polymer granulates were premixed in the mixer and then fed into the extruder. The extruder homogenized them and a masterbatch of PCL/n-HAp was produced before the principal process of forming fibers. Fibers were extruded from the melted components at a temperature of 170°C and were spun with a take-up velocity of 247m/min. The obtained fibers were then carded using a laboratory carding machine and arranged in a 2D sheet with a random fiber orientation. The fibers in the web were then bonded together using a needle-punching process. For both samples, the same amount of needling and needling depth was used. Finally, two types of non-woven scaffolds were obtained using this method: (1) pure PCL (mPCL) and (2) PCL modified with n-HAp (mPCL/n-HAp).

2.1.3. Hybrid Nano/Micro-Fibrous Scaffolds

Hybrid bilayer scaffolds consisting of a top nanofiber layer and a bottom microfiber layer were obtained by electrospinning nPCL or nPCL/n-HAp nanofibers directly onto the micro-fibrous mPCL or mPCL/n-HAp nonwoven scaffolds. The same electrospinning process parameters for nanofibrous scaffold fabrication were applied to produce a hybrid bilayer scaffolds. The nanofibers were collected on a rotary drum wrapped with a mPCL or mPCL/n-HAp nonwovens.

Two types of hybrid materials were obtained: (1) a non-modified hybrid scaffold (H-PCL) and (2) a n-HAp modified hybrid scaffold (H-PCL/n-HAp).

2.2. Scaffold Characterization

In order to determine the diameter and morphology of the nano- and micro-fibrous materials, scaffold samples were covered by a sputtered gold coating and analyzed by a scanning electron microscope (JEOL JSM 5500). The average diameter of the fibers was determined by performing measurements on 60 fibers.

A capillary flow porometer (PMI, USA) was applied to evaluate the pore size distribution of nano- and micro-fibrous scaffolds. In a PMI porometer, a non-reacting gas flows through a dry sample and then through the same sample after it has been wet with isopropyl alcohol. The change in flow rate is evaluated as a function of pressure for both dry and wet processes[33]. Three samples from each material were used. For each sample, the pore size measurement was repeated three times.

Attenuated total reflectance-Fourier transform infrared spectroscopy (ATR-FTIR) was used to characterize the nano- and micro-fibrous scaffolds. FTIR spectra were recorded using the FTS Digilab 60 BioRad spectrophotometer (in the 400–4000cm^{-1} range). All spectra were acquired through the accumulation of 64 scans at a resolution of 4cm^{-1}. One sample from each material was used in the study.

Water contact angles indicating the wetting ability of the nano- and micro-fibrous scaffolds were measured using drop shape analysis (DSA 10, KRUSS). A single droplet of doubly distilled water (20μL) was applied to the scaffold surface and contact angle measurements were taken at room temperature. Five measurements were done at different locations and the average value of the contact angle was obtained using standard deviation.

The tensile properties of the nanofibrous, microfibrous, and hybrid scaffolds were analyzed using a Zwick-Roell Z 2.5 testing machine at a cross-head speed of 10 mm/min. The rectangular samples were 10cm long and 2cm wide. At least three specimens were tested for each sample and the mean values of tensile strength were reported. The thickness of the scaffolds was determined using the thickness gage TILMET 73. The thickness test was performed on eight samples of each type of fibrous scaffolds. The values of the average thickness were determined.

The adhesion tests were done using a Zwick-Roell Z 2.5 testing machine at a cross-head speed of 10 mm/min. The rectangular samples were approximately 2.0 × 2.0cm and 0.5cm high. Double-sided tape was placed on the surface of the compression plates and coated with a thin layer of superglue. The samples were then placed between the plates and the hybrid scaffold was compressed up to −30 N and allowed to sit for 1 min to allow the glue to dry[21]. Then, the tensile test was performed until the two layers were separated. Three specimens for each test were considered. The maximum load and average adhesion strength were reported.

To determine the bioactivity of nano-/micro-fibrous scaffolds, the effect of the presence of n-HAp particles in the hybrid scaffold on apatite formation after immersion in simulated body fluid (SBF) was evaluated. In comparison, non-modified hybrid nano-/micro-fibrous scaffolds were also analyzed. Hybrid scaffolds were incubated for 7 days in 14mL of 1.59 SBF (pH 7.4; 37°C). The SBF was prepared following Kokubo's protocol[34]. The surface morphology of the samples was examined by SEM.

All the data in this article has been presented as mean ± standard deviation and has been analyzed using Student's t test for the calculation of the significance level of the data. Differences were considered significant when $p \leq 0.05$.

3. Results and Discussion

Interconnected nano- and micro-fibrous nonwovens were obtained using electrospinning and needle-punching processes. SEM images revealed a beadless, porous structure of the electrospun scaffolds [**Figure 1(a), (b)**]. Some aggregated n-HAp particles in the case of nPCL/n-HAp nanofibers were observed. The fiber diameter histogram [**Figure 1(c)**] revealed that most of the fibers were between 300 and 850nm with the highest frequency occurring in the 500nm range for modified nPCL/n-HAp nanofibers and in the 650nm range for pure nPCL samples. It was found that the addition of n-HAp particles into the PCL solution can reduce the fiber diameter. It seems likely that the addition of n-HAp into the PCL solution can cause an increase in net charge density during electrospinning. An increase in net charge should increase the electrostatic force within the jet and result in higher stretching of the jet, hence the decrease in fiber diameter. On the other hand, n-HAp particles agglomerate easily, therefore, the measured fiber diameter for both types of nanofibrous scaffolds will be in a similar range (300–1150nm). The SEM images of the micro-fibrous scaffolds obtained using the needle-punching process are shown in **Figure 1(d)** and **(e)**. The mPCL and mPCL/n-HAp fibers appeared randomly distributed without preferential orientation. The fiber's surface was affected by the mechanical intertwining of the fibers by needles. In addition, in the case of the mPCL/n-HAp sample, n-HAp agglomerates were seen on the fiber's surface. The fiber's diameter distribution is presented in **Figure 1(f)**. The diameter of fibers varied from 20 to 80μm for pure mPCL nonwoven and from 20 to 100μm for the modified mPCL/ n-HAp scaffold. The average diameter of the fibers was 37.6lm ± 17.7lm for pure mPCL and 49.9μm ± 21.7μm for composite mPCL/n-HAp fibers.

Figure 2 shows the pore size distribution of the electrospun nano-fibrous and needle-punched micro-fibrous scaffolds. The results revealed a very narrow distribution of pore size for nPCL and nPCL/n-HAp scaffolds centered at 1.2 and 2.0μm, respectively [**Figure 2(a)**]. The results obtained confirmed a general view that the electrospinning technique allows one to obtain scaffolds with a highly porous network of interconnected pores[35]. However, the average pore size in electrospun nanofibrous scaffolds is insufficient for bone tissue engineering[17]. The relatively small pore size compared to the cellular diameter (5–20μm) could not allow cell migration within the scaffolds[19]. In the case of micro-fibrous scaffolds,

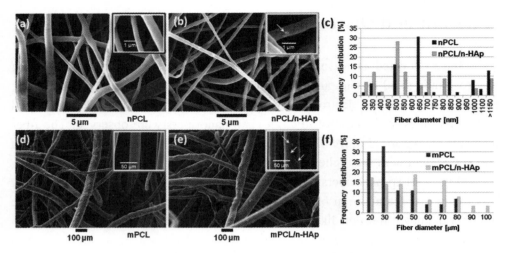

Figure 1. SEM images and fiber diameter distributions for the fibrous scaffolds: a-c nano-fibrous nPCL and nPCL/n-HAp scaffold fabricated by electrospinning and d-f micro-fibrous mPCL and mPCL/n-HAp scaffold fabricated using the needle-punching process.

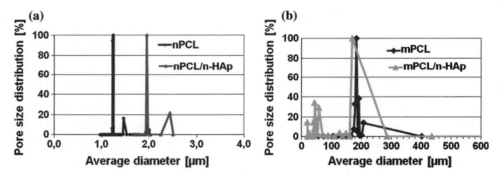

Figure 2. Pore size distributions for fibrous scaffolds: a nanofibrous scaffold fabricated by electrospinning and b microfibrous scaffold fabricated using the needle-punching process.

pore size distribution was much wider [**Figure 2(b)**]. The main pore fraction for n-HAp modified micro-fibrous samples was in the range of 160−280lm, smaller pores in the range of 20−60μm were also present. In the case of pure mPCL samples, major pore fractions were in the range of 180−200μm and 200 −400μm.

Figure 3 shows FTIR spectra of nano- and micro-fibrous scaffolds as well as of pure n-HAp powder. Several characteristic bands of PCL were observed for both types of nano- and micro-fibrous samples at $1727 cm^{-1}$ (C=O stretching), $1293 cm^{-1}$ (C−O and C−C stretching), $1240 cm^{-1}$ (C−O−C asymmetric stretching),

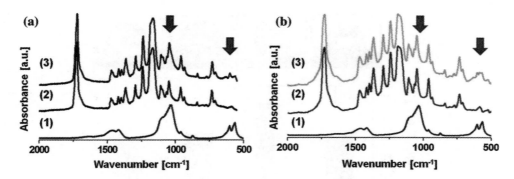

Figure 3. FTIR spectra of fibrous scaffolds: a nano-fibrous scaffold fabricated by electrospinning (1) n-HAp, (2) nPCL, and (3) nPCL/n-HAp; b micro-fibrous scaffold fabricated using the needlepunching process (1) n-HAp, (2) mPCL, and (3) mPCL/n-Hap.

and 1175cm^{-1} (C−O−C symmetric stretching)[36]. In the case of n-HAp modified scaffolds, some bands characteristic of n-HAp were identified. Although the main ν3 PO_4^{3-} stretching mode at about 1033cm^{-1} was overlapped by the PCL spectrum, the ν4 vibrational band was detected for nPCL/n-HAp and mPCL/n-HAp samples at 605 and 566cm^{-1}[37]. The results confirmed a successful incorporation of n-HAp particles both into the nano- and micro-fibrous scaffold.

The contact angle measurement was performed in order to understand the influence of n-HAp on micro- and nano-fibrous scaffolds' surfaces wetting ability. Water contact angle values of the prepared scaffold are presented in **Figure 4(a)**. All fibrous scaffolds revealed contact angles of over 100°, which means that water does not spread on the surface of these materials due to their hydrophobic properties. In addition, the incorporation of n-HAp nanoparticles within the nPCL nanofibrous scaffold (nPCL/n-HAp) increased the hydrophobicity of its surface and the highest value of the contact angle (about 135°) was observed using nPCL/n-HAp samples. The static contact angle measured is regulated by the surface chemistry and the surface roughness of the sample. The smoother the surface, the smaller the contact angle. The addition of n-HAp into the electrospun scaffolds increase the surface roughness resulting from the presence of extra crystallites on the fiber's surface and thus increases the contact angle. In the case of the nonwovens prepared with the traditional needle-punching technique, the addition of n-HAp has no effect on the change of the contact angle value.

Representative stress-strain curves and average tensile strengths of obtained

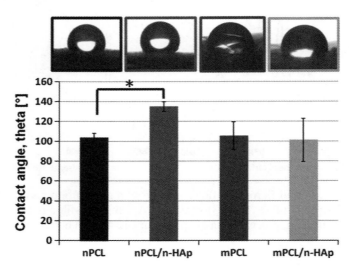

Figure 4. The contact angle of nano- and micro-fibrous scaffolds. The data are expressed as the mean ± SD (n = 5). The asterisk indicates a statistically significant difference.

nano-fibrous, micro-fibrous, and hybrid scaffolds are shown in **Figure 5**. Nano- and micro-fibrous samples showed different tensile behavior [**Figure 5(b), (d)**]. The stress and strain values of micro-fibrous scaffolds were lower compared to nano-fibrous scaffolds, indicating an inferior deformability and flexibility. The addition of n-HAp powder resulted in a decrease of mechanical properties of obtained nano- and micro-fibrous scaffolds, probably due to the agglomeration of n-HAp particles. Inorganic nanoparticles generally agglomerate easily and cannot be intermixed well. The addition of n-HAp particles creates weak links in the PCL matrix. These weak links become stress concentrators in a continuum matrix and thus reduce the tensile strength of the composite fibers. The tensile strength decrease ranges from 2.2MPa for an unmodified nano-fibrous scaffold to 1.3MPa for a composite nano-fibrous scaffold and, in the case of microfibrous scaffolds from 8.7kPa, for an unmodified scaffold to 5.4kPa for composite scaffold [**Figure 5(a), (c)**]. It should be noted that the higher porosity of micro-fibrous scaffolds is responsible for their lower tensile strength.

Hybrid scaffolds were fabricated using the same process parameters as those applied before for individual scaffold components. The typical morphology of the hybrid nano-/micro-fibrous scaffolds obtained is presented in **Figure 6**. Homogenous deposition of the randomly orientated nano-fibrous layer on the micro-fibrous nonwovens was observed in the case of both types of samples (n-HAp

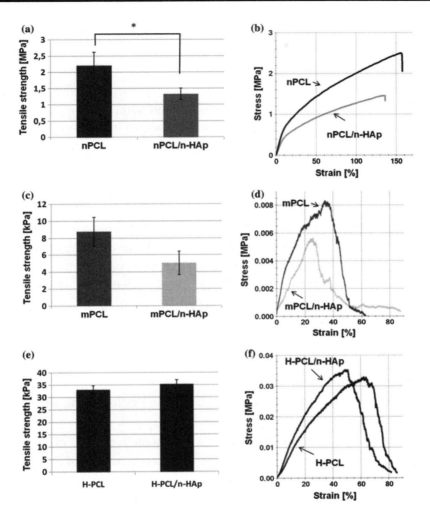

Figure 5. Tensile strength and typical stress/strain curves of (a, b) nanofibrous, (c, d) microfibrous, and (e, f) nano-/micro-hybrid scaffolds. The data are expressed as the mean ± SD (n = 3). The asterisk indicates a statistically significant difference.

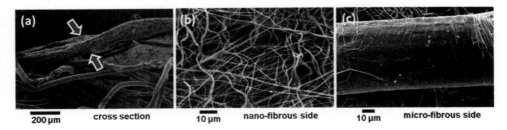

Figure 6. Microstructure of H-PCL hybrid scaffold a cross section, b nano-fibrous side of scaffold, and c micro-fibrous side of scaffold.

modified and non-modified). A micro-fibrous 3D porous structure served as a support for the nanofibrous mesh. The cross section revealed that nanofibers formed a layer with the thickness of about 115μm.

In **Figure 7**, the pore size distribution for both hybrid materials is presented. The unmodified hybrid H-PCL scaffolds show a bimodal distribution of pore sizes, centered around 3.0μm and in the range of 20–58μm. In the case of modified hybrid H-PCL/n-HAp, the main fraction of pores is in the range of 26–32μm. Capillary flow porometry assumes a cylindrical pore or capillary of varying diameter through the thickness of the scaffold and measures only its smallest diameter (throat diameter)[38]. The results indicate that, in hybrid bilayered scaffolds composed of nano- and micro-fibers, it is possible to obtain three types of pore size ranges: (1) pores larger than 200μm (pores in which cells could grow), (2) smaller pores dedicated to the diffusion of nutrients or/and metabolites of bone-forming cells (20–40μm), and (3) pores enabling better neovascularization and cell attachment (2–5μm).

The mechanical properties of the microfibrous scaffolds were improved remarkably by introducing a nanofibrous layer. The hybrid scaffolds had much higher tensile strength than mPCL and mPCL/n-HAp scaffolds [**Figure 5(e), (f)**].

Figure 8(a) and **Figure 8(b)** shows a representative curve of the adhesion tests. The first linear part of the curve is the recovery from the −30N compression step. At around 0N, the slope of the curve changed, corresponding to the point where the scaffold was not compressed any more, *i.e.*, the start of the adhesion test.

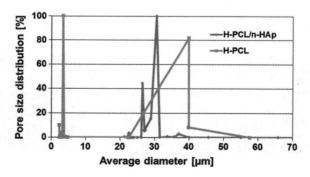

Figure 7. Pore size distribution for hybrid nano-/micro-fibrous scaffold.

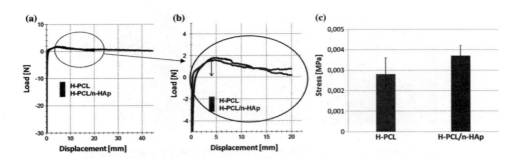

Figure 8. Representative curves of the adhesion tests of the H-PCL and H-PCL/n-HAp hybrid scaffolds (a, b), adhesion strength of hybrid scaffolds (c). There were no significant differences among the H-PCL and H-PCL/n-HAp samples.

The load increased up to a peak and then gradually decreased. At this stage, one of the layers was completely separated from an adjacent layer. Adhesion strength values and standard deviation associated with H-PCL and H-PCL/n-HAp samples are shown in **Figure 8(c)**. The adhesion load was around 2N for both types of hybrid scaffolds. Non-modified H-PCL scaffolds had an adhesion strength of 2.8kPa ± 0.8kPa and modified H-PCL/n-HAp scaffolds had an adhesion strength of 3.7kPa ± 0.5kPa. There were no significant differences among the adhesion strength values of H-PCL and H-PCL/n-HAp samples.

The bone-bonding ability of hybrid materials was evaluated by examining the ability of apatite to form on its surface in a SBF with ion concentrations nearly equal to those of human blood plasma. The macroscopic images of hybrid bilayer scaffolds after immersion in SBF for 7 days are shown in **Figure 9**. Obviously, no apatite-like materials were precipitated within incubation of a pure (not modified) H-PCL hybrid scaffold [**Figure 9(a)**, **(b)**]. However, with the introduction of n-HAp nanoparticles into the hybrid H-PCL/n-HAp scaffold, the formation of calcium phosphate occurred and an apatite-like mineral layer completely covered the surface of the sample [**Figure 9(c)**, **(d)**]. The surface of apatite layer deposited by immersion in SBF was formed by globular structures. Preliminary results showed that the presence of n-HAp particles within the nano- and micro-fibrous hybrid scaffold provokes the nucleation of calcium phosphate in SBF.

In this study, bioactive hybrid scaffolds were fabricated through direct spinning of nanofibers on micro-fibrous nonwovens. Both scaffold layers were modified with n-HAp in order to ensure the bioactive character of the final hybrid

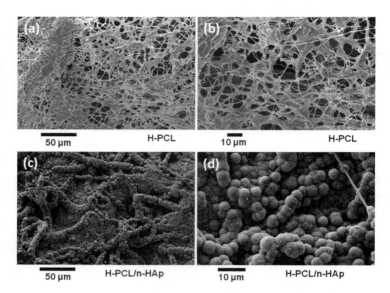

Figure 9. Microstructure of H-PCL (a, b) and H-PCL/n-HAp (c, d) scaffolds after 7 days of immersion in SBF.

scaffold. The presence of n-HAp was confirmed by FTIR and SEM studies. The materials usually proposed for bone tissue engineering made of ECM-mimicking nanofibrous electrospun mats cannot be used as 3D scaffolds due to their very small pore size. It is well known that pore size distribution, porosity, and pore interconnectivity are crucial parameters for tissue engineering scaffolds as they provide the optimal spatial and nutritional conditions for the cells and determine the successful integration of the natural bone tissue and the material, as well as enabling neovascularization and cell attachment. By combining nano-fibers with micro-fibers, a new scaffold material with a 3D structure and different pore size ranges was obtained. Such bilayer material can be used to separate and allow for the independent proliferation of two distinct tissue types: soft tissue on one side and bone tissue on the other side. The bioactive character of these hybrid scaffolds was confirmed by an SBF mineralization study. After only 7 days, the hybrid scaffold surface was already completely covered by apatite deposits, proving the biocompatibility of these materials.

4. Conclusions

The main objective of this study was to obtain new hybrid nano-/micro-fibrous materials which could be prospectively applied as three-dimensional porous scaffolds for bone tissue engineering purposes. In this study, it was determined that the combined use of two techniques, namely electrospinning and needle-punching processes, could be beneficial for the fabrication of 3D bioactive scaffolds with a desired pore size and a final nano-/micro-fibrous structure. Such scaffolds can be successfully used in the body to separate soft tissue from bone tissue in order to allow for new bone growth.

Acknowledgements

This study was supported by the Polish Ministry of Science and Higher Education (Project Iuventus Plus II No. IP2011044671) and by the Polish National Science Center (Project No. N507550938).

Open Access: This article is distributed under the terms of the Creative Commons Attribution License which permits any use, distribution, and reproduction in any medium, provided the original author(s) and the source are credited.

Source: Rajzer I. Fabrication of bioactive polycaprolactone/hydroxyapatite scaffolds with final bilayer nano-/micro-fibrous structures for tissue engineering application[J]. Journal of Materials Science, 2014, 49(16):5799−5807.

References

[1] Liu C, Xia Z, Czernuszka T (2007) Design and development of three-dimensional scaffolds for tissue engineering. Trans IChemE A 85(A7):1051–1064.

[2] Tuzlakoglu K, Reis RL (2009) Biodegradable polymeric fiber structures in tissue engineering. Tissue Eng B 15(1):17–27.

[3] Tuzlakoglu K, Santos MI, Neves N, Reis RL (2011) Design of nano- and microfiber combined scaffold by electrospinning of collagen onto starch-based fiber meshes: a man-made equivalent of natural extracellular matrix. Tissue Eng A 17(3–4):463–473.

[4] Jang J-H, Castano O, Kim H-W (2009) Electrospun materials as potential platforms for bone tissue engineering. Adv Drug Deliv Rev 61:1065–1083.

[5] Ngiam M, Liao S, Patil AJ, Cheng Z, Chan CK, Ramakrishna S (2009) The fabrication of nano-hydroxyapatite on PLGA and PLGA/collagen nanofibrous composite

scaffolds and their effects in osteoblastic behavior for bone tissue engineering. Bone 45:4–16.

[6] Rajzer I, Menaszek E, Kwiatkowski R, Chrzanowski W (2014) Bioactive nanocomposite PLDL/nano-hydroxyapatite electrospun membranes for bone tissue engineering. J Mater Sci Mater Med 25(5):1239–1247. doi:10.1007/s10856-014-5149-9.

[7] Prabhakaran MP, Ghasemi-Mobarakeh L, Ramakrishna S (2011) Electrospun composite nanofibers for tissue regeneration. J Nanosci Nanotechnol 11(4):3039–3057.

[8] Stodolak-Zych E, Szumera M, Blazewicz M (2013) Osteoconductive nanocomposite materials for bone regeneration. Mater Sci Forum 730–732:38–43.

[9] Fabbri P, Bondioli F, Messori M, Bartoli C, Dinucci D, Chiellini F (2010) Porous scaffolds of polycaprolactone reinforced with in situ generated hydroxyapatite for bone tissue engineering. J Mater Sci Mater Med 21:343–351.

[10] Stodolak-Zych E, Fraczek-Szczypta A, Wiechec A, Blazewicz M (2012) Nanocomposite polymer scaffolds for bone tissue regeneration. Acta Phys Pol A 121(2): 518–521.

[11] Lu Z, Roohani-Esfahani S-I, Kwok PCL, Zreiqat H (2011) Osteoblasts on rod shaped hydroxyapatite nanoparticles incorporated PCL film provide an optimal osteogenic niche for stem cell differentiation. Tissue Eng A 17(11–12):1651–1661.

[12] Rajzer I, Menaszek E (2012) Cell differentiation on electrospun poly-(epsilon-caprolactone) membranes modified with hydroxy-apatite. J Tissue Eng Regen Med 6(Supplement 1):194.

[13] Bianco A, Federico ED, Moscatelli I, Camaioni A, Armentano I, Campagnolo L, Dottori M, Kenny JM, Siracusa G, Gusmano G (2009) Electrospun poly-(e-caprolactone)/Ca-deficient hydroxyapatite nanohybrids: microstructure, mechanical properties and cell response by murine embryonic stem cells. Mater Sci Eng C 29:2063–2071.

[14] Sanzana ES, Navarro M, Ginebra MP, Planell JA, Ojeda AC, Montecinos HA (2013) Role of porosity and pore architecture in the in vivo bone regeneration capacity of biodegradable glass scaffolds. J Biomed Mater Res A 102:1767–1773. doi:10.1002/jbm.a.34845.

[15] Pham QP, Sharma U, Mikos AG (2006) Electrospun poly(e-caprolactone) microfiber and multilayer nanofiber/microfiber scaffolds: characterization of scaffold and measurement of cellular infiltration. Biomacromolecules 7:2796–2805.

[16] Lee JB, Jeong SI, Bae MS, Yang DH, Heo DN, Kim CH, Eben Alsberg E, Kwon IK (2011) Highly porous electrospun nanofibers enhanced by ultrasonication for improved cellular infiltration. Tissue Eng A 17(21–22):2695–2702.

[17] Tuzlakoglu K, Bolgen N, Salgado AJ, Gomes ME, Piskin E, Reis RL (2005) Nano- and micro-fiber combined scaffolds: a new architecture for bone tissue engineering. J Mater Sci Mater Med 16:1099–1104.

[18] Shalumon KT, Chennazhi KP, Tamura H, Kawahara K, Nair SV, Jayakumar R (2012) Fabrication of three-dimensional nano, micro and micro/nano scaffolds of porous poly(lactic acid) by electrospinning and comparison of cell infiltration by Z-stacking/three-dimensional projection technique. IET Nanobiotechnol 6(1):16–25.

[19] Kim SJ, Jang DH, Park WH, Min B-M (2010) Fabrication and characterization of 3-dimensional PLGA nanofiber/microfiber composite scaffolds. Polymer 51(6):1320–1327.

[20] Srouji S, Kizhner T, Suss-Tobi E, Livne E, Zussman E (2008) 3-D Nanofibrous electrospun multilayered construct is an alternative ECM mimicking scaffold. J Mater Sci Mater Med 19:1249–1255.

[21] Vaquette C, Cooper-White J (2013) A simple method for fabricating 3-D multilayered composite scaffolds. Acta Biomater 9:4599–4608.

[22] Vaquette C, Cooper-White J (2012) The use of an electrostatic lens to enhance the efficiency of the electrospinning process. Cell Tissue Res 347:815–826.

[23] Chung S, Ingle NP, Montero GA, Kim SH, King MW (2010) Bioresorbable elastomeric vascular tissue engineering scaffolds via melt spinning and electrospinning. Acta Biomater 6:1958–1967.

[24] Santos MI, Tuzlakoglu K, Fuchs S, Gomes ME, Peters K, Unger RE, Piskin E, Reis RL, Kirkpatrick CJ (2008) Endothelial cell colonization and angiogenic potential of combined nano- and micro-fibrous scaffolds for bone tissue engineering. Biomaterials 29:4306–4313.

[25] Rajzer I, Grzybowska-Pietras J, Janicki J (2011) Fabrication of bioactive carbon nonwovens for bone tissue regeneration. Fibres Text East Eur 84(1):66–72.

[26] Kasoju N, Bhonde RR, Bora U (2009) Fabrication of a novel micro-nano fibrous nonwoven scaffold with Antheraea assama silk fibroin for use in tissue engineering. Mater Lett 63:2466–2469.

[27] Rajzer I, Rom M, Błaz_ewicz M (2010) Production and properties of modified carbon fibers for medical applications. Fiber Polym 11(4):615–624.

[28] Pielichowska K, Blazewicz S (2010) Bioactive polymer/hydroxyapatite (nano)-composites for bone tissue regeneration. Adv Polym Sci 232(1):97–207

[29] Bye FJ, Bissoli J, Black L, Bullock AJ et al (2013) Development of bilayer and trilayer nanofibrous/microfibrous scaffolds for regenerative medicine. Biomater Sci 1:942–951.

[30] Ślósarczyk A, Paszkiewicz Z, Zima A (2010) The effect of phosphate source on the sintering of carbonate substituted hydroxyapatite. Ceram Int 36(2):577–582.

[31] Rajzer I, Fabia J, Graczyk T, Piekarczyk W (2013) Evaluation of PCL and PCL/n-HAp fibres processed by melt spinning. Eng Biomater 118:2–4.

[32] Rajzer I, Rom M, Fabia J, Sarna E, Janicki J (2010) Fabrication and characterization

of bioactive PLA fibers for bone tissue engineering. In: 23rd European conference on biomaterials, Tampere, Finland, p. 448. ISBN 978-1-61782-086-1.

[33] Li D, Frey MW, Joo YL (2006) Characterization of nanofibrous membranes with capillary flow porometry. J Membr Sci 286:104–114.

[34] Kokubo T, Takadama H (2006) How useful is SBF in predicting in vivo bone bioactivity? Biomaterials 27(15):2907–2915.

[35] Wang N, Burugapalli K, Song W, Halls J, Moussy F, Zheng Y, Ma Y, Wu Z, Li K (2013) Tailored fibro-porous structure of electrospun polyurethane membranes, their size-dependent properties and trans-membrane glucose diffusion. J Membr Sci 427:207–2017.

[36] Gautam S, Dinda AK, Mishra NC (2013) Fabrication and characterization of PCL/gelatin composite nanofibrous scaffold for tissue engineering applications by electrospinning method. Mater Sci Eng C 33:1228–1235.

[37] Rajzer I, Kwiatkowski R, Piekarczyk W, Binias´ W, Janicki J (2012) Carbon nanofibers produced from electrospun PAN/HAp precursors as scaffolds for bone tissue engineering. Mater Sci Eng C 32(8):2562–2569.

[38] Szentivanyi A, Chakradeo T, Zernetsch H, Glasmacher B (2011) Electrospun cellular microenvironments: understanding controlled release and scaffold structure. Adv Drug Deliv Rev 63:209–220.

Chapter 15

Fast-Curing Epoxy Polymers with Silica Nanoparticles: Properties and Rheo-Kinetic Modelling

A. Keller[1,2], K. Masania[2], A. C. Taylor[1], C. Dransfeld[2]

[1]Department of Mechanical Engineering, Imperial College London,
South Kensington Campus, London SW7 2AZ, UK
[2]Institute of Polymer Engineering, University of Applied Sciences and Arts,
Northwestern Switzerland FHNW, Klosterzelgstrasse 2,
5210 Windisch, Switzerland

Abstract: Fast-curing epoxy polymers allow thermoset parts to be manufactured in minutes, but the curing reaction is highly exothermic with heat flows up to 20 times higher than conventional epoxies. The low thermal conductivity of the polymer causes the mechanical and kinetic properties of parts to vary through their thickness. In the present work, silica nanoparticles were used to reduce the exotherm, and hence improve the consistency of the parts. The mechanical and kinetic properties were measured as a function of part thickness. The exothermic heat of reaction was significantly reduced with the addition of silica nanoparticles, which were well dispersed in the epoxy. The silica nanoparticles increased the Young's modulus linearly from 3.6 to 4.6GPa with 20wt% of silica, but the fracture energy was found to increase less than for many slow-curing epoxy resins, with values of 176–211J m^{-2} being measured. Although there was no additional toughening, shear band yielding was observed. Further, the addition of silica nanoparticles increased the molecular weight between crosslinks, indicating the relevance of detailed cure kinetics when studying fast-curing epoxy resins. A model was developed to describe the increase in viscosity and degree of cure of the unmodified and the silica-modified epoxies. A heat transfer equation was used to predict the temperature and resulting properties through the thickness of a plate, as well as the effect of the addition of silica nanoparticles. The predictions were compared to the experimental data, and the agreement was found to be very good.

1. Introduction

Epoxy polymers have many applications, including in adhesives, coatings and fibre composite materials. Conventionally, such thermoset polymers are cured relatively slowly, but the demand for increased manufacturing throughput is driving cycle times down from hours or tens of minutes to several or a few minutes. Some of this reduction in cycle time can be partly achieved by reducing the time taken to fill the mould. For example, for liquid composite moulding processes, the time required for fibre impregnation can be greatly reduced by using through-thickness impregnation (e.g. by compression resin transfer moulding (CRTM)[1]). This approach reduces the impregnation length compared to standard in-plane infusion processes, and therefore reduces the impregnation time by orders of magnitude. However, curing is generally the rate-determining step in the manufacturing process, and so fast-curing epoxies are being developed.

Although the rheology and cure kinetics of slow-curing epoxies have been studied and modelled[2]–[5], the use of fast-curing systems gives new challenges [6][7]. To achieve the short cure times the mould is pre-heated, so the degree of cure and viscosity of the epoxy evolve as a function of time whilst the mould is being filled, and so these must be considered as variables for flow modelling. The fast and highly exothermic nature of the curing also causes difficulties in measuring accurate experimental data. Yang et al.[6] found that the total heat of reaction from isothermal measurements decreases at higher temperatures because the heat flow is not measured correctly in the initial stage of curing. To overcome this, Prime et al.[7] extrapolated isothermal curing curves from dynamic measurements for fast-curing resins.

This exothermic heat also has a significant impact on the manufacturing and properties of epoxy polymer or composite parts[8][9]. The poor thermal conductivity of the epoxy results in a large temperature increase in the centre of the part (especially when thick), which gives a higher degree of cure in the centre than at the edges. Hence the glass transition temperature and the properties of the epoxy will vary through the thickness. The addition of ceramic particles will reduce the temperature exotherm and cure shrinkage, by reducing the volume of epoxy and increasing the thermal conductivity, as reported in[10]. For composite manufacturing, any such particles must be small such that they are not filtered out during infusion processes, for example silica nanoparticles[11].

The addition of particles can also improve the properties of the epoxy polymer. Although epoxies have high modulus, strength and temperature stability due to their highly cross-linked matrix structure, this structure makes the polymer relatively brittle. The fracture toughness of the epoxy can be increased by adding particles such as silica nanoparticles[12] or core-shell rubber nanoparticles[13]. Such silica nanoparticles give increases in stiffness, fracture energy[12][14][15] and cyclic-fatigue resistance[16]. It has been shown that these silica nanoparticles remain well dispersed in slow-curing epoxies[17].

This suggests that silica nanoparticles may be advantageous when manufacturing components using fast-curing epoxies, so the present work will investigate of the effects of silica nanoparticles on fast-curing resins by studying the curing behaviour, the rheological behaviour and the mechanical response. The rheology and kinetics will be modelled, and the variation in the glass transition temperature through the thickness of plates will be predicted and compared with experimental results.

2. Materials

The epoxy resin used was a diglycidyl ether of bisphenol A (DGEBA) epoxy with an epoxy equivalent weight (EEW) of 181.5g eq^{-1}, "XB 3585" from Huntsman Advanced Materials, Switzerland. The curing agent was a mixture of diethylenetriamine and 4,4'-isopropylidenediphenol, "XB 3458" from Huntsman Advanced Materials, and was used at a stoichiometric ratio of 100:19 by weight of epoxy to hardener.

The silica (SiO$_2$) nanoparticles were supplied as a masterbatch, "Nanopox F400" from Evonik Hanse, Germany, with 40wt% of silica nanoparticles predispersed in DGEBA (EEW = 295g eq^{-1}) with a mean diameter of 20nm[18]. Formulations with up to 20wt% of silica nanoparticles (20N) were used for the mechanical and fracture tests, with some additional measurements with up to 35.8wt% of silica nanoparticles (35.8N).

For comparison, a conventional slower-curing epoxy was also used. This was "HexFlow RTM6", from Hexcel, UK, which comprises a tetrafunctional epoxy resin, tetraglycidyl 4-4' diaminodiphenylmethane, and two amine hardeners: 4,4'-methylene-bis(2,6-diethylanaline) and 4,4'-methylene-bis(2,6-diisopropylaniline).

3. Experimental Methods

3.1. Thermal Mechanical Characterisation

Two differential scanning calorimeters were used for the thermal characterisation. Firstly, measurements were performed using a TA Q1000 with a cooler unit that enabled measurements below room temperature. The glass transition temperatures, T_g, of the uncured and cured polymers were measured using a sample mass of 2mg and heating rate of 10°C min^{-1}. Modulated measurements (±0.30°C every 15s) were conducted to determine the glass transition temperature and the residual heat of reaction for the partially cured resin. Secondly, isothermal and dynamic measurements to determine the total heat of reaction were conducted using a Mettler DSC 1 using 5mg of resin in "tzero" aluminium pans as this DSC allows manual opening of the pre-heated furnace. All the measurements were conducted using a nitrogen sample purge flow to reduce oxidation of the resin.

The molecular weight between cross-links, M_c, was calculated using the relation proposed by Nielsen[19] as

$$M_c = \frac{3.9 \times 10^4}{T_g - T_{g0}} \quad (1)$$

where T_{g0} is the glass transition temperature of the linear polymer, and was taken as 55°C from Bellenger and Verdu[20]. The validity of this method has been shown previously in[21] and it was assumed that the silica nanoparticles do no interact with the polymer network, as was shown by the constant T_g of silica nanoparticle-modified slow-curing epoxies[12].

3.2. Rheological Study

A PAAR Physica MCR 302 plate-plate rheometer or a TA AR 200 EX plate-plate rheometer was used to measure the change in viscosity of the unmodified and the silica nanoparticle-modified epoxies. Disposable aluminium plates of 25mm diameter were used, and the gap between the plates was set to 1mm. A strain of 1% and an angular frequency of $10s^{-1}$ were used.

3.3. Manufacturing

Bulk epoxy plates of 3, 4 and 6mm thickness were cast in 8-mm-thick aluminium moulds, which were pre-heated to 80°C. The plates were cured in an oven at 80°C for 12min. This slightly slower cure-cycle was preferred to prevent the very strong exothermic reaction and decomposition, which may occur for 6mm plates manufactured at 100°C.

3.4. Microstructure

The morphology of the unmodified and silica nanoparticle-modified epoxies was observed using atomic force microscopy (AFM). Height and phase images were obtained using a MultiMode scanning probe microscope from Veeco, USA, with a NanoScope IV controller and an E scanner. The samples were prepared at room temperature using a PowerTome XL cryo-microtome form RMC Products, USA.

3.5. Modulus and Yield Behaviour

The Young's modulus, E_t, of the bulk unmodified and silica nanoparticle-modified epoxies were determined using uniaxial tensile tests according to BS ISO 572-2[22].

Dumbbell specimens (type 1BA) were machined from the epoxy plates using a water-jet cutter. The tests were performed using an Instron 5584 universal testing machine, with a gauge length of 25mm and a displacement rate of 1mm min^{-1}. An Instron 2620–601 dynamic extensometer was used to measure the strain. A minimum of five samples were tested for each formulation.

3.6. Compression Properties

Plane strain compression (PSC) tests were performed to determine the compressive properties of the polymers as proposed by Williams and Ford[23], and as discussed in[12]. Polished specimens with dimensions of 40 × 40 × 3mm^3 were compressed between two parallel dies of 12mm width using an Instron 5585H at a displacement rate of 0.1mm min^{-1}.

An unmodified epoxy sample was interrupted post-yield, *i.e.* in the strain softening region, then sectioned and polished. This sample was examined using optical microscopy between crossed polarisers to confirm that shear band yielding was present.

3.7. Fracture Properties

Single-edge-notched bending (SENB) tests were performed according to ASTM D5045[24] to determine the planestrain fracture toughness, KC, and the fracture energy, GC, of the unmodified and of the silica nanoparticle-modified epoxies. Sample dimensions of 60 × 12× 6mm with a V-notch of 4mm in depth were water-jet cut. A sharp pre-crack was obtained by tapping a new razor blade, cooled with liquid nitrogen, into the notch. An Instron 5584 was used to perform the tests at a constant displacement rate of 1mm min^{-1}.

4. Cure Kinetic Modelling

4.1. Determination of Glass Transition Temperature

Differential scanning calorimetry (DSC) with a heating rate of 10°C min^{-1} was used to obtain the glass transition temperature, T_g. The value for the uncured epoxy, T_{g0}, was measured to be $-27.0°C \pm 0.5°C$, and for the cured epoxy, $T_{g\infty}$, to be $121°C \pm 1°C$. These values remained constant with the addition of silica nanoparticles, indicating no kinetic effect, and hence there appears to be no chemical interaction between the particles and the epoxy, as previously reported[12].

4.2. Determination of Total Heat of Reaction

The total heat of reaction, ΔH_{tot}, was determined using dynamic measurements with heating rates, $dT\,dt^{-1}$, from 10 to 30°C min^{-1} at 5°C min^{-1} intervals between room temperature until the material was fully cured. The mean value was determined from the three highest ΔH_{tot} readings. For the unmodified epoxy ΔH_{tot} was measured to be 494J g^{-1} ± 4J g^{-1}.

Measurements were performed for 10wt%, 20wt%, 25wt% and 35.8wt% of silica (where 35.8wt% is the maximum possible amount of silica in the epoxy when mixed with the curing agent). The addition of silica nanoparticles reduces the mass of the epoxy matrix, leading to a linear reduction (m = -4.84J g^{-1} wt%) of heat flow proportional to the wt% of silica nanoparticles.

4.3. Isoconversion

Isothermal cure curves were determined from the dynamic measurements using the model-free kinetics (MFK) method proposed by Vyazovkin[25][26]. This method has previously shown excellent agreement with measured data and was found to be superior compared to other studied isoconversion methods[25][27][28].

Vyazovkin[26] defined the time to reach a degree of cure, t_α, for a given temperature, T_0, as follows:

$$t_\alpha = \frac{J[E_a, T(t_\alpha)]}{\exp\left(\frac{-E_a}{RT_0}\right)} \qquad (2)$$

where E_a is the activation energy, R is the universal gas constant and

$$J[E_a, T(t_\alpha)] \equiv \int_{t_{\alpha-\Delta\alpha}}^{t_\alpha} \exp\left[\frac{-E_a}{RT_i(t)}\right] dt \qquad (3)$$

where E_a minimises

$$\Phi(E_a) = \sum_{i=1}^{n} \sum_{j \neq i}^{n} \frac{J[E_a, T_i(t_\alpha)]}{J[E_a, T_j(t_\alpha)]} \qquad (4)$$

4.4. Isothermal Measurements

The fast-curing nature of the epoxy causes several difficulties when conducting isothermal measurements using DSC. One method is to equilibrate the sample and the furnace at room temperature, and then to heat as fast as possible to the desired temperature. This method was found to be unsuitable for this fast-curing epoxy, as the resin cured partially before the heat flow was correctly measured at the set temperature, even with heating rates of up to 500°C min^{-1}.

An alternative method is to preheat the furnace of the DSC, and then to place the sample inside. Approximately, 3–4s of data were lost while the DSC chamber closed and began recording the heat flow. This method could be used with the Mettler DSC 1. Despite opening the device, the DSC was able to maintain a constant furnace temperature.

The heat flow was measured for temperatures from 60°C to 110°C, at 10°C intervals. Measurements for the unmodified epoxy are shown in **Figure 1**. Data were recorded until a proper baseline was reached. The reaction rates peak at the

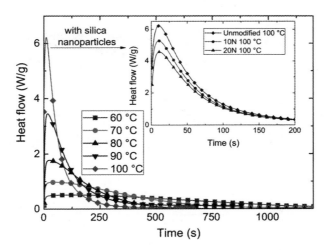

Figure 1. Heat flow for isothermal measurements of the unmodified epoxy and (inset) measurements at 100°C for the unmodified, 10 and 20wt% silica-modified epoxies.

same degree of cure ($\alpha = 0.1$) for all isothermal measurements up to 100°C. For higher temperatures, the peak was observed at a higher degree of cure as the DSC was not able to resolve the initial part of the curing reaction correctly due to the reaction progressing too quickly.

The maximum heat flow shows a relatively high magnitude compared to standard-curing epoxies. At 60°C a maximum heat flow of 0.508W g^{-1} is reached, increasing to 6.21W g^{-1} at 100°C (the upper processing temperature). In comparison, the HexFlow RTM6, a commonly used epoxy in the aerospace industry, has a maximum heat flow of 0.353W g^{-1} at a cure temperature of 180°C, *i.e.* a factor of 20 lower than the epoxy in this study. The measurements were repeated for the epoxies containing 10 and 20wt% of silica nanoparticles, which again showed a decreased heat flow and total heat of reaction. The heat flow for isothermal conditions at 100°C is shown in **Figure 1** (inset), and the reduction in heat flow can be seen clearly.

4.5. Comparison of Isoconversion and Measurement

The measured data and the isoconversion curves using the MFK approach from Eq. (2) were compared for temperatures from 80°C to 110°C, as shown in **Figure 2**. The degree of cure progression shows good agreement between the

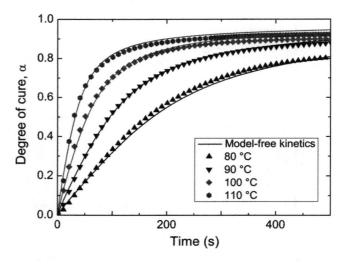

Figure 2. Comparison between isothermal measurements (points) and modelled curves (lines) using the model-free kinetics approach from dynamic measurements for the unmodified epoxy.

measured and modelled data, especially in the initial stage of the curing reaction. However, the model over predicts the degree of cure over the whole temperature range. The gap between the measurement and the model towards the end of the conversion curves becomes larger with increasing temperature. This is due to the loss of data in the initial stage of the measurement before the DSC was able to record the heat flow correctly. This was further verified by noting that with increasing temperature more data were lost.

4.6. Kinetic Modelling

The degree of cure, α, can be calculated for every time, t, during the cure reaction using

$$d\alpha = \frac{\int_{t_1}^{t_2} \frac{dH}{dt}}{\Delta H_{tot}} \qquad (5)$$

where $dH\,dt^{-1}$ is the measured heat flow in W g^{-1}, and ΔH_{tot} is the total heat of reaction in J g^{-1} from "Determination of total heat of reaction" section.

The heat flow and total heat of reaction decreased in proportion to the amount of silica, leading to identical conversion curves for all formulations that were studied. The reaction rate, dα/dt, is proportional to the heat flow,

$$\frac{d\alpha}{dt} = \frac{\frac{dH}{dt}}{\Delta H_{tot}} \qquad (6)$$

As described by Bailleul et al.[29], the reaction rate can be modelled as a function of the temperature and the degree of cure using

$$\frac{d\alpha}{dt} = k(T)G(\alpha) \qquad (7)$$

where

$$k(T) = k_1 e^{(-E_1(\frac{T_{ref}}{T} - 1))} \qquad (8)$$

and

$$G(\alpha) = \sum_{i=0}^{m} G_i \alpha^i \qquad (9)$$

where k_1 is the frequency factor of the cure reaction, E_1 is the activation energy, T_{ref} is a randomly chosen temperature and G_i is a polynomial function. Ruiz and Trochu[30] have extended the Bailleul model by a third term to consider the effects of glass transition temperature on the reaction rate by

$$k_3(T,\alpha) = (\alpha_{max}(T) - \alpha)^n \qquad (10)$$

with

$$n = f(T) \qquad (11)$$

where α_{max} is the maximum degree of cure for isothermal curing and n is the power exponent which depends on temperature, leading to the following model:

$$\frac{d\alpha}{dt} = k_1 e^{(-E_1(\frac{T_{ref}}{T}-1))} \sum_{i=0}^{m} G_i \alpha^i (\alpha_{max}(T) - \alpha)^n \qquad (12)$$

The polynomial term is normalised to 1 at the maximum reaction rate. The reaction rate at this point can be modelled solely by Eq. (8).

The maximum degree of cure, α_{max}, was determined from the isothermal and ensuing modulated measurements. The values of α_{max} determined from the dynamic modulated measurements were generally higher compared to the isothermal measurements. This difference was caused by the loss of data in the initial part of the measurement during the isothermal runs.

Therefore, to avoid this error due to onset of cure before data recording and subsequent baseline selection to measure heat flow, models for α_{max} were developed using the modulated measurements,

$$\alpha_{max} = \frac{T_c - T_{g0}}{(T_c - T_{g0})(1-\mu) + (T_{g\infty} - T_{g0})\mu} \qquad (13)$$

where T_c is the cure temperature, T_{g0} is the glass transition temperature of the uncured resin, $T_{g\infty}$ is the glass transition temperature of the cured resin, and μ is a fitting parameter.

The parameters of the Arrhenius term were determined by normalising the polynomial term to 1 at the point of the maximum reaction rate, occurring at a degree of cure of 0.1.

$$\frac{d\alpha}{dt} = k_1 e^{\left(-E_1(\frac{T_{ref}}{T}-1)\right)} 1 \qquad (14)$$

The reference temperature, T_{ref}, was set to 350K, which lies in the middle of the measured range. The natural logarithm of Eq. (14) leads to a linear function. The activation energy, E_1, was obtained from the slope of the linear fit and the frequency factor, k_1, equals the exponential function of the intercept.

With the Arrhenius and the α_{max} terms solved, the parameters for the polynomial term were determined using

$$\frac{\frac{d\alpha}{dt}}{k_1 e^{(-E_1(\frac{T_{ref}}{T}-1))}} = \sum_{i=0}^{m} G_i \alpha^i (\alpha_{max}(T) - \alpha)^n \tag{15}$$

The polynomial term was found to be a 4th order polynomial divided by a second-order polynomial, using a least square fit, as

$$G(\alpha) = \frac{G_1 \alpha^4 + G_2 \alpha^3 + G_3 \alpha^2 + G_4 \alpha + G_5}{G_6 \alpha^2 + G_7 \alpha} \tag{16}$$

and reaction exponent n, using

$$n = n_s T_c + n_i \tag{17}$$

The model from Eq. 13, using the parameters shown in **Table 1**, gives good agreement with the dynamic measurements, as shown in **Figure 3(a)**, with R^2-values between 0.995 and 0.999. The model for the isothermal measurements showed a less good agreement than that for the dynamic measurements. For higher temperatures (80°C–100°C), the initial curing is modelled well. However, the model overshoots the measured data as the curing reaction starts before the heat flow is measured by the DSC. This effect increases at higher temperatures, which is reflected in the model, illustrated in **Figure 3(b)** with R^2-values between 0.981 and 0.997.

4.7. Glass Transition Modelling

The glass transition temperature, T_g, was modelled as a function of the degree of cure, α, using the DiBenedetto model[31]

Table 1. Fitting parameters for the cure kinetic model.

Name	Symbol	Value
Polynomial parameter	G_1	−1351
Polynomial parameter	G_2	−2678
Polynomial parameter	G_3	10194
Polynomial parameter	G_4	4338
Polynomial parameter	G_5	−5.208
Polynomial parameter	G_6	587
Polynomial parameter	G_7	4570
Reaction exponent factor slope	n_s	0.018
Reaction exponent factor intercept	n_i	−5.2367
Frequency factor	k_1 (s^{-1})	0.00309
Activation energy	E_1	22.28
Reference temperature	T_{ref} (K)	350
T_g of the uncured resin	T_{g0} (K)	246.15
T_g of the cured resin	$T_{g\infty}$ (K)	394.4
Parameter for α_{max} model	μ	0.32

$$T_g = T_{g0} + \frac{(T_{g\infty} - T_{g0})\lambda\alpha}{1-(1-\lambda)\alpha} \qquad (18)$$

where T_{g0} and $T_{g\infty}$ are the glass transition temperature of the uncured and the fully cured material, respectively, from "Determination of glass transition temperature" section, and λ is a fitting parameter. The results are shown in **Figure 4(a)**.

Measurements were obtained using a modulated DSC set-up. The T_g was measured directly from the reversible heat flow, and α was determined from the residual heat of reaction from the irreversible heat flow.

4.8. Heat Transfer Modelling

During the curing of an epoxy polymer, exothermic heat is created during the crosslinking reaction. With fast-curing epoxies, this exothermic heat leads to a large overshoot in the resin temperature compared to the mould temperature. Hence, the curing reaction cannot be assumed to occur at the mould temperature. The heat conduction model was solved in one dimension to account for a thickness

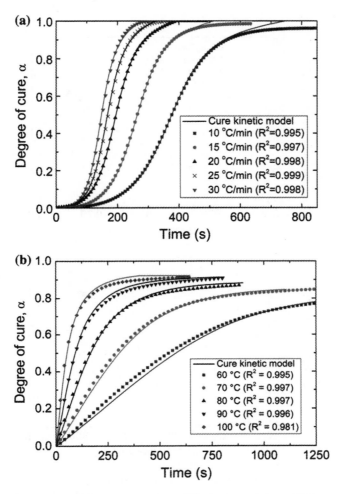

Figure 3. Cure kinetic model for a dynamic measurements and b isothermal measurements of the unmodified epoxy.

dependant temperature overshoot to give

$$\rho_r H_{tot} \frac{d\alpha}{dt} + k_{xx} \frac{\partial^2 T}{\partial x^2} = \rho C_p \frac{\partial T}{\partial t} \qquad (19)$$

where ρ_r is the resin density, ρ is the density of the composite resin, H_{tot} is the total heat of reaction from "Determination of total heat of reaction" section, $d\alpha\ dt^{-1}$ is the modelled reaction rate from Eq. (12), k_{xx} is the thermal conductivity and C_p is the specific heat capacity of the epoxy resin.

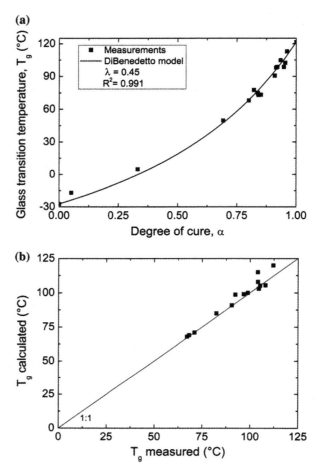

Figure 4. a Glass transition temperature, T_g, as a function of degree of cure, a, compared to the DiBenedetto model[31] (line), and b comparison of measured and modelled T_g.

4.9. Finite Element Approach

Equation (19) was discretised using the finite element method with linear elements according to[8][32], leading to an equation in the form

$$[C]\{\dot{T}\}+[K]\{T\}-\{F\}=0 \qquad (20)$$

where [C] is the thermal capacity matrix, {T} is the temperature vector, [K] is the thermal conductivity matrix and {F} is the thermal load vector due to chemical reaction of the resin, with

$$[C] = \int_1 \rho C_p [N]^T [N] dx \qquad (21)$$

$$[K] = \int_1 k_{xx} [\dot{N}]^T [\dot{N}] dx \qquad (22)$$

$$\{F\} = \int_1^T [N]^T \rho_r H_{tot} \frac{d\alpha}{dt} dx \qquad (23)$$

where [N] is the shape function. A finite difference scheme was used for time discretisation[32],

$$([C] + \theta \Delta t [K])\{T\}^{n+1} = ([C] - (1-\theta)\Delta t [K])\{T\}^n + \Delta t (\theta \{F\}^{n+1} + (1-\theta)\{F\}^n) \qquad (24)$$

where θ is a parameter to define the art of the scheme, chosen as 0.5 for a semi-implicit scheme, and Δt is the time step. For small time steps, the thermal load vector can be approximated by

$$\{F\}^{n+1} \approx \{F\}^n \qquad (25)$$

which leads to

$$([C] + \theta \Delta t [K])\{T\}^{n+1} = ([C] - (1-\theta)\Delta t [K])\{T\}^n + \Delta t \{F\}^n \qquad (26)$$

4.10. Initial Conditions

The polymer plates were cast in 8-mm-thick aluminium moulds. The values for the density, thermal conductivity, k, and specific heat capacity, C_p, of the materials which were used to solve Eq. (26), are summarised in **Table 2**, where the total heat of reaction was discussed in "Determination of total heat of reaction" section.

Table 2. Values used for heat transfer modeling.

	Aluminium	Unmodified (density of epoxy in formulation)	10 N (density of epoxy in 10 N formulation)	20 N (density of epoxy in 20 N formulation)
Density (kg m^{-3})	2700	1194	1247	1301
Density of epoxy (kg m^{-3})	–	1194	1185	1176
Thermal conductivity, k (W m^{-1} °C)	135	0.20 [33]	0.23 [33]	0.26 [33]
Specific heat capacity, C_p (J kg^{-1} °C)	1000 [37]	1700	1700	1700

After a convergence analysis, ten nodes per mm were used in the finite element analysis, where the time step was varied depending on the cure time, to produce approximately 3000 data points. The initial temperatures of the mould and epoxy were set to identical values, assuming that the resin heats up quickly during casting. Then, through internal heat generation during curing, the thermal heat vector {F} was calculated to produce exothermic energy, resulting in an increase in the resin temperature and subsequently the mould temperature via conduction. The heat transfer between the mould and oven via convection was considered to be small and therefore neglected.

The predicted temperature distribution across the aluminium mould and the resin (a 6-mm-thick unmodified epoxy plate in this case) is shown in **Figure 5**. At time t = 0s, a stable temperature distribution was given as per boundary conditions. Over time, the internal heat was generated and conducted into the aluminium mould, which was predicted to increase in temperature by about 7°C during the approximately 60s that were described. Large temperature variations across the thickness of the epoxy plate were predicted, with a significantly higher temperature in the centre of the epoxy plate. Thereafter, the temperature decreased back to oven temperature, and the heat was dissipated slowly because convection between the oven and mould was not described in the model.

The specific heat capacity did not change significantly with the addition of 20wt% silica nanoparticles (1700 vs. 1600J kg^{-1} °C) and hence was kept constant. The values of k increased from 0.2 for the unmodified epoxy to 0.23 with 10wt% and 0.26 for 20wt% silica nanoparticles according to[33].

4.11. Comparison to Experimental Data

Equation (26) was solved using discretisation to predict the progression of

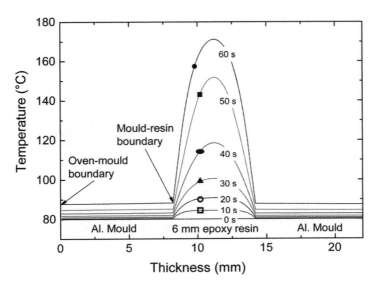

Figure 5. Modelled temperature distribution across the thickness of the aluminium mould and 6-mm epoxy resin plate, cured at an oven temperature of 80°C.

the temperature, degree of cure and glass transition temperature during cure. The resin temperature can overshoot the mould temperature significantly if a high mould temperature is used and/or thick plates are manufactured. For example, the temperature of a 3-mm plate manufactured with a mould temperature of 100°C peaked at a temperature of 176°C. This temperature peak occurs at the very beginning of the curing reaction, and results from the kinetics as the maximum reaction rate occurred at the initiation of cure, *i.e.* after 20s for a 3-mm plate cured at 100°C. Further, when the plate thickness was increased, the heat generated in the centre of the plate could not be dissipated easily due to the low thermal conductivity of the epoxy.

A comparison between the measured and modelled temperature progression is shown in **Figure 6** for four cases. There is a good agreement between the modelled and experimental temperature evolution for the 3-mm plates produced using a 50°C mould, see **Figure 6(a)**. The temperature difference of approximately 2°C has no significant influence on the kinetics, as can be seen in **Table 3** where there is good agreement between the measured and calculated T_g values.

A comparison of the exothermic temperature of a 4-mm-thick plate cured at 80°C, measured 0.8mm from the edge, and a 3-mm plate cured at 100°C,

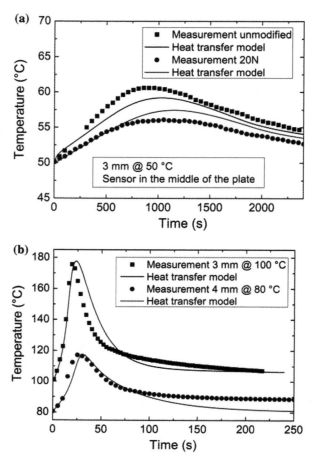

Figure 6. Comparison of experimental (points) and modelled temperature progression (line) comparing a the effect of addition of silica nanoparticles and b variation of plate thickness and cure temperature, on the overall temperature exotherm.

Table 3. Comparison of measured and modelled glass transition temperature, T_g.

Material	Thickness (mm)	Position	Mould temperature (°C)	T_g measured (°C)	T_g calculated (°C)
Unmodified	3	Edge	50	71	71
10 N	3	Edge	50	68	69
20 N	3	Edge	50	67	68
Unmodified	4	Edge	60	82	78
Unmodified	4	Edge	80	90	91
Unmodified	4	Centre	80	92	99
Unmodified	6	Edge	80	105	103
10 N	6	Edge	80	99	100
20 N	6	Edge	80	97	99
Unmodified	6	Centre	80	112	120
10 N	6	Centre	80	104	115
20 N	6	Centre	80	104	108

measured in the centre, is shown in **Figure 6(b)**. In both cases, a large temperature overshoot was measured and predicted. The temperature progression is replicated well at the edge of plates, as can be seen for the 4-mm plate at 80°C. Measurements taken in the centre of the plates, such as for the 3-mm plate at 100°C, show a slight delay after the temperature peaks, which leads to an overestimation of T_g. This is probably a result of the assumed constant thermal conductivity, which will increase slightly with increasing temperature.

As can be seen in **Figure 6(a)**, the peak temperature is reduced with the addition of silica nanoparticles due to the reduced total heat of reaction. For a 3-mm plate manufactured with a mould temperature of 100°C, the calculated peak temperature is reduced from 173°C to 163°C with 10wt% and to 154°C with 20wt% of silica nanoparticles. This gives a more controllable manufacturing process with a reduced risk of epoxy decomposition and a lower T_g.

The T_g was calculated using the DiBenedetto model (Eq. 18), and was predicted to vary over the thickness of the plates. For example, a 6-mm plate manufactured at 80°C has a T_g difference of 7°C between the edge and the centre, see **Table 3**. For the experimental data, at least two measurements were conducted for each position, and the repeatability was found to be very good (within ±1°C).

The higher values of T_g in the centre of the plate can be explained by the higher peak temperature. At the edges, the heat is transported away through the mould faster than in the centre of the plate. With increasing plate thickness, there is an increase in the temperature gradient over the thickness (an example is shown in **Figure 5**) as well as an increase in the maximum temperature.

A comparison of the experimentally determined and modelled values for T_g is shown in **Figure 4(b)**, and the values are summarised in **Table 3**. Measurements were taken from plates that were manufactured using different curing cycles, nanoparticle content, and positions through the thickness. The results generally agree well with the modelled data. As discussed above, the largest differences were for the centre of the 6-mm plates.

5. Rheological Modelling

5.1. Shear rate Dependency

The shear rate dependency was measured at room temperature, and a little shear thinning was observed when measurements were made between 100 and 1000s^{-1}. The viscosity reduced from 3.75Pas at a shear rate of 100s^{-1} to 3.25Pas at 1000s^{-1}. The time between these two measurements was 100s, during which the resin viscosity increased by 0.5Pas resulting from the curing reaction, at a constant shear rate of 1s^{-1}.

At the higher temperatures required for fast curing, the shear rate dependency became negligible as the curing reaction progressed very quickly, leading to a rapid increase in viscosity. For example, the viscosity increased from 0.5 to 10Pas at 100°C in 20s for the unmodified formulation.

5.2. Isothermal Measurements

Viscosity measurements for the unmodified and modified epoxies were obtained from 50°C to 100°C, at 10°C intervals, using a plate–plate rheometer. Example results for 60°C and 90°C are shown in **Figure 7**. The first measurement point occurs at between 30 and 60s, which is the time taken from the moment when the resin first touches the preheated plate until the rheometer starts collecting data. A measurement at 60°C was also performed using 35.8wt% of silica nanoparticles for comparison, which resulted in a substantial increase in the initial viscosity, from 0.26Pas for the unmodified resin to 1.49Pas for the 35.8wt% silica nanoparticle-modified epoxy. This viscosity increase was believed to occur due to a reduction of polymer chain mobility (non-linear effect) and would be expected to result in an increased fibre impregnation time during composite processing.

The times taken for the viscosity to reach 1 and 10Pas at a given temperature, thus describing a hypothetical process window for fibre composite processing, for the unmodified epoxy, 10wt% and 20wt% silica nanoparticle-modified epoxies, are compared in **Figure 8**. At 60°C, it takes about 170s for the unmodified epoxy to reach 1Pas, and at 90°C this viscosity is reached in 84s. At any temperature,

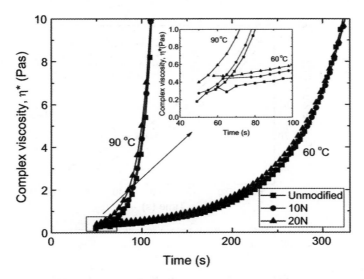

Figure 7. Isothermal measurements at 60 and 90°C. The initial viscosity increases with addition of silica nanoparticles (inset).

the time taken to reach 1Pas is roughly 7s less for the 10wt% and 10−15s less for the 20wt% silica nanoparticle epoxy compared to the unmodified epoxy, due to the increased viscosity. The time taken to reach 10Pas was nevertheless similar for all three formulations, as the cure reaction kinetics dominated the evolution of the viscosity.

The time to reach a certain viscosity value can be modelled by fitting the following equation:

$$t = e^{\frac{T-C}{B}} \qquad (27)$$

where T is the temperature, t is the time, B and C are fitting constants as summarised in **Table 4**.

5.3. Rheological Modelling

The model used or the rheological modelling is based on the Kiuna *et al.* approach[4] where the differential equation which describes the curing of the resin is

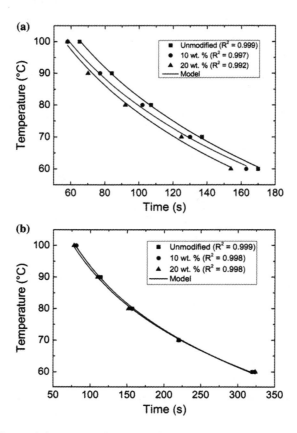

Figure 8. Comparison of time to reach a viscosity of 1Pas (a) and 10Pas (b) for the unmodified and modified epoxies.

Table 4. Parameters for the time to reach viscosities of 1Pas and 10Pas.

	$B_{1\ Pas}$	$C_{1\ Pas}$	$B_{10\ Pas}$	$C_{10\ Pas}$
Unmodified	−41.4	273	−29.1	228
10 N	−38.5	257	−29.0	227
20 N	−39.2	258	−28.2	222

$$\frac{d\alpha}{dt} = \frac{k(T)}{g'(\alpha_2)} \qquad (28)$$

The model is based on the assumption that the viscosity can be modelled with the change in cure and where α_2 represents the dimensionless viscosity.

$$\alpha_2 = f(\tau) = \ln\left(\frac{\eta(T)}{\eta_0(T)}\right) \qquad (29)$$

where η_0 is the viscosity of the uncured material and s is the elapsed dimensionless cure time,

$$\tau = k(T)t \qquad (30)$$

and

$$k(T) = \frac{1}{t_1} \qquad (31)$$

where t_1 is the time when α_2 becomes 1. Combining the derivative of the dimensionless viscosity, Eq. (28), with Eq. (29), leads to the following differential equation

$$\frac{d\eta}{dt} + \left(-\frac{k(T)}{g'(\alpha_2)} - \frac{d\eta_0}{dt}\frac{1}{\eta_0(T)}\right)\eta = 0 \qquad (32)$$

For isothermal cure, the equation can be solved for the viscosity, which leads to

$$\eta(t,T) = \exp\left(\frac{k(T)}{g'(\alpha_2)}t\right)\eta_0 \qquad (33)$$

The initial viscosity, η_0, and the advance of curing show an exponential dependency for different isothermal temperatures,

$$\eta_0(T) = A_1 \exp\left(\frac{E_1}{RT}\right) \qquad (34)$$

and

$$k(T) = A_2 \exp\left(\frac{E_2}{RT}\right) \qquad (35)$$

where R is the universal gas constant and where A_1, E_1, A_2 and E_2 are fitting parameters.

The next step is to determine the best fit for $g'(\alpha_2)$, for which an exponential approach was used

$$\alpha_2 = y_0 + A_3 \exp\left(\frac{\tau}{E_3}\right) \tag{36}$$

where y_0, A_3 and E_3 again are fitting parameters.

By substituting Eqs. (34–36) into Eq. (33), with the addition of a term to adjust for the shift in gel time, $(A_4 \Delta T + E_4)$, the model for isothermal conditions becomes

$$\eta(t,T) = \exp\left(\frac{A_2 \exp\exp\left(\frac{E_2}{RT}\right)}{E_3 \frac{1}{A_3} \exp\exp\left(A_2 \exp\exp\left(\frac{E_2}{RT}\right)\frac{t}{E_3}\right)} t(A_4 T + E_4)\right) A_1 \exp\left(\frac{E_1}{RT}\right) \tag{37}$$

A comparison between measurements and the model for different isothermal temperatures of the unmodified epoxy is shown in **Figure 9**. The model agrees well with the measured viscosity progression during the initial stage of the curing process. The modelling was repeated for the 10wt%, 20wt% and 35.8wt% silica nanoparticle-modified epoxies by modifying the values for A_1 and E_1, which represent the initial viscosity, as well as A_4 and E_4 to take into account the time shift. Intermediate silica contents can be interpolated from the values for these four formulations using the third-order polynomials shown in **Table 5**.

The calculated mean R^2 values of 0.978, 0.9774 and 0.965 describe the quality of fit for the unmodified, 10wt% and 20wt% silica nanoparticle-modified epoxies, respectively, and thus the models represent the experimental data well.

To take the non-isothermal conditions during cure into account, either Eq.

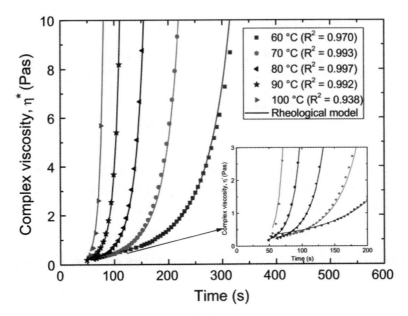

Figure 9. Rheological model (lines) compared to experimental measurements (points) for the unmodified epoxy and (inset) low viscosity range.

Table 5. Parameters used for the rheological model.

Parameter	Unmodified epoxy	10 N	20 N	35.8 N
A_1	1.32×10^{-9}	6.81×10^{-10}	3.62×10^{-10}	1.37×10^{-10}
E_1	52,931	56,049	58,094	64,000
A_2	7518	7518	7518	7518
E_2	$-38,710$	$-38,710$	$-38,710$	$-38,710$
A_3	2.7	2.7	2.7	2.7
E_3	2.2	2.2	2.2	2.2
A_4	0.003	0.0046	0.006	0.007
E_4	-0.409	-1.047	-1.519	-1.732

The nanoparticle content-dependant equations to modify parameters are shown below (N is the wt% of silica)

$A_1 = -2.5684 \times 10^{-14}N^3 + 2.3459 \times 10^{-12}N^2 - 8.4316 \times 10^{-11}N + 1.3156 \times 10^{-9}$

$E_1 = 0.36941N^3 - 16.444N^2 - 439.28N + 52931$

$A_4 = -5.5238 \times 10^{-8}N^3 + 6.5714 \times 10^{-7}N^2 - 1.5895 \times 10^{-4}N + 3.00 \times 10^{-3}$

$E_4 = 1.40495 \times 10^{-5}N^3 + 4.06514 \times 10^{-4}N^2 - 6.92301 \times 10^{-2}N - 0.409$

(32) can be used, or numerical integration of the time-dependent mathematical terms can be used to give

$$\eta(t,T) = \exp\left(\dfrac{\sum_{i}^{n} A_2 \exp\exp\left(\dfrac{E_2}{RT}\right) t_i (A_4 T + E_4)}{E_3 \dfrac{1}{A_3 \exp\left[\sum_{i}^{n} A_2 \exp\exp\left(\dfrac{E_2}{RT}\right) \dfrac{t_i}{E_3}\right]}} \right) A_1 \exp\exp\left(\dfrac{E_1}{RT}\right) \quad (38)$$

A comparison of three heating rates for the unmodified epoxy is shown in **Figure 10**, where the overall agreement was found to be good, with a slight delay of the viscosity increase (latency).

6. Thermal and Mechanical Properties of the Epoxy Polymers

6.1. Thermal Properties

The molecular weight between cross-links, Mc, can be calculated, as shown in **Table 6** for the 6-mm plates that were manufactured at 80°C. The observed variation through the thickness of the epoxy is apparent, where values of 780 and

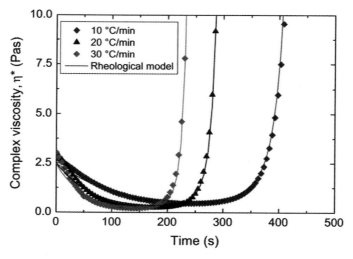

Figure 10. Comparison of the developed rheological model (lines) to dynamic measurements (points) at different heating rates for the unmodified epoxy.

Table 6. Measured glass transition temperature and calculated molecular weight between crosslinks, M_c, of the epoxy polymers cured at 80°C.

	Thickness (mm)	Position	T_g measured (°C)	Molecular weight, M_c (g mol^{-1})
Unmodified	6	Edge	105	780
10 N	6	Edge	99	886
20 N	6	Edge	97	929
Unmodified	6	Centre	112	684
10 N	6	Centre	104	796
20 N	6	Centre	104	796

684g mol^{-1} were measured for the edge and centre of the unmodified epoxy, respectively. The value of Mc increased from 684 to 796g mol^{-1} with the addition of 20wt% silica nanoparticles.

6.2. Morphology

AFM was conducted on the unmodified epoxy and on the epoxies with 10 or 20wt% silica nanoparticles. The resultant phase images are shown in **Figure 11**. The planed surface of the unmodified sample was flat and featureless, see **Figure 11(a)**, as expected for a homogenous thermoset polymer. The microtoming direction is shown on each image, and the scratches parallel to this direction are artefacts of the planning process. The images in **Figure 11(b), (c)**, show that there is a good dispersion and no significant size variation in the 20 nm diameter silica nanoparticles. The area disorder, $\overline{AD}$, was calculated for images of side length L = 2.5µm and L = 5µm for the 10wt% silica nanoparticle-modified epoxies using the methodology described in[34]. A value of $\overline{AD} = 0.4342$ and indicates that the particles in the test material are better dispersed than random, confirming that the silica is well dispersed.

6.3. Mechanical Properties

The tensile Young's modulus, E_t, values of the unmodified and modified epoxies are summarised in **Table 7**. The modulus for the unmodified epoxy was measured to be 3.58GPa. The addition of the nanoparticles increased the modulus

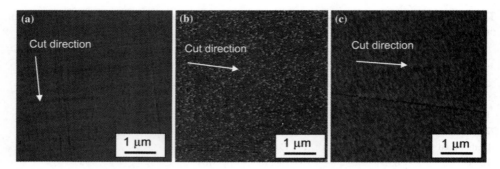

Figure 11. Atomic force microscope phase images of the unmodified epoxy (a), 10N (b) and 20N (c) formulations. The light yellow dots are the silica nanoparticles (Color figure online).

Table 7. Mechanical properties of the epoxy polymers.

	Tensile modulus, E_t (GPa)	Compressive modulus, E_c (GPa)	Compressive yield stress, σ_{yc} (MPa)	Compressive yield strain, ε_{yc} (–)	Fracture energy, G_C (J m^{-2})	Fracture toughness, K_C (MPa$\sqrt{m}$)
Unmodified	3.58 ± 0.11	2.14 ± 0.7	102 ± 11	0.10 ± 0.00	177 ± 35	0.85 ± 0.09
10 N	4.13 ± 010	2.18 ± 0.1	102 ± 1	0.08 ± 0.00	176 ± 15	0.91 ± 0.04
20 N	4.62 ± 0.15	2.13 ± 0.1	104 ± 1	0.08 ± 0.01	211 ± 22	1.05 ± 0.05

as expected approximately linearly to 4.13GPa with the addition of 10wt% of silica, and to 4.62GPa with the addition of 20wt% of silica nanoparticles.

The true stress versus true strain curves were obtained using plane-strain compression tests for the unmodified and silica-modified epoxies, see **Figure 12**. All the samples demonstrate strain softening after yield. The strain soft-ening was followed by a strain hardening region until the test samples broke. The values for the compressive modulus, E_c, compressive yield stress, σ_{yc}, the true yield strain, σ_{yc}, fracture stress, σ_f, and true fracture strain, σ_f, are given in **Table 7**. The fracture stress increased with the silica nanoparticle content, whereas the true fracture strain remains relatively constant. A sample of the unmodified epoxy was unloaded during the strain softening part of the response. The sample was sectioned, polished, and cross-polarised optical microscopy images were then taken. In **Figure 12** (inset), shear band yielding can be clearly observed in the partially loaded and polished section of the compressed region, hence shear band yielding in the process zone of the crack tip would be expected, initiating at the particles in the case of the silica nanoparticle-modified epoxies.

Single-edge-notch bend tests were conducted to investigate the effect of the

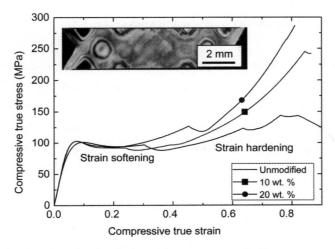

Figure 12. Compressive true stress versus true strain of the unmodified and silica nanoparticle-modified epoxies, and (inset) a cross-polarised optical image of a section from the unmodified epoxy interrupted in the strain softening portion of the curve.

silica content on the fracture energy and fracture toughness. The results are summarised in **Table 7**, and show that a small difference was measured in the fracture energy (177–211 J m^{-2}) or the fracture toughness (0.85–1.05 MPa$\sqrt{m}$), with the addition of up to 20wt% of silica nanoparticles. This indicates that this fastcuring epoxy polymer was not readily toughenable using the silica nanoparticles.

7. Discussion

The kinetics of the epoxy polymer were studied and modelled with and without the addition of silica nanoparticles. The combination of the kinetic and heat transfer models shows excellent agreement for the temperature progression and T_g measurements. The main challenge faced with such a fast-curing epoxy was to obtain proper isothermal measurements using DSC. Several possibilities were tried and a good set-up was achieved. As was observed from temperature measurement of the bulk epoxy plates, the temperature does not remain constant during isothermal curing. No influence of the silica nanoparticles was found on the kinetics itself, but the addition of the particles does reduce the total heat of the reaction (by a reduction in resin mass), which leads to a reduced exotherm during cure. This can prevent the material from damage or decomposition during the manufacturing of thick parts and/or the use of fast cure cycles, and hence leads to a more uniform T_g distribution over the thickness.

Rheological modelling was performed to predict the evolution of the viscosity during fibre impregnation. The influence of the silica nanoparticles on the initial viscosity was found to be small up to 20wt% silica nanoparticles (a relatively high loading), making them very useful for heat reduction, without adversely affecting processing. A very high silica content (35.8wt%) does lead, however, to a viscosity increase, which would cause difficulties during fabric impregnation in a composite material.

The lack of toughening was unexpected and inconsistent with published work for slow-curing epoxies[12][14][15]. Two major mechanisms have been widely recognised to provide toughening in particle-modified epoxies: namely (i) plastic shear bands and (ii) debonding of the matrix from the particles and subsequent plastic void growth of the epoxy. PSC tests show that the unmodified and silica-modified polymers do indeed strain soften and the polished sections provide evidence of shear bands. Likewise, observation of the plan view of fractured samples using cross-polarised light confirmed that shear bands were present in all samples. Scanning electron microscopy could not identify particle debonding and plastic void growth, hence the expected toughening would be modest due to energy absorption via shear band yielding alone, without particle debonding and void growth in the crack tip plastic zone. The level of interfacial adhesion can be quantified using the model of Vörös and Pukánszky[35][36] to compute the proportionality constant, k, for interfacial stress transfer by assuming that the particles carry a load proportional to their volume fraction using stress averaging[12]. Interestingly, as the molecular weight between crosslinks was calculated to decrease with increasing silica nanoparticle content, the yield strength of the polymer network would be expected to decrease, with a superposition of stress increase interfacial stress transfer due to the silica nanoparticles in the measured compressive yield strength. Nevertheless, a value of k = 1.05 was calculated, which indicates that the silica nanoparticles are relatively well bonded to the epoxy and disregarding the effect of reduction in bulk polymer yield properties (where values of the range 0.46–1.88 were previously reported for k[12]).

The difference between the unmodified polymer and that with the addition of silica nanoparticles seems to be the degree to which the polymer exotherms (and its corresponding temperature overshoot, see **Figure 6**). This may be demonstrated by observing the results shown in **Table 3**. The mould temperature, the corresponding measured T_g of the cured epoxy and the calculated T_g of the epoxy were recorded for different cure cycles from the centre and edge of the epoxy plates. Our findings indicate that (i) thinner epoxy plates have a lower exotherm and as expected T_g is consistently lower throughout the plate. It may be noted that the amine hardener starts to decompose at temperatures above 200°C as would be the case for a 6-mm epoxy plate manufactured using a 100°C mould temperature. (ii) The T_g at the edge of the plate is consistently lower than in the centre of the plate, as heat from the exothermic reaction is conducted away from the sample due to the thermal mass of the mould. (iii) The addition of silica particles results in a lower T_g (112°C for the unmodified epoxy compared with 104°C for 20wt%. silica nanoparticles), which indicates that the epoxy network likely has a lower crosslink density.

8. Conclusions

The thermal, mechanical and fracture properties of a silica nanoparticle-modified fast-curing epoxy have been measured, and a kinetic and rheological model was successfully developed. There is a significant exotherm during curing which causes a variation in properties across the thickness of cast plates. Hence the heat conduction equation was used to model the resin temperature during curing, enabling the degree of cure and the resultant T_g to be predicted. Comparison of the calculated T_g values with experimental data shows good agreement, and demonstrates the accuracy of the kinetic models. The addition of silica nanoparticles had no influence on the curing reaction itself, but it reduces the total heat of reaction, and hence reduces the exothermic reaction during cure. This makes the manufacturing of thick parts and the use of fast cure cycles more controllable.

The influence of the silica nanoparticles on the initial viscosity was found to be small with up to 20wt% of silica. No additional toughening was obtained by the addition of silica nanoparticles. Shear band yielding was present in the samples, but there was no additional energy dissipation caused by the silica, and nanoparticle debonding and void growth were not observed.

Acknowledgements

The authors would like to thank the EPSRC Centre for Innovative Manufacturing in Composites Grant Reference EP/1033513/1, Project Number RGS 109687, and the Commission for Technology and Innovation in Switzerland Grant 15828.1 PFEN-IW for funding this study. The authors would also like to thank Huntsman Advanced Materials, Switzerland, and Evonik Hanse, Germany, for supplying materials, plus Anton Paar Switzerland AG, Switzerland, for supporting the rheological measurements. The contributions of Chong, F. Leone, J. Studer, R. Stra̋ssle and P. Tsotra are gratefully acknowledged. Data supporting this publication can be obtained on request from adhesion.group@imperial.ac.uk.

Open Access: This article is distributed under the terms of the Creative Commons Attribution 4.0 International License (http://creativecommons.org/licenses/by/4.0/), which permits unrestricted use, distribution, and reproduction in any medium, provided you give appropriate credit to the original author(s) and the source, provide a link to the Creative Commons license, and indicate if changes were made.

Source: Keller A, Masania K, Taylor A C, *et al.* Fast-curing epoxy polymers with silica nanoparticles: properties and rheo-kinetic modelling[J]. Journal of Materials Science, 2016, 51(1):1–16.

References

[1] Bhat P, Merotte J, Simacek P, Advani SG (2009) Process analysis of compression resin transfer molding. Compos A 40(4):431–441.

[2] Karkanas PI, Partridge IK (2000) Cure modeling and monitoring of epoxy/amine resin systems. I. Cure kinetics modeling. J Appl Polym Sci 77(7):1419–1431.

[3] Yousefi A, Lafleur PG, Gauvin R (1997) Kinetic studies of thermoset cure reactions: a review. Polym Compos 18(2):157–168.

[4] Kiuna N, Lawrence CJ, Fontana QPV, Lee PD, Selerland T, Spelt PDM (2002) A model for resin viscosity during cure in the resin transfer moulding process. Compos A 33(11):1497–1503.

[5] Hardis R, Jessop JLP, Peters FE, Kessler MR (2013) Cure kinetics characterization and monitoring of an epoxy resin using DSC, Raman spectroscopy, and DEA. Com-

pos A 49:100–108.

[6] Yang LF, Yao KD, Koh W (1999) Kinetics analysis of the curing reaction of fast cure epoxy prepregs. J Appl Polym Sci 73(8):1501–1508.

[7] Prime RB, Michalski C, Neag CM (2005) Kinetic analysis of a fast reacting thermoset system. Thermochim Acta 429(2): 213–217.

[8] Guo Z-S, Du S, Zhang B (2005) Temperature field of thick thermoset composite laminates during cure process. Compos Sci Technol 65(3):517–523.

[9] Zhang J, Xu Y, Huang P (2009) Effect of cure cycle on curing process and hardness for epoxy resin. Express Polym Lett 3(9): 534–541.

[10] Nelson J, Hine AM, Goetz DP, Sedgwick P, Lowe RH, Rexeisen E, King RE, Aitken C, Pham Q (2013) Properties and applications of nanosilica-modified tooling prepregs. SAMPE Journal 49:7–17.

[11] Sprenger S, Kothmann MH, Altstaedt V (2014) Carbon fiber-reinforced composites using an epoxy resin matrix modified with reactive liquid rubber and silica nanoparticles. Compos Sci Technol 105:86–95. doi:10.1016/j.compscitech.2014.10.003.

[12] Hsieh TH, Kinloch AJ, Masania K, Taylor AC, Sprenger S (2010) The mechanisms and mechanics of the toughening of epoxy polymers modified with silica nanoparticles. Polymer 51(26): 6284–6294. doi:10.1016/j.polymer.2010.10.048.

[13] Giannakopoulos I, Masania K, Taylor AC (2011) Toughening of epoxy using core-shell particles. J Mater Sci 46(2):327–338. doi:10.1007/s10853-010-4816-6.

[14] Hsieh TH, Kinloch AJ, Masania K, Sohn Lee J, Taylor AC, Sprenger S (2010) The toughness of epoxy polymers and fibre composites modified with rubber microparticles and silica nanoparticles. J Mater Sci 45(5):1193–1210. doi:10.1007/s10853-009-4064-9.

[15] Bray DJ, Dittanet P, Guild FJ, Kinloch AJ, Masania K, Pearson RA, Taylor AC (2013) The modelling of the toughening of epoxy polymers via silica nanoparticles: the effects of volume fraction and particle size. Polymer 54(26):7022–7032. doi:10.1016/j. polymer.2013.10.034.

[16] Blackman BRK, Kinloch AJ, Sohn Lee J, Taylor AC, Agarwal R, Schueneman G, Sprenger S (2007) The fracture and fatigue behaviour of nano-modified epoxy polymers. J Mater Sci 42(16):7049–7051. doi:10.1007/s10853-007-1768-6.

[17] Kinloch AJ, Masania K, Taylor AC, Sprenger S, Egan D (2008) The fracture of glass-fibre reinforced epoxy composites using nanoparticle-modified matrices. J Mater Sci 43(3):1151–1154. doi:10.1007/s10853-007-2390-3.

[18] Kinloch AJ, Mohammed RD, Taylor AC, Sprenger S, Egan D (2006) The interlaminar toughness of carbon-fibre reinforced plastic composites using 'hybrid-toughened' matrices. J Mater Sci 41(15):5043–5046. doi:10.1007/s10853-006-0130-8.

[19] Nielsen LE (1969) Cross-linking effect on physical properties of polymers. J Macromol Sci 3:69–103. doi:10.1080/1558372 6908545897.

[20] Bellenger V, Verdu J, Morel E (1987) Effect of structure on glass transition temperature of amine crosslinked epoxies. J Polym Sci 25(6):1219–1234.

[21] Pearson RA, Yee AF (1989) Toughening mechanisms in elastomer-modified epoxies. 3. The effect of cross-link density. J Mater Sci 24(7):2571–2580. doi:10.1007/BF01174528.

[22] BS-EN-ISO-527-2, Plastics: determination of tensile properties. Part 2: test conditions for moulding and extrusion plastics, 2012, BSI, London.

[23] Williams J, Ford H (1964) Stress-strain relationships for some unreinforced plastics. J Mech Eng Sci 6(4):405–417. doi:10.1243/ JMES_JOUR_1964_006_055_02.

[24] ASTM-D5045-99, Standard test methods for plane-strain fracture toughness and strain energy release rate of plastic materials, 2007, American Society for Testing and Materials, West Conshohocken.

[25] Vyazovkin S, Wight CA (1999) Model-free and model-fitting approaches to kinetic analysis of isothermal and nonisothermal data. Thermochim Acta 340:53–68.

[26] Vyazovkin S (2006) Model-free kinetics. J Therm Anal Calorim 83(1):45–51.

[27] Kandelbauer A (2009) Wuzella, G, Mahendran, A, Taudes, I, Widsten, P, Model-free kinetic analysis of melamine–formaldehyde resin cure. Chem Eng J 152(2):556–565.

[28] Rivero G, Pettarin V, Vázquez A, Manfredi LB (2011) Curing kinetics of a furan resin and its nanocomposites. Thermochim Acta 516(1):79–87.

[29] Bailleul J-L, Delaunay D, Jarny Y (1996) Determination of temperature variable properties of composite materials: methodology and experimental results. J Reinf Plast Compos 15(5):479–496.

[30] Ruiz E, Trochu F (2005) Thermomechanical properties during cure of glass-polyester RTM composites: elastic and viscoelastic modeling. J Compos Mater 39(10):881–916.

[31] DiBenedetto AT (1987) Prediction of the glass transition temperature of polymers: a model based on the principle of corresponding states. J Polym Sci 25(9):1949–1969.

[32] Lewis RW, Nithiarasu P, Seetharamu K (2004) Fundamentals of finite element method for heat and fluid flow. Wiley, Chichester.

[33] Wong CP, Bollampally RS (1999) Thermal conductivity, elastic modulus, and coefficient of thermal expansion of polymer composites filled with ceramic particles for electronic packaging. J Appl Polym Sci 74(14):3396–3403.

[34] Bray DJ, Gilmour SG, Guild FJ, Hsieh TH, Masania K, Taylor AC (2011) Quantifying nanoparticle dispersion: application of the Delaunay network for objective analysis of sample micrographs. J Mater Sci 46(19):6437–6452. doi:10.1007/s10853-011-5615-4.

[35] Vörös G, Pukánszky B (1995) Stress distribution in particulate filled composites and its effect on micromechanical deformation. J Mater Sci 30(16):4171–4178. doi:10.1007/BF00360726.

[36] Pukánszky B, Vörös G (1996) Stress distribution around inclusions, interaction, and mechanical properties of particulate-filled composites. Polym Compos 17(3):384–392. doi:10.1002/pc. 10625.

[37] Ashby MF, Jones DR, Heinzelmann M (2006) Werkstoffe: Eigenschaften, Mechanismen und Anwendungen. Elsevier, Spektrum Akad, Verlag.

Chapter 16

Graphene as a Lubricant on Ag for Electrical Contact Applications

Fang Mao[1], Urban Wiklund[2], Anna M. Andersson[3], Ulf Jansson[1]

[1]Department of Chemistry-Ångström Laboratory, Uppsala University, Box 538, 751 21 Uppsala, Sweden
[2]Department of Engineering Sciences, Uppsala University, Box 534, 751 21 Uppsala, Sweden
[3]ABB AB, Corporate Research, 721 78 Västerås, Sweden

Abstract: The potential of graphene as a solid lubricant in sliding Ag-based electrical contacts has been investigated. Graphene was easily and quickly deposited by evaporating a few droplets of a commercial graphene solution in air. The addition of graphene reduced the friction coefficient in an Ag/Ag contact with a factor of ~10. The lubricating effect was maintained for more than 150,000 cycles in a pin-on-disk test at 1N. A reduction in friction coefficient was also observed with other counter surfaces such as steel and W but the life time was strongly dependent on the materials combination. Ag/Ag contacts exhibited a significantly longer life time than steel/Ag and W/Ag contacts. The trend was explained by an increased affinity for metal-carbon bond formation.

1. Introduction

Electrical contacts are important in modern technology. From a materials science point of view, the design of such contacts is a complex problem, in particular for a sliding contact. In general, a contact material must have a low resistivity, a low contact resistance, a high corrosion resistance, and also be reasonable inexpensive. For a sliding contact, additional materials properties are required. The material cannot be too soft and must exhibit a low wear rate. In addition, a low friction coefficient between the sliding surfaces is required. All these properties are difficult to combine in one single material and the development of new, more reliable contact materials is therefore a true challenge.

One of the most widely used contact materials is Ag. This noble metal exhibits a low resistivity and low contact resistance. The disadvantages of Ag, are the rather high materials costs and the fact that it form surface compounds such as sulfides which may be detrimental for the performance. Most important of all, in a sliding contact application, it is too soft and the friction coefficient between two sliding Ag surfaces is far too high (>1). Consequently, there is a need to modify the Ag surfaces to reduce the friction coefficient by e.g., adding a surface coating on the Ag contact. One example for such coating is AgI which can be deposited by electrochemical techniques or by exposure to an I2 solution[1]-[3]. Ag versus AgI-coated Ag contacts typically exhibit a friction coefficient of 0.3 but have a limited life time as the AgI coatings have a rather high wear rate[3]. Hence, other methods to produce low friction surfaces on Ag contacts with long lifetimes are needed.

One of the emerging lubricating materials is graphene, which has been widely studied for its remarkable mechanical, electrical, optical, and thermal properties[4]-[7]. The tribological properties, especially as a lubricating additive, have attracted intense research attention as well[8]-[13]. Recently, however, Berman et al. demonstrated that the addition of few-layer graphene flakes to a steel surface can significantly reduce the friction coefficient from ~1 to about 0.15 against steel in a pin-on-disk test[14][15]. It is conceivable that graphene layers (GL) also could drastically reduce the friction coefficient for a sliding Ag/Ag contacts but this has yet not been demonstrated. Ag interacts weakly with graphene and forms weak Ag −C bonds. It is therefore possible that the tribological behavior for a graphene-coated Ag surface can be quite different compared to a graphene-coated steel surface where stronger interactions with graphene and also stronger Fe−C bonds are expected. Furthermore, it is possible to design a contact where a Ag surface is sliding against a counter surface of another metal or alloy. In this case, the graphene can be expected to exhibit different effects on the tribological properties depending on the metal-graphene interactions.

The aim of this study is to investigate the potential use of graphene in sliding Ag-based contacts. Also, trends of metal-graphene interactions are studied by using Ag/Ag, Ag/steel, and Ag/W contacts. Me–C bond strength is known to vary among the transition metals where W shows the strongest bonds and Ag the weakest. Fe-based alloys such as steel are therefore expected to exhibit intermediate bond strength to carbon. We have investigated the tribological behavior of these materials combinations with added graphene and characterized the contacts with Raman spectroscopy and X-ray photoelectron spectroscopy (XPS).

2. Materials and Methods

Tribological studies of flat Ag samples with and without added graphene were performed using a ball-on-disk tribometer (from VIT) with rotation geometry at room temperature. The flat Ag samples from Alfa Aesar (99.95% purity) were polished and its roughness was measured using an optical profiler WYKO NT1100 (from Veeco/WYKO) to R_q = ~18nm. The counter materials were a silver-coated cylindrical Cu rod with a hemispherical tip of 9mm diameter with R_q = ~60nm, bearing steel balls of 6mm diameter with R_q = ~10nm, and tungsten balls of 8mm diameter with R_q = ~50nm. All the specimens were cleaned by sonication in acetone, and then in isopropanol and followed by flushing in dry N_2 to clean up any contaminants left from the sample preparation and polishing steps. All of the tribological tests were carried out at a sliding speed of 0.02m/s with contact track of 2.5mm radius. The friction coefficient was continually recorded during each test. Two different normal loads were applied for the tribological tests; 2N normal load was used for lifetime testing of lubricating GL, while more detailed comparisons between different balls against Ag with added graphene were performed using a normal load of 1N. A graphene-containing ethanol solution (1mg/L) from Graphene Supermarket Inc. was used a graphene source. The solution contained monolayer graphene with an average flake size of 550nm. Before the tribological tests, three droplets of the graphene solution were added on the highly polished silver plate surface and allowed to evaporate in ambient atmosphere (humidity 30%).

The surface morphology was studied using a scanning electron microscopy (SEM; Merlin, Zeiss) with a field emission gun as the electron source and an acceleration voltage of 5kV. After the tribological tests, the contact tracks were examined using Raman spectroscopy, using a red laser light ($\lambda = 633$nm) in a Renishaw Invia-Raman spectroscope. The chemical bonds in the contact tracks were studied with XPS using a Physical Systems Quantum 2000 spectrometer with monochromated Al Kα radiation. The analysis was performed with an analysis spot of 50µm without any pre-sputtering. The instrument was calibrated against Au, Ag, and Cu references. The composition ratio of Ag/C on the surface of the contact tracks was estimated using XPS areas and sensitivity factors given by the Physical Electronics Software MultiPak V6.1A. The imaging of the contact tracks was performed with an Olympus optical microscope (from Leitz). The roughness and height profile of the sample surface of the contact tracks were determined with an optical profiler WYKO NT1100 (from Veeco/WYKO).

The contact resistances of the Ag surface and the graphene-coated Ag were measured using a custom-made set-up, based on a four-point resistance method, measuring the voltage drop as a current flows from the probe tip to the test surface. The terminal on the probe is placed as close to the contact point as possible to ensure the distance for shared path of current and voltage is as short as possible to reduce the resistance contribution from probe. The path of the current and voltage divides immediately after the contact point to ensure the contact resistance measurement is not affected by the sheet resistance of the measured film or the internal resistance of the wires. The normal load during contact resistance measurements was varied from 1 to 5N. All measurements were carried out against a commercial Au-coated probe K60.05.33 (from Fixtest, Germany) with a hemispherical tip with Φ 3.3mm.

3. Results

The lubricating effect of graphene was evaluated by measurements of friction coefficients of a clean Ag surface and a graphene-modified Ag surface using three different counter materials Ag, steel, and W. In the following the graphene-free systems are denoted Ag/Ag, steel/Ag, and W/Ag while the graphene-modified systems are denoted Ag/GL/Ag, steel/GL/Ag, and W/GL/Ag, respectively. During the initial experiments, we observed a variation in results also for the same set of counter materials. A general observation was that the graphene solution was susceptible to aging and that the lubricating effect of the applied graphene solution decreased by time. Furthermore, some variation in e.g., friction coefficient was also observed between different experiments. Thus the results presented below shows a representative set of observations where the general trends between materials will be demonstrated.

A typical set of friction curves for the different materials pairs with and without added graphene are shown in **Figure 1**. The optical micrographs and profiles of the tracks for each tribotest are also included as insets. As can be seen, a dramatic reduction of the friction coefficient is observed after addition of graphene, in particular for the Ag/GL/Ag system. In addition, the roughness of tracks is also affected by the addition of graphene.

As shown in **Figure 1(a)**, Ag/Ag performed poorly with a high and fluctuating friction coefficient (~1.15). The optical micrograph shows that the Ag/Ag pair suffered severe adhesive material transfer between the surfaces, which is also illustrated by the profile of the track surface. With addition of graphene, however, the friction coefficient was reduced remarkably to around 0.2–0.25 initially. Typically, the friction was reduced with time and reached a value of 0.1–0.15 after 1000 laps. However, a friction coefficient as low as 0.05 was observed in some experiments. In the Ag/Ag pair, the high tendency for adhesion gives a strong interface, similar in strength to the two mating materials, and promotes material transfer and a large area of contact. All of these effects result in a high roughness of the contact track and a high friction coefficient for the Ag/Ag pair. However, GL weaken the 'interface', where it is deposited, and thus provide a preferred shear plane, resulting in less adhesion, less material transfer, a smoother track, and a lower friction coefficient in the Ag/GL/Ag pair.

The steel/Ag pair initially shows a very low friction coefficient (0.15), due to the low roughness of steel ball, resulting in a delayed onset of adhesion. After 50 cycles, the friction coefficient gradually starts to increase to ~0.85. With addition of graphene, the friction coefficient was reduced from about 0.85 (steel/Ag) to

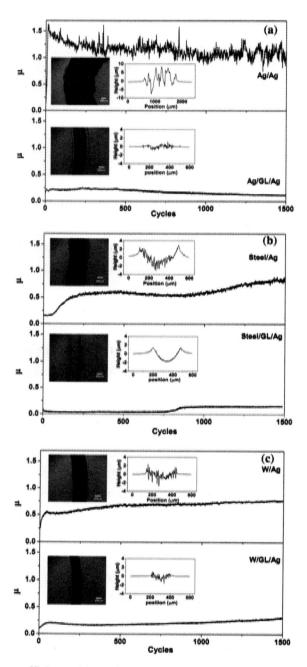

Figure 1. Friction coefficients (μ), optical micrographs (scale bar 200μm) and height profiles of the tracks from the pin-on-disk tests (load: 1N) for a Ag, b steel, and c W against Ag plate with and without graphene deposition. All tribological tests were manually stopped after 1500 cycles.

0.15 (steel/GL/ Ag) after 1500 cycles. The example in **Figure 1(b)** shows an experiment where the friction in the steel/GL/Ag pair initially was below 0.1 and increased to about 0.15. The optical micrographs and height profiles of the tracks illustrates a rough track surface and ridges piling up to a similar volume as the groove in the steel/Ag track, indicating a combination of adhesive material transfer and plastic deformation of the Ag surface in the steel/Ag pair. With addition of graphene, however, almost no adhesive material transfer is visible in the steel/GL/Ag track, instead plastic deformation dominates completely.

As shown in **Figure 1(c)**, graphene also improves the tribological behavior of the W/Ag pair. The addition of graphene decreased the friction coefficient from 0.75 (W/Ag) to 0.45 (W/GL/Ag). The height profile of W/Ag track shows a combination of adhesive transfer and plastic deformation, similar as steel/Ag track. However, with addition of graphene, unlike the steel/GL/Ag pair, some adhesive material transfer is evident in the W/GL/Ag track. The results in **Figure 1** show that graphene was a less effective lubricant with W as a counter surface than with Ag and steel as counter surfaces.

The results in **Figure 1**, from tests carried out at a 1N load, clearly show a lubricating effect with graphene strongly influenced by the metal in the counter surface. To further study this effect, a lifetime study was performed. In a first experiment, a Ag/GL/Ag pair was tested at 1N (black curve in **Figure 2**). This

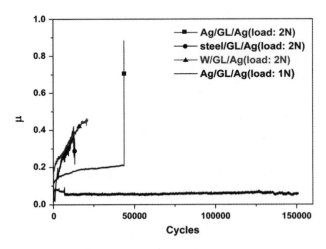

Figure 2. Lifetime testings of lubricating graphene in the tribological pairs of different metal counter surfaces (Ag, Steel, and W) against the graphene-coated Ag plate.

system showed a friction coefficient about 0.06. The test was terminated after about 150,000 laps without loss of lubricating effect. Berman et al. have observed that the lubricating effect of graphene is load dependent[15]. A second set of experiments were therefore carried out at 2N. In this case, the Ag/GL/Ag pair showed a slightly higher friction coefficient for more than 40,000 laps, followed by a rapid increase in friction probably due to a loss of lubrication. In contrast, both the steel/GL/Ag and W/GL/Ag pairs showed a considerably lower lifetime in our experimental set-up. The lifetime of the steel/GL/Ag pair was determined to about 2700 cycles (not even discernible in **Figure 2**) when the friction coefficient starts to fluctuate and increase gradually to high values. The corresponding life time for the W/GL/Ag pair was determined to be only about 500 cycles.

SEM and Raman spectroscopy were used to characterize the surfaces after addition of graphene on Ag. As shown in **Figure 3(a)**, after evaporation of the ethanol, grayish flakes on the surface can be seen in the SEM images. It is clear that the size of the deposited flakes varies with massive amount of flakes less than 1μm in diameter. It also shows that the silver surface is not fully covered by the deposited flakes. Raman spectroscopy confirms that the flakes indeed are graphene [see **Figure 3(b)**]. The characteristic peaks of graphene were observed at ~1330cm^{-1} (D peak), ~1600cm^{-1} (G peak) and ~2650cm^{-1} (2D peak). The D peak is due to a breathing mode of sp^2 atoms in rings, which is activated by disordered structures, e.g., edges or defects from partial oxidation of graphene[16]–[19]. The D peak is quite strong, indicating a large amount of disordered structures, e.g.,

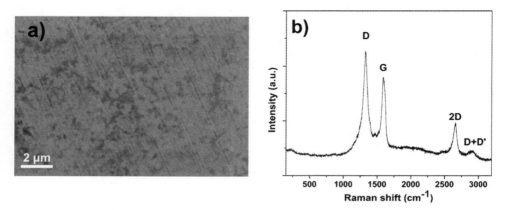

Figure 3. A SEM image and b Raman spectrum of asdeposited graphene flakes on Ag plate surface before pin-on-disk test. Scale bar for SEM image is 2μm.

edges or partial oxidation in graphene flakes. The Raman results suggest that the flakes consist of few-layer graphene, very similar to those observed by Berman et al. on steel surfaces[15].

To further study the lubricating effect of the GL, Raman spectroscopy was also carried out inside the track after the completion of the dry sliding tribological tests. The Raman spectra and the SEM images of the tracks are shown in **Figure 4**. The track of the Ag/Ag pair clearly suggests an adhesive material transfer as observed in **Figure 1**. In contrast, the SEM image of the Ag/GL/Ag pair shows a microscopic plowing pattern with plenty of grayish flakes remaining in the track. The existence of GL in the tracks after 1500 cycles was also confirmed by the characteristic Raman spectrum, which is almost identical to that from as-deposited flakes in **Figure 3**. The SEM images from the steel/Ag and steel/GL/Ag pairs also confirm that graphene has a strong impact of the tribological behavior. The track in the steel/Ag pair shows adhesive material transfer mixed with plowing while the steel/GL/Ag pair only exhibits minute plowing with remains of grayish flakes. The Raman spectrum from the steel/Ag track shows a number of peaks e.g., at ~560, 650, and 1320cm^{-1}. These peaks can be attributed to metal oxides such as Fe_2O_3 which has been transferred from the steel ball to the Ag surface during the tribological test[20]. In contrast, the Raman spectrum from the steel/GL/Ag track shows clear peaks from graphene, similar to the Ag/GL/Ag case, but no metal oxide peaks. This shows that the presence of graphene has a strong influence on the tribological behavior also with steel as a counter surface. A completely different behavior was observed with W as a counter material. The Raman spectrum from the W/Ag track shows a strong peak W-O at ~900cm^{-1} suggesting an extensive formation of tungsten oxides in the track[21]. Upon addition of graphene, the W/GL/Ag track shows very few graphene flakes in SEM. Furthermore, the W-O peak is still clearly seen in the Raman spectrum but only very weak and broad D and G peaks are observed. Such peaks are typical for amorphous carbon and suggest that the graphene has been highly damaged or completely destroyed during the sliding test.

XPS analysis can give supplementary information about graphene coverage inside the tracks through the C/Ag composition ratio. As shown in **Figure 5**, the composition ratio of C/Ag inside the tracks of the three bare metals/Ag was almost similar to 1, which means that there were some carbon contaminations on the

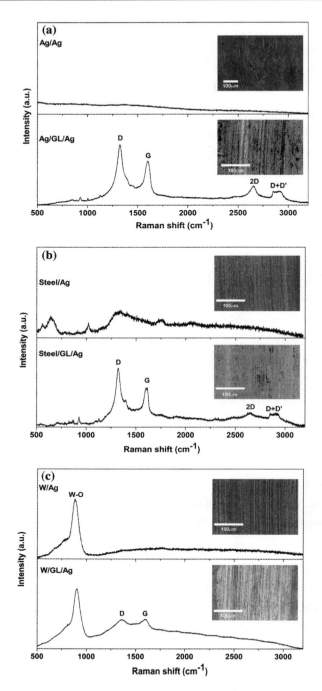

Figure 4. Raman spectra and SEM images for the tracks on the Ag plate with and without graphene deposition after pin-on-disk tests using different counters: a Ag ball; b steel ball; and c W ball. Scale bar for SEM images are 100μm.

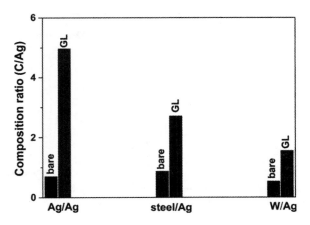

Figure 5. Composition ratio of C/Ag inside the tracks on the Ag plate after pin-on-disk tests using different counter surfaces.

bare track surface after the tribological tests. This originates from the carbon-containing contaminants adsorbed on the surfaces. In contrast, the three metals/GL/Ag pairs, exhibit significantly higher C/Ag ratios in the tracks. The Ag/GL/Ag track exhibits a higher ratio than steel/GL/Ag and considerably larger than that for the W/GK/Ag track. This result is consistent with the Raman and SEM results that plenty of graphene remains in Ag/GL/Ag track and less graphene left in the W/GL/Ag track.

To understand the reasons for different tribological behaviors in different metal-graphene interfaces, XPS was used to analyze the chemical bonding in the tracks of different metals/GL/Ag pairs. As shown in **Figure 6(a)**, Ag, C, and O signals were detected in all tracks. The C1s peak could originate from a mixture of remaining graphene flakes and other carbon contaminations. The O can be attributed to partially oxidized graphene or oxidized metal particles formed during the sliding tests. Some organic contaminants can also contribute. It is interesting to notice that W peaks were also observed in the XPS spectrum from the track of the W/GL/Ag pair, while no Fe peaks were seen in the spectrum from the steel/GL/Ag track. This indicates that W has been transferred to the Ag surface from the W ball during sliding while no such transfer occurs from the steel ball. High resolution spectra from the W4f peak indicate that the W is oxidized (not shown here).

High resolution spectra of the C1s peak obtained after 1500 cycles are shown in **Figure 6(b)**. In the track of Ag/GL/Ag, a C1s peak is observed at 284.8eV,

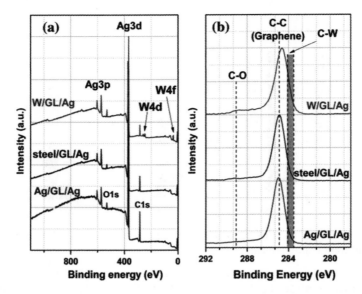

Figure 6. A Survey XPS spectra and b high resolution XPS spectra of C1s peaks for the track surface on the graphene-coated Ag plate after pin-on-disk tests using different metal counters.

which can be attributed to mainly C–C bonds in graphene[22]. Also, the C1s peaks show weak contribution around 289eV which can be attributed to a type of C–O bonds. The relative intensity of the C–O contribution is slightly higher for the steel/GL/Ag par and significantly higher for the W/GL/Ag pair. This suggests a somewhat stronger oxidation of the GL with a steel counter surface. Furthermore, for the W/GL/Ag pair, the C1s peak is shifted with 0.2eV to a lower binding energy together with an increase in the intensity of C–O feature. This shift suggests a strong interaction of carbon with a metal such as W. The peak shift is not an indication of the formation of a hexagonal WC phase since the C1s peak in this compound is observed at 283.5eV. However, the C–W binding energy is strongly dependent on the type of W–C coordination and, for example, a binding energy of 284.1eV has been observed in W_2C[23]. It can be concluded, however, that the peak shift and increased C–O intensity is in good agreement with a decomposition of the graphene flakes in the W/GL/Ag track and the formation of amorphous carbon and some type of W-C compound.

The contact resistance of graphene-coated Ag surfaces was also evaluated using a four-point resistance method. A general observation was that the addition of graphene slightly reduced the contact resistance compared to the pure Ag surface, but at a similar level, as shown in **Figure 7**. The graphene-coated Ag surface exhibited a contact resistance of 0.8mΩ, compared to 1.18mΩ for pure Ag at a normal load of 1N. The deviations decrease with increasing load, e.g., 0.41mΩ for graphene/Ag and 0.59mΩ for Ag using 5N. Graphene as a zero-overlap semimetal (with both holes and electrons as charge carriers) with very high electrical conductivity could be the main contribution for the good contact property of graphene-coated Ag surface[24].

4. Discussion

Our results clearly show that graphene is an excellent lubricant in Ag/GL/Ag contacts. In contrast, graphene showed a lubricating effect with tungsten as a counter surface but the friction coefficient was higher and the lifetime of the graphene was much shorter. The steel/GL/Ag pair showed an intermediate behavior. There are two main mechanisms which may contribute to these results: (i) the general trend in the Me–C bond strength and the Me-graphene interactions and (ii) the mechanical deformation mechanisms in the metal/GL/Ag contacts.

Our XPS and Raman results clearly suggest that graphene on W in the track is highly damaged and probably partly decomposed. The shift of the XPS C1s peak suggests a strong C-W interaction between the metal and damaged graphene and/or partial formation of a C-W compound. No direct indications of such decomposition were observed on Ag or steel. It is clear that chemical interactions

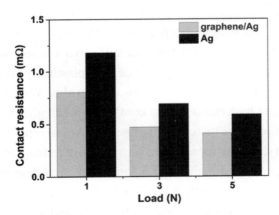

Figure 7. Contact resistance of Ag and graphene-coated Ag (graphene/Ag) surface for different loads.

between metals and carbon are strongly dependent on the position of the metal in the periodic table. In general, the Me−C bond strength decreases with increasing number of d-electrons in the metal. In our study, the W−C bond strength is rather high and several tungsten carbide phases are known with the hexagonal WC phase an illustrative example. The Fe−C bonds are weaker and only metastable iron carbides are known. Finally, the Ag−C interactions are very weak and no stable or metastable silver carbides are known. Consequently, from a thermodynamic point of view, a driving force exists to form carbides by a chemical reaction between W and graphene while no such driving force exists for Fe and Ag. Secondly, the kinetics of decomposition of graphene can be strongly affected by Me−C interactions. A change of charge transfer can dramatically change the stability of a molecule due to changes in the bonding/antibonding states. Theoretical calculations on adsorption of graphene have shown mainly weak adsorption (physisorption) on late transition metals such as Cu, Ag, and Au while a much stronger interactions including chemisorptions can be observed on Co and Ni[25]–[27]. Hence, from a pure chemical point of we should expect strong interactions between graphene and W and to some extent also Fe. This can explain the general trends observed in the tribological properties described above.

An alternative explanation to the observed trends in friction could possibly be variations in the mechanical deformation mechanisms in the metal/GL/Ag contacts. The tracks after tribological tests indeed look very different; some shallow and some extending to large depths. But when comparing the different tribological behaviors, it is important to keep the vast differences in mechanical strength in mind and to separate the mechanical plastic deformation from the more interesting tribological mechanisms behind the friction coefficients. Ag is a very soft metal known for very limited strain hardening. In any contact with considerably harder materials, like steel or W, only the Ag will deform plastically. This means that the steel balls, the W balls and even the Ag tips used in this work will all be pressed into the silver counter material until the contact area has grown sufficiently large to make the contact pressure match the hardness of Ag. In other words, the contact pressure will be very similar for all three (or six) pairs tested here, despite the fact that the radii of the spherical bodies differ somewhat. And despite the fact that the specific shapes of the tracks will be different. This holds both initially, at first contact, and during the subsequent mechanical deformation occurring during the tribological testing. In other words, the much harder steel and W will merely serve to shape the surface of the track in Ag, on which the crucial tribological mechanisms will take place, *i.e.*, material transfer, graphene retention or decomposition, oxidation, etc.

However, once material transfer commence, the shape of the components will degrade. This is very clearly demonstrated by the Ag/Ag pair where the initially spherical against flat geometry is completely lost due to excessive material transfer. A strong intermetallic bond, as in the Ag/Ag pair, is of course very effective in initiating material transfer. All other pairs tested, and especially those with interfaces containing graphene, will be less prone to material transfer. But once it is initiated, the surfaces will begin to degrade and the effectiveness of any graphene present in the surface will be reduced further.

Consequently, we conclude that the most likely explanation for the observed trend in tribological behavior is variations in the chemical interactions between metal and graphene.

5. Conclusion

In this study, the potential use of GL as a solid lubricant on sliding Ag electrical contact was investigated. It has been shown that small amount of graphene flakes, which can easily and quickly be deposited by evaporating a commercial solution in air, can dramatically reduce friction in dry sliding Ag/Ag contacts. With the lubricating effect of the graphene flakes, the friction coefficient was reduced by a factor of ~10 (e.g., from 1.15 to 0.12). Moreover, the track was much smoother, showing almost no signs of adhesive material transfer. The lifetime of the lubricating graphene was very long and the contact resistance for graphene/Ag surface was similar to pure Ag.

The lubricating effect was dependent on the counter surface. The Ag/GL/Ag contact exhibited the lowest friction coefficient and longest lifetime (>40,000 cylces at 2N). In contrast, the steel/GL/Ag and W/GL/Ag exhibited higher friction coefficients and shorter lifetimes. The life-time of the W/GL/Ag pair was only 500 cycles. Metal-graphene interaction is believed to be the main reason for the differences in friction reduction, where W-graphene shows the strongest bonds and

Ag the weakest. Fe-based alloys such as steel exhibits intermediate bond strength to graphene. Therefore, the friction reduction with graphene lubrication increased in the sequence Ag ball > steel ball > W ball. Alternative models to explain the trend, such as variations in hardness between the counter surfaces could be excluded due to the softness of the Ag surface. Overall, the deposited GL holds a great promise as an effective solid lubricant to significantly reduce friction in sliding Ag electrical contacts, especially in such a simple, quick, energy-saving, and cost-efficient way.

Acknowledgements

The authors wish to acknowledge the financial support of the KIC InnoEnergy and the Swedish Centre for Smart Grids and Energy Storage (SweGRIDS). We also want to thank Pedro Berastegui for his technical support. Ulf Jansson also acknowledges Knut och Alice Wallenberg (KaW) foundation for support. Urban Wiklund also acknowledges the Swedish Foundation for Strategic Research (via the program Technical advancement through controlled tribofilms) for the financial support.

Open Access: This article is distributed under the terms of the Creative Commons Attribution 4.0 International License (http://creativecommons.org/licenses/by/4.0/), which permits unrestricted use, distribution, and reproduction in any medium, provided you give appropriate credit to the original author(s) and the source, provide a link to the Creative Commons license, and indicate if changes were made.

Source: Mao F, Wiklund U, Andersson A M, et al. Graphene as a lubricant on Ag for electrical contact applications[J]. Journal of Materials Science, 2015, 50(19): 6518–6525.

References

[1] Arnell S, Andersson G (2001) Silver iodide as a solid lubricant for power contacts. In: Proceedings of the forty-seventh IEEE holm conference on electrical contacts, 239–244. doi:10.1109/holm.2001.953217.

[2] Lauridsen J, Eklund P, Lu J, Knutsson A, Odén M, Mannerbro R, Andersson AM, Hultman L (2012) Microstructural and chemical analysis of AgI coatings used as a solid lubricant in electrical sliding contacts. Tribol Lett 46(2):187–193. doi:10.1007/s11249-012-9938-3.

[3] Sundberg J, Mao F, Andersson A, Wiklund U, Jansson U (2013). Solution-based synthesis of AgI coatings for low-friction applications. J Mater Sci 48(5):2236–2244. doi:10.1007/s10853-012-6999-5.

[4] Gao Y, Hao P (2009) Mechanical properties of monolayer graphene under tensile and compressive loading. Physica E 41(8):1561–1566. doi:10.1016/j.physe.2009.04.033.

[5] Geim AK (2009) Graphene: status and prospects. Science 324(5934):1530–1534. doi:10.1126/science.1158877.

[6] Lee C, Wei X, Kysar JW, Hone J (2008) Measurement of the elastic properties and intrinsic strength of monolayer graphene. Science 321(5887):385–388. doi:10.2307/20054532.

[7] Stankovich S, Dikin DA, Dommett GHB, Kohlhaas KM, Zimney EJ, Stach EA, Piner RD, Nguyen ST, Ruoff RS (2006) Graphene-based composite materials. Nature 442(7100):282–286. doi:10. 1038/nature04969.

[8] Choudhary S, Mungse HP, Khatri OP (2012) Dispersion of alkylated graphene in organic solvents and its potential for lubrication applications. J Mater Chem 22(39):21032–21039. doi:10.1039/c2jm34741e.

[9] Kandanur SS, Rafiee MA, Yavari F, Schrameyer M, Yu Z-Z, Blanchet TA, Koratkar N (2012) Suppression of wear in graphene polymer composites. Carbon 50(9):3178–3183. doi:10.1016/j. carbon.2011.10.038.

[10] Lee C, Wei X, Li Q, Carpick R, Kysar JW, Hone J (2009) Elastic and frictional properties of graphene. Phys Status Solidi B 246(11–12):2562–2567. doi:10.1002/pssb.200982329

[11] Lin J, Wang L, Chen G (2011) Modification of graphene platelets and their tribological properties as a lubricant additive. Tribol Lett 41(1):209–215. doi:10.1007/s11249-010-9702-5.

[12] Sandoz-Rosado EJ, Tertuliano OA, Terrell EJ (2012) An ato-mistic study of the abrasive wear and failure of graphene sheets when used as a solid lubricant and a comparison to diamond-like-carbon coatings. Carbon 50(11):4078–4084. doi:10.1016/j.car bon.2012.04.055.

[13] Berman D, Erdemir A, Sumant AV (2014) Graphene as a protective coating and superior lubricant for electrical contacts. Appl Phys Lett 105(23):231907. doi:10.1063/1.4903933.

[14] Berman D, Erdemir A, Sumant AV (2013) Few layer graphene to reduce wear and friction on sliding steel surfaces. Carbon 54:454–459. doi:10.1016/j.carbon.2012.11.061.

[15] Berman D, Erdemir A, Sumant AV (2013) Reduced wear and friction enabled by graphene layers on sliding steel surfaces in dry nitrogen. Carbon 59:167–175. doi:10.1016/j.carbon.2013.03. 006.

[16] Ferrari AC, Robertson J (2000) Interpretation of Raman spectra of disordered and amorphous carbon. Phys Rev B 61(20): 14095–14107.

[17] Pócsik I, Hundhausen M, Koós M, Ley L (1998) Origin of the D peak in the Raman spectrum of microcrystalline graphite. J Non-Cryst Solids 227–230, Part 2 (0): 1083–1086. doi:10.1016/S0022-3093(98)00349-4.

[18] Ferrari AC (2007) Raman spectroscopy of graphene and graphite: disorder, electron–phonon coupling, doping and nonadiabatic effects. Solid State Commun 143(1–2):47–57. doi:10.1016/j.ssc. 2007.03.052.

[19] Ferrari AC, Meyer JC, Scardaci V, Casiraghi C, Lazzeri M, Mauri F, Piscanec S, Jiang D, Novoselov KS, Roth S, Geim AK (2006) Raman spectrum of graphene and graphene layers. Phys Rev Lett. doi:10.1103/PhysRevLett.97.187401.

[20] Muralha VSF, Rehren T, Clark RJH (2011) Characterization of an iron smelting slag from Zimbabwe by Raman microscopy and electron beam analysis. J Raman Spectrosc 42(12):2077–2084. doi:10.1002/jrs.2961.

[21] Voevodin AA, O'Neill JP, Zabinski JS (1999) Tribological performance and tribochemistry of nanocrystalline WC/amorphous diamond-like carbon composites. Thin Solid Films 342(1–2): 194–200. doi:10.1016/S0040-6090(98)01456-4.

[22] Ferrah D, Penuelas J, Bottela C, Grenet G, Ouerghi A (2013) X-ray photoelectron spectroscopy (XPS) and diffraction (XPD) study of a few layers of graphene on 6H-SiC(0001). Surf Sci 615:47–56. doi:10.1016/j.susc.2013.04.006.

[23] Aizawa T, Hishita S, Tanaka T, Otani S (2011) Surface reconstruction of W2C(0001). J Phys-Condens Mat 23(30):305007.

[24] Castro Neto AH, Guinea F, Peres NMR, Novoselov KS, Geim AK (2009) The electronic properties of graphene. Rev Mod Phys 81(1):109–162.

[25] Xu Z, Buehler MJ (2010) Interface structure and mechanics between graphene and metal substrates: a first-principles study. J Phys 22(48):485301.

[26] Giovannetti G, Khomyakov PA, Brocks G, Karpan VM, van den Brink J, Kelly PJ (2008) Doping graphene with metal contacts. Phys Rev Lett 101(2):026803.

[27] Andersen M, Hornekær L, Hammer B (2012) Graphene on metal surfaces and its hydrogen adsorption: a meta-GGA functional study. Phys Rev B 86(8):085405.

Chapter 17

Growth of Carbon Nanofibers from Methane on a Hydroxyapatite-Supported Nickel Catalyst

Ewa Miniach, Agata Śliwak, Adam Moyseowicz, Grażyna Gryglewicz[*]

Department of Polymer and Carbonaceous Materials, Faculty of Chemistry, Wrocław University of Technology, Gdańska 7/9, 50-344 Wrocław, Poland

Abstract: Carbon nanofibers (CNFs) were grown using catalytic chemical vapor deposition (CCVD) with methane as the carbon source and a hydroxyapatite-supported nickel catalyst (Ni/HAp). The catalyst, which contained approximately 14wt% Ni, was prepared using the incipient wetness method with an aqueous nickel nitrate solution. Temperature-programmed reduction and X-ray diffraction were used to characterize the active phase of Ni/HAp. Three variables were evaluated to optimize the CNF growth process, including the temperature and the time of catalyst reduction as well as the reaction time, at 650°C. Regardless of the applied CCVD process conditions, herringbone bamboo-like CNFs were grown during methane decomposition over Ni/HAp, which was confirmed using transmission electron microscopy. A high CNF yield of nearly $10\,g_{CNF}\,g_{cat}^{-1}$ was achieved at 650°C after a reaction time of 3h when the catalyst was subjected to a reduction at the same temperature for 2h under a hydrogen flow prior to synthesis. As the reduction temperature increased from 450°C to 650°C, both the yield and diameters of the CNFs increased. The beneficial effects of including hydrogen in the reaction mixture on the catalytic performance of Ni/HAp and the purity of the grown CNFs were demonstrated.

1. Introduction

Carbon nanofibers (CNFs), which include carbon nanotubes (CNTs), have been synthesized since the 1960s. However, the most significant discovery regarding CNTs was in 1991 when Iijima reported that highly graphitized carbon, which was formed from the arc discharge of graphite electrodes, contained several coaxial tubes and a hollow core[1]. Since this study was reported, the synthesis and application of CNFs/CNTs have attracted increasing interest due to their variety of extraordinary structures, which provide unique mechanical and electromagnetic properties, chemical inertness, high conductivity, surface properties, and ease of structure control[2]. Catalytic chemical vapor deposition (CCVD) using hydrocarbons is the most extensively used method to produce CNFs due to its high efficiency and selectivity at low costs. Many variables affect CNF growth, such as the catalyst composition, reducibility of the catalyst, nature of the metal/support interaction, particle size of the active metal, synthesis temperature, carbon source, and composition of the reaction gas mixture[3]–[6]. The search for an efficient catalytic system that can provide good stability and dispersion of a catalytic active metal as well as high yield of well-defined carbon nanostructures remains the subject of much research[7][8].

Methane decomposition over transition metals such as Ni, Fe or Co, is a promising approach for CNF synthesis due to the abundance and low cost of natural gas. Among the catalysts, Ni is characterized by exhibiting the highest catalytic activity in the production of CNFs[9][10]. In most studies, Ni was deposited on widely used supports, such as Al_2O_3, SiO_2, TiO_2, MgO[11][12], and zeolites[13] in addition to activated carbon[14][15] and carbon nanofibers/nanotubes[16][17]. Herein, we report the use of hydroxyapatite (HAp) as a support for a Ni catalyst to produce CNFs using methane decomposition. Only a few studies on the use of HAp as a catalyst support to produce CNF/HAp composites using a mixture of methane and nitrogen have been reported for potential applications in the dental and medical fields[18]–[20]. CNF-containing HAp-based composites are candidates for use in implants due to their enhanced mechanical properties[21]–[23]. The chemical inertness of HAp is an important property for a material that will be used as a catalyst support. HAp does not exhibit any catalytic activity during the high-temperature treatment[20] or create spinels that would lead to a reduction in the catalyst reducibility, which is often observed for alumina-supported nickel catalysts[24][25]. Therefore, HAp appears to be a good support for Ni catalysts for the synthesis of CNF using CCVD.

In this study, selected variables of the CCVD process were studied to obtain CNFs with high yield and purity using a Ni/HAp catalyst and methane as the carbon source. We focused on the catalyst reduction conditions in the CCVD process because they are rarely reported in the literature. Our results demonstrate that the optimization of the temperature and time of catalyst reduction is crucial for high-yield CNF synthesis. For optimal reduction conditions, the kinetics of CNF growth were studied to evaluate the susceptibility of Ni/HAp to deactivation.

2. Experimental

2.1. Catalyst Preparation

Hydroxyapatite (HAp) with a purity grade of 97% (Sigma-Aldrich, particle size <200nm) was used as the support. The HAp-supported Ni catalyst was prepared using the incipient wetness method with an aqueous solution of $Ni(NO_3)\cdot 6H_2O$ (ACROS Organics, 99%). The amount of Ni precursor was adjusted to achieve a Ni weight loading of 14wt% in the catalyst. The sample was further dried at 110°C for 2h and subsequently calcined in a nitrogen atmosphere at 350°C for 4h. The Ni content in the catalyst was determined to be 13.4wt% using atomic absorption spectroscopy (AAS). The prepared catalyst is referred to as Ni/HAp.

2.2. Synthesis of CNFs

CNFs were synthesized using CCVD with methane as the carbon precursor. First, 200mg of the Ni/HAp catalyst was loaded in a quartz boat and placed in a quartz tube in the central heat zone of a conventional horizontal furnace. Prior to CNF growth, the catalyst was reduced under hydrogen flowing at a rate of 150 ml min at various temperatures (450°C–650°C) and times (10–120min). After the reduction, a mixture of CH_4 and H_2 (1:1, v/v, 150ml min) was passed through the quartz tube at 650°C. The reaction time was varied from 3 to 180min. The resulting CNFs were subsequently cooled to ambient temperature in a nitrogen atmosphere. Both the catalyst reduction and the CNF synthesis were performed at atmospheric pressure. The purity grades of CH_4, H_2, and N_2 were 99.995%, 99.999%, and 99.999%, respectively. The yield (Υ) of CNFs was calculated based on the catalyst weight increase during the reaction using the following equation:

$$\Upsilon = \frac{m_t - m_{cat}}{m_{cat}} \tag{1}$$

where Υ is the yield of CNF expressed in g_{CNF} per g_{cat} [$g_{CNF}\, g_{cat}^{-1}$], m_t is the total weight of the catalyst and the deposited carbon at the end of the reaction, and m_{cat} is the initial mass of the catalyst. Additionally, the yield (Υ_{Ni}) was expressed per gram of metal ($g_{CNF}\, g_{Ni}^{-1}$) according to the following equation:

$$\Upsilon_{Ni} = \frac{m_t - m_{cat}}{0.134 \cdot m_{cat}} \tag{2}$$

The difference in the CNF yield for two separate runs, which were performed using the same experimental conditions, was not higher than 7% of the average yield. The as-received CNFs were treated with 5% HCl for 2h at room temperature in a sonication bath to remove the nickel and hydroxyapatite. Subsequently, the samples were washed with distilled water, filtered, and dried at 110°C for 2h.

2.3. Catalyst Characterization

The nickel content in the Ni/HAp catalyst was determined by AAS using a SOLAAR S4 spectrometer. Temperature-programmed reduction (TPR) was performed on a ChemBET 3000 analyzer (Quantachrome) using a reducing mixture of H_2 (5% v/v)-Ar (100ml·min^{-1}). The catalyst sample was heated at a rate of

20°C·min^{-1} from ambient temperature to 980°C. The hydrogen consumption was determined using CuO as a standard for the TPR analysis. X-ray diffraction (XRD) analysis was performed on an Ultima IV Rigaku diffractometer using CuKa radiation (λ = 1.54056Å). The average sizes of the NiO and Ni crystallites were calculated using Scherrer's equation.

2.4. CNF Characterization

The morphologies of the as-grown and purified CNFs, HAp, and as-prepared Ni/HAp catalyst were studied using scanning electron microscopy (SEM) on an EVO LS15 Zeiss microscope. The arrangement of graphene layers in the CNFs was determined by transmission electron microscopy (TEM) using a FEI Tecnai G^2 20 X-TWIN micro-scope, which was operated at an acceleration voltage of 200kV. The sample was prepared by ultrasonic dispersion in ethanol, and a few drops of the suspension were placed onto a copper micro-grid covered with a perforated carbon film. The average CNF diameter was determined based on the measurement of 50 nanofibers for each sample. Thermogravimetric analysis (TGA) was performed on the purified CNFs using a TGA/DSC1 Mettler Toledo (thermobalance).

3. Results and Discussion

3.1. Catalyst Characterization

Figure 1 shows the X-ray diffraction patterns of the as-prepared (a) and reduced Ni/HAp (b) catalysts. For the as-prepared Ni/HAp catalyst, we observed diffraction peaks arising from different planes of the NiO phase [*i.e.*, 2h = 37.2° (111), 43.2° (200), 62.9° (220), 75.5° (311), and 79.5° (222)] [**Figure 1(a)**] [ICSD No:01-1239]. The diffraction peaks at 2h = 25.9° (002), 29.0° (210), 31.8° (211), 32.2° (112), 32.9° (300), 34.0°, 39.8° (310), 46.7° (222), 49.5° (213), 50.5° (321), and 53.1° (004) were due to the crystalline $Ca_5(PO_4)_3(OH)$ phase. The average size of the NiO crystallites, which was determined using XRD, was 22nm for the fresh Ni/HAp catalyst. After catalyst reduction at 650°C for 2h, the XRD pattern [**Figure 1(b)**] showed diffraction peaks arising from metallic Ni [*i.e.*, 2θ = 44.4° (111), 51.8° (200), and 76.4° (220)] [ICSD No:04-0850]. The Ni crystallite size was

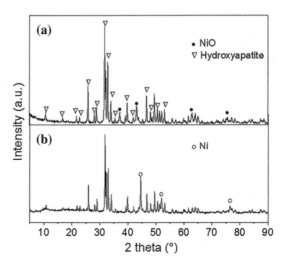

Figure 1. XRD pattern of the fresh Ni/HAp catalyst (a) and reduced Ni/HAp catalyst (b).

bigger than that of NiO (34 vs. 22nm), suggesting sintering and aggregation of the nickel phase during the heat treatment in a hydrogen atmosphere. The strong influence of the metal particle size on the CNF yield has been previously reported[7][26][27]. Excessively small metal particles will not form CNFs because the quantity of metal is not sufficient to enable CNF formation according to the proposed mechanism of CNF growth during CCVD. In contrast, excessively large metal particles are inactive because the carbon diffusion through these particles is hindered[28]. Chen et al.[7] reported the relationship between the NiO particle size of a hydrotalcite-supported catalyst and the yield of CNFs produced from methane decomposition. The most suitable NiO particle sizes, which constitute the catalytic centers for CNF growth after the hydrogen treatment, were determined to be approximately 34nm. Moreover, based on XRD analysis, the average NiO crystal size of the as-prepared catalyst was comparable to the Ni crystal size of the reduced catalyst, which was determined by chemisorption[7]. Therefore, the synthesized Ni/HAp catalyst with an average Ni crystallite size of 34nm appears to be a promising catalytic system for CNF growth. Some studies[26][27] have reported that Ni particles with sizes of 10–60nm are characteristic of highly active catalysts for the production of CNFs via methane decomposition.

The morphology of the as-prepared Ni/HAp catalyst was very similar to that of the support (Fig. S1). The HAp powder consists of agglomerated spherical nanoparticles, and their sizes are preserved in the Ni/HAp catalyst. SEM-EDX investigation of the reduced Ni/HAp catalyst revealed that nickel was homogeneously distributed on the surface of the support particles (Fig. S2).

The reducibility of the Ni/HAp catalyst, which was characterized by the TPR profile, is shown in **Figure 2**. Only one wide peak with a maximum at 470°C was observed due to the reduction of NiO to metallic Ni. A small shoulder on the curve at approximately 380°C may correspond to the reduction of very fine NiO particles. The reduction of NiO began at 200°C and ended at 670°C. The hydrogen consumption determined by TPR was 2.08 mmol·g^{-1}.

3.2. Effect of the Reduction Temperature

Based on the TPR results, a temperature range of 450 to 650°C was selected to determine the effect of the Ni/HAp reduction temperature on the yield, morphology, and structure of the grown CNFs. The CNFs were synthesized at 650°C using methane diluted with hydrogen in a ratio of 1:1 (v/v). The reduction time was 2h. The CNF yields, which depend on the reduction temperature of the Ni/HAp catalyst, are listed in **Table 1**. As the reduction temperature increased from 450°C to 650°C, the CNF yield increased from 3.9 to 5.9 $g_{CNF} \, g_{cat}^{-1}$, which may be due to the increasing number of Ni active sites on the catalyst surface. The latter

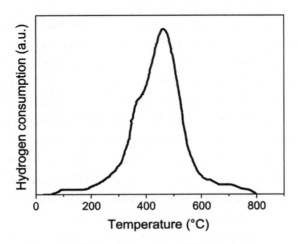

Figure 2. TPR profile of the Ni/HAp catalyst.

Table 1. Effect of the Ni/HAp reduction temperature on CNF growth.

Temperature (°C)	Reaction yield	
	($g_{CNF}\, g_{cat}^{-1}$)	($g_{CNF}\, g_{Ni}^{-1}$)
450	3.9	27.6
550	4.2	30.3
600	5.8	41.6
650	5.9	42.5

result is supported by the TPR profile of Ni/HAp (**Figure 2**), which indicates that approximately half of the NiO was reduced to the metallic form at 450°C. Pronounced CNF growth was observed up to 600°C, but further CNF growth was only slightly higher at 650°C. As expected, this growth was followed by an increase in the amount of Ni, which contributed to the CNF growth and yielded 42.5 $g_{CNF}\, g_{Ni}^{-1}$ at 650°C compared to 27.6 $g_{CNF}\, g_{Ni}^{-1}$ at 450°C. An enhancement in CNF growth from methane over a Ni−Cu−MgO catalyst with an increase in the reduction temperature from 600°C to 1000°C has also been reported by Wang et al.[29].

Figure 3 shows the TEM images of the as-grown CNFs, which were synthesized over the Ni/HAp catalyst and reduced at different temperatures. Regardless of the reduction temperature, the graphene layers were stacked at an angle to the fiber axis, which indicates the presence of their herring-bone structural type. The herringbone bamboo-like CNFs were primarily grown on the surface of the Ni/HAp catalyst. CNFs with a herringbone-type graphene layer arrangement exhibit open edges on their outer surface, which result in high chemical activity and may be beneficial in medical and other applications. The presence of Ni particles at the ends of the nanofibers suggests a tip-growth mechanism due to weak catalyst-support interactions[8]. The herring-bone type of CNFs is a typical structure that forms during CCVD when using methane over a Ni catalyst[30]. The diameters of the grown CNFs were determined by TEM. As shown in **Figure 4**, the CNF diameters increased as the reduction temperature increased. At 550°C, CNFs with a diameter of 25−45nm were grown with a predominant contribution of narrower nanofibers. At 600°C, a wide distribution of CNF diameters in the range of 25−55nm was observed. CNFs with a diameter of 36−55nm were obtained when the Ni/HAp catalyst was reduced at 650°C. The shift in the CNF diameter to larger values as the reduction temperature increases was due to an increase in the Ni crystallite size. CNFs with larger diameters were also grown by increasing the

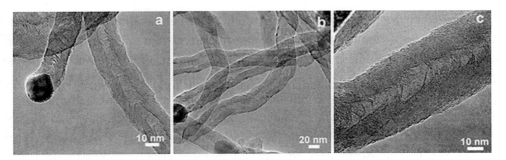

Figure 3. TEM images of CNFs synthesized with a catalyst reduction at 550°C (a), 600°C (b), and 650°C (c).

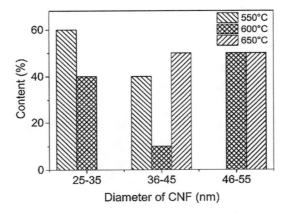

Figure 4. Diameter distribution of CNFs as a function of the reduction temperature of Ni/HAp.

reaction temperature[16][31].

An increase in the reduction temperature can lead to the migration of nickel particles on the support surface and formation of larger agglomerates, which results in larger CNF diameters. However, the catalyst, operating conditions, and reactant gas composition affect the CNF diameters[32]. In our study, the most suitable temperature for Ni/HAp reduction was determined to be 650°C, which results in maximum reduction of the NiO phase.

3.3. Effect of the Reduction Time

In general, a higher amount of NiO reduced to a metallic form corresponds to a higher CNF yield. The extent of catalyst reduction depends on both the temperature and reduction time. The effect of the reduction time on CNF growth was investigated at 650°C, where the maximum conversion of NiO to Ni was observed. After catalyst reduction, the CNFs were synthesized at the same temperature for 1h using a mixture of methane and hydrogen at a ratio of 1:1 (v/v). **Figure 5** shows the effect of the reduction time on the CNF growth over Ni/HAp. An increase in the catalyst reduction times led to a higher CNF yield, which is due to the increased amounts of reduced metallic nickel. A maximum CNF weight gain ($5.9\,g_{CNF}\,g_{cat}^{-1}$) was obtained at 2h of reduction. A sharp increase in CNF growth was observed for up to 30min of reduction, and then, the growth rate subsequently decreased. The plateau in the plot of the yield as a function of the reduction time indicates that 2h is an optimal time for activation of all of the catalytic Ni centers on the HAp surface. Longer reduction times most likely promote the formation of large agglomerates of Ni, which are unable to catalyze the decomposition of methane.

Figure 6 shows the SEM images of the obtained CNFs at different Ni/HAp reduction times. After 10min of reduction, the CNFs were heterogeneously dispersed on the catalyst surface and characterized by a broad size distribution, which ranged from short to several micrometers in length [**Figure 6(a)**]. An increase in the reduction time to 60min produced a strong entangled network of nanofibers that were more than a dozen micrometers in length [**Figure 6(b)**]. A dense network of entangled nanofibers was observed when Ni/HAp was reduced at 650°C for 2h. The obtained results indicate that a reduction time of 2h is optimal for CNF

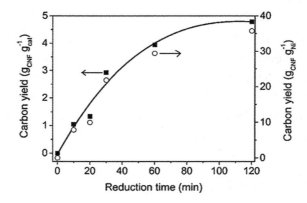

Figure 5. Relationship between the CNF yield and the reduction time of Ni/HAp at 650°C.

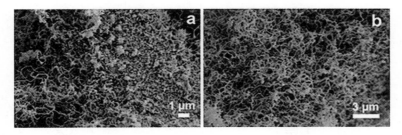

Figure 6. SEM images of the CNFs obtained after 10min (a) and 60min (b) of Ni/Hap reduction.

production over the Ni/HAp catalyst.

3.4. Kinetics of CNF Growth

The temperature of CCVD is crucial to the growth of CNFs[13][17]. For a given catalyst, the temperature must be higher than the decomposition temperature of the carbon precursor on the catalyst surface. An increase in the synthesis temperature results in an enhanced CNF yield but only to a certain extent. Excessive temperatures promote the uncontrolled deposition of carbon as well as the migration of metallic catalyst nanoparticles across the surface, resulting in the formation of large nanoparticles that do not promote CNF growth. An increase in reaction temperature up to 600°C–650°C increases methane conversion to filamentous carbon and hydrogen over Ni catalysts[33], which is consistent with results for Ni catalysts that exhibit the highest activity toward methane decomposition at 650°C[34]. In our study, the kinetic study of the CNF growth over Ni/HAp was performed at 650°C using a mixture of CH_4 and H_2. Prior to CNF growth, the Ni/HAp catalyst was reduced with hydrogen under the previously determined optimal conditions (*i.e.*, at 650°C for 2h). **Figure 7** shows the CNF yield as a function of the synthesis time.

An increase in the CNF productivity from 0.15 to 9.92 $g_{CNF}\ g_{cat}^{-1}$ was observed when the reaction time was increased from 3 to 180min. During the first few minutes of the process, only a small portion of Ni contributed to the CNF growth (1.12 $g_{CNF}\ g_{Ni}^{-1}$). After 180min, a large amount of Ni was involved in CNF growth (74.03 $g_{CNF}\ g_{Ni}^{-1}$). The highest rate of CNF growth was observed during the first hour of synthesis. At longer reaction times, the growth of the CNFs

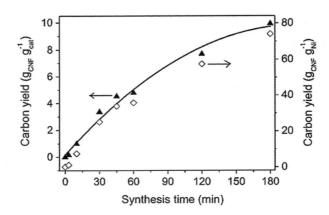

Figure 7. CNF growth as a function of the synthesis time over the Ni/HAp catalyst at 650°C.

slightly decreased. The high yield of CNFs after 3h of synthesis (nearly $10\,g_{CNF}\,g_{cat}^{-1}$) was due to the large amount of the nickel active phase, which results from the hydrogen in the reaction gas mixture preventing deposition of amorphous carbon on the catalyst surface. The presence of hydrogen in the reaction mixture plays a crucial role in the formation of CNFs due to the decomposition of hydrocarbons[31][35]. Hydrogen can either accelerate or suppress the formation of carbon[36], which affects the yield and morphology of the resulting CNFs as well as catalyst deactivation. This effect can be interpreted in multiple ways. Hydrogen, which is present in the feed stream along with methane, may be responsible for the decomposition of nickel carbides to form catalytically active metal particles[37] or remove the graphite overlayer that encapsulates the catalyst active surface[38]. In our study, hydrogen undoubtedly facilitates CNF growth by maintaining the catalytic activity of Ni, which leads to a high CNF yield. A continuous weight gain of CNFs with synthesis time may be related to hydrogen activation of new catalytic centers that were not activated during the catalyst reduction step. The two processes can occur simultaneously (*i.e.*, the reduction of NiO nanoparticles to create new active sites for CNF growth and the decomposition of methane on previously activated Ni centers, which both lead to an increase in the length of nanofibers). This result is supported by SEM investigations of the CNFs obtained using different synthesis times (**Figure 8**).

After a reaction time of 3min, the catalyst surface was not entirely coated with CNFs, which were several micrometers in length [**Figure 8(a)**]. Much longer

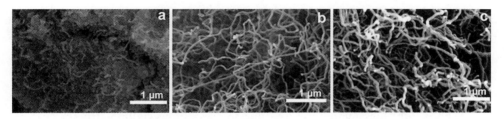

Figure 8. SEM images of CNFs synthesized at 650°C for 3min (a), 60min (b), and 180min (c).

and entangled CNFs appeared after a longer synthesis time [**Figure 8(b)**]. The increasing number of nanofibers on the support surface suggests the creation of new active sites during CCVD, which is supported by the higher CNF yield expressed as $g_{CNF}\,g_{Ni}^{-1}$. After a reaction time of 180min, the entire surface was coated with a dense layer of strongly entangled nanofibers that were 10 μm in length [**Figure 8(c)**], providing a yield of approximately $10\,g_{CNF}\,g_{cat}^{-1}$. It is important to note that by adjusting the reaction time, CNF/HAp composites with adjustable CNFs contents and lengths can be tailored for specific requirements.

The beneficial effect of hydrogen on the CNF growth over Ni/HAp using methane as a carbon precursor can also be demonstrated by comparing the obtained CNF yields with those reported by Ashok et al.[20]. The Ni/HAp catalyst with 15wt% of Ni after 4h of synthesis at 650°C in a methane stream (without hydrogen) produced $0.76\,g_{CNF}\,g_{cat}^{-1}$ of CNFs. In our study, a mixture of methane and hydrogen resulted in a CNF yield that was several times greater to as high as $9.92\,g_{CNF}\,g_{cat}^{-1}$. Few studies of Ni catalysts based on different supports with high performance in hydrocarbon decomposition to produce CNFs have been reported (e.g., Ni/Al$_2$O$_3$ with 90wt% of metal loading)[39]. This catalyst enabled a CNF yield of $20\,g_{CNF}\,g_{cat}^{-1}$ after 2h of reaction. However, the difficulties associated with the thermal instability of Al$_2$O$_3$ and its low mechanical strength have been reported by Boukha et al.[25]. Moreover, alumina can create spinels with nickel, which prevents the formation of nickel growth centers[25]. Based on these negative features, alumina can be replaced by other promising Ni catalyst supports, such as the studied hydroxyapatite, which has excellent mechanical properties and does not react with nickel.

It is interesting to note that although notably high yields of CNFs were obtained, amorphous carbon was not formed during the CCVD process, which was confirmed by SEM analysis. This result suggests that the presence of hydrogen in the reaction mixture efficiently prevents the deposition of undesirable carbon. The TGA analysis revealed a similar oxidation behavior of the resulting CNFs regardless of the synthesis time (**Figure 9**).

The weight loss, which may be due to the non-crystalline structure of carbonaceous materials, such as amorphous carbon, occurred between 300°C and 400°C. In this temperature range, no oxidation was observed even for the CNFs synthesized for 3h. This result demonstrates that a highly ordered graphitic structure without amorphous carbon is characteristic of CNFs that are synthesized over Ni/HAp using a CH_4 and H_2 mixture.

4. Conclusions

This study demonstrates that the use of hydroxyapatite as a support for a nickel catalyst results in the production of herringbone bamboo-like CNFs in a high yield with no amorphous carbon by CCVD using a mixture of methane and hydrogen as the reactant gas. At a reaction temperature of 650°C, the growth of CNFs and their diameter distribution strongly depended on the catalyst reduction temperature. An increase in the reduction temperature of Ni/HAp from 450°C to 650°C increased the CNF growth due to the larger amount of reduced metallic nickel. Moreover, reduction at higher temperatures produced CNFs with larger diameters due to agglomeration of nickel particles during the high-temperature treatment.

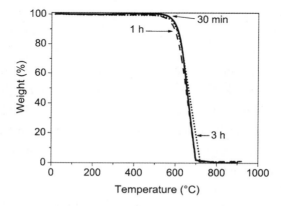

Figure 9. TGA curves of CNFs synthesized with different reaction times.

This result suggests that the diameter distribution of the grown CNFs can be tailored by adjusting the reduction temperature of the catalyst. A notably high CNF yield up to approximately $10\,\mathrm{g_{CNF}\,g_{cat}^{-1}}$, was obtained at 650°C with a reaction time of 3h when Ni/HAp was reduced at 650°C for 2h. At this temperature, all of the reducible NiO in the Ni/HAp catalyst was converted to the metallic form, which was confirmed by TPR. Due to the inertness of hydroxyapatite and the presence of hydrogen in the reaction mixture, the Ni/HAp catalyst was resistant to deactivation and maintained its catalytic activity during CNF growth over long CCVD processing times. Due to the graphitic edges that were exposed on the outer surface of the herringbone CNFs, this structure has a high potential for modification, which may create new applications for CNF/HAp composites.

Acknowledgements

This work was financed by a statutory activity subsidy from the Polish Ministry of Science and Higher Education for the Faculty of Chemistry of Wrocław University of Technology. The project was supported by Wrocław Centre of Biotechnology, under The Leading National Research Centre (KNOW) programme for years 2014–2018.

Compliance with Ethical Standards

Conflict of interest: The authors declare that they have no conflict of interest.

Open Access: This article is distributed under the terms of the Creative Commons Attribution 4.0 International License (http://creativecommons.org/licenses/by/4.0/), which permits unrestricted use, distribution, and reproduction in any medium, provided you give appropriate credit to the original author(s) and the source, provide a link to the Creative Commons license, and indicate if changes were made.

Electronic supplementary material: The online version of this article (doi:10.1007/s10853-016-9839-1) contains supplementary material, which is available to authorized users.

Source: Miniach E, Śliwak A, Moyseowicz A. Growth of carbon nanofibers from methane on a hydroxyapatite-supported nickel catalyst[J]. Journal of Materials Science, 2016, 51(11):5367–5376.

References

[1] Iijima S (1991) Helical microtubules of graphitic carbon. Nature 354:56–58.

[2] Feng L, Xie N, Zhong J (2014) Carbon nanofibers and their composites: a review of synthesizing, properties and applications. Materials 7:3919–3945.

[3] Zhou JH, Sui ZJ, Li P, Chen D, Dai YC, Yuan WK (2006) Structural characterization of carbon nanofibers formed from different carbon-containing gases. Carbon 44:3255–3266.

[4] Tessonier JP, Rosenthal D, Hansen TW, Hess Ch, Schuster ME, Blume R, Girsdies F, Pfander N, Timpe O, Su DS, Schlogl R (2009) Analysis of the structure and chemical properties of some commercial carbon nanostructures. Carbon 47:1779–1798.

[5] Moyseowicz A, Śliwak A, Gryglewicz G (2016) Influence of structural and textural parameters of carbon nanofibers on their capacitive behavior. J Mater Sci 51:3431–3439. doi:10.1007/s10853-015-9660-2.

[6] Yu L, Qin Y, Sui L, Zhang Q, Cui Z (2008) Two opposite growth modes of carbon nanofibers prepared by catalytic decomposition of acetylene at low temperature. J Mater Sci 43:883–886. doi:10.1007/s10853-007-2191-8.

[7] Chen D, Christensen KO, Ochoa-Fernandez E, Yu Z, Totdal B, Latorre N, Monzon A, Holmen A (2005) Synthesis of carbon nanofibers: effects of Ni crystal size during methane decomposition. J Catal 229:82–96.

[8] Vander Wal RL, Ticich TM, Curtis VE (2001) Substrate-support interactions in metal catalyzed carbon nanofiber growth. Carbon 39:2277–2289.

[9] Yamada Y, Hosono Y, Murakoshi N, Higashi N, Ichi-oka H, Miyake T et al (2006) Carbon nanofiber formation on iron group metal loaded on SiO_2. Diam Relat Mater 15(4–8):1080–1084.

[10] Romero A, Garrido A, Nieto-Marquez A, Sanchez P, de Lucas A, Valverde JL (2008) Synthesis and structural characteristics of highly graphitized carbon nanofibers produced from the catalytic decomposition of ethylene: influence of the active metal (Co, Ni, Fe) and the zeolite type support. Microporous Mesoporous Mater 110(2–3):318–329.

[11] Takenaka S, Shigeta Y, Tanabe E, Otsuka K (2003) Methane decomposition into hydrogen and carbon nanofibers over supported Pd-Ni catalysts. J Catal 220:468–477.

[12] Li Y, Li D, Wang G (2011) Methane decomposition to CO_x—free hydrogen and na-

no-carbon material on group 8–10 base metal catalysts: a review. Catal Today 162:1–48.

[13] De Lucas A, Garrido A, Sanchez P, Romero A, Valverde JL (2005) Growth of carbon nanofibers from Ni/Y zeolite based catalysts: effects of Ni introduction method, reaction temperature, and reaction gas composition. Ind Eng Chem Res 44:8225–8236.

[14] Rinaldi A, Abdullah N, Ali M, Furche A, Bee Abd Hamid S, Sheng Su S, Schlogl R (2009) Controlling the yield and structure of carbon nanofibers grown on a nickel/activated carbon catalyst. Carbon 47:3023–3033.

[15] Gryglewicz G, Śliwak A, Beguin F (2013) Carbon nanofibers grafted on activated carbon as an electrode in high-power supercapacitors. ChemSusChem 6:1516–1522.

[16] Romero A, Garrido A, Nieto-Marquez A, Raquel de la Oa A, de Lucas A, Valverde JL (2007) The influence of operating conditions on the growth of carbon nanofibers on carbon nanofiber-supported nickel catalyst. Appl Catal A 319:246–258.

[17] Zhao Y, Li Ch, Yao K, Liang J (2007) Preparation of carbon nanofibers over carbon nanotube-nickel catalyst in propylene decomposition. J Mater Sci 42:4240–4244. doi:10.1007/ s10853-006-0676-5.

[18] Li H, Zhao N, Liu Y, Liang C, Shi C, Du X, Li J (2008) Fabrication and properties of carbon nanotubes reinforced Fe/hydroxyapatite composites by in situ chemical vapor deposition. Compos A 39(7):1128–1132.

[19] Li H, Wang L, Liang C, Wang Z, Zhao W (2010) Dispersion of carbon nanotubes in hydroxyapatite powder by in situ chemical vapor deposition. Mater Sci Eng B 166(1):19–23.

[20] Ashok J, Naveen Kumar S, Subrahmanyam M, Venugopal A (2008) Pure H2 production by decomposition of methane over Ni supported on hydroxyapatite catalyst. Catal Lett 121:283–290.

[21] Kaya C (2008) Electrophoretic deposition of carbon nanotube-reinforced hydroxyapatite bioactive layers on Ti–6Al–4 V alloys for biomedical applications. Ceram Int 34:1843–1847.

[22] Mukherjee S, Kundu B, Sen S, Chanda A (2014) Improved properties of hydroxyapatite–carbon nanotube biocomposite: mechanical, in vitro bioactivity and biological studies. Ceram Int 40:5635–5643.

[23] Prodana M, Duta M, Ionita D, Bojin D, Stan MS, Dinischiotu A, Demetrescu I (2015) A new complex ceramic coating with carbon nanotubes, hydroxyapatite and TiO2 nanotubes on Ti surface for biomedical applications. Ceram Int 41:6318–6325.

[24] Piao L, Li Y, Chen J, Chang L, Lin JYS (2002) Methane decomposition to carbon nanotubes and hydrogen on an alumina supported nickel aerogel catalyst. Catal Today 74(1–2):145–155.

[25] Boukha Z, Jimenez-Gonzalez C, de Rivas B, Gonzalez-Velasco JR, Gutierrez-Ortiz JI, Lopez-Fonseca R (2014) Synthesis, characterization and performance evaluation

of spinel-derived Al_2O_3 catalysts for various methane reforming reactions. Appl Catal B 158–159:190–201.

[26] Ermakova MA, Ermakov DY, Kuvshinov GG, Plasova LM (1999) New nickel catalysts for the formation of filamentous carbon in the reaction of methane decomposition. J Catal 187:77–84.

[27] Ermakova MA, Ermakov DY, Kuvshinov GG (2000) Effective catalysts for direct cracking of methane to produce hydrogen and filamentous carbon part I. Nickel catalysts. Appl Catal A 201:61–70.

[28] Chinthaginjala JK, Seshan K, Lefferts L (2007) Preparation and application of carbon-nanofiber based microstructured materials as catalyst supports. Ind Eng Chem Res 46(12):3968–3978.

[29] Wang H, Terry R, Baker K (2004) Decomposition of methane over a Ni-Cu-MgO catalyst to produce hydrogen and carbon nanofibers. J Phys Chem B 108:20273–20277.

[30] Takenaka S, Kobayashi S, Ogihara H, Otsuka K (2003) Ni/ SiO2 catalyst effective for methane decomposition into hydrogen and carbon nanofiber. J Catal 217:79–87.

[31] Singh C, Shaffer MSP, Windle AH (2003) Production of controlled architectures of aligned carbon nanotubes by an injection chemical vapour deposition method. Carbon 41(2):359–368.

[32] Jong KPD, Geus JW (2000) Carbon nanofibers: catalytic synthesis and applications. Cat Rev Sci Eng 42:481–510.

[33] Zhang W, Ge Q, Xu H (2011) Influences of reaction conditions on methane decomposition over non-supported Ni catalyst. J Nat Gas Chem 20:339–344.

[34] Zhang W, Ge Q, Xu H (2010) Influences of precipitate rinsing solvents on Ni catalyst for methane decomposition to COx—free hydrogen. J Phys Chem A 114:3818–3823.

[35] Vieira R, Ledoux MJ, Pham-Huu C (2004) Synthesis and characterization of carbon nanofibers with macroscopic shaping formed by catalytic decomposition of C_2H_2/H_2 over nickel catalyst. Appl Catal A 274:1–8.

[36] Ci L, Wei J, Wei B, Liang J, Xu C, Wu D (2001) Carbon nanofibers and single-walled carbon nanotubes prepared by the floating catalyst method. Carbon 39:329–335.

[37] Yang KL, Yang RT (1986) The accelerating and retarding effects of hydrogen on carbon deposition on metal surfaces. Carbon 24:687–693.

[38] Kim MS, Rodriguez NM, Baker RTK (1991) The interaction of hydrocarbons with copper-nickel and nickel in the formation of carbon filaments. J Catal 131:60–73.

[39] Zavarukhin SG, Kuvshinov GG (2004) The kinetic model of formation of nanofibrous carbon from CH_4-H_2 mixture over a high-loaded nickel catalyst with consideration for the catalyst deactivation. Appl Catal A 272:219–227.

Chapter 18

High-Temperature Reactivity and Wetting Characteristics of Al/ZnO System Related to the Zinc Oxide Single Crystal Orientation

Joanna Wojewoda-Budka[1], Katarzyna Stan[1], Rafal Nowak[2], Natalia Sobczak[2,3]

[1]Institute of Metallurgy and Materials Science, Polish Academy of Sciences, 25 Reymonta St., 30-059 Cracow, Poland
[2]Foundry Research Institute, 73 Zakopianska St., 30-418 Cracow, Poland
[3]Motor Transport Institute, 80 Jagiellonska St., 03-301 Warsaw, Poland

Abstract: Wettability and reactivity of the Al/ZnO couple at a temperature of 1273 K in vacuum have been studied with application of modified sessile drop method. The results obtained for the ZnO single crystal substrates of various orientations evidenced non-wetting by liquid alu-minium, showing high contact angles of 100°–107°. Detailed transmission electron microscopy characterization unambiguously proved that the creation of two interfacial layers revealing highly epitaxial growth took place at high-temperature interaction between liquid Al drop and ZnO substrate. The layers of columnar grains consisted of reactively formed Al_2O_3 and $ZnAl_2O_4$, the second one growing at the expense of the alumina. The alumina was identified as either cubic γ-, tetragonal δ- or monoclinic λ-Al_2O_3, depending on subtle changes in chemistry of the surface top layer of the ZnO substrates.

1. Introduction

Light-weight Al−Al$_2$O$_3$ composites produced in situ due to redox reaction between liquid Al and oxides such as SiO$_2$, mullite, kaolinite, CeO$_2$, NiO, MgO and ZnO are attractive alternative for conventional light alloys used in aviation and ground transport industry[1]–[6]. That is why, there is a strong need to extend the knowledge on both reactivity and wettability of these systems of interest. The sessile drop experiment is of particular usefulness in such studies and was successfully applied previously to Al/SiO$_2$[1], Al/NiO[3][4], Al/MgO[5], Al/ZnO[6]–[8] and Al/Y$_2$O$_3$[9] systems. The last two systems reveal similarities, since the reactively formed aluminium oxide crystallites are large and surrounded by the metallic channels composed of either Al(Mg) or Al(Zn). These two systems can find an application when the higher content of alumina is desired, and larger amount of metal is predictably reduced, as one way to control the quantity of the alloying element is by using MeO, which produces Me in a vapour form by the redox reaction with Al. Zinc and magnesium oxides fulfil this requirement. The experiment performed at 1273K leads to the formation of reaction product region (RPR) composed of alumina particles interpenetrated by the metallic channels being either Al(Mg) or Al(Zn). In the RPR, the alumina content can be increased, while the zinc or magnesium is kept low in these metallic channels because of its diffusion into the drop followed by further vaporization from the liquid Al(Zn) or Al(Mg) metal. So far, several researchers have used zinc oxide for in situ reactive synthesis of Al−Al$_2$O$_3$ composites[9]–[13].

The structural investigation of Al/ZnOSC sessile drop couples reported in works[7][8] revealed the formation of not only large α-Al$_2$O$_3$ crystals interpenetrated by the Al(Zn) channels but also an additional thin layer located next to the ZnO substrate. In study[6], the examination performed on polycrystalline ZnO substrate reacting with molten aluminium established the occurrence of thin layer of about 250nm specified as metastable δ-Al$_2$O$_3$. In the next paper, the application of the same sample preparation procedure but using the single crystal ZnO substrate of ⟨0001⟩ orientation (further denoted as ZnO$^{<0001>}$) was reported. The creation of the thin alumina layer at the ZnO-side interface was also detected using the transmission electron microscopy (TEM). At that moment, its growth was presumed to be an effect of cooling[8]. In order to confirm this statement, the Al/ZnO couple was produced by capillary purification (CP) procedure combined with drop pushing procedure (DP) described in[14]. Such a unique procedure comprises squeezing of the liquid aluminium from the capillary on the substrate surface at the test temperature of 1273K, followed by pushing it into another location of the substrate after appropriate contact time. Several advantages arose from the experiment performed in such a way. At first, it allowed non-contact heating of both Al and ZnO materials and, therefore, to avoid the effect of heating history due to the interaction between them during contact heating to the test temperature, commonly used in wettability tests. Next, it caused the removal of the primary oxide film from the Al drop directly in the high-temperature chamber under vacuum. And finally, it helped to exclude or at least significantly reduce the effect of the cooling history, as the droplet was no more in contact with the substrate during the cooling. Three phases were detected within the Al/ZnO$^{<1-100>}$ couple: locally distributed large crystals of α-Al_2O_3, the alumina of unknown type and $ZnAl_2O_4$ spinel placed between the unknown Al_2O_3 and ZnO. Both alumina and spinel took form of columnar grains, which placed one above other formed thin and continuous layer of about 200nm in thickness[7]. The above results as well as incomplete literature data require systematic examination of the Al/ZnO contact system comprising of both the reactivity and wettability strongly correlated with the substrate surface structure, which was the aim of the research presented here. The systematic studies with application of the pushing drop procedure and various crystallographic orientations of the ZnO substrate were performed first time ever.

2. Experimental

Aluminium of 99.999% purity and ZnO single crystal substrates were subjected to the sessile drop wettability tests at 1273K using experimental facility described in details in work[14]. All samples were ultrasonically cleaned in the isopropanol before the test. Zinc oxide single crystals ZnO^{SC} of <1-100>, <1-120>, and <0001> orientations and of two variants of atomic termination of the ZnO surface (denoted as $ZnO^{<0001>Zn}$ or $ZnO^{<0001>O}$ for Zn-terminated and O-terminated surfaces, respectively) are grown by hydrothermal method under high pressure with the surface roughness less than 5Å.

As the experimental conditions have significant influence on the wetting test results[15], two testing procedures were applied in the sessile drop wettability tests:

(1) CP and (2) drop pushing in order to expose the Al/ZnO interface at high temperature directly in vacuum chamber after appropriate interaction time and uncover freshly formed reaction products for further SPM and TEM investigation. It consisted of squeezing the aluminium droplet through an alumina capillary at the test temperature and its deposition on the ZnO substrate. After 20min of Al–ZnO interaction, the drop was transferred to a new location of the same ZnO substrate for another 5min contact and then finally pushed away from the substrate to the alumina support placed next to the ZnO single crystal[14]. After wettability tests, the Al/ZnO couples were cooled at the rate of ~10K/min. All the manipulations were undertaken under dynamic vacuum produced by the continuously working turbomolecular pump.

The TEM lamellas were obtained using focused ion beam technique (Quanta 3D, FEI) and in situ lift out technique, which guaranteed the precise probing of the place of analyses. The TEM techniques comprising of bright-field (BF), high-resolution observation and also selected area electron diffraction (SAED) were carried out using TECNAI G2 FEG super TWIN (200kV) micro-scope. Additionally, high-angle annular dark-field detector allowed the observation in the scanning transmission mode and an energy dispersive X-ray spectroscopy system manufactured by EDAX company integrated with the microscope let to carry out the local chemical analyses as well as to collect the maps of element distribution in the particular location in the sessile drop sample.

3. Results and Discussion

3.1. Wettability Tests

The contact angle measurements of aluminium on ZnO^{SC} substrates of <1-100>, <1-120> and <0001> (Zn or O faced) orientations were carried out at 1273K for 20min. The obtained results [**Figure 1(a)** and **Table 1**] indicate that interaction of liquid aluminium with all ZnO substrates exhibits non-wetting behaviour ($\theta > 90°$), reaching the contact angle values of $\theta = 100°-107°$.

The wetting kinetics of the drop pushed away to another location at the same substrate has been also studied for 5min. As can be seen in **Figure 1(b)** for the

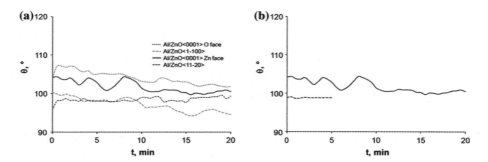

Figure 1. Wetting kinetics of Al/ZnO^{SC} system measured at first location of the Al drop on the ZnO substrate at 1273K lasting for 20min, in which ZnO substrates of various orientations, <0001> of two variants O or Zn faced, <1–100> and <1–120>, were applied (a). Comparison of the wetting angle changes for the drop placed at first (20min at 1273K) and second (5min at 1273K) location of the same ZnO <0001>^{Zn} substrate (b).

Table 1. Approximate thickness values of alumina and spinel layers formed due to the interaction between aluminium and ZnO single crystal of various orientation after 5 and 20min in 1273K as well as the type of formed Al$_2$O$_3$.

Couple	Testing conditions		Contact angle (°)				Time of interaction (min) at 1273 K	Thickness (nm) of Al$_2$O$_3$	Thickness (nm) of ZnAl$_2$O$_4$	Type of Al$_2$O$_3$
	Time (min)	Temp (K)	T = 0 min		T = 20 min					
			θ left	θ right	θ left	θ right				
Al/ZnO$^{(0001)Zn}$	20 + 5	1273	105.1	103.4	103.6	97.5	20	150	600	δ or γ
	5	1273	96.3	101.6	95.0	103.3	5	350	400	δ or γ
Al/ZnO$^{(0001)O}$	20 + 5	1273	101.0	101.3	102.0	101.7	20	300	500	δ
	5	1273	95.9	100.0	91.8	95.1	5	400	400	λ
Al/ZnO$^{(1-100)\,Zn}$	20 + 5	1273	101.9	99.0	96.0	93.1	20	200	550	γ, δ, λ
	5	1273	1*	1*	1*	1*	5	150	200	γ, δ, λ
Al/ZnO$^{(11-20)Zn}$	20 + 5	1273	90.6	101.1	100.6	98.6	20	180	1000	δ or γ
	5	1273	2*	2*	2*	2*	5	350	280	δ or γ

1* sticking of the drop to the pusher making impossible to calculate the contact angles
2* drop too big to push it to another place on substrate

ZnO <0001> substrate, no change in the contact angle value took place in comparison with the first position of the drop on the substrate. The contact angle formed on ZnO <0001> surface was lower by 6°, reaching the value of 99°. The dissimilar wetting behaviour of the Al drop in the first and second positions on the same substrate might be related to the effect of liquid metal saturation with oxygen and zinc during first 20min contact, resulting in a suppressed interaction between the fresh substrate surface and the Al(O, Zn) drop in its second position after it was moved.

The tendency to lower the contact angle value with time of interaction can be associated with either the evaporation from the drop or due to chemical reaction, which takes place at the drop/substrate interface under experimental conditions used. First phenomenon can be excluded by the measurements of the diameter (d_{max}) and height (h_{max}) of the drop at the beginning and at the end of the contact angle measurements. During the performed experiment, d_{max} increased with decreasing h_{max}, while the corresponding drop volume changed up to 0.2% only, which clearly evidenced that although the metal evaporation takes place, its effect on the drop size is negligible. Therefore, it was presumed that lowering of the contact angle is caused by the formation of the new compound at the drop/substrate interface.

The literature data suffer from very limited information concerning both the wettability and reactivity in Al/ZnO system followed by the detailed microstructure and phase composition characterization. The presented results can be compared with those reported by Sobczak et al.[1][16] showing high reactivity accompanied with a lack of wettability in the Al/ZnO couple. High contact angles ($\theta = 111$ for single crystalline substrate and $\theta = 140$ for polycrystalline substrate) were measured at 1273K. At that temperature, the Al_2O_3 reaction product, suggested by the electron microscopy characterization coupled with EDS analysis, is known to be wetted by molten Al, as reported in the analysis of literature data on contact angle measurements of Al/Al_2O_3 system performed by Sobczak et al.[17]. Since the removal of oxide film from Al drop did not improve the wettability in Al/ZnO couple, its non-wetting behaviour was explained either by secondary roughening of initially smooth ZnO single crystal surface or by secondary oxidation of the Al drop surface[1]. The first explanation came from the observation of the fine alumina precipitates detected at the substrate surface around the Al drop and probably created by evaporative-reactive deposition caused by enlarged transport of Al vapour under UHV and its reaction with ZnO. The idea of secondary oxidation of the drop surface was based on the observations of fine alumina precipitates at the drop surface, particularly densely localized close to the triple line. Their formation was explained by enhanced transfer of oxygen from the ZnO substrate caused by copious evaporation of Zn. It should be highlighted also that in previous reports, no information about the type of reactively formed alumina was provided, while the results obtained were compared with experimental data showing wetting in the Al/Al_2O_3 couple at 1273K when α-Al_2O_3 was used. Following more recent reports by Shen et al.[18]–[21] concerning the influence of the α-Al_2O_3 single crystal orientation on wetting its surface by molten aluminium in the temperature range of 800°C–1500°C, it can be found that this effect is very strong. The wetting of molten Al on differently oriented single crystals R<0112>, A<1120> and C<0001> showed that the adhesion of the molten aluminium on the R and A surfaces is much stronger in comparison with the C surface.

Therefore, in this study, the detailed structural investigation was performed using two advanced techniques of materials characterization, i.e. the scanning probe micro-scopy (SPM) technique (in order to examine the subtle changes in substrate surface topography after high-temperature contact with liquid Al), followed by the transmission electron microscopy examination of the microstructure and phase composition of the compounds present at the drop/substrate interface.

3.2. Substrate Surface Topography after Interaction with Al Drop

Figure 2(a) shows the image of the ZnO$^{<11-20>}$ surface exposed at high temperature after interaction with aluminium. There are five characteristic areas that can be distinguished. Regions 1 and 2 correspond to the place of the first contact of the Al drop with ZnO substrate. Region 3 represents the triple line region formed after 20min of interaction. Region 4 is just outside of the contact place, while the fifth region is far away from it. During dropping the aluminium from the capillary, the volume of the squeezed drop slightly increases. Each ring marked with arrows visible in **Figure 2(a)** corresponds to the new volume.

Figures 2(b)-(f) shows the series of measurements performed consequently for all mentioned above regions. First region [**Figure 2(b)**] reveals the dramatic change of the substrate structure although the contact time was extremely short and lasted only few seconds. It resulted in the immediate reduction of the ZnO by liquid Al. The presence of the triple line, of about 100nm depth, can be also noted here. The region in contact with the Al is composed of many craters along which the liquid aluminium can infiltrate the monocrystalline substrate and reactively formed products can appear.

The examination of region 2 [**Figure 2(c)**] confirmed the presence of large

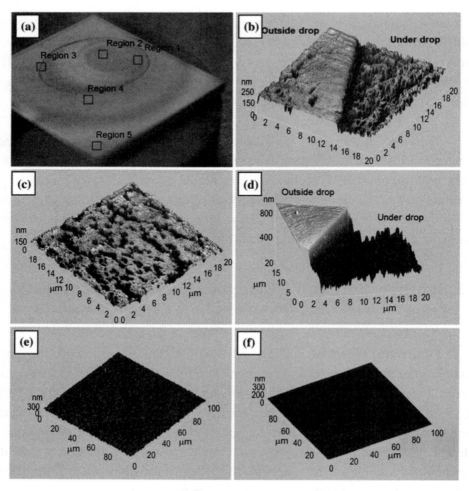

Figure 2. SPM images of the ZnO$^{<11-20>}$ surface exposed at high temperature after interaction with aluminium: general view of the sample surface with particular areas of interest (a). Region 1 and 2 denotes the first contact of the Al drop with ZnO substrate (b, c), region 3 represents the triple line region formed after 20min of interaction (d), region 4-surface area close but beyond the contact with the drop (e) and region 5 is far away from it (f).

channels through which the transfer of liquid Al to the reaction front and reactively formed zinc to the drop took place. The maximum depth of these channels was 140nm. Region 3 visualized in **Figure 2(d)** is the surface topography after extra dropping of aluminium. This structure is different from the previous one because of lack of large craters and the occurrence of fine columnar crystals within this area. Also the RPR is located much lower than the place outside the drop. The dissolution of the ZnO substrate in the drop results in the formation of the characteristic crater about 700nm deep in the triple line area. Furthermore, in region 4 [**Figure 2(e)**], representing the area beside the drop/substrate contact, the secondary roughening of the surface can be observed contrary to the flat surface of the zinc oxide (R_a = 2nm) measured far away from the place of Al/ZnO contact [**Figure 2(f)**]. This phenomenon takes place at 1273K due to chemical reaction between Al vapour and ZnO substrate surface, and it can contribute to the formation of high apparent contact angles as explained in[1].

3.3. Microstructure and Phase Composition of the Products Reactively Formed at the Drop/Substrate Interface

The surface of the ZnO substrates after high-temperature interaction with molten aluminium was next observed with scanning electron microscopy (**Figure 3**). **Figure 3(a)** shows the overall view of the sample obtained in DP. The places where the aluminium drop contacted the substrate for 20min (the first position) and 5min (the second position) are well distinguishable as the circles of brighter contrasts. The local existence of large crystals, previously identified as α-Al_2O_3[7], can be noticed [**Figure 3(b)**]. They appeared due to the interaction of ZnO substrate with so-called "daughter" droplets, which remained after pushing off the "mother" drop.

In all cross-sectioned Al/ZnO couples, the formation of two layers revealing strong epitaxial growth was observed in the following order: ZnO/$ZnAl_2O_4$/Al_2O_3.

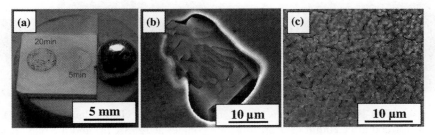

Figure 3. Overall view of the Al/ZnO sessile drop sample on alumina support after using drop pushing procedure. Al drop was removed from ZnO substrate onto alumina support (a). SEM-SE top view image of the substrate surface with visible large dendrites of α-Al_2O_3 locally distributed on the substrate surface corresponding to the first position of the drop contacting ZnO for 20min (b) and area of contact close to large dendrites (c).

Both the spinel and aluminium oxide took the shape of long, columnar crystals, of which the acquisition of either the electron diffraction or high-resolution image was not a trivial task because of their mutual overlapping.

Figure 4 shows the microstructure of the cross-sectioned area within the places where Al drop reacted for particular time-the first lasting 20min and the second for another 5min (total of 20min + 5min) with <0001>-oriented ZnO- and Zn-faced substrates. As it can be noticed in **Figure 4(a), (e)**, a layer of columnar grains can be distinguished above the ZnO$^{<0001>Zn}$. Observations in scanning transmission mode [**Figure 4(f), (g)**] revealed various contrast within this layer. The EDS analysis showed the presence of alumina and $ZnAl_2O_4$ spinel. This last compound was observed between Al_2O_3 and ZnO. For the Zn-terminated <0001> orientation of ZnO single crystal, the approximate thicknesses of reactively formed layers after interaction at the first location of the drop on the substrate (20min time) were 150 and 600nm, and in the second location (another 5min), they were 350- and 400-nm thick, for the alumina and spinel, respectively. In contrary to this,

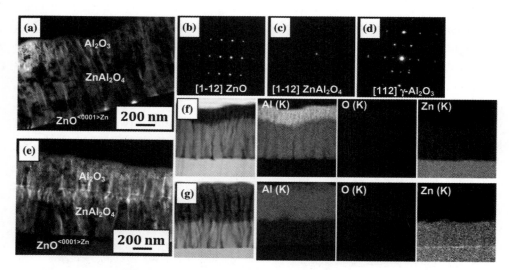

Figure 4. TEM-BF image of the RPR/ZnO$^{<0001>Zn}$ with visible layers of Al_2O_3 and $ZnAl_2O_4$ formed after 20min of drop/substrate contact at 1273K (a) and the one created at second location of the drop staying in contact with the substrate for another 5min time (e) of interaction with Al at 1273K. b-d shows the SAED patterns of the ZnO, $ZnAl_2O_4$ and Al_2O_3 crystals, respectively. f and g present the STEM images together with the map of element distribution Al, O and Zn recorded for the RPR formed at first and second location of the drop on the ZnO$^{<0001>Zn}$ substrate, respectively.

when the top layer of the single crystal consisted of the oxygen atoms (ZnO$^{<0001>O}$), the thicknesses of both products after 20min of the interaction were 300 nm (Al$_2$O$_3$) and 500nm (ZnAl$_2$O$_4$), while at the second location of the drop (5min of reaction at 1273K), they were similar and both of about 400 nm (**Figure 5**).

Particular attention was paid to the determination of the type of formed aluminium oxide. The analysis of the SEAD patterns showed that the crystal structure of Al$_2$O$_3$, formed after 20min interaction of aluminium with ZnO$^{<0001>Zn}$, was either γ- or δ-Al$_2$O$_3$ phase, while for the ZnO$^{<0001>O}$, the columnar grains of tetragonal phase delta prevailed. At the second location of the drop, the drop/substrate interaction was different for the applied two variants of the single crystal. In the case of the top surface layer composed of zinc atoms, the appearance of γ- or δ-Al$_2$O$_3$ was observed, wherein the orientation of the gamma oxide was the same as that of spinel [**Figures 4(b)-(d)**]. Conversely, the monoclinic λ-Al$_2$O$_3$ phase was mostly formed for the samples with the top layer of ZnO single crystal built of oxygen atoms [**Figures 5(a)-(c)**].

Interaction of molten aluminium and ZnO of <1−100> orientation [**Figure 6(a), (b)**] resulted in the appearance of the aluminium oxide and the spinel of 200 and 550nm in thickness (after 2min) and 150 and 200nm (after 5min). However, the most spectacular difference in the thickness of these two layers was observed for the ZnO substrate of <1−120> orientation [**Figure 4(c), (d)**], where after 20min of interaction, the aluminium oxide thickness was only 180 nm, while that of spinel increased up to 1000nm. After 5min of interaction, the widths were similar and reached 350nm (aluminium oxide) and 280nm (spinel). The reaction of

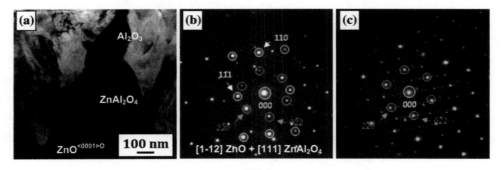

Figure 5. TEM-BF image of the RPR/ZnO$^{<0001>O}$ with Al$_2$O$_3$ and ZnAl$_2$O$_4$ crystals formed after 5min of interaction of ZnO with Al drop at 1273K (a). b and c show the SAED patterns of the ZnO together with ZnAl$_2$O$_4$ (b) and ZnAl$_2$O$_4$ with Al$_2$O$_3$ crystals (c).

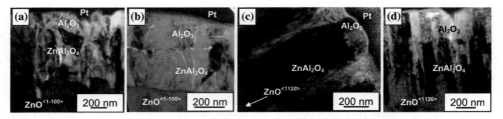

Figure 6. TEM-BF images of the RPR/ZnO$^{<1-100>}$ cross section with Al$_2$O$_3$ and ZnAl$_2$O$_4$ crystals formed after 20min (a) and 5min (b) of interaction with Al at 1273 K together with BF images of the RPR/ZnO$^{<1-120>}$ with Al$_2$O$_3$ and ZnAl$_2$O$_4$ crystals formed after 20min (c) and 5min (d) of interaction time.

<1−120>-oriented ZnO single crystal and aluminium at 1273K for both 20 and 5min led to the formation of γ- or δ-Al$_2$O$_3$ but the gamma alumina dominated for the second interaction (**Figure 7**). On the other hand, the research performed on sample Al/ZnO$^{<1-100>}$ has shown the presence of all three of the mentioned variations of metastable alumina γ-, δ- or λ-Al$_2$O$_3$. The comparison of thickness of both alumina and spinel layers as well as the type of formed Al$_2$O$_3$ has been presented in **Table 1**.

As it can be easily noticed, subtle changes of the top surface chemistry of the ZnO substrate influence the type of created alumina phase as well as the thickness of both alumina and spinel. The largest thickness was obtained in the case of spinel formed after the reaction with ZnO of <11−20> orientation. However, it should be also pointed that these phase composition changes do not cause the changes in the contact angle values, which are almost the same in all the cases and are in fact the contact angle between the aluminium and metastable alumina. The morphology of the alumina takes shape of longitudinal columnar grains in all the cases—also visible under SPM. The potential of the C4 morphology creation (co-continuous ceramic composite composed of mutually interpenetrated metallic and ceramic phases) was previously attributed to the volume mismatch between the substrate (ZnO) and reactively formed alumina that was believed to be α-Al$_2$O$_3$, being as large as 40%[22]. Such a type of morphology occurs after long time of interaction, when the reactively formed product is dominated by alumina interpenetrated with Al(Zn). Short time of interaction results in creation of two products: alumina and spinel in the form of layers one above the other. The question that arises is which route of creation of the products among the two proposed below takes place in this study? The analysis of experimental results suggests that the

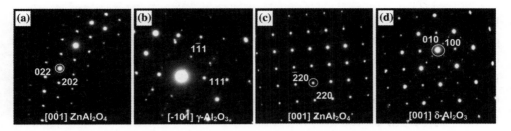

Figure 7. SAED patterns of the ZnAl$_2$O$_4$ and Al$_2$O$_3$ crystals formed in the RPR of the Al/ZnO$^{<1-120>}$ couple after 20min of interaction with aluminium at 1273K.

reactively changed interfaces can be formed by two possible sequences (single slash denotes stable and double slash denotes unstable interface):

$$Al//ZnO \rightarrow Al/Al_2O_3//ZnO$$
$$\rightarrow Al/Al_2O_3/ZnAl_2O_4/ZnO \quad (1)$$

or

$$Al//ZnO \rightarrow Al//ZnAl_2O_4/ZnO$$
$$\rightarrow Al/Al_2O_3/ZnAl_2O_4/ZnO \quad (2)$$

The results in **Table 1** confirm that the growth of the spinel (especially pronounced in the case of <11–20> orientation) is favoured during the first contact of the aluminium with zinc oxide, when the drop is not saturated with the oxygen and zinc. After interaction in the first location, the chemical composition of the drop changes and in its second location, the drop becomes less reactive in contact with fresh ZnO surface. Therefore, the reactively formed alumina and spinel are of more or less of the same thickness. Taking into account the time of interaction at the first location (20min of contact plus another 5min, when the drop is placed and kept at the second location before it is pushed away) and the second one (5min only) and comparing the microstructures, it can be concluded that alumina is formed as the first layer and later the spinel layer grows at the expense of Al$_2$O$_3$.

4. Conclusions

For the first time, the effects of heating and cooling history on the interface transformation in highly reactive Al/ZnO system were dramatically reduced by combined application of two testing procedures allowing non-contact heating of a couple of materials as well as exposing of the interface at high temperature directly in the UHV chamber during the sessile drop wettability tests. Under conditions of this study, the liquid Al did not wet the ZnO substrate ($\theta > 90°$) independently from its crystallographic orientation.

Detailed characterization of exposed substrate-side interface formed between liquid Al and ZnO single crystals of different crystallographic orientations showed that non-wetting behaviour in highly reactive Al/ZnO system may be related with the formation of non-wettable reaction product since the α-Al_2O_3 phase, widely accepted to be wettable by liquid Al at 1273K and expected in previous reports as a main reaction product in the Al/ZnO couple, was not found in this study. It was evidenced that the high-temperature interaction in Al/ZnO couple led to the creation of two reactively formed layers corresponding to the metastable Al_2O_3 and $ZnAl_2O_4$ spinel, both characterized with highly epitaxial growth. The thicknesses of those layers varied with respect to the drop content and time of interaction: longer time promoted the spinel growth at the expense of the alumina. The alumina was identified as either cubic γ-, tetragonal δ- or monoclinic λ-Al_2O_3.

Moreover, the type of created alumina was affected by the crystallographic orientation and chemistry of the top surface layer of the ZnO substrates showing dissimilar alumina formed even for the substrates of the same <0001> orientation but of various Zn- or O-terminated surfaces.

Acknowledgements

The Ministry of Science and Higher Education of Poland within the Project Iuventus Plus No. IP2011061071 has supported this study. Samples were prepared in the Centre for High-Temperature Studies at the Foundry Research Institute in Cracow. The SEM and TEM studies were performed in the Accredited Testing Laboratories at the Institute of Metallurgy and Materials Science of the Polish Academy of Sciences in Cracow.

Compliance with Ethical Standards

Conflict of interest: The authors declare that they have no conflict of interest.

Open Access: This article is distributed under the terms of the Creative Commons Attribution 4.0 International License (http://creativecommons.org/licenses/by/4.0/), which permits unrestricted use, distribution, and reproduction in any medium, provided you give appropriate credit to the original author(s) and the source, provide a link to the Creative Commons license, and indicate if changes were made.

Source: Wojewoda-Budka J, Stan K, Nowak R. High-temperature reactivity and wetting characteristics of Al/ZnO system related to the zinc oxide single crystal orientation[J]. Journal of Materials Science, 2016, 51(4):1–9.

References

[1] Sobczak N (2005) Wettability and reactivity between molten aluminium and selected oxides. Solid State Phenom 101–102:221–226.

[2] Reddy BSB, Das K, Das S (2007) A review on the synthesis of in situ aluminium based composites by thermal, mechanical and mechanical–thermal activation of chemical reactions. J Mater Sci 42:9366–9378. doi:10.1007/s10853-007-1827-z.

[3] Sobczak N, Oblakowski J, Nowak R, Kudyba A, Radziwill W (2005) Interaction between liquid aluminium and NiO single crystals. J Mater Sci 40:2313–2318. doi:10.1007/s10853-005-1951-6.

[4] Wojewoda-Budka J, Stan K, Onderka B, Nowak R, Sobczak N (2014) Microstructure, chemistry and thermodynamics of Al/NiO couples obtained at 1000 C. J Alloys Compd 615:178–182.

[5] Nowak R, Sobczak N, Sienicki E, Morgiel J (2011) Structural characterization of reaction product region in Al/MgO and Al/ $MgAl_2O_4$ systems. Solid State Phenom 172–174:1273–1278.

[6] Wojewoda-Budka J, Sobczak N, Morgiel J, Nowak R (2010) Reactivity of molten aluminium with polycrystalline ZnO sub-strate. J Mater Sci 45(16):4291–4298. doi:10.1007/s10853-010-4379-6.

[7] Wojewoda-Budka J, Sobczak N, Morgiel J, Nowak R (2011) TEM studies of the temperature and crystal orientation influence on the microstructure and phase composition of the reaction products in Al/ZnO system. Solid State Phenom 172–174:

1267–1272.

[8] Wojewoda-Budka J, Sobczak N, Stan K, Nowak R (2013) Microstructural characterization of the reaction product region formed due to the high temperature interaction of ZnO(0001) single crystal with liquid aluminium. Arch Metall Mater 58(2):349–353.

[9] Wojewoda-Budka J, Sobczak N, Onderka B, Morgiel J, Nowak R (2010) Interaction between liquid aluminium and yttria substrate-microstructure characterization and thermodynamic considerations. J Mater Sci 45(8):2042–2050. doi:10.1007/s10853-009-4135-y.

[10] Yu P, Deng CJ, Ma NG, Ng DHL (2004) A new method of producing uniformly distributed alumina particles in Al-based metal matrix composite. Mater Lett 58:679–682.

[11] Durai TG, Das K, Das S (2007) Synthesis and characterization of Al matrix composites reinforced by in situ alumina particulates. Mater Sci Eng A 445–446:100–105.

[12] Chen G, Sun GX (1998) Study on in situ reaction-processed Al-Zn/α-Al_2O_3(p) composites. Mater Sci Eng A 244:291–295.

[13] Maleki A, Panjepour M, Niroumand B, Meratian M (2010) Mechanism of zinc oxide-aluminium aluminothermic reaction. J Mater Sci 45:5574–5580. doi:10.1007/s10853-010-4619-9.

[14] Sobczak N, Nowak R, Asthana R, Purgert R (2010) Wetting in high-temperature materials processing: the case of Ni/MgO and NiW10/MgO. Scripta Mater 62(12):949–954.

[15] Eustathopoulos N, Sobczak N, Passerone A, Nogi K (2005) Measurement of contact angle and work of adhesion at high temperature. J Mater Sci 40:2271–2280. doi:10.1007/s10853-005-1945-4.

[16] Sobczak N (2006) In: Gupta N, Hunt WH (eds) Solidification processing of metal matrix composites. TMS Publications, Ohio, pp 133–146.

[17] Sobczak N, Sobczak J, Asthana R, Purgert R (2010) The mystery of molten metal. China Foundry 7(4):425–437.

[18] Shen P, Fujii H, Matsumoto T, Nogi K (2003) Wetting of (0001) α-Al_2O_3 single crystals by molten Al. Scripta Mater 48:779–784.

[19] Shen P, Fujii H, Matsumoto T, Nogi K (2003) The influence of surface structure on wetting of α-Al_2O_3 by aluminium in a reduced atmosphere. Acta Mater 51:4897–4906.

[20] Shen P, Fujii H, Matsumoto T, Nogi K (2004) Effect of substrate crystallographic orientation on wettability and adhesion in several representative systems. J Mater Process Technol 155–156: 1256–1260.

[21] Shen P, Fujii H, Matsumoto T, Nogi K (2005) Influence of substrate crystallographic orientation on the wettability and adhesion of α-Al_2O_3 single crystals by liquid Al and Cu. J Mater Sci 40:2329–2333. doi:10.1007/s10853-005-1954-3.

[22] Sobczak N (2007) In: Sobczak J (ed) Innovations in foundry (in Polish), Part II. Foundry Research Institute, Krakow, pp 187–198.

Chapter 19

Improving Shear Bond Strength of Temporary Crown and Fixed Dental Prosthesis Resins by Surface Treatments

Seung-Ryong Ha[1], Sung-Hun Kim[2], Jai-Bong Lee[2], Jung-Suk Han[2], In-Sung Yeo[2]

[1]Department of Dentistry, Ajou University School of Medicine, Suwon, Republic of Korea

[2]Department of Prosthodontics and Dental Research Institute, School of Dentistry, Seoul National University, Seoul, Republic of Korea

Abstract: This study evaluated the effect of surface treatments on the bond strength of repaired temporary resins. One-hundred flat-surfaced cylindrical specimens (Ø 7mm × 12mm) of each temporary resin (2 bis-acryl resins and 2 poly-methyl-methacrylates) were prepared. The specimens were randomly divided into 10 groups (n = 10), according to the types of surface treatments: untreated, adhesive treated, silanated, silane + adhesive treated, hydrofluoric acid etched, laser treated, sandblasted, sandblasting + adhesive treated, sandblasting + silanated, and tribochemical silica coating + silanated. Each resin material of the same brand with cylindrical shape (Ø 3mm × 3mm) was polymerized onto the resin surfaces, and specimens were stored for 24h in distilled water. The shear bond strengths were measured and failure modes were examined. All data were analyzed with a one-way ANOVA and multiple comparison Scheffé post hoc test ($\alpha = 0.05$). For bis-acryl resins, the highest shear bond strength was observed in sandblasted group and the lowest was observed in the control group. Results show that the repair bond strength was improved for bis-acryl resin by 23% than that of the control group due to the increase in surface roughness by sandblasting. However, chemical treatment did not improve repair bond strength. The surface treatment of bis-acryl resins with sandblasting seems to be promising for the improvement of repair bond strength.

1. Introduction

Temporary crowns and fixed dental prostheses (FDPs) are often required to provide long-term stability and tooth protection while complementary treatments are provided[1]. The long-term maintenance of the temporary restorations with procedures such as endodontic therapy, orthodontics, chemotherapy, tissue grafting, and implant surgery is frequently useful[1]. The temporary restorations must meet not only esthetic and biologic needs, but also mechanical needs such as resistance to dislodging forces and functional loads[2]. However, complications, such as fractures, could occur with more extensive temporary restorations that are intended for long-term uses. The restorations are subject to various forces in oral conditions: compressive force at the load application; and tensile and shear force at the load resistance[3]. Fracture of a long-span temporary restoration is more likely to happen compared to a short-span restoration because the fracture resistance is inversely proportional to the cube of the restoration length. Fracture of the temporary restorations could cause economic loss and discomfort to both clinicians and patients[4].

In clinical practice, most repairs of temporary restorations are accomplished using an auto-polymerizing resin of the same brand that was used in original temporary restorations. A durable repairing system for the fractured temporary restorations is desired to avoid frequent fracture. Attempts to improve bond strength of restorative materials involve mechanical and chemical means. Many methods have been introduced for modification of a filling composite resin surface: sandblasting, roughening with diamond instrument, abrasive papers, and acid etching[5]–[7]. Several studies have demonstrated that the application of chemical agent, such as bonding agent or silane coupling agent, enhances the bond strength[6]–[11].

Koumjian and Nimmo[12] reported that transverse strength was reduced by 85% after repair of a temporary resin. Their study proposed that it might be more convenient to prepare a new temporary restoration than to repair the restoration. However, repair of the fractured temporary restoration could be cost- and time-efficient treatment option in clinical situations. For a successful repair, bond strength comparable to the strength of the original material is required. Therefore, it is important to measure the repair bond strengths of various surface treatments on the temporary restorations. However, there have been only few studies on the effect of surface treatments on the repair bond strength of the temporary crown and FDP materials[13]–[15].

The present study was conducted to determine the effect of different surface treatments on the shear bond strength of the temporary crown and FDP materials. The null hypothesis to be tested was that there is no difference in shear bond strength among various surface treatments on the repaired temporary crown and FDP materials.

2. Materials and Methods

2.1. Preparation of the Specimens

The materials investigated in the present study were 2 bis-acryl resins and 2 auto-polymerizing polymethyl methacrylate (PMMA) resins (**Table 1**). The bis-acryl resin was dispensed from a cartridge in a dispensing gun through a mixing tip. A small amount of the resin was extruded and discarded, and then the resin was placed into a customized polytetrafluoroethylene (PTFE) mold of 7mm inner diameter and 12mm height [**Figure 1(a)**]. The PMMA resins were mixed according to the manufacturers' instructions. The PMMA powder was saturated and mixed with liquid monomer, using a metal spatula for 20s and immediately placed into the PTFE mold. A vinyl strip and a glass plate were located onto the mold to form flat end surfaces, and hand pressure was applied to extrude excess material. The specimens were allowed to polymerize for 60min at $23°C \pm 1°C$. Four-hundred specimens were fabricated (100 specimens for each material).

Table 1. Materials, manufacturers, lot numbers, and main compositions of the temporary crown and FDP materials investigated in this study.

Material	Manufacturer	Lot No.	Composition	Characteristics
Protemp 3 Garant	3M ESPE, Seefeld, Germany	B 319023 C 318795	DMA, SA, strontium glass	Bis-acryl composite resin
Luxatemp	DMG, Hamburg, Germany	513917	DMA, UDMA, GMA, silica, glass powder	
Vertex	Vertex-Dental/Dentimex, Zeist, Netherlands	YR493L10 YR274P02	MMA, cross-linker, accelerator, polymer	Polymethyl methacrylate resin
Jet	Lang, IL, USA	14425075/01AC 6000906AI/02AB	MMA, DMT, DEP, polymer	

DMA dimethacrylate, *SA* silicic acid, *UDMA* urethane dimethacrylate, *GMA* glycol methacrylate, *MMA* methyl methacrylate, *DMT* N,N-dimethyl-*p*-toluidine, *DEP* diethyl phthalate

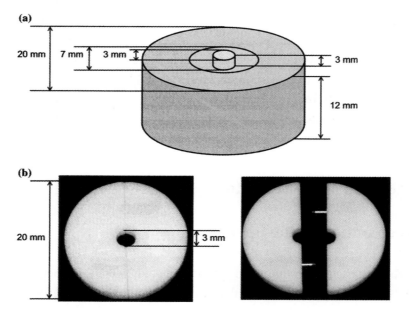

Figure 1. Polytetrafluoroethylene mold used in this study: a for preparation of the specimens; b for repair.

2.2. Surface Treatment of Specimens

The specimens were randomly assigned to one of the ten groups (n = 10 for each group). Groups of specimens with the abbreviations and the preparation methods are presented in **Table 2**. Each specimen was modified by various surface treatments as follows:

Table 2. Test groups for repairing temporary crown and FDP materials.

Group abbreviation	Surface treatment	Mechanical treatment	Chemical treatment
CON	Control	None	None
ADH	Adhesive	None	Unfilled bis-GMA resin application
SIL	Silane	None	Silane application
SI/A	Silane + adhesive	None	Silane application, followed by unfilled bis-GMA resin application
HFA	Hydrofluoric acid	Etching with 4 % hydrofluoric acid	None
LAS	Laser	Roughening with Er,Cr:YSGG laser	None
SAN	Sandblasting	Sandblasting with 50 µm Al_2O_3 particles	None
SA/A	Sandblasting + adhesive	Sandblasting with 50 µm Al_2O_3 particles	Unfilled bis-GMA resin application
SA/S	Sandblasting + silane	Sandblasting with 50 µm Al_2O_3 particles	Silane application
TS/S	Tribochemical silica coating + silane	Tribochemical silica coating with 30 µm silicatized sand	Silane application

CON control, *ADH* adhesive, *GMA* glycol methacrylate, *SIL* silane, *SI/A* silane + adhesive, *HFA* hydrofluoric acid, *LAS* laser, *Er* erbium, *Cr* chromium, *YSGG* yttrium–scandium–gallium–garnet, *SAN* sandblasting, *SA/A* sandblasting + adhesive, *SA/S* sandblasting + silane, *TS/S* tribochemical silica coating + silane

- Group 1 (CON): No surface treatment.

- Group 2 (ADH): Adhesive monomer (Adper Scotch-bond Multi-Purpose adhesive, Lot No. 6PN, 3M ESPE, St. Paul, MN, USA) was applied twice, thinned with oil-free compressed air, and then cured for 20s using a quartz halogen curing light (Elipar Trilight, 3M ESPE, Seefeld, Germany). The light intensity of the lamp was measured regularly with a radiation meter and maintained at the same level for all tests.

- Group 3 (SIL): Silane coupling agent (Porcelain primer, Lot No. 0700000153, Bisco, Schaumberg, IL, USA) was applied as a single coat and was allowed to dry for 5min.

- Group 4 (SI/A): Silane coupling agent was applied under the same conditions as above. Any residual solvent was evaporated with oil-free compressed air for 10s. Then, adhesive monomer was applied under the same conditions as in Group 2.

- Group 5 (HFA): A thin layer of 4% hydrofluoric acid gel (Porcelain etchant, Lot No. 0600000878, Bisco) was applied for 120s. The specimen was rinsed with water for 120s and dried with oil-free compressed air for 10s.

- Group 6 (LAS): The specimen was irradiated with the Er,Cr:YSGG laser (Waterlase MD, Lot No. 6200218, Biolase technology, San Clemente, CA, USA) under water cooling (30% water, 30% air) at 2.25W, 30Hz. The optic fiber was used in a non-contact mode, in back and forth motions to assure a controlled irradiation of the surface. The specimen was rinsed with water for 20s and dried with oil-free compressed air for 10s.

- Group 7 (SAN): Sandblasting with 50 lm aluminum oxide (Al_2O_3) particles was applied using an airborne-particle abrasion device (S-U-PROGRESA 200, Schuler-Dental, Germany) from a distance of approximately 10mm at a pressure of 2 bar for 10s. The specimen was rinsed with water for 20s and dried with oil-free compressed air for 10s.

- Group 8 (SA/A): Sandblasting process was applied using the same device under the same conditions as above. Then, adhesive monomer was applied under the same conditions as in Group 2.

- Group 9 (SA/S): Sandblasting was done under the same conditions as in Group 7. Then, silane coupling agent was applied under the same conditions as in Group 3.

- Group 10 (TS/S): Tribochemical silica coating was achieved using an intraoral blaster (3M ESPE, St. Paul, MN, USA) from a distance of approximately 10mm with 30lm silicatized sand (RocatecTM-Soft, 3M ESPE) for 10s. The specimen was rinsed with water for 20s and dried with oil-free compressed air for 10s. Then, silane coupling agent was applied under the same conditions as above.

2.3. Scanning Electron Microscope Examination

Additional specimens of each resin were prepared for examination with a scanning electron microscope (SEM). Each specimen modified by mechanical treatment was examined using the SEM (FE-SEM, S-4700, Hitachi, Tokyo, Japan) at ×500 and ×2000 magnification to observe the topographic patterns.

2.4. Repair of the Specimens

A PTFE mold with an opening of 3mm diameter and 3mm height was used for the repair of the specimens [**Figure 1(b)**]. The mold was positioned on the modified surface of each specimen, and its opening was filled with each fresh resin of the same brand to complete the repair procedure. The specimen was allowed to polymerize for 60min at 23°C ± 1°C, and then the PTFE mold was gently removed from the specimen. The tested specimens received an identification number and were stored individually in distilled water at 37°C for 24h before mechanical testing.

2.5. Shear Bond Strength Test

All specimens were moved from the storage container directly onto the testing apparatus. The specimens were inserted into a shear test jig, and the jig was secured in a universal testing machine (Instron, Model 3345, Instron, Canton, MA, USA). Then, shear load was applied to the adjacent bonding interface with a crosshead speed of 0.5mm/min until fracture occurred using knife-edge rod (**Figure 2**). Tests were carried out at the temperature of 23°C ± 1°C.

The bond strength values were calculated using the formula:

$$\sigma = L/A$$

where σ is the bond strength (in MPa), L is the load at failure (in N), and A is the repaired area (in mm^2).

2.6. Failure Mode Analysis

The interfacial fractured surfaces of each test group were examined using a stereoscopic microscope (945, Meiji 2000, Meiji Techno, Saitama, Japan). The specimens were classified according to fracture patterns: adhesive failure, mixed failure (combination of cohesive and adhesive failure), and cohesive failure.

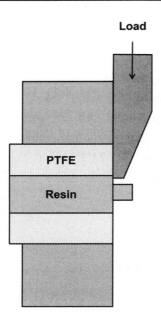

Figure 2. Schematic diagram of shear bond test jig.

2.7. Statistical Analysis

Statistical analysis was performed by a one-way ANOVA and multiple comparison Scheffé post hoc tests with the statistical software (SPSS 22, SPSS Inc., Chicago, IL, USA). The test was performed at a significance level of 0.05.

3. Results

The mean shear bond strength values and standard deviations of all groups are demonstrated in **Table 3**. The results of one-way ANOVA are presented in **Tables 4–7**.

3.1. Bis-acryl Resin

For Protemp 3 Garant, mean shear bond strength values ranged from 32.4 to 39.2MPa. Group SAN obtained the highest mean shear bond strength with the value of 39.2MPa, followed by Group SA/A with the value of 35.9MPa. These

Table 3. Mean shear bond strength (MPa) with standard deviations in parenthesis of temporary crown and FDP materials.

Group abbreviation	Protemp 3 Garant	Luxatemp	Vertex	Jet
CON	32.4 (2.1)^{b}	29.1 (2.0)^{c}	25.9 (1.8)^{e}	25.0 (2.0)^{f,g,h}
ADH	34.8 (3.2)^{a,b}	32.9 (2.2)^{c}	29.5 (2.5)^{d}	26.1 (2.2)^{f,g}
SIL	35.6 (2.7)^{a,b}	31.6 (3.2)^{c}	26.6 (1.6)^{d,e}	26.8 (1.9)^{f}
SI/A	33.6 (4.2)^{a,b}	31.3 (4.5)^{c}	26.4 (1.8)^{d,e}	24.0 (1.5)^{f,g,h}
HFA	34.0 (7.2)^{a,b}	34.2 (5.9)^{c}	27.2 (1.9)^{d,e}	24.1 (1.0)^{f,g,h}
LAS	33.7 (2.5)^{a,b}	31.7 (3.9)^{c}	26.1 (1.9)^{e}	22.8 (2.1)^{h}
SAN	39.2 (1.9)^{a}	35.8 (3.4)^{c}	26.5 (1.0)^{d,e}	24.6 (1.6)^{f,g,h}
SA/A	35.9 (3.5)^{a,b}	32.3 (3.9)^{c}	26.7 (2.0)^{d,e}	23.6 (1.3)^{g,h}
SA/S	34.7 (2.5)^{a,b}	33.0 (3.8)^{c}	25.7 (1.1)^{e}	23.8 (1.0)^{f,g,h}
TS/S	34.0 (2.2)^{a,b}	32.7 (3.7)^{c}	26.5 (1.2)^{d,e}	23.6 (1.1)^{g,h}

Protemp 3 Garant, Luxatemp, Vertex, and Jet data are analyzed separately

CON control, *ADH* adhesive, *SIL* silane, *SI/A* silane + adhesive, *HFA* hydrofluoric acid, *LAS* laser, *SAN* sandblasting, *SA/A* sandblasting + adhesive, *SA/S* sandblasting + silane, *TS/S* tribochemical silica coating + silane

Same superscripted lowercase letters in each temporary crown and FDP material indicate no significant differences (Scheffé test: $P > 0.05$)

Table 4. Statistical analysis of shear bond strength of Protemp 3 Garant.

	Sum of squares	df	Mean square	F	P
Treatment	307.20	9	34.13	2.76	0.007
Error	1114.84	90	12.39		
Total	1422.03	99			

Table 5. Statistical analysis of shear bond strength of Luxatemp.

	Sum of squares	df	Mean square	F	P
Treatment	286.82	9	31.87	2.23	0.027
Error	1288.78	90	14.32		
Total	1575.60	99			

two methods did not differ significantly. On the other hand, Group CON exhibited the lowest value with the mean value of 32.4MPa. The mean shear bond strength was 21% higher in the Group SAN than in the Group CON ($P < 0.05$, Scheffé test).

Table 6. Statistical analysis of shear bond strength of Vertex.

	Sum of squares	df	Mean square	F	P
Treatment	98.99	9	11.00	3.70	0.001
Error	267.33	90	2.97		
Total	366.32	99			

Table 7. Statistical analysis of shear bond strength of Jet.

	Sum of squares	df	Mean square	F	P
Treatment	134.07	9	14.90	5.74	0.000
Error	233.70	90	2.60		
Total	367.77	99			

For Luxatemp, mean shear bond strength values ranged from 29.1 to 35.8MPa. Group SAN revealed the highest mean shear bond strength with the value of 35.8MPa, but Group CON showed the lowest with the value of 29.1MPa. Although there was no significant difference in the mean shear bond strengths among the groups, it is noted that the Group SAN showed 23% higher value, on average, than the Group CON.

3.2. Polymethyl Methacrylate

For Vertex, mean shear bond strength values ranged from 25.7 to 29.5MPa. Group ADH had the highest mean shear bond strength value and Group SA/S had the lowest. The shear bond strength of the Group ADH was significantly higher than those of the Groups SA/S, CON, and LAS ($P < 0.05$, Scheffé test).

For Jet, mean shear bond strength values ranged from 22.8 to 26.8MPa. As displayed in **Table 7**, the mean shear bond strengths differed significantly among groups ($P = 0.000$, one-way ANOVA). Group SIL showed significantly higher mean shear bond strength value when compared to the Groups LAS, SA/A, or TS/S ($P < 0.05$, Scheffé test). In addition, a significant difference was noted between Groups ADH and LAS. However, when compared to Group CON, there was no statistical difference in mean shear bond strength values.

3.3. Surface Morphology

SEM images showed that the topographic patterns differed among the specimens of which were etched with hydrofluoric acid, roughened with the Er,Cr:YSGG laser, or abraded with airborne Al_2O_3 particles (**Figures 3–6**). The surfaces of untreated specimens appeared relatively smooth [**Figures 3(a), 4(a), 5(a), 6(a) and 3(b), 4(b), 5(b), 6(b)**]. Hydrofluoric acid gel dissolved the fillers of the bis-acryl resins and produced porous irregular surfaces [**Figures 3(d), 4(d)**]. However, SEM image of the PMMA resin surface treated with hydrofluoric acid showed no substantial difference from that of the untreated control group [**Figures 5(d), 6(d)**]. The appearance of laser-treated specimen was very different from that of untreated specimen. Microcracks, fissures, grooves, and concavities were present on the surface of laser-treated specimen [**Figures 3(f), 4(f), 5(f), 6(f)**]. SEM images of the sandblasted surfaces showed a micromechanical retention system and demonstrated visible changes in the topographic pattern. Their surface roughness was significantly increased [**Figures 3(h), 4(h), 5(h), 6(h)**].

3.4. Failure Mode

Figure 7 presents the failure mode for all groups in the present study. Predominantly, cohesive failures were found in all groups of each resin. The cohesive failures were observed in each repaired resin, not repairing resin.

4. Discussion

As revealed by one-way ANOVA on the testing results of each material, the shear bond strengths of the temporary crown and FDP materials were affected by various surface treatments. Thus, the null hypothesis that there is no difference in shear bond strength among various surface treatments on the repaired temporary crown and FDP materials should be rejected. Adequate surface treatments should be carefully selected and utilized for each temporary restoration system due to the differences in chemical compositions of the temporary crown and FDP materials. For selection of the optimal surface treatment for every clinical situation, it is critical to know the bond strengths resulted from different surface treatments.

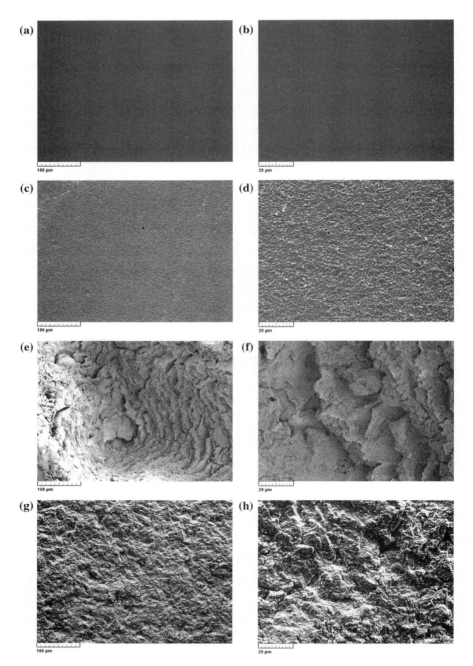

Figure 3. Scanning electron micrographs (the left sides magnification ×500 and the right sides ×2000) of Protemp 3 Garant specimen surfaces, where a, b control; c, d etching with 4% hydrofluoric acid; e, f roughening with Er,Cr:YSGG laser; and g, h sandblasting with 50μm Al_2O_3.

Chapter 19

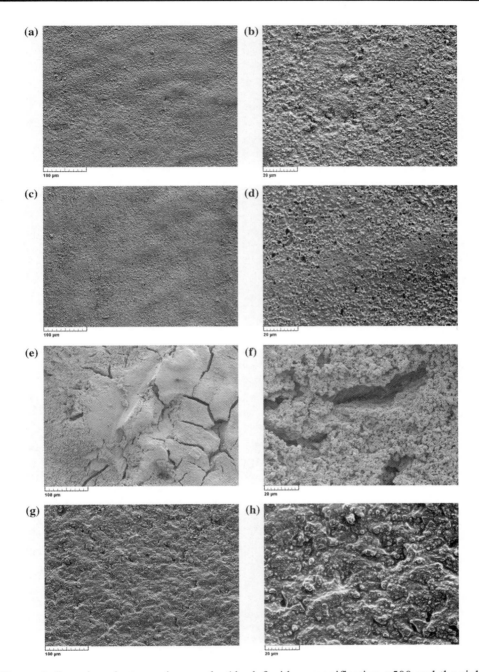

Figure 4. Scanning electron micrographs (the left sides magnification ×500 and the right sides ×2000) of Luxatemp specimen surfaces, where a, b control; c, d etching with 4% hydrofluoric acid; e, f roughening with Er,Cr:YSGG laser; and g, h sandblasting with 50μm Al_2O_3.

409

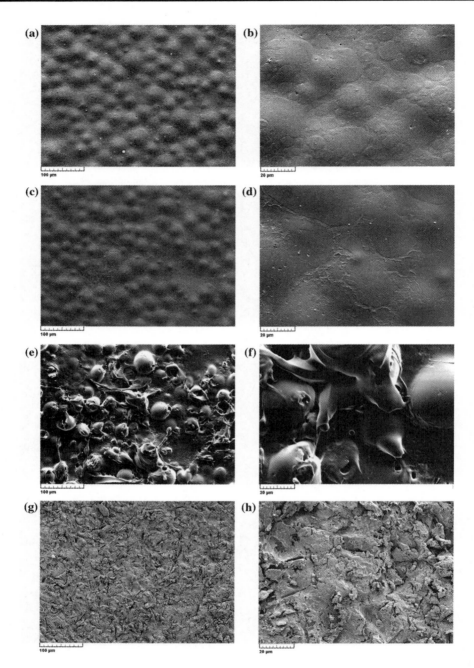

Figure 5. Scanning electron micrographs (the left sides magnification ×500 and the right sides ×2000) of Vertex specimen surfaces, where a, b control; c, d etching with 4% hydrofluoric acid; e, f roughening with Er,Cr:YSGG laser; and g, h sandblasting with 50μm Al_2O_3.

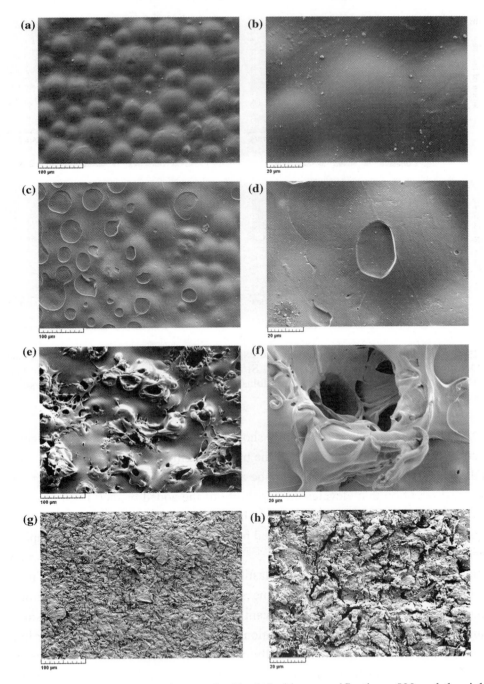

Figure 6. Scanning electron micrographs (the left sides magnification ×500 and the right sides ×2000) of Jet specimen surfaces, where a, b control; c, d etching with 4% hydrofluoric acid; e, f roughening with Er,Cr:YSGG laser; and g, h sandblasting with 50μm Al_2O_3.

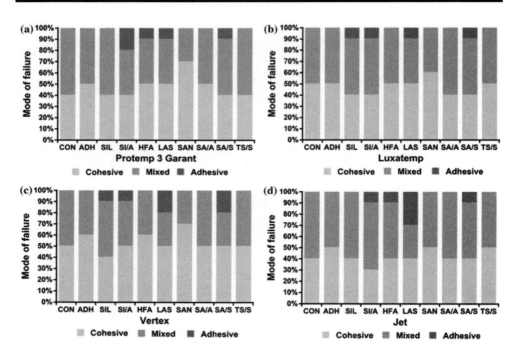

Figure 7. Failure mode distribution of the experimental and control groups. Group codes: CON control, ADH adhesive, SIL silane, SI/A silane + adhesive, HFA hydrofluoric acid, LAS laser, SAN sandblasting, SA/A sandblasting + adhesive, SA/S sandblasting + silane, TS/S tribochemical silica coating + silane.

The results of the current study showed that sandblasting alone significantly increased the shear bond strength of the bis-acryl temporary crown and FDP materials. The effect of sandblasting may be attributed to an increase in the micromechanical retention that elevates the capability of the added bis-acryl resin to interlock mechanically onto the old resin. These results are in close agreement with many investigations reporting improved bond strengths when the filling composite resin was sandblasted before repair[5][7][9][11]. The findings of the current study demonstrate the importance of sandblasting and micromechanical retention in the bis-acryl temporary resin repair. Some studies have also shown that micromechanical retention is the most significant factor in the filling composite resin repair[5][16][17]. However, some investigations have reported the reduced repair bond strength after sandblasting in the filling composite resin repair[18]–[20]. Possible causes of these reductions in bond strength are supposed in several studies. Surface debris or air inclusion on the repair site, exposure of filler components following sandblasting, and viscosity of filling composite resin can be all attributed to the reduction in bond strength[21].

The laser system used in the current study was Er,Cr:YSGG laser. This produces water vapor which increases pressure until a thermally induced mechanical ablation occurs[22][23]. In the current study, laser treatment led to the formation of microcracks, fissures, grooves, and concavities. Although the surface of laser-treated specimen in the SEM images was rougher than that of the control group, laser treatment did not result in the increase of bond strength compared to the control group. The roughened surface containing cliffs, microcracks, and other destructive topographic pattern may affect the results. Moreover, the presence of smear layer or surface debris following laser treatment could reduce the bond strength in each resin. This may suppose that the surface roughness is not a single critical factor contributing to the repair bond strength.

It has been reported that strong acid might dissolve filler on the filling composite resin surface, leaving gaps or pores, and create surface irregularities that allow micromechanical retention[24][25]. This effect of strong acid is dependent on the type, percentage, and size of the filler[24]. However, some studies have reported that etching with hydrofluoric acid did not increase the adhesion of resin to some filling composite resins[25][26]. In a study by Kula et al.[27], immersion of the filling composite resin in acidic medium decomposed the inorganic filler particles, resulting in impaired adhesion between composite layers. Swift et al.[24] investigated the dissolution of the filler and softening of the resin after etching of the composites with a 9.6% hydrofluoric acid for shorter duration. The study showed either decreased or increased bond strength, depending on the kind of filling composite resin being repaired[24]. The results of the present study showed that the application of hydrofluoric acid did not significantly improve the bond strength. This finding could be explained by variations in compositions of the temporary materials. In terms of bis-acryl resin tested in this study, it was evident that hydrofluoric acid increased the shear bond strength. However, this effect could not be proved due to the high standard deviation in this group. The SEM images revealed the differences in the modified surfaces of the tested materials. According to the images, the surface treated with hydrofluoric acid [**Figures 3(c), 4(c)**] appeared to be slightly smoother than that treated with sandblasting. It suggests that the acid may have slightly eroded the bis-acryl resin surface. In addition, this surface treatment did not show the numerous surface irregularities shown by the specimens treated with laser, but no significant differences in shear bond strength between them were noted. The surface topography indicated that the laser-treated specimen has a combination of micro- and macro-mechanical retention systems, but the sandblasted surface only demonstrated a micromechanical retention system and increased surface roughness.

It has been reported that the utilization of adhesive monomers significantly increases the repair bond strength in the filling composite resins[16][28][29]. Several possible mechanisms of the adhesive monomer during the filling composite resin repair include chemical bond formation to the surface fillers and to the matrix and micromechanical interlocking formed by infiltration of the monomer into microcracks in the matrix[30]. Many adhesive monomers consist of chloro-phosphate esters of bis-GMA resin. Since the phosphate groups are polar, they may play a role in the affinity of inorganic filler particles by bonding to silane and hydrogen. This may form covalent bond to the unreacted methacrylate groups on the matrix[28]. In addition, the adhesive monomers enable the achievement of better wetting of the surface[25]. A solvent and a surfactant are often added, and the wetting properties of the adhesive monomers are increased by their low viscosity[19][28][31]. Adhesives promote penetration of solvent systems and of monomers into the composite surface, depending on the degree of hydration and the chemical affinity of materials, and create a non-polymerized oxygen inhibition layer that could ultimately promote adhesion of new composites[32][33]. However, the study of Hagge et al.[34] showed that the shear bond strength values of the flowable composite resin were significantly higher in surface treatment with sandblasting alone than with the combination of sand-blasting and adhesive monomer. The results of the present study correspond well with those of the earlier studies in the composite resin. The current study showed that the use of the adhesive monomer did not enhance the repair bond strengths of all tested temporary resins except for Vertex.

When the material has no specific groups to bond to the silane coupling agent or when little filler remains on the surface, the effect of silane could be useless[25]. The surface could be treated with tribochemical silica coating to achieve a chemical bonding with the silane. Through this treatment, it is possible to deposit a mixture of silica particles and alumina on the surface[35]. These particles could form covalent bonds through its hydroxyl groups with hydrolyzed silanol groups in the silane. This makes the surface more reactive to the methacrylate groups of the resin[25][36]. Silane coupling agent improves the wettability of the filler and adhesive monomer that facilitates their infiltration into the irregularities created by sand-blasting[25][36]. However, in the present study, silanization of specimens after tribochemical silica coating or sandblasting did not increase the shear bond strength significantly. The failure of silane coupling agents to increase the shear bond strength may propose that mechanical retention is the single most important factor contributing to bond strength. The repair procedure should not alter the original color of the temporary restorations. Moreover, the procedure needs to be easy, rapid, and inexpensive to perform. In the present study, roughening the surface of bis-acryl resin by sandblasting showed a greater improvement on the repair strength than using the chemical treatment. Thus, it appears that the application of chemical agent is unnecessary for repairing temporary restorations.

The values of strengths obtained in the present study seem to be higher than those in clinical situations because the repairs were carried out only a few hours after polymerization of the original temporary materials. Furthermore, the repaired surface was stored largely untouched until the surface modification procedures. Bond strength between the original material and newly added resin is dependent on unreacted C=C double bonds[25]. The resins often have incomplete C=C double-bond conversion after being polymerized[37]. As the material ages, more cross-linking decreases the capability of fresh monomer to infiltrate into the matrix, and fewer and fewer unreacted C=C double bonds remain[38].

The results of in vitro testing cannot be postulated in the clinical situation, as the design of the present study did not consider factors in the oral environment, such as dynamic forces of mastication or fatigue loading. The repaired surface area used in this study was about $7mm^2$ (1.52 p), but fractured surface area of clinically used temporary restorations is usually of a smaller size. It should be noted that this was a comparative study where all variables were controlled except for the surface treatment. Hence, it should be kept in mind that the shear bond strength is only one of many behaviors in response to a particular stress and that strength is just one property of temporary crown and FDP materials. In addition, the present study design offered no data on the long-term stability of the repaired specimens. Further investigations are necessary to evaluate the effect of the thermal cycling on the repair bond strength of temporary crown and FDP materials. Moreover, it is necessary to determine the repair bond strength after long-term use of the materials.

Finally, the influence of changing the application condition of sandblasting needs further investigation.

5. Conclusions

The surface treatment of bis-acryl resins with sandblasting seems to be promising for the improvement of repair bond strength.

Compliance with Ethical Standards

Conflict of Interest: The authors declare that they have no potential conflict of interest to this work.

Open Access: This article is distributed under the terms of the Creative Commons Attribution 4.0 International License (http://creativecommons.org/licenses/by/4.0/), which permits unrestricted use, distribution, and reproduction in any medium, provided you give appropriate credit to the original author(s) and the source, provide a link to the Creative Commons license, and indicate if changes were made.

Source: Ha S R, Kim S H, Lee J B, *et al.* Improving shear bond strength of temporary crown and fixed dental prosthesis resins by surface treatments[J]. Journal of Materials Science, 2016, 51(3):1463–1475.

References

[1] Gratton DG, Aquilino SA (2004) Interim restorations. Dent Clin North Am 48(vii):487–497. doi:10.1016/j.cden.2003.12.007.

[2] Gergauff AG, Holloway JA (2000) Provisional restorations. In: Rosenstiel SF, Land MF, Fujimoto J (eds) Contemporary fixed prosthodontics, 3rd edn. Mosby, St. Louis, pp 380–416.

[3] el-Ebrashi MK, Craig RG, Peyton FA (1970) Experimental stress analysis of dental restorations. VII. Structural design and stress analysis of fixed partial dentures. J Prosthet Dent 23:177–186.

[4] Kim SH, Watts DC (2007) In vitro study of edge-strength of provisional poly-

mer-based crown and fixed partial denture materials. Dent Mater 23:1570–1573. doi:10.1016/j.dental.2007. 06.023.

[5] da Costa TR, Serrano AM, Atman AP, Loguercio AD, Reis A (2012) Durability of composite repair using different surface treatments. J Dent 40:513–521. doi:10.1016/j.jdent.2012.03.001.

[6] Hickel R, Brushaver K, Ilie N (2013) Repair of restorations—criteria for decision making and clinical recommendations. Dent Mater 29:28–50. doi:10.1016/j.dental.2012.07.006.

[7] Ozcan M, Corazza PH, Marocho SM, Barbosa SH, Bottino MA (2013) Repair bond strength of microhybrid, nanohybrid and nanofilled resin composites: effect of substrate resin type, surface conditioning and ageing. Clin Oral Investig 17:1751–1758. doi:10.1007/s00784-012-0863-5.

[8] Baur V, Ilie N (2013) Repair of dental resin-based composites. Clin Oral Investig 17:601–608. doi:10.1007/s00784-012-0722-4.

[9] Bacchi A, Consani RL, Sinhoreti MA, Feitosa VP, Cavalcante LM, Pfeifer CS, Schneider LF (2013) Repair bond strength in aged methacrylate- and silorane-based composites. J Adhes Dent 15:447–452. doi:10.3290/j.jad.a29590.

[10] El-Askary FS, El-Banna AH, van Noort R (2012) Immediate vs delayed repair bond strength of a nanohybrid resin composite. J Adhes Dent 14:265–274. doi:10.3290/j.jad.a22716.

[11] Melo MA, Moyses MR, Santos SG, Alcantara CE, Ribeiro JC (2011) Effects of different surface treatments and accelerated artificial aging on the bond strength of composite resin repairs. Braz Oral Res 25:485–491.

[12] Koumjian JH, Nimmo A (1990) Evaluation of fracture resistance of resins used for provisional restorations. J Prosthet Dent 64:654–657.

[13] Hammond BD, Cooper JR 3rd, Lazarchik DA (2009) Predictable repair of provisional restorations. J Esthet Restor Dent 21:19–24. doi:10.1111/j.1708-8240.2008.00225.x discussion 25.

[14] Chen HL, Lai YL, Chou IC, Hu CJ, Lee SY (2008) Shear bond strength of provisional restoration materials repaired with light-cured resins. Oper Dent 33:508–515. doi:10.2341/07-130.

[15] Patras M, Naka O, Doukoudakis S, Pissiotis A (2012) Management of provisional restorations' deficiencies: a literature review. J Esthet Restor Dent 24:26–38. doi:10.1111/j.1708-8240.2011. 00467.x.

[16] Turner CW, Meiers JC (1993) Repair of an aged, contaminated indirect composite resin with a direct, visible-light-cured com-posite resin. Oper Dent 18:187–194.

[17] Kimyai S, Oskoee SS, Mohammadi N, Rikhtegaran S, Bahari M, Oskoee PA, Vahedpour H (2015) Effect of different mechanical and chemical surface treatments on the repaired bond strength of an indirect composite resin. Lasers Med Sci 30:

653–659. doi:10. 1007/s10103-013-1391-5.

[18] Eliades GC, Caputo AA (1989) The strength of layering technique in visible light-cured composites. J Prosthet Dent 61:31–38.

[19] Eli I, Liberman R, Levi N, Haspel Y (1988) Bond strength of joined posterior light-cured composites: comparison of surface treatments. J Prosthet Dent 60: 185–189.

[20] Pounder B, Gregory WA, Powers JM (1987) Bond strengths of repaired composite resins. Oper Dent 12:127–131.

[21] Gregory WA, Pounder B, Bakus E (1990) Bond strengths of chemically dissimilar repaired composite resins. J Prosthet Dent 64:664–668.

[22] Mohammadi N, Savadi Oskoee S, Abed Kahnamoui M, Bahari M, Kimyai S, Rikhtegaran S (2013) Effect of Er, Cr:YSGG pretreatment on bond strength of fiber posts to root canal dentin using a self-adhesive resin cement. Lasers Med Sci 28:65–69. doi:10.1007/s10103-012-1063-x.

[23] Perussi LR, Pavone C, de Oliveira GJ, Cerri PS, Marcantonio RA (2012) Effects of the Er, Cr:YSGG laser on bone and soft tissue in a rat model. Lasers Med Sci 27:95–102. doi:10.1007/s10103-011-0920-3.

[24] Swift EJ Jr, LeValley BD, Boyer DB (1992) Evaluation of new methods for composite repair. Dent Mater 8:362–365.

[25] Ozcan M, Alander P, Vallittu PK, Huysmans MC, Kalk W (2005) Effect of three surface conditioning methods to improve bond strength of particulate filler resin composites. J Mater Sci Mater Med 16:21–27. doi:10.1007/s10856-005-6442-4.

[26] Bouschlicher MR, Reinhardt JW, Vargas MA (1997) Surface treatment techniques for resin composite repair. Am J Dent 10:279–283.

[27] Kula K, Nelson S, Kula T, Thompson V (1986) In vitro effect of acidulated phosphate fluoride gel on the surface of composites with different filler particles. J Prosthet Dent 56:161–169.

[28] Shahdad SA, Kennedy JG (1998) Bond strength of repaired anterior composite resins: an in vitro study. J Dent 26:685–694.

[29] Oztas N, Alacam A, Bardakcy Y (2003) The effect of air abrasion with two new bonding agents on composite repair. Oper Dent 28:149–154.

[30] Brosh T, Pilo R, Bichacho N, Blutstein R (1997) Effect of combinations of surface treatments and bonding agents on the bond strength of repaired composites. J Prosthet Dent 77:122–126.

[31] Puckett AD, Holder R, O'Hara JW (1991) Strength of posterior composite repairs using different composite/bonding agent combinations. Oper Dent 16:136–140.

[32] Teixeira EC, Bayne SC, Thompson JY, Ritter AV, Swift EJ (2005) Shear bond strength of self-etching bonding systems in combination with various composites

used for repairing aged composites. J Adhes Dent 7:159–164.

[33] Lastumaki TM, Kallio TT, Vallittu PK (2002) The bond strength of light-curing composite resin to finally polymerized and aged glass fiber-reinforced composite substrate. Biomaterials 23:4533–4539.

[34] Hagge MS, Lindemuth JS, Jones AG (2002) Shear bond strength of bis-acryl composite provisional material repaired with flow-able composite. J Esthet Restor Dent 14:47–52.

[35] Ozcan M (2002) The use of chairside silica coating for different dental applications: a clinical report. J Prosthet Dent 87:469–472.

[36] Ozcan M, Barbosa SH, Melo RM, Galhano GA, Bottino MA (2007) Effect of surface conditioning methods on the microtensile bond strength of resin composite to composite after aging conditions. Dent Mater 23:1276–1282. doi:10.1016/j.dental.2006.11. 007.

[37] Floyd CJ, Dickens SH (2006) Network structure of Bis-GMA- and UDMA-based resin systems. Dent Mater 22:1143–1149. doi:10.1016/j.dental.2005.10.009.

[38] Chay SH, Wong SL, Mohamed N, Chia A, Yap AU (2007) Effects of surface treatment and aging on the bond strength of orthodontic brackets to provisional materials. Am J Orthod Dentofacial Orthop 132:577.e7–577.e11. doi:10.1016/j.ajodo. 2004.01.024.

Chapter 20

In Vitro Biocompatibility of Anodized Titanium with Deposited Silver Nanodendrites

Mariusz Kaczmarek[1,*], Karolina Jurczyk[2], Jeremiasz K. Koper[3], Anna Paszel-Jaworska[4], Aleksandra Romaniuk[4], Natalia Lipińska[4], Jakub Żurawski[5], Paulina Urbaniak[6], Jarosław Jakubowicz[3], Mieczysława U. Jurczyk[7]

[1]Department of Immunology, Chair of Clinical Immunology, Poznan University of Medical Sciences, Rokietnicka 5D, 60-806 Poznan, Poland
[2]Department of Conservative Dentistry and Periodontology, Poznan University of Medical Sciences, Bukowska 70, 60-812 Poznan, Poland
[3]Institute of Materials Science and Engineering, Poznan University of Technology, Jana Pawla II 24, 61-138 Poznan, Poland
[4]Department of Clinical Chemistry and Molecular Diagnostics, Poznan University of Medical Sciences, Przybyszewskiego 49, 60-355 Poznan, Poland
[5]Department of Immunobiochemistry, Chair of Biology and Environmental Sciences, Poznan University of Medical Sciences, Rokietnicka 8, 60-806 Poznan, Poland
[6]Department of Cell Biology, Poznan University of Medical Sciences, Rokietnicka 5D, 60-806 Poznan, Poland
[7]Division Mother's and Child's Health, Poznan University of Medical Sciences, Polna 33, 60-535 Poznan, Poland

Abstract: Engineers searching new dental biomaterials try to modify the structure of the material in order to achieve the best performance as well as increased migration and proliferation of cells involved in the osseointegration of the implant. In this work we show in vitro test results of the Ti, which was anodically oxidized at high voltages with additionally deposited silver in the form of nanodendrites. The in vitro cytocompatibility of these materials was evaluated and compared with a conventional microcrystalline titanium. During the studies, established cell line of human gingival fibroblasts (HGF) and osteoblasts were cultured in the presence of tested materials, and its survival rate and proliferation activity were examined. Titanium samples modified with silver has a higher degree of biocompatibility in comparison with the unmodified reference material. Cells in contact with studied material showed a higher relative viability potential, stable level of proliferation activity, and lower rate of mortality. Biocompatibility tests carried out indicate that the anodically oxidized titanium at high voltages with additionally deposited nanosilver could be a possible candidate for dental implants and other medicinal applications.

1. Introduction

Titanium (Ti) and its alloys are most useful and most often investigated metallic biomaterials. Titanium possesses high strength to density ratio, relatively low Young modulus value, very good corrosion resistance, and biocompatibility. For providing fast osseointegration and long-term usage in the human body, the implant surface should be modified, *i.e.*, it should be rough or porous, oxidized, and covered by biocompatible coating including calcium-phosphate compounds[1]–[3]. Surfaces showing, micro-, and nanoirregularities are useful in biocompatibility improvements[4].

Among many surface treatment technologies applied for Ti, the electrochemical one is very useful, giving surface roughening and new chemical and phase composition, which improve surface biocompatibility[5]–[7]. Anodic oxidation results in the formation of rough titanium oxide, which improves osseointegration. The oxidation process can be done in standard conditions as well as can be supported by spark-discharge process in the plasma electrolytic oxidation (PEO), done at high voltages[8]–[10]. As the results, formation of different size pores or cavities as well as nanotubes are possible[11]–[13]. The anodically oxidized titanium-based dental implants are commonly available[14]. By carefully choosing the oxidation conditions it is possible to control the oxide thickness, which is correlated with its color[15].

During and after implantation, there is a risk of bacterial attack in the wound tissue. To avoid this inconvenient postoperative effect, an introduction of antibacterial agent into implant surface layer is highly recommended and possible. The commonly known antibacterial agents are silver ions, however it should be noticed, that high silver ion concentration could prevent osseointegration. Cytotoxicity of silver (Ag) nanoparticles at a concentration of $8\mu g/cm^2$ to E. coli was reported previously[16]. The large surface area to volume ratio of Ag nanoparticles provides good antibacterial behavior[17].

The surface, which is oxidized and has porous topography, is an excellent template for controlled introduction of silver ions. Previously[18] we show preferential deposition mechanism of Ag nanodendrites, which are deposited inside the pits in the oxide layer. Thus we suggest that the surface pits positively effect on both, osteoblast cells and antibacterial nanoparticle attachment.

The aim of this study was to determine of bio-compatibility in vitro of the Ti, which was anodically oxidized at high voltages with additionally deposited Ag in the form of nanodendrites. Biocompatibility of the tested material was referred in relation to human osteoblasts and gingival fibroblasts. They are the major cellular elements that determine the osseointegration and the acceptance of implant in the oral cavity.

2. Materials and Methods

2.1. Sample Preparation

The commercially pure titanium (CP-Ti) with purity >99.6% (Goodfellow) was used for electrochemical treatment and biocompatibility test. The original φ 10mm Ti rod was cut into a form of small tablets (10mm diameter and 5mm height), grinded up to a 1000 sand paper and then polished in Al_2O_3 suspension to a mirror-like surface. The samples were anodically oxidized in home-made Teflon electro-chemical cell. The platinum (Pt) electrode was applied as reference electrode. The oxidation process proceeded under Atlas Sollich high voltage potentiostat (300V/3A) control. The structural and morphological changes with using broad anodic oxidation voltages were studied in our previous works[19][20]. In this work for biocompatibility test, samples optimally oxidized at 210V versus OCP (open circuit potential), at constant time of 30min were chosen. As the electrolyte, solution of 2M H_3PO_4 with addition of 1wt% HF was used. After the oxidation process, the surfaces were rinsed in water and dried under a stream of nitrogen. For obtaining antibacterial characteristics, the surface was electro-chemically modified by deposition of Ag nanocrystals, in the form of nanodendrites (nanotrees) using the electrolytic deposition process[18]. The aqueous solution of 0.01M $AgNO_3$ + 0.01M HNO_3 composition was served as a substrate for the Ag ions. The Ti samples were immersed in the electrolyte and additionally an Ag plate ($8cm^2$) was used to support the transport of the Ag ions and Ag refilling into the electrolyte. An SCE electrode was used as the reference electrode. EG&G electrochemical cell was applied and the electrolyte inside was mixed using magnetic stirring (250rpm). The process was controlled by a Solartron 1285 potentiostat. The Ag nanodendrites depositing parameters were as follows: potential −1V versus OCP, time 60s, temperature 20°C. After deposition, the samples were rinsed in distilled water and dried in a stream of nitrogen. For the purpose of this study, materials CP-Ti, anodically oxidized Ti, anodically oxidized with deposited nano-Ag are named as A0, C1, and C2, respectively.

2.2. Materials Characterization

Structure was determined using Panalytical Empyrean XRD with CuKa1 radiation, equipped with crystallographic database. Surface topography was determined using Tescan SEM model Vega 5135 equipped with PGT model Prism 200 Avalon EDS, Leica DCM 3D confocal microscope with EPI 20X-L objective, Quesant Q-scope 250 AFM. The AFM tapping mode was applied during surface scanning. The Nanoandmore probes (with pre-mounted Nanosensors Supersharp-SiliconTM) were used in the surface scanning.

2.3. In Vitro Biocompatibility Studies

2.3.1. Cell Cultures

Established line of human gingival fibroblasts HGF-1 (ATCC® CRL-2014™) and human osteoblast U-2 OS (ATCC®HTB-96™) were used for the study. Cell lines were derived from ATCC collection. HGF-1 cells were cultured in DMEM/ Ham's F12 media (mixed in 1:1) with L-glutamine and 15mM HEPES, supplemented with 10% fetal bovine serum (FBS) and 1% antibiotic solution (10,000 U penicillin, 10mg/ml streptomycin, 25mg/ml amphotericin B). U-2 OS cells were cultured in DMEM medium only supplemented as described above. Cells were cultured under standard conditions, in plastic plates in an incubator at 37°C temperature, in atmosphere of 5% CO_2 and increased humidity level of 95%. When the cells reached confluent monolayers (~90% of cells), the culture media were removed and cells were washed with phosphate-buffered saline (PBS). After removing PBS, the cells were harvested from the surface of plates using a 0.25% Trypsin-EDTA. The cells were counted using a Fuchs-Rosenthal's hematologic camera. Thus prepared cells were placed on the surface of the tested samples.

2.3.2. In Vitro Evaluation

Before testing, the samples of the material were sterilized using an autoclave at 120°C for 15min. Sterile samples were placed in 24-well culture plates, pre-filled with 1 ml of culture medium. Approximately 3×10^4 of cells were placed directly on the surface of studied material samples. Cultures were grown for 72 and 96h. To ensure sterile conditions during the analyses, a chamber with laminar air flow and disposable sterile equipment were used.

2.3.3. Imaging of Samples Using Fluorescence Microscopy

HGF-1 fibroblasts and U-2 OS osteoblasts growing on the tested materials were imaged with a fluorescent microscope after 96h. Cells were stained with a thiazole orange (TO). This fluorescent dye penetrates the cell membrane of living cells and bind RNA. Microscopic images of cells growing on the tested samples were archived within 10min with a fluorescence microscope using appropriate color filters, at a magnification of 40, 100, 400, and 1000 fold.

2.3.4. Cell Viability Assay (MTT Cytotoxicity Test)

To evaluate the cytotoxicity of the tested materials the MTT assay was performed. Cytotoxicity level was assessed by determining the percentage of dead cells as well as the degree of inhibition of their growth. During the test, the water-soluble MTT tetrazolium salts (3-(4,5-dimethylthiazol-2-yl)-2,5-diphenyltetra-zolium bromide) are reduced to a blue-purple insoluble formazan crystals. The reduction reaction occurs in the presence of the active mitochondrial dehydrogenase, only in living cells. Finally, the formazan crystals are extracted from the cells with a solubilizing solution (10% SDS in 0.01M HCl). The intensity of the color reaction is directly proportional to the number of living cells. During analyses cells were cultured directly on material samples located in the complete culture media supplemented as described above in 24-well culture plate. After 72 and 96h of culture on the surface of the tested samples 100μl of fresh culture media and 10μl of a solution of MTT (5mg/ml thiazolyl blue Tetrazolium Bromide) were applied. Samples were then incubated for 4h under standard conditions. Finally, the formazan crystals were released from the cells by adding 100μl of a solubilizing solution. After overnight incubation, the absorbance of solutions using a microtiter reader (Multiscan, Labsystems) at two wavelengths: $\lambda = 570$ and 690nm. The viability of cells growing on the tested material was expressed in relation to the viability of cells growing directly on culture plate surface without any material (control samples) as well as growing simultaneously on a sample of the reference material (A0 samples). Results were presented as a Relative Viability of Cells value (RVC), which was calculated from the formula:

$$\mathrm{RVC}[\%] = \left[\frac{(a-b)}{(c-b)}\right] \times 100$$

a absorbance of the tested sample; b absorbance of the blank control (reaction without the cells); c absorbance of the control grown on the reference material.

2.3.5. Evaluation of a Cell Cycle

After 72 and 96h of culture, cells covering the material samples were transferred to new plastic tubes. Next, the cells were detached from the surface of the samples with trypsin. Harvested cells were resuspended in a fresh DMEM medium and centrifuged at 1000rpm for 5min at room temperature. After discarding, the supernatant cells were washed once with PBS. Finally, to the cell pellets a solution of propidium iodide (PI) and RNase was added, and then incubated for 1h at 37°C in the dark. Cells stained with PI were evaluated using the flow cytometer FACS-Canto (BD Biosciences). Histograms obtained during the evaluation were analyzed by FACS Diva software. During the analysis, the mean fluorescent intensity (MFI) of stained cells was measured. The percentage of cells in S-phase of the cell cycle corresponds to proliferative activity of the cells grown on the studied material. In addition, on the basis of the cytometric histogram results, the estimated percentage of dead cells as well as cells in the G2/M phase, preceding the process of mitosis, were determined.

3. Results

3.1. Structure Properties

Study of Ti surface topography and its properties after different processing conditions of high voltage anodic oxidation was described in previous manuscripts[19][20]. For the in vitro tests, the surface was oxidized at 210V for 30min in 2M H_3PO_4 + 1wt% HF electrolyte (**Figure 1**), which showed optimum properties, useful for implant applications[19][20]. The surface treated at these conditions has 3D scaffold morphology with interconnected pores. The surface pores have elongated in plane channel-type form, which were formed by the fusion of the circular pores lying close to each other during anodization. The roughness (**Table 1**) and morphology of the Ti after anodic oxidation predisposes the surface for implant application. The differences in both Sv and Sz roughness parameters, measured using different techniques (**Table 1**), are the consequence of different area analysis in AFM and confocal microscopes, however Sa roughness is comparable, independent of the analyzed area.

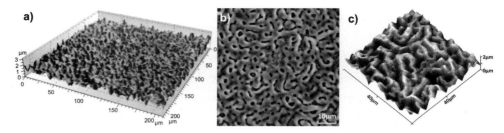

Figure 1. Confocal (a), SEM (b), and AFM (c) pictures of anodically oxidized titanium.

Table 1. 3D roughness parameters measured on anodically oxidized surface.

3D roughness parameter	Method of measurements	
	Confocal	AFM
Sv (µm)	1.452	1.085
Sz (µm)	3.445	1.976
Sa (µm)	0.248	0.240

The electrochemical silver deposition, done on the oxidized Ti at conditions presented in experimental paragraph, resulted in the formation of highly developed nanoparticles in the form of dendrites (**Figure 2**). The nanodendrites start to grow from bottom of the surface pores [**Figure 2(a), (b)**], formed in the anodicoxidation process. These surface pores support preferential ions flow from the electrolyte, giving formation of nanodendrites. The Ag nanodendrites are composed of mainly rod-type stem, from which branched arms oriented at about 60° propagate [**Figure 2(b)**]. The Ag nanodendrites have uniform geometry and a relatively large specific surface area, which should be useful in antibacterial action. The EDS analysis of the deposited Ag nanodendrites [**Figure 2(c)**] showed the presence of Ag on the sample surface. The chemical composition of the nanoparticles was measured in many different points, however the large spot of the X-ray beam, results not only in particle characterization, but Ti substrate detection too. Moreover phosphorus content was also detected as the effect of phosphorus implantation from the electrolyte during anodic oxidation. The phosphorus content can act positively in bone through hydroxyapatite formation. In the EDS spectrum the peaks corresponding to Ag are the main peaks. The Ag nanodendrites have a stem diameter of ~100–150nm and the length in the range of 5–100µm, whereas the length of the arms is in the range of 0.2–20µm. For comparison, Wei *et al.*[21] use silicon template to produce Ag dendrites of 1–2µm in stem diameter and 10–50µm

Figure 2. SEM pictures of anodically oxidized titanium with deposited silver nanodendrites (a, b—different magnifications) and EDS analysis of the silver particles (c).

in length. The dendritic silver structures preferentially grow along (111) and (200) directions[21].

The XRD spectra shown in **Figure 3** describe Ti surface changes after electrochemical treatment. The Ti after anodic oxidation shows typical α-Ti-type structure [**Figure 3(a)**]. No crystalline form of Ti oxides was found, which means that amorphous oxide formation proceeds during applied treatment conditions. The etching of Ti, with the removal of surface oxide is also possible (the electrolyte contains 1% HF, which strongly acts as oxide etching). The electrochemical silver deposition results are shown in [**Figure 3(b), (c)**]. The spectrum (b) presents dominant Ti peaks masking the small Ag peaks (relatively very low Ag amount deposited on Ti template) and for comparison the spectrum for deposited silver particles only, after their removal from the surface (c) is shown. The deposited material clearly represents silver (c). As it is seen on the spectrum (b), additional phase TiN was formed during silver deposition process, as the result of Ti and electrolyte ($AgNO_3 + HNO_3$) reaction.

3.2. In Vitro Evaluation Results

3.2.1. Fluorescent Imaging of Samples

Both, HGF-1 fibroblasts and U-2 OS osteoblasts cultured on different types of materials showed different growth patterns, according to the type and composition of the samples as well as their texture. The growth of both cell types on the

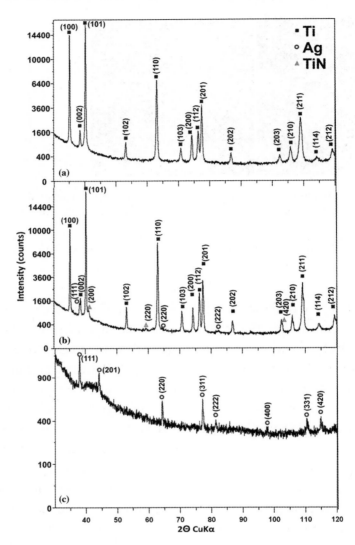

Figure 3. XRD of the titanium after anodic oxidation (a), anodic oxidation and silver nanoparticles deposition (b), and for comparison only for silver nanoparticles removed from the surface (c).

surface of the control material A0 was homogeneous and showed a parallel and orderly pattern. Particularly fibroblasts were characterized by irregular growth, with the shape of multidirectional network. Their growth on the surface of the anodically oxidized Ti samples with deposited nano-Ag (C2) appeared to be more intense and more regular than on the surface of the anodically oxidized Ti samples without nano-Ag deposits (C1). Osteoblasts also showed much more intensive

Chapter 20

growth on the surface of C2 sample as compared with sample C1. However, their growth on the surface of the C1 material was clearly limited [**Figure 4(a), (c)**]. Moreover within the cell nuclei numerous nucleolus organizer regions (NOR) were visible (**Figure 5**).

3.2.2. Cell Viability Assay (MTT Cytotoxicity Test)

First the spontaneous ability to reduce the soluble tetrazolium salt by the tested samples was evaluated (**Figure 6**). During MTT tests two independent cultures for each cell line were made. In each culture, cells were grown in triplicates directly on samples of the test materials. The cultures were performed for 72 and 96h. As a control of spontaneous growth of cells, the cells were grown directly on

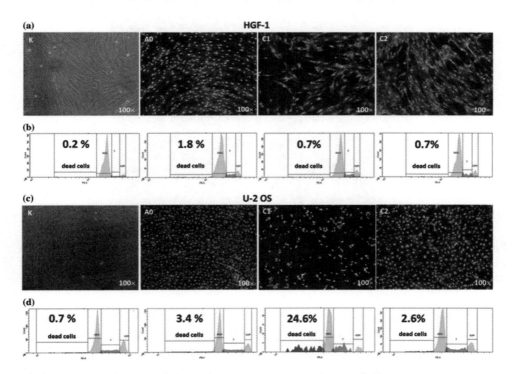

Figure 4. Fluorescent imaging of human gingival fibroblasts (HGF-1) and human osteoblasts (U-2 OS) growing on samples of the tested materials using fluorescent microscopy system (a, c). Histograms visualizing the phases of the cell cycle, obtained during flow cytometry analysis, with particular indication to the percentage of dead cells (b, d). Samples: K—cells grown on culture wells without tested materials, A0—CP-Ti, C1—anodically oxidized Ti, C2—anodically oxidized with deposited nano-Ag.

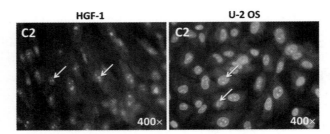

Figure 5. Nucleolus organizer regions within the cell nuclei of HGF-1 fibroblasts and U-2 OS osteoblasts cultured on the surface of the C2 material sample—white arrows; cells during division—red arrows (TO staining, 4009 magnification).

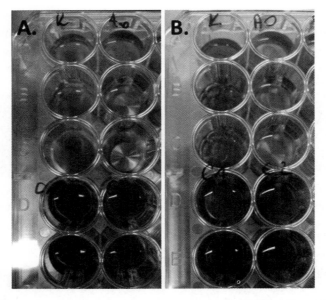

Figure 6. Samples of the A0, C1, and C2 materials placed in the wells with culture media with the addition of MTT solution. Samples before reduction of MTT tetrazolium salt (a) and after reduction (b).

culture plates in the wells without the tested material samples (control). Simultaneously, cells were cultured on the surface of the reference material (control of cell growth on a reference material, A0). Viability of cells growing on C1 and C2 samples was compared to viability of cells growing in wells without test samples (control) as well as to cells growing on a reference material. MTT tests showed differences in viability potential of HGF-1 and U-2 OS cell lines (**Figure 7**). Tested cells showed similar relative viability potential (RVC) in contact with tested samples compared to the control, despite natural diversities between the

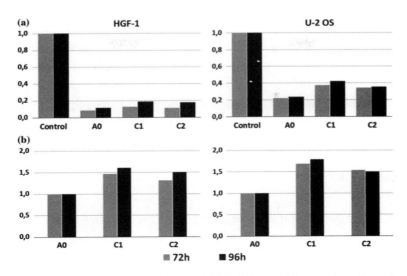

Figure 7. Viability of HGF-1 fibroblasts and U-2 OS osteoblasts cultured on the C1 and C2 samples, relative to cells growing directly on the surface of the culture plates (control) as well as to cells growing on the reference material (A0), in the 72 and 96h of culture.

HGF-1 and U-2 OS lines. Their potential was smaller than potential of control cells. Both cell lines showed higher viable potential in contact with C1 and C2 samples compared to cells growing on A0, however potential of U-2 OS was stronger. Moreover, RVC of both the cell lines increased simultaneously in the course of time.

3.2.3. Evaluation of the Cell Cycle

The cell cycle of HGF-1 and U-2 OS cells growing on the surface of the tested samples was estimated, using the flow cytometer. The results of the evaluation were visualized in the form of histograms, which allowed to determine the percentage of cells in particular phases of the cell cycle, as well as percentage of dead cells, which died in the course of culture on the surface of tested materials [**Figure 4(b), (d)**]. HGF-1 fibroblasts cultured on C1 and C2 samples showed decreased proliferative activity in relation to HGF-1 cultured on the A0 material. Both cell lines cultured on the C1 and C2 samples compared to those on A0 were characterized by a much lower mortality rate, which was comparable to mortality observed among cells directly growing on a culture plate, without contact with tested materials. U-2 OS cells revealed higher proliferative potential as compared to the HGF-1 cells, independently from time and tested material. Studied osteoblasts cultured on C1 and C2 samples showed lower percentage of cells in the S-phase of the cell cycle as compared to the control, but higher as compared to cells on the surface of the A0 samples. However, during the test the U-2 OS osteoblasts cultured on the C1 samples showed high percentage of dead cells (**Figure 8**).

4. Discussion

The main task is to adjust the structure of the material to achieve increased migration and proliferation of cells involved in the osseointegration of the implant. It is important to create material with high durability and resistance to the envi-

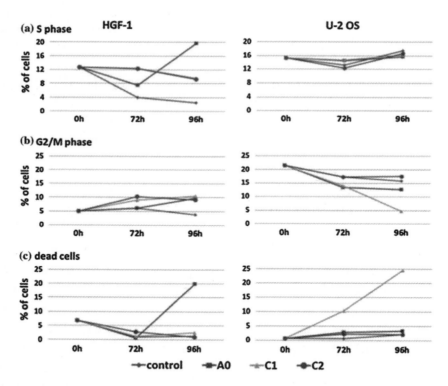

Figure 8. Proportional participation of HGF-1 and U-2 OS cells cultured on the-surface of the C1 and C2 material samples in particular phases of the cell cycle. Percentage of cells in the S-phase (proliferative activity; a); percentage of cells in the G2/M phase (percentage of cells in the premitotic phase; b); percentage of dead cells (apoptotic and/or necrotic cells; c).

ronment surrounding the tissue and a low susceptibility to colonization by pathogenic microorganisms. However, till now materials having characteristics identical to natural has not been created. It has been proven that the porous materials in the most accurate way fill these assumptions[5][9].

The anodic oxidation process done at high voltages is very useful in surface biofunctionalization, *i.e.*, formation of porous, rough, and biocompatible oxide[8]–[10][19][20]. As we have shown in this work, the PEO followed by spark discharge, which proceeds on the Ti surface, results in the formation of highly connected channels in the surface of the oxide layer. The channels support not only bone cell nucleation and growth, but vascularization process too. In this way formation of strong bond between implant material and bone is provided. Thus the process done at high voltages significantly overcomes the conventional low voltage oxidation process[5]. Additionally inside the channel structure, the oxide thickness at pore bottom is lowest, which positively acts on silver ions flow from electrolyte into the surface of highest electrical conductivity (lowest oxide thickness)[18]. The Ag nanodendrites grow inside the pores and forms biocompatible layer with enhanced bacterial killing properties. Moreover the Ag nanoparticles fixed inside the pores are protected against their removing during implant handling.

The main reason for implantation failure is local inflammation caused by reaction of tissue on the introduction of a foreign body. Cascade of negative events in the organism may be accompanied by the activity of pathogenic microorganisms. Disrupted relationship between host and microflora resulting in disease of the oral structures, mainly include dental caries, gingivitis or periodontitis or peri-implantitis[22]. Peri-implantitis around osseointegrated implants initiated by bacterial infection affect the soft and hard tissues and can lead to bone damage[23]. Some bacteria attached to implanted restorative materials form the biofilm that protects them against antibiotic treatment, leading to antibiotic-resistant periprosthetic infection[24].

Nanostructured biomaterials exhibit exceptional mechanical and surface features[25]. They can mimic the bones, by what are considered the next-generation materials[26]. The shape, spatial and chemical structure, roughness, and surface energy have impact on the adhesion and proliferation of cells. However, the rough surface of the dental implants can promote the colonization by pathogens[27]. Therefore, the next challenge for the dental implantology is an indication of materials, which will limit the risk of colonization with pathogenic microorganisms. Due to proven antibacterial properties Ag nanoparticles are intensively investigated as a material used in dental applications, mainly as dental fillings or implants[28][29]. Strong antibacterial features of Ag are associated with the ability of this metal to interact with sulfhydryl groups of proteins as well as with DNA[30]. Ag nanoparticles have activity against Gram-positive and Gram-negative bacteria, fungi, and viruses[31][32].

However, the use of Ag nanoparticles carries the risk of cytotoxicity causing tissue destruction and local inflammation. Certain Ag compounds exhibited nanotoxicity for HGF[33]. Ag nanoparticles by induction of oxidative stress can be also mutagenic for cells[34]. However, the use of Ag nanoparticles coated on a titanium core seems to be very useful in dental technology. Mei *et al.*[35] used processes of anodization and Ag plasma immersion ion implantation for production of Ti nanotubes containing Ag nanoparticles. The technology used has confirmed the high biocompatibility of Ti and the antimicrobial effect of the Ag particles, enhanced by the large depth of the deposited Ag. Other group indicated an anti-biofilm effect of Ag nanoparticles immobilized on Ti against Staphylococcus epidermidis. This activity was related with inhibition of bacteria adhesion and down regulation of transcription of the intercellular adhesion operon (icaAD)[36]. Liu *et al.* proved that Ag nanoparticles promoted proliferation and maturation of preosteoblasts and induced osteogenesis of rat bones, accompanied with suppression of bacterial survival[37].

In our study we examined the biocompatibility of the samples of anodized Ti modified with deposited Ag nanodendrites. Material samples were evaluated based on contact with the HGF and osteoblasts (U-2 OS) cell lines in vitro. Evaluation of fibroblasts and osteoblasts growth served as a model showing the behavior of cells of the soft and hard tissue in contact with the tested material. Cytocompatibility of tested nanosilver-titanium composites was compared to conventional microcrystalline Ti, which served as a reference material. Studies on dental implants are mainly focused on assessing the interaction between materials and bone cells as well as soft tissue, because the process of osseointegration may be disturbed by anomalies and infections of soft tissue[38]. To evaluate the cytocompatibility of the tested samples, in vitro characterization tests in static condition were performed. Samples were tested for cytotoxicity with MTT assay. Proliferative activity of cells directly growing on samples was determined with cytofluorimeter (determination of the percentage of cells in S-phase of the cell cycle stained with PI). Furthermore, cells growing on the tested samples were stained with thiazole orange and visualized in a fluorescence microscope. Our analyses indicated that the nanocrystalline Ti modified with Ag has a higher degree of biocompatibility in comparison with the unmodified reference material, which was micro-crystalline Ti. Cells used in the tests showed a higher relative viability potential in contact with tested materials, than cells in contact with the surface of control samples. Both, fibroblasts and osteoblasts showed numerous NOR within the cell nuclei what may serve as an evidence of increased proliferative activity of tested cells. This observation was supported by numerous cells in the process of mitosis. Moreover, cytometric analysis confirmed these results, particularly when tested cells showed the stable level of proliferation activity and lower rate of mortality.

Modification of the titanium through the anodic oxidation process at high voltages with additional deposition of nanosilver can significantly change the properties of the reference material, which can improve its biocompatibility, but also can introduce new quality. The development of such material is a major goal of modern materials science. Anodized titanium modified with deposited silver nanodendrites, the material presented in this report significantly brings us closer to this target.

Acknowledgements

The work has been financed by Polish Ministry of Science and Higher Education within statutory activity.

Compliance with Ethical Standards

Conflict of interest Authors declare no conflict of interest.

Open Access: This article is distributed under the terms of the Creative Commons

Attribution 4.0 Inter-national License (http://creativecommons.org/licenses/by/4.0/), which permits unrestricted use, distribution, and reproduction in any medium, provided you give appropriate credit to the original author(s) and the source, provide a link to the Creative Commons license, and indicate if changes were made.

Source: Kaczmarek M, Jurczyk K, Koper J K. In vitro biocompatibility of anodized titanium with deposited silver nanodendrites[J]. Journal of Materials Science, 2016, 51(11):5259–5270.

References

[1] Sul YT (2003) The significance of the surface properties of oxidized titanium to the bone response: special emphasis on potential biochemical bonding of oxidized titanium implant. Biomaterials 24:3893–3907.

[2] Oh H-J, Lee J-H, Kim Y-J, Suh S-J, Lee J-H, Chi Ch-S (2008) Surface characteristics of porous anodic TiO_2 layer for biomedical applications. Mater Chem Phys 109:10–14.

[3] Adamek G, Jakubowicz J (2010) Mechanoelectrochemical synthesis and properties of porous nano-Ti-6Al-4V alloy with hydroxyapatite layer for biomedical applications. Electrochem Commun 12:653–656.

[4] Webster TJ, Ejiofor JU (2004) Increased osteoblast adhesion on nanophase metals: Ti, Ti6Al4V, and CoCrMo. Biomaterials 25:4731–4739.

[5] Jakubowicz J (2008) Formation of porous TiOx biomaterias in H_3PO_4 electrolytes. Electrochem Commun 10:735–739.

[6] Jakubowicz J, Adamek G, Jurczyk MU, Jurczyk M (2012) 3D surface topography study of the biofunctionalized nanocrystalline Ti-6Zr-4Nb/Ca-P. Mater Charact 70:55–62.

[7] Lee J-H, Kim S-E, Kim Y-J, Chi Ch-S, Oh H-J (2006) Effects of microstructure of anodic titania on the formation of bioactive compounds. Mater Chem Phys 98:39–43.

[8] Yang B, Uchida M, Kim H-M, Zhang X, Kokubo T (2004) Preparation of bioactive metal via anodic oxidation treatment. Biomaterials 25:1003–1010.

[9] Huang P, Xu K-W, Han Y (2005) Preparation and apatite layer formation of plasma electrolytic oxidation film on titanium for biomedical application. Mater Lett 59:185–189.

[10] Huang P, Wang F, Xu K, Han Y (2007) Mechanical properties of titania prepared by plasma electrolytic oxidation at different voltages. Surf Coat Technol 201:

5168–5171.

[11] Yang W-E, Hsu M-L, Lin M-Ch, Chen Z-H, Chen L-K, Huang H-H (2009) Nano/submicron-scale TiO_2 network on titanium surface for dental implant application. J Alloys Compd 479:642–647.

[12] Macak JM, Sirotna K, Schmuki P (2005) Self-organized porous titanium oxide prepared in Na_2SO_4/NaF electrolytes. Electrochim Acta 50:3679–3684.

[13] Tsuchiya H, Macak JM, Taveira L, Balaur E, Ghicov A, Sirotna K, Shmuki P (2005) Self-organized TiO_2 nanotubes prepared in ammonium fluoride containing acetic acid electrolytes. Electrochem Commun 7:576–580.

[14] Le Guehennec L, Soueidan A, Layrolle P, Amouriq Y (2007) Surface treatments of titanium dental implants for rapid osseointegration. Dent Mater 23:844–854.

[15] Yang Ch-L, Chen F-L, Chen S-W (2006) Anodization of the dental arch wires. Mater Chem Phys 100:268–274.

[16] Rai M, Yadav A, Gade A (2009) Silver nanoparticles as a new generation of antimicrobials. Biotechnol Adv 27:76–83.

[17] Chen X, Schleusener HJ (2008) Nanosilver: a nanoproduct in medical application. Toxicol Lett 176:1–12.

[18] Jakubowicz J, Koper JK, Adamek G, Połomska M, Wolak J (2015) Silver nano-trees deposited in the pores of anodically oxidized titanium and Ti scaffold. Int J Electrochem Sci 10:4165–4172.

[19] Koper JK, Jakubowicz J (2014) Correlation of wettability with surface structure and morphology of the anodically oxidized titanium implants. J Biomater Tissue Eng 4:459–464.

[20] Koper JK, Jakubowicz J (2015) Corrosion resistance of porous titanium surface prepared at moderate and high potentials in H_3PO_4/HF electrolytes. Prot Metal Phys Chem Surf 51:295–303.

[21] Wei Y, Chen Y, Ye L, Chang P (2011) Preparation of dendritic-like Ag crystals using monocrystalline silicon as template. Mater Res Bull 46:929–936.

[22] Allaker RP, Memarzadeh K (2014) Nanoparticles and the control of oral infections. Int J Antimicrob Agents 43:95–104.

[23] Mombelli A (2000) Microbiology and antimicrobial therapy of peri-implantitis. Periodontology 2002(28):177–189.

[24] Stewart PS, Costerton JW (2001) Antibiotic resistance of bacteria in biofilms. Lancet 358:135–138.

[25] Allaker RP, Ren G (2008) Potential impact of nanotechnology on the control of infectious diseases. Trans R Soc Trop Med Hyg 102:1–2.

[26] Webster TJ, Ejiofor JU (2004) Increased osteoblast adhesion on nanophase metals: Ti,

Ti6Al4V, and CoCrMo. Biomaterials 25:4731–4739.

[27] Burgers R, Hahnel S, Reichert TE, Rosentritt M, Behr M, Gerlach T *et al* (2010) Adhesion of Candida albicans to various dental implant surfaces and the influence of salivary pellicle proteins. Acta Biomater 6:2307–2313.

[28] Garcia-Contreras R, Argueta-Figueroa L, Mejia-Rubalcava C, Jimenez-Martinez R, Cuevas-Guajardo S, Sanchez-Reyna PA *et al* (2011) Perspectives for the use of silver nanoparticles in dental practice. Int Dent J 61:297–301.

[29] Jurczyk K, Adamek G, Kubicka MM, Jurczyk M (2015) Nanostructured titanium-10wt% 45S5 bioglass-Ag composite foams for medical applications. Materials 8:1398–1412.

[30] Lansdown AB (2006) Silver in health care: antimicrobial effects and safety in use. Curr Probl Dermatol 33:17–34.

[31] Sadhasivam S, Shanmugam P, Yun K (2010) Biosynthesis of silver nanoparticles by Streptomyces hygroscopicus and antimicrobial activity against medically important pathogenic microorganisms. Colloids Surf B Biointerfaces. 81:358–362.

[32] Lu L, Sun RW, Chen R, Hui CK, Ho CM, Luk JM *et al* (2008) Silver nanoparticles inhibit hepatitis B virus replication. Antivir Ther. 13:253–262.

[33] Park EJ, Yi J, Kim Y, Choi K, Park K (2010) Silver nanoparticles induce cytotoxicity by a Trojan-horse type mechanism. Toxicol In Vitro 24:872–878.

[34] Mei N, Zhang Y, Chen Y, Guo X, Ding W, Ali SF *et al* (2012) Silver nanoparticle-induced mutations and oxidative stress in mouse lymphoma cells. Environ Mol Mutagen 53:409–419.

[35] Mei S, Wang H, Wang W, Tong L, Pan H, Ruan C *et al* (2014) Antibacterial effects and biocompatibility of titanium surfaces with graded silver incorporation in titania nanotubes. Biomaterials 35:4255–4265.

[36] Qin H, Cao H, Zhao Y, Zhu C, Cheng T, Wang Q *et al* (2014) In vitro and in vivo anti-biofilm effects of silver nanoparticles immobilized on titanium. Biomaterials 35:9114–9125.

[37] Liu Y, Zheng Z, Zara JN, Hsu C, Soofer DE, Lee KS *et al* (2012) The antimicrobial and osteoinductive properties of silver nanoparticle/poly (DL-lactic-co-glycolic acid)-coated stainless steel. Biomaterials 33:8745–8756.

[38] Hench LL (1991) Bioceramics: from concept to clinic. J Am Ceram Soc 74:1487–1510.

Chapter 21

Influence of Structural and Textural Parameters of Carbon Nanofibers on Their Capacitive Behavior

Adam Moyseowicz, Agata Śliwak, Grażyna Gryglewicz

Department of Polymer and Carbonaceous Materials, Faculty of Chemistry, Wrocław University of Technology, Gdańska 7/9, 50-344 Wrocław, Poland

Abstract: Herringbone, platelet, and tubular carbon nanofibers (CNFs) were synthesized by catalytic chemical vapor deposition using methane, propane, and ethylene as carbon precursors. Alumina-supported nickel and iron catalysts were used for the syntheses. The resultant CNFs were characterized by scanning electron microscopy, transmission electron microscopy, and nitrogen sorption at 77K. The performance of a CNF-based supercapacitor working in 6mol L^{-1} KOH was analyzed using cyclic voltammetry, galvanostatic charge/discharge, and electrochemical impedance spectroscopy techniques. The Brunauer-Emmett-Teller (BET) surface area of the CNFs ranged between 150 and 296m^2 g^{-1}. An increase in the CNF diameter was accompanied by a decrease in the BET surface area. A comparison of the porous textures and the structure types of the CNFs demonstrated that the performance of the CNF-based supercapacitor is enhanced primarily by the exposed edges of the graphitic layers on the CNF surface, followed by the specific surface area. Among the studied CNFs, the highest capacitance value, 26F g^{-1} at 0.2A g^{-1}, was obtained for the platelet-type CNFs. Tubular CNFs exhibited the lowest capacitance value, which increased from 4 to 33F g^{-1} at 0.2A g^{-1} upon air treatment at 450°C. The presence of exposed graphitic edges on the air-treated CNT surface and an increase in the specific surface area are considered to be responsible for the enhancement of the capacitor performance.

1. Introduction

The increasing demand for electrical devices has stimulated the intensive development of electric double-layer capacitors (EDLCs) and lithium-ion batteries (LiBs), which are used in a wide range of industry and daily life applications[1]–[3]. Carbon materials, including activated carbons, carbon aerogels, carbon nanofibers (CNFs), and graphene materials, have been extensively investigated for efficient and long-lasting energy storage in EDLCs and LiBs[3]–[7]. In an EDLC, electrical energy is stored by the electrostatic accumulation of charge in the electric double-layer at the electrode/electrolyte interface, thus making it possible to achieve fast charge-discharge rates and high power values. EDLC performance is strongly related to the electrode material, and can be enhanced by selecting carbons with a high surface area and good electrical conductivity[8]. Generally, the higher the surface area of the carbon material accessible to the electrolyte ion, the higher capacitance value is obtained[9]–[13]. However, the capacitance becomes almost constant for carbons with BET surface area higher than $1800 m^2 g^{-1}$[14][15].

CNFs have been reported to form unique nanostructures, depending on the arrangement of the graphene layers along the fiber axis. Typically, three structural types of CNFs are distinguished based on the angle of the graphene layers with respect to the fiber axis; herringbone, platelet, and tubular CNFs[16]. It has been widely accepted to describe tubular CNFs with graphene layers parallel to the filament axis as carbon nanotubes (CNTs).

The exceptional physicochemical properties of CNFs, including CNTs, such as a large area of exposed surface for electrolyte ions, high electrical conductivity, and chemical stability, underpin their potential applications for supercapacitors[1][17]. However, pure CNFs/CNTs as capacitor electrode materials can only supply relatively low capacitance values, *i.e.*, from 5 to 30F g^{-1}, due to their moderate surface area (up to $400 m^2 g^{-1}$)[18]–[21]. It has been reported that CNTs with structural defects exhibit the higher gravimetric capacitance than purified and defect-free ones[18][22].

The capacitance of CNFs/CNTs can be enhanced through development of their surface area by chemical activation[23] and air treatment[24] or the introduction of oxygen and nitrogen functional groups into their surface to provide pseudocapacitance[25]-[27]. Their superior conductive properties make CNFs/CNTs very attractive as percolating additives in activated carbon-based electrodes of supercapacitors[28]-[31], conducting supports of metallic oxides[32][33] and excellent components of carbon/carbon composites[17][34][35]. A very high electrical conductivity of 10^3–10^4 S cm^{-1} has been reported for multi-walled CNTs[36].

Kim et al.[18] studied the capacitive behavior of well-defined CNF surfaces. They reported that the capacitance values for CNFs with exposed graphitic edges, such as herringbone and platelet, are several times higher than for tubular CNFs. This finding was explained by more efficient charging of the graphitic edges at the surface of the carbon materials under electrochemical polarization compared with the basal plane surface. Unfortunately, the impact of the specific surface area on the capacitance of CNFs has been omitted. The diameter of CNFs may be another factor influencing the electrochemical performance[18].

The aim of this work was to determine which feature of CNFs has the predominant impact on their capacitive behavior. The interplay of the graphitic alignment, the specific surface area, and the nanofiber diameter in influencing the performance of the CNF-based electrochemical capacitor was discussed. For this purpose, CNFs of various structures (herringbone, platelet, and tubular) were synthesized by the CVD method using different carbon precursors and catalysts. The capacitive behavior of the resulting CNFs was tested in a two-electrode cell by cyclic voltammetry, galvanostatic charging/discharging, and electrochemical impedance spectroscopy (EIS) techniques in 6L^{-1} KOH as electrolyte.

2. Experimental

2.1. Preparation of Alumina-Supported Catalysts

Two catalysts, an alumina-supported nickel catalyst and an alumina-supported iron catalyst, were prepared by the incipient wetness method. Alumina (particle size < 50nm) was obtained from Sigma-Aldrich. The preparation of Ni/ Al_2O_3 using an aqueous solution of $Ni(NO_3)_2·6H_2O$ (Across, 98%) was reported in previous work[25]. The Fe/Al_2O_3 catalyst was prepared using an aqueous solution of $Fe(NO_3)_3·H_2O$ (Sigma-Aldrich, 98%). The amount of Ni and Fe precursor was adjusted to achieve 10wt% of the metal in the catalyst. The as-prepared samples were dried at 110°C and subjected to calcination in the air at 350°C for 4h.

2.2. CCVD Process

The CNFs of different structures were synthesized by CCVD using Ni/Al_2O_3 and Fe/Al_2O_3 as catalysts and methane, propane, and ethylene as carbon precursors. The herringbone CNFs were synthesized over Ni/Al_2O_3 catalyst using different carbon sources, *i.e.*, methane (HCNF1) and propane (HCNF2). The same catalyst was applied for the growth of platelet CNFs (PCNF) from propane. The Fe/Al_2O_3 catalyst was used for the synthesis of tubular CNFs with ethylene as a carbon precursor. The CCVD process conditions, including the temperatures of catalyst reduction and CNF synthesis, are given in Table S1. The CCVD processes were performed in a conventional horizontal furnace. Two hundred milligrams of the catalyst was spread in the bottom of a quartz boat and placed in the center of the quartz tube. Prior to the CNF growth, the Ni/Al_2O_3 catalyst was reduced for 2h under a hydrogen flow (150ml min^{-1}) at 550°C. Subsequently, a mixture of methane and hydrogen at a volume ratio of 1:1 (150ml min^{-1}) was introduced into the reactor for 1 h at 650°C and cooled to room temperature in a nitrogen atmosphere (150ml min^{-1}). The CCVD processes using propane were performed under the same conditions, including the reduction step, except for the synthesis temperature, which was 500°C for HCNF2 and 450°C for PCNF. The synthesis of tubular CNFs (labeled as CNT) was performed using the Fe/Al_2O_3 catalyst without the reduction step. The mixture of C_2H_4 and H2 (1:3, v/v, 150ml min^{-1}) was introduced into the reactor for 1h at 650°C.

The removal of the catalysts from the as-received CNFs was performed by hydrofluoric acid treatment for 2h at room temperature. Then, the CNFs were filtered, washed with distilled water, and finally dried at 110°C for 2h.

2.3. Oxidative Treatment

Tubular CNFs were treated with air (330ml min^{-1}) in a quartz boat in the horizontal reactor at 450°C (CNT450) for 1h. Afterwards, the reactor was cooled to room temperature under a nitrogen flow (150ml min^{-1}).

2.4. CNF Characterization

The CNFs were observed with an EVO LS13 Zeiss scanning electron microscope (SEM). High-resolution trans-mission electron microscope (HRTEM) images were obtained using a FEI Tecnai G^2 20X-TWIN microscope, operating at an acceleration voltage of 200kV. A few drops of CNF suspension in methanol were dropped onto a copper microgrid with a holy carbon thin film. The diameters of CNFs were estimated by counting about 100 nanofibers on the TEM images. The porous texture characteristics of the materials were determined by N2 sorption at 77K by using a NOVA 2000 gas sorption analyzer (Quantachrome). Prior to measurements, the sample was outgassed overnight at 300°C. The specific surface area (SBET) was calculated from the Brunauer-Emmett-Teller (BET) equation. The amount of nitrogen adsorbed at a relative pressure of $p/p_0 = 0.96$ was employed to determine the total pore volume (V_T). The micropore volume (V_{DR}) was estimated from the Dubinin-Radushkevich equation. The mesopore volume (V_{mes}) was determined as the difference between the total pore volume and the micropore volume. The pore size distribution (PSD) was obtained by means of the quenched solid density functional theory (QSDFT) method. The elemental compositions of the pristine and air-treated CNTs were determined by X-ray photoelectron spectroscopy (XPS) using a PHI 5000 VersaProbe.

2.5. Electrochemical Measurements

The electrodes were composed of 90wt% of CNFs and 10wt% of polyvinylidene fluoride (PVDF) as a binder. The electrodes were in the form of pellets with a geometric surface area of 0.9cm^2 and a thickness of approximately 0.2mm. Two-electrode symmetric capacitors were assembled in a Swagelok system with pellets of comparable mass (8mg–12mg). The measurements were performed in 6L^{-1} KOH aqueous solution using gold current collectors to avoid corrosion and to preserve comparable experimental conditions, using a potentiostat-galvanostat VMP3 Biologic in a voltage range of 0–0.8V. The electrochemical properties of CNFs were determined by cyclic voltammetry at a voltage scan rate of 1–100mV s^{-1} and galvanostatic cycling at current densities in the range 0.2–20A g^{-1}. The specific capacitance was expressed in farads per mass of active material in one electrode. EIS measurements were performed under open circuit potential in a frequency range from 100kHz to 10mHz at amplitude of 5mV.

The specific capacitance values (C, F g^{-1}) were calculated from the galvanostatic discharge curves and the CV curves using Eqs. (1) and (2), respectively.

$$C = \frac{It}{m\Delta V} \tag{1}$$

$$C = \frac{\left(\int IdV\right)}{vm\Delta V} \tag{2}$$

where I (A) is the response current, t (s) is the discharge time, v (V s^{-1}) is the scan rate, ΔV (V) is the potential window and m (g) is the mass of the active material in one electrode.

3. Results and Discussion

3.1. Structure of CNFs

HRTEM images of the as-grown carbon nanostructures produced by the decomposition of methane, propane, and ethylene over the alumina-supported Ni and Fe catalysts are shown in **Figure 1**. The CNFs were obtained with a yield in the range of 1.7–5.1gCNF/gcat (Table S1). SEM images of the synthesized CNFs are shown in Figure S1. Various types of CNF structure were obtained depending on the catalyst, synthesis temperature, and carbon source. CNFs with herringbone structure without hollow core were synthesized over nickel catalyst using both methane and propane as carbon precursors, *i.e.*, HCNF1 and HCNF2, respectively [**Figure 1(a), (b)**]. These CNFs have graphene layers aligned at an angle of less than 90° to the fiber axis. The diameter of HCNF1 varied from 20 to 60nm. HCNF2 was

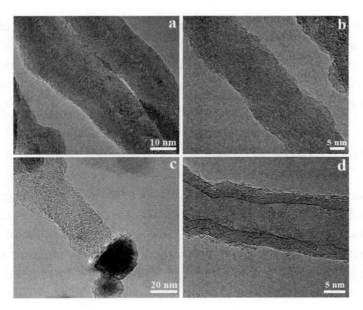

Figure 1. HRTEM images of the pristine HCNF1 (a), HCNF2 (b), PCNF (c), and CNT (d).

characterized by thinner nanofibers with a diameter not exceeding 40nm. The significant differences in diameter between HCNF1 and HCNF2 are related to the temperature applied for their synthesis. Methane requires a higher temperature than propane due to its higher decomposition energy to carbon and hydrogen (37.8 and 26.0kJ^{-1} H$_2$ for methane and propane, respectively)[37]. Increased reaction temperatures favor the migration of catalyst particles on the support surface, which results in their aggregation, leading to the growth of thicker CNFs[38]. It is well known that the diameter of nanofibers is controlled by the size of the catalyst particle responsible for their growth[39]. For propane, the decrease in temperature from 500°C to 450°C led to a change in the CNF structure from herringbone (HCNF2) to platelet (PCNF), maintaining the same range of CNF diameters. In the platelet structure, the graphene layers are aligned perpendicularly to the fiber axis [**Figure 1(c)**]. Our study clearly showed that both the diameter and the structure of grown CNFs are strongly dependent on the temperature of the CCVD process. The synthesis temperature has an indirect influence on the CNF structure through various diffusion rates of carbon into the catalyst particles and different orientations of the graphene layers precipitated on the metal nanoparticle[38]. The CNTs obtained over Fe/Al$_2$O$_3$ catalyst using ethylene as a carbon source were the thinnest among the synthesized nanocarbons, with diameters of

10–20nm, **Figure 1(d)**. A residual amount of amorphous carbon was observed on the CNT surface. The parallel orientation of the graphene layers and the lower diameter arose from the changes to the catalyst and the carbon precursor. The use of an iron catalyst promotes the growth of CNTs, whereas nickel favors the formation of herringbone CNFs[40][41].

3.2. Porous Texture of CNFs

Considering the strong relationship between the porosity development of active electrode material and the amount of charge accumulated in the electric-double layer, the porous texture of CNFs was determined by N_2 sorption at 77K. **Figure 2(a)** shows the nitrogen adsorption-desorption isotherms for synthesized CNFs. The calculated textural parameters are given in **Table 1**. The diameters of CNFs estimated based on the TEM examination are also included. The presence of

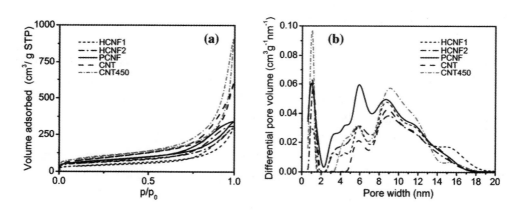

Figure 2. Adsorption-desorption isotherms of nitrogen at 77K (a) and QSDFT pore size distribution curves (b) of synthesized CNFs and CNTs.

Table 1. Porous texture characteristics of CNFs and CNTs determined by N_2 sorption at 77K.

Sample	CNF diameter (nm)	S_{BET} (m² g⁻¹)	V_T (cm³ g⁻¹)	V_{mes} (cm³ g⁻¹)	V_{DR} (cm³ g⁻¹)
HCNF1	20–60	150	0.417	0.366	0.051
HCNF2	15–40	223	0.442	0.359	0.083
PCNF	20–40	251	0.486	0.398	0.088
CNT	10–20	296	0.805	0.691	0.114
CNT450	<20	380	0.918	0.773	0.145

a large hysteresis loop on the adsorption-desorption isotherms for all studied nanocarbons reveals their mesoporous nature. This feature is clearly displayed in the PSD curves [**Figure 2(b)**]. It was revealed that all CNFs/CNTs samples contained mesopores with a width <20nm. The CNFs, including CNTs, had BET surface areas between 150 and 296m^2 g^{-1} (**Table 1**). The total pore volume varied from 0.417 to 0.805cm^3 g^{-1} with a mesopore contribution of 0.77–0.88. It was revealed that the differences in the textural parameters are related to the diameter of the CNFs. HCNF1 was characterized by the largest diameters, ranging from 20 to 60nm, and the lowest development of porosity among the studied nanocarbons. In the case of CNFs with the same structure type but smaller diameter (HCNF2), increased porosity development was observed. The BET surface area of HCNF1 was substantially higher than for HCNF2 (223 and 150m^2 g^{-1}, respectively). PCNF, which was characterized by a diameter comparable to HCNF2, exhibited a slightly higher specific surface area (251m^2 g^{-1} for PCNF and 223m^2 g^{-1} for HCNF2, respectively). The difference in the BET surface area between PCNF and HCNF2 could be explained by their different roughness of the outer surface due to structural defects and imperfections on the surface of the nanofibers[42][43]. However, TEM observation does not provide a clear evidence for this finding. The values of other textural parameters of PCNF were also higher than for HCNF2. CNT was characterized by both the highest BET surface area (296m^2 g^{-1}) and total pore volume (0.805cm^3 g^{-1}) and the smallest diameter (approximately 10–20nm) among the synthesized carbon nanomaterials. The results clearly show a tendency of increasing porosity development with decreasing CNF diameters.

3.3. CNF Performance in Capacitors

Figure 3 shows the voltammograms of CNFs recorded using a two-electrode cell at different scan rates in 6L^{-1} KOH. At a scan rate of 10mV s^{-1}, the CV curves were close to a rectangular shape, indicating an ideal capacitor behavior, including quick charge propagation and fast charge/discharge kinetics [**Figure 3(a)**]. Excellent charge propagation in an electrical double layer was observed for all samples, even at the higher scan rate of 100mV s^{-1} [**Figure 3(b)**]. This result confirms that CNFs exhibit very good electrical conductivity properties. However, the capacitance values were relatively low, ranging from 4 to 26F g^{-1} at 10mV s^{-1} and from 3 to 22F g^{-1} at 100mV s^{-1}, due to the poorly developed surface area of the CNFs. The lowest values were obtained for tubular CNFs despite their having

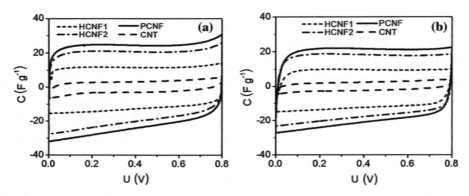

Figure 3. Cyclic voltammograms of CNFs recorded in 6L^{-1} KOH at scan rates of 10mV s^{-1} (a) and 100mV s^{-1} (b).

the most developed surface area among the studied nanomaterials. The capacitance of the CNF-based capacitors increased in the sequence: CNT < HCNF1 < HCNF2 < PCNF. Notably, considerably higher values were recorded for herringbone and platelet CNFs, with open graphitic edges on their surface, compared to the tubular type, for which the basal planes are exposed to electrolyte ions. The results obtained are in line with the finding of Kim et al.[18] that the edge surfaces of CNFs promote charge storage. Kim et al. synthesized also a series of CNFs with different graphitic layers alignment. Moreover, platelet CNFs were subjected to graphitization at 2800°C, which resulted in the closure of opened graphitic edges and formation of domelike basal planes. Lower capacitance values were reported for both tubular CNF and graphitized PCNF compared with herringbone- and platelet-type CNFs. In our work, we have also revealed that an increase in the BET surface area of CNFs was not followed by an increase in the capacitance value, high-lighting the importance of the structural alignment of the graphene layers in the nanostructured carbon materials.

For the further characterization of CNF-based capacitors, galvanostatic charging/discharging was applied in the range of a current density between 0.2 and 20A g^{-1}. The variations in the capacitance value with increasing current load for the tested samples in two-electrode cells are presented in **Figure 4**. The galvanostatic measurements confirmed the tendency shown by cyclic voltammetry. The capacitance value ranged from 4 for CNT to 26F g^{-1} for PCNF at 0.2A g^{-1}. PCNF and HCNF2 exhibited superior capacitive performance (26 and 21F g^{-1} at 0.2A g^{-1} and 21 and 18F g^{-1} at 1A g^{-1}, respectively) compared with HCNF1 and CNT.

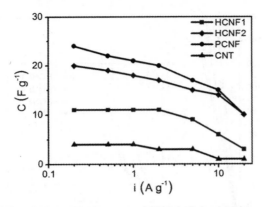

Figure 4. Specific capacitance versus current load for the CNF-based capacitors working in 6 L^{-1} KOH.

The higher capacitances of PCNF and HCNF2 are attributed to porosity development (251 and 223m^2 g^{-1}, respectively). HCNF1 exhibited lower capacitance values (20 at 0.2A g^{-1} and 11F g^{-1} at 1A g^{-1}) because of its lower surface area (150m^2 g^{-1}) as consequence of the wider nanofiber diameters. The worst capacitive behavior was again demonstrated by CNT. Although CNT had a larger specific surface area than PCNF (296 vs. 251m^2 g^{-1}), its electrical charge storage capability was limited (4 vs. 26F g^{-1}). This result is probably related to the exposure of the basal plane surface to electrolyte ions, which is more favorable for conducting electrons than for accumulating them. The results suggest that the capacitance of CNFs is controlled more by their surface structure than by their specific surface area.

The impedance spectroscopy (EIS) technique was employed to assess the electrochemical frequency behavior of CNF-based electrodes. **Figure 5** shows Nyquist plots for the CNFs measured in the range of 0.01–100kHz in a two-electrode cell. Nyquist plots are commonly used to analyze EIS data and reflect conductive properties of carbons. The equivalent series resistance (ESR) determined from the Z' axis intercept of the Nyquist plot was found to be comparable for all CNFs (0.31–0.61Ω); this is a combinational resistance of electrode materials, electrolyte and contact at the active material/current collector interface. A nearly vertical line at low-frequency region observed for all the studied materials indicates a very good diffusion at the interface between the electrolyte and electroactive material, confirming their excellent conductive properties[44]. However, the Nyquist plots at high-frequency region revealed a substantial difference in the

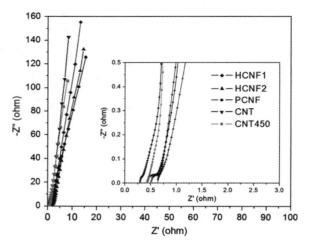

Figure 5. Nyquist plots of different CNFs measured in a two-electrode cell in 6L^{-1} KOH solution. The inset shows the expanded highfrequency region of the plots.

behavior between CNFs with open graphitic edges and CNT. In contrast to CNT, a semicircle is observed for a platelet and herring-bone CNF-based electrode, which indicates a higher interfacial electron-transfer resistance, probably because of exposed graphitic edges. A low interfacial electron-transfer resistance observed for the CNT-based electrode favors the capacitance retention at high current densities. An increase of the current load from 10 to 20A g^{-1} resulted in a drop of the capacitance value by only 8% for CNT, but much more for HCNF2 (29%), PCNF (33%), and HCNF1 (50%).

Our preliminary study showed a good linear correlation between the capacitance and the BET surface area for CNFs with exposed graphene layer edges in their structure, *i.e.*, herringbone and platelet CNFs (Figure S2). Nevertheless, this relationship should be supported by the examination of more CNF samples. No such relationship was observed by for CNFs produced from methane and acetylene on an alumina-supported nickel catalyst[19].

3.4. Air Treatment of CNT

CNT was heated with air at 450°C to remove residual amorphous carbon from the surface. A weight loss of 3.6wt% was observed during air treatment. HRTEM examination indicated that the external graphene layers of CNT450 were slightly deformed and even interrupted, although the core structure remained unchanged [**Figure 6(a)**]. Moreover, the air treatment resulted in a noticeable reduction in the diameter of the air-treated CNT. The BET surface area increased from 296 to 380m^2 g^{-1} due to air treatment. Both HRTEM and PSD results obtained for the air-treated CNFs explain their enhanced micro- and mesoporosity compared with the pristine CNTs, which is due to opening of basal external graphene layers and following reduction of nanofiber diameters during air treatment [**Table 1**; **Figure 2(b)**].

The results of electrochemical measurements showed better capacitance behavior for air-treated CNT than pristine CNT [**Figure 6(b), (c)**]. The almost ideal rectangular shape of the CV curve confirms the very good conductivity properties of the air-treated CNT despite some defects in the external basal planes. At a scan

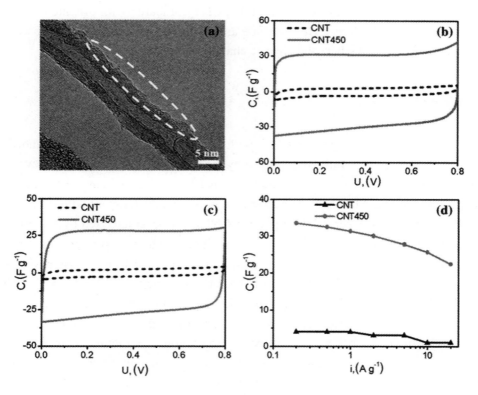

Figure 6. HRTEM image of CNT air-treated at 450°C (a). Deformation of the external graphene layers is indicated by a circle. Cyclic voltammograms of CNT and air-treated CNT capacitors operating in 6L^{-1} KOH at scan rates of 10mV s^{-1} (b) and 100mV s^{-1} (c) and specific capacitance versus current load (d) in twoelectrode cells.

rate of 10mV s^{-1}, the capacitance value of CNT increased from 4 to 30F g^{-1} for CNT450, **Figure 6(b)**. Moreover, at a much higher scan rate of 100mV s^{-1}, the capacitance behavior of CNT450 remained nearly unchanged with very good charge propagation, **Figure 6(c)**. The results of galvanostatic charge/discharge measurements demonstrated a similar tendency to the results of cyclic voltammetry, **Figure 6(d)**. Moreover, the comparison of the EIS spectra for CNT and CNT450 samples (**Figure 5**) could suggest an insignificant impact of air treatment on the conductive properties of CNT. The results of the HRTEM examination of CNT450 support this finding, revealing only a small imperfection on the external surface and preserved core structure. It should be added that the contribution of oxygen functional groups to the overall capacitance, due to pseudo-faradaic redox reactions, can be neglected. The XPS analysis revealed that the oxygen content in CNT450 was very low, equal 2.4 at% (Figure S3). Therefore, the enhancement of the capacitance properties of the air-treated CNT compared with pristine CNT can be explained by the larger BET surface area and the graphitic edges exposed on the nanofiber surface where the electric charge can be stored.

4. Conclusions

Various structure types of CNFs with different porous textures and nanofiber diameters were successfully synthesized by the CCVD method. The study revealed that the capacitance of CNFs is related to their BET surface area and the alignment of graphitic layers. Moreover, an increase in porosity development with decreasing CNF diameter was observed. The platelet CNFs were characterized by the highest capacitance values (26F g^{-1} at 0.2A g^{-1}), though not the highest surface area, among the studied nanostructured carbon materials. In turn, despite the most developed surface area, CNT exhibited the lowest capacitance values (4F g^{-1} at 0.2A g^{-1}). The results clearly showed that exposed graphitic layer edges are beneficial for charge accumulation in an EDLC. This finding was confirmed by the enhanced capacitance of air-treated CNT as a result of the formation of structural defects on the surface during oxidative treatment, which was accompanied by an increase in the BET surface area. The obtained results prove that the capacitance of CNFs is determined by both the graphene layer alignment and the specific surface area, the latter being related to the CNF diameter. This result suggests that the performance of CNF-based capacitors can be improved by using thinner CNFs with exposed graphitic edges on their surface.

Acknowledgements

This work was financed by a statutory activity subsidy from the Polish Ministry of Science and Higher Education for the Faculty of Chemistry of Wrocław University of Technology.

Open Access: This article is distributed under the terms of the Creative Commons Attribution 4.0 International License (http://creativecommons.org/licenses/by/4.0/), which permits unrestricted use, distribution, and reproduction in any medium, provided you give appropriate credit to the original author(s) and the source, provide a link to the Creative Commons license, and indicate if changes were made.

Source: Moyseowicz A, Śliwak A, Gryglewicz G. Influence of structural and textural parameters of carbon nanofibers on their capacitive behavior[J]. Journal of Materials Science, 2016, 51(7):3431–3439.

References

[1] Inagaki M, Konno H, Tanaike O (2010) Carbon materials for electrochemical capacitors. J Power Sources 195:7880–7903.

[2] Pandolfo AG, Hollenkamp AF (2006) Carbon properties and their role in supercapacitors. J Power Sources 157:11–27.

[3] Zhong H, Xu F, Li Z, Fu R, Wu D (2013) High-energy super-capacitors based on hierarchical porous carbon with an ultrahigh ion-accessible surface area in ionic liquid electrolyte. Nanoscale 5:4678–4682.

[4] Goriparti S, Miele E, De Angelis F, Di Fabrizio E, Zaccaria RP, Capiglia C (2014) Review on recent progress of nanostructured anode materials for Li-ion batteries. J Power Sources 257:421–443.

[5] Yang X, Huang H, Li Z, Zhong M, Zhang G, Wu D (2014) Preparation and lithium-storage performance of carbon/silica composite with a unique porous bicontinous nanostructure. Carbon 77:257–280.

[6] Yang X, Huang H, Zhang G, Li X, Wu D, Fu R (2015) Carbon aerogel with 3-D continuous skeleton and mesopore structure for lithium-ion batteries application. Mater

Chem Phys 149–150: 657–662.

[7] Choi H, Yoon H (2015) Nanostructured electrode materials for electrochemical capacitor applications. Nanomater 5:906–936.

[8] Frackowiak E, Beguin F (2001) Carbon materials for the electrochemical storage of energy in capacitors. Carbon 39:937–950.

[9] Shi H (1996) Activated carbons and double layer capacitance. Electrochim Acta 41:1633–1639.

[10] Lozano-Castello D, Cazorla-Amoros D, Linares-Solano A, Shiraishi S, Kurihara H, Oya A (2003) Influence of pore structure and surface chemistry on electric double layer capacitance in non-aqueous electrolyte. Carbon 41:1765–1775.

[11] Gryglewicz G, Machnikowski J, Lorenc-Grabowska E, Lota G, Frackowiak E (2005) Effect of pore size distribution of coal-based activated carbons on double layer capacitance. Electrochim Acta 50:1197–1206.

[12] Fernández JA, Tennison S, Kozynchenko O, Rubiera F, Stoeckli F, Centeno TA (2009) Effect of mesoporosity on specific capacitance of carbons. Carbon 47:1598–1604.

[13] Yang W, Feng Y, Xiao D, Yuan H (2015) Fabrication of microporous and mesoporous carbon spheres for high-performance supercapacitor electrode materials. Int J Energy Res. doi:10.1002/er.3301

[14] Raymundo-Piñero E, Kierzek K, Machnikowski J, Béguin F (2006) Relationship between the nanoporous texture of activated carbons and their capacitance properties in different electrolytes. Carbon 44:2498–2507.

[15] Barbieri O, Hahn M, Herzog A, Kötz R (2005) Capacitance limits of high surface area activated carbons for double layer capacitors. Carbon 43:1303–1310.

[16] Rodriguez NM, Chambers A, Baker RTK (1995) Catalytic engineering of carbon nanostructures. Langmuir 11:3862–3866.

[17] Lota G, Fic K, Frackowiak E (2011) Carbon nanotubes and their composites in electrochemical applications. Energy Environ Sci 4:1592–1605.

[18] Kim T, Lim S, Kwon K, Hong SH, Qiao W, Rhee CK *et al* (2006) Electrochemical capacitances of well-defined carbon surfaces. Langmuir 22:9086–9088.

[19] Hulicova-Jurcakova D, Li X, Zhu Z, de Marco R, Lu GQ (2008) Graphitic carbon nanofibers synthesized by the chemical vapor deposition (CVD) method and their electrochemical performances in supercapacitors. Energy Fuels 22:4139–4145.

[20] Portet C, Yushin G, Gogotsi Y (2007) Electrochemical performance of carbon onions, nanodiamonds, carbon black and multiwalled nanotubes in electrical double layer capacitors. Carbon 45:2511–2518.

[21] McDonough JR, Choi JW, Yang Y, La Mantia F, Zhang Y, Cui Y (2009) Carbon nanofiber supercapacitors with large areal capacitances. Appl Phys Lett 95:243109.

[22] Frackowiak E, Jurewicz K, Delpeux S, Beguin F (2001) Nanotubular materials for supercapacitors. J Power Sources 97–98:822–825.

[23] Frackowiak E, Delpeux S, Jurewicz K, Szostak K, Cazorla-Amoros D, Béguin F (2002) Enhanced capacitance of carbon nanotubes through chemical activation. Chem Phys Lett 361:35–41.

[24] Seo M-K, Park S-J (2010) Influence of air-oxidation on electric double layer capacitances of multi-walled carbon nanotubes electrodes. Curr Appl Phys 10:241–244.

[25] Śiwak A, Grzyb B, Ćwikła J, Gryglewicz G (2013) Influence of wet oxidation of herringbone carbon nanofibers on the pseudo-capacitance effect. Carbon 64:324–333.

[26] Ye J-S, Liu X, Cui HF, Zhang W-D, Sheu F-S, Lim TM (2005) Electrochemical oxidation of multi-walled carbon nanotubes and its application to electrochemical double layer capacitors. Electrochem Commun 7:249–255.

[27] Jurewicz A, Babeł K, Pietrzak R, Delpeux S, Wachowska H (2006) Capacitance properties of multi-walled carbon nanotubes modified by activation and ammoxidation. Carbon 44:2368–2375.

[28] Portet C, Taberna PL, Simon P, Flahaut E (2005) Influence of carbon nanotubes addition on carbon-arbon supercapacitor performances in organic electrolyte. J Power Sources 139:371–378.

[29] Gryglewicz G, Śliwak A, Béguin F (2013) Carbon nanofibers grafted on activated carbon as an electrode in high-power supercapacitors. ChemSusChem 6:1516–1522.

[30] Huang N, Kirk DW, Thorpe SJ, Liang C, Xu L, Li W *et al* (2015) Effect of carbon nanotube loadings on supercapacitor characteristics. Int J Energy Res 39:336–343.

[31] La Mantia F, Huggins RA (2013) Oxidation process on conducting carbon additives for lithium-ion batteries. J Appl Electrochem 43:1–7.

[32] Yu G, Xie X, Pan L, Bao Z, Cui Y (2013) Hybrid nanostructured materials for high-performance electrochemical capacitors. Nano Energy 2:213–234.

[33] Śliwak A, Gryglewicz G (2014) High-voltage asymmetric super-capacitors based on carbon and manganese oxide/oxidized carbon nanofiber composite electrodes. Energy Technol 2:819–824.

[34] Noked M, Okashy S, Zimrin T, Aurbach D (2012) Composite carbon nanotube/carbon electrodes for electrical double-layer super capacitors. Angew Chem Int Ed 51:1568–1571.

[35] Sivakkumar SR, Pandolfo AG (2014) Carbon nanotubes/amorphous carbon composites as high-power negative electrodes in lithium ion capacitors. J Appl Electrochem 44:105–113.

[36] Al-Saleh MH, Sundararaj U (2009) A review of vapour grown carbon nanofibers/polymer conductive composites. Carbon 47: 2–22.

[37] Muradov N (2001) Hydrogen via methane decomposition: an application for decarbonization of fossil fuels. Int J Hydrogen Energy 26:1165–1175.

[38] Romero A, Garrido A, Nieto-Márquez A, de la Osa AR, de Lucas A, Valverde JL (2007) The influence of operating conditions on the growth of carbon nanofibers on carbon nanofiber-supported nickel catalysts. Appl Catal A 319:246–258.

[39] De Yong KP, Geus JW (2000) Carbon nanofibers: catalytic synthesis and applications. Catal Rev 42:481–510.

[40] Martin-Gullon I, Vera J, Conesa JA, González JL, Merino C (2006) Differences between carbon nanofibers produced using Fe and Ni catalysts in a floating catalyst reactor. Carbon 44: 1572–1580.

[41] Yu Z, Chen D, Rønning M, Tøtdal B, Vrålstad T, Ochoa-Fernández E, Holmen A (2008) Large-scale synthesis of carbon nanofibers on Ni–Fe–Al hydrotalcite derived catalysts. Appl Catal A 338:147–158.

[42] Zhou JH, Sui ZJ, Li P, Chen D, Dai YC, Yuan WK (2006) Structural characterization of carbon nanofibers formed from different carbon-containing gases. Carbon 44: 3255–3262.

[43] Ramos A, Cameán I, García AB (2013) Graphitization thermal treatment of carbon nanofibers. Carbon 59:2–32.

[44] Kavian R, Vicenzo A, Bestetti M (2011) Growth of carbon nanotubes on aluminium foil for supercapacitors electrodes. J Mater Sci 46:1487–1493. doi:10.1007/s10853-010-4950-1.

Chapter 22
Investigation of Ammonium Diuranate Calcination with High-Temperature X-Ray Diffraction

R. Eloirdi, D. Ho Mer Lin, K. Mayer, R. Caciuffo, T. Fanghänel

European Commission, Joint Research Centre, Institute for Transuranium Elements, P.O. Box 2340, 76125 Karlsruhe, Germany

Abstract: The thermal decomposition of ammonium diuranate (ADU) in air is investigated using in situ high-temperature X-ray diffraction (HT-XRD), thermogravimetry and differential thermal analysis. Data have been collected in the temperature range from 30°C to 1000°C, allowing the observation of phase transformation and the assessment of the energy changes involved in the calcination of ADU. The starting material $2UO_3 \cdot NH_3 \cdot 3H_2O$ undergoes a process involving several endothermic and exothermic reactions. In situ HT-XRD shows that amorphous UO_3 is obtained after achieving complete dehydration at 300°C, and denitration at about 450°C. After cooling from heat treatment at 600°C, a crystalline UO_3 phase appears, as displayed by ex situ XRD. The self-reduction of UO_3 into orthorhombic U_3O_8 takes place at about 600°C, but a long heat treatment or higher temperature is required to stabilise the structure of U_3O_8 at room temperature. U_3O_8 remains stable in air up to 850°C. Above this temperature, oxygen losses lead to the formation of U_3O_{8-x}, as demonstrated by subtle changes in the diffraction pattern

459

and by a mass loss recorded by TGA.

1. Introduction

Ammonium diuranate $(NH_4)_2U_2O_7$ (ADU), once used to create coloured glazes in ceramics, is the most prominent chemical compound among the uranium ore concentrates, often referred to as "yellow cake": ADU plays also an important role in the fabrication of uranium oxide fuel. For this reason, ADU is of great interest for nuclear forensics, a relatively young discipline that aims at providing hints on the history and intended use of nuclear materials of unknown origin, on the basis of certain measurable parameters that, as fingerprints, are characteristics of the material, of the place where it was produced and of the route followed for its transformation[1]–[6].

The term "ammonium diuranate" is actually a misnomer, but it remains of common use. Controversies over the composition and structure of ADU have sparked numerous studies[7]–[13]. Its morphology has also attracted a great deal of attention, as it is well established that the characteristics of precursor materials are inherited into the final product[14][15]. Laboratory-scale preparations of ADU are usually based on the reaction of UO_3 or uranyl nitrate with water and ammonia[8][16], or by the reaction between hydrated UO_3 with liquid or gaseous ammonia[11][16]. On the other hand, industrial productions are based on two routes, either via uranyl nitrate with aqueous or gaseous ammonia, or via the reaction of uranyl fluoride with ammonium hydroxide[15]. Industrial processes often include drying the wet concentrate at 120°C–400°C or calcinations at 400°C–850°C, resulting in the possibility of mixtures. Since the properties of the final product depend on the starting material, it is important to understand the decomposition steps and the intermediate products of the ammonium diuranate. Some studies have been previously reported, however the method of preparation and used techniques were different. For instance, in the report by Price and Stuart[17], nitrate-free ADU was prepared, and the thermal decomposition was monitored by TG with accompanying infrared spectroscopy data. As mentioned by Woolfrey[16], the thermal decomposition of ammonium uranates studied by some authors was not in complete agreement. In more recent publications, scanning electron microscopy and ex situ XRD were used to study calcined ADU, but only for a few selected temperatures[18], deduced from to exothermic and endothermic peaks observed by TG/DTA analyses and characteristic of the intermediate products of the reaction.

In this study, we have applied high-temperature X-ray diffraction (HT-XRD), thermogravimetry and differential temperature analyses (TG/DTA) as complementary tools for retracing and understanding the process leading to the decomposition of ADU under air. The investigated temperature spanned the range between 30°C and 1000°C. Although the XRD analysis does not permit an accurate determination of oxygen positions in heavy metal oxides, it enables one to follow the changes in the crystallographic structure, which can be correlated with the variation of mass detected by TG and the energy transfer measured by DTA. To the best of our knowledge, no studies of ADU decomposition based on in situ HT-XRD have been reported so far. The results obtained are compared with those gathered by ex situ XRD, focusing on the different phases appearing along the heat treatment. Finally, we discuss the formation of U_3O_8 above 600°C and its stability with temperature, an interesting issue due to the use of U_3O_8 as a preferred storage form[19]. Also, the main purpose of this paper is to see the potential of HT-XRD associated to TG/DTA as a tool for nuclear forensic investigations.

2. Experimental

2.1. Synthesis of ADU

Ammonium hydroxide (Merck, 25% w/v solution) was added manually in drop-wise fashion to 100g/L of uranyl nitrate hexahydrate (purity 99%) ($UO_2(NO_3)_2 \cdot 6H_2O$) which was constantly stirred (~250rpm) at 68°C[20]. The initial pH of uranyl nitrate solution was about 2, and the final pH after the completion of NH_4OH addition was about 11. The yellow precipitate was subsequently filtered and washed three times with ultrapure water (18MX) and was allowed to dry overnight at room temperature. A few grams of material were obtained.

2.2. Instrument

TG/DTA analyses were carried out under flow of oxygen (4N) at a heating rate of 10°C/min on a temperature range of 30°C till 1000°C. The apparatus used was NETZSCH Simultaneous Analyzer STA 449 Jupiter. In situ HT-XRD patterns were acquired with a Bruker D8 powder diffractometer installed in a glove-box and mounted in a Bragg-rentano configuration with a curved Ge monochromator (111), a Cu X-ray tube (40kV, 40mA), a Vantec detector and an Anton Paar HTK2000 heating chamber equipped with a Pt heating plate. Data were collected over a full diffraction angle range $2\theta = 10°-100°$, in steps of 0.017°, with a counting time per step of 2s. The explored temperature range spanned the interval 30°C–1000°C, with the sample kept under static synthetic air (21vol% O_2 + 79vol% N_2) with a starting pressure of 800×10^2Pa. Prior to measurement, MgO was used to check the calibration of the furnace chamber. Temperature errors were determined to fall within 5% in the whole investigated interval. Ex situ XRD patterns were acquired with a Bruker D8 diffractometer similar to the in situ HT-XRD apparatus but using a Lynxeye detector and a Si wafer (111) as a sample holder.

3. Results

3.1. X-Ray Diffraction Patterns of Starting and Final Material

The XRD pattern of the starting material measured at 30°C is shown in **Figure 1(a)**. All diffraction peaks can be matched to those reported in the ICDD database (PDF number 00-044-0069) for the hexagonal structure of $2UO_3 \cdot NH_3 \cdot 3H_2O$. The precipitation of the ADU obtained by adding a large excess of ammonia, leads to a well-defined composition[11]. The oxidation state +VI of the reference material is in agreement with the yellow colour of the ADU sample [inset of **Figure 1(a)**].

Figure 1(b) illustrates the room temperature XRD pattern of an ADU sample after heat treatment in air up to 1000°C during the in situ HT-XRD investigation. This pattern can be fitted with the orthorhombic structure of U_3O_8 (Space Group C2 mm, No. 38), PDF number 01-074-2101[21].

3.2. Thermogravimetry and Differential Thermal Analyses of ADU

The TG/DTA curves of the ADU sample are reported in **Figure 2**. For a

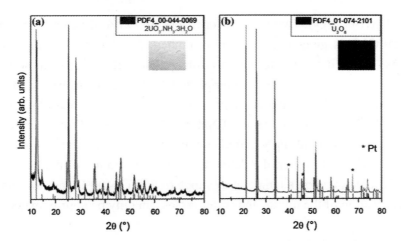

Figure 1. a X-ray diffraction pattern of ADU sample (starting material) matched to reference material ($2UO_3 \cdot NH_3 \cdot 3H_2O$) shown in red-PDF (00-044-0069) The inset shows a picture of the sample; the yellow colour is characteristic of the ADU material. b X-ray diffraction pattern of an ADU sample obtained after a heat treatment up to 1000°C in air, matched to that of the reference material (U_3O_8) shown in blue colour (PDF 01-074-2101). The dark colour of the sample (inset) is characteristic of U_3O_8 (Color figure online).

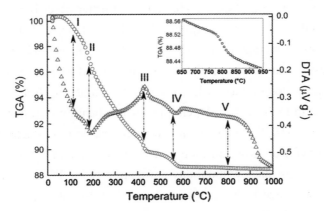

Figure 2. Thermogravimetry curve (circles) and differential thermal curve (triangles) of the ADU sample. Inset details of the DTA curve in the range 650°C–950°C.

better description, the data is denoted in five parts (I-V) that correspond largely to the appearance of endothermic and exothermic peaks. The shoulder and the peak, noted as I and II, respectively, occur at about 100°C and 200°C and correspond to the removal of physically and chemically adsorbed water molecules. Because of the superposition of the peaks, the TG curve analysis can be at best semi-quantitative. Thus, for the starting material sample whose XRD pattern is assigned to

$2UO_3 \cdot NH_3 \cdot xH_2O$, TG/DTA reveals a mass loss of 7wt% of water, taking place at about 300°C. This gives rise to 2.5mol number of water in the sample which is quite close to the data reported in literature[9]. The difference could be explained by the temperature considered for complete dehydration. Considering that the denitration begins at 300°C and completes at about 450°C, the mass loss of NH_3 of about 3wt% corresponds to 0.5mol of NH_3 for 1mol of UO_3. From literature[16], we know that ammonia is strongly bound and that below 250°C, and it is not taken out from the uranates. However, the dehydration and denitration of the sample cannot be fully differentiated on a temperature range of 250°C–400°C and, for this reason, the quantification of the nitrate and water losses can only be approximated.

The exothermic peak (III) occurring at about 420°C can be related to the denitration of the sample. After denitration and formation of UO_3, a constant mass is maintained over 100°C before reduction to U_3O_8. Such behaviour has been reported in a study of UO_3 with amorphous structure[22]. This is followed by an endothermic peak (IV) occurring slightly above 550°C which corresponds to the self-reduction of UO_3 into U_3O_8[17]. This will be further described by the HT-XRD study. The crystallisation of UO_3 coincides with a mass loss of about 1.5wt%, and is assigned to its reduction into U_3O_8 as seen later by HT-XRD. However, this mass loss leads to a stoichiometry of $UO_{2.73}$ instead of $UO_{2.67}$ expected for U_3O_8. In fact, the exact stoichiometry of U_3O_8 is difficult to obtain. Indeed, the O/U ratio deviates significantly from 8/3 depending on the heating, temperature, time and thermal history[23]. Also the oxygen content cannot be precisely determined with the used TG/DTA device or X-ray diffraction. However, at about 800°C (peak V), a further weight loss takes place without any correlation in the DTA curve (inset **Figure 2**). This could be related to the transition from U_3O_8 to U_3O_{8-x}, a structural change that does not affect significantly the energy of the system.

Above 850°C, a drastic drop in the DTA curve occurs, which may correspond to the decomposition of the sample without mass loss, as demonstrated by the constant TG curve in this high temperature range. Above 900°C, there is a decrease in the DTA curve, representative of an endothermic process completed at a temperature higher than 1000°C. A similar DTA peak was reported earlier in literature[24] and was attributed to the conversion of U_3O_{8-x} to U_8O_{21}[25].

3.3. High-Temperature X-Ray Diffraction of ADU

In order to study the decomposition of ADU and to understand the evolution of its structure with temperature, diffraction patterns have been measured with the sample kept at various temperatures up to 1000°C. The results are shown in **Figure 3**. A qualitative analysis of the diffraction profiles immediately reveals a transformation from a partially crystalline structure at room temperature to an amorphous state between 150°C and 500°C, until eventually a highly crystalline phase abruptly appears at 600°C. As described in "Thermogravimetry and Differential Thermal Analyses of ADU" Section, ADU is reactive at a temperature range of 100°C–600°C, where a handful of endothermic and exothermic peaks appear in the TG/DTA curve. This is in contrast with HT-XRD which shows mainly the appearance of weak and broad intensity peaks relating to the presence of disorder in the material. When combined, **Figures 2** and **3** provide complementary information on the process of ADU decomposition in air.

In the following, the different XRD patterns observed with in situ HT-XRD are compared to those observed with ex situ XRD made after isothermal treatment in TG/DTA. In **Figure 4**, XRD patterns obtained from both instruments are placed side by side. As an example, pattern (a) is the profile measured in situ with HT-XRD at 150°C, whereas the pattern (a') is measured at room temperature on the sample that had been heated at 150°C in air. At this temperature, we do expect the loss of physically absorbed water molecules, which still induces disorder in the

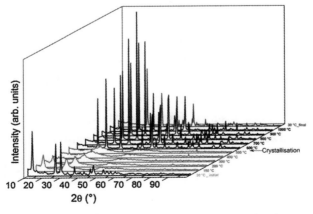

Figure 3. In situ HT-XRD study of the decomposition of ADU in air, measured from 30°C to 1000°C, in addition to the normal XRD pattern measured after cooling to 30°C.

material as seen by the broader peaks in HT-XRD. The cooling leads to a better-defined pattern very close to the one of the starting material [**Figure 1(a)**]. The decrease in the intensity of few peaks could be due to the loss of the physically absorbed water. It has been reported that dehydration of minerals occurs in several steps and each accompanied by structural changes and a decrease in inter-lamellar spacing[17].

The HT-XRD pattern (b) measured at 400°C displays even broader peaks than at a lower temperature, thus indicating a rather amorphous material. After cooling, such sample (pattern b') still shows similarities with the sample cooled from 150°C (pattern a'), with a further decrease in intensity and linewidth of the peaks. Based on the information given by the TG/DTA at this temperature, the loss of chemically absorbed water is completed, while the process of denitration is still on-going. The HT-XRD pattern obtained at 500°C (Figure 3) is similar to the one reported for 400°C (pattern b). Thus, after complete denitration, expected for the sample measured at 500°C, the resulting UO_3 is amorphous as reported in the literature for one of its known structures[19][22]. Indeed, it has been reported in literature that UO_3 has one amorphous and six crystalline modifications (α, β, δ, ε, γ, ξ)[23].

At 600°C (pattern c), the sudden crystallisation of the sample is revealed by the appearance of sharp peaks. These peaks could be easily assigned to U_3O_8 with an orthorhombic structure (Space Group C2 mm, No. 38). The result is in agreement with the one reported in Ref.[26], where the structure of U_3O_8 was determined by a combination of HT-XRD and neutron diffraction. The in situ HT-XRD sample measured at 600°C, 700°C and 800°C (pattern d) are similar. The sample measured at 600°C is metastable. The diffraction profile recorded for this sample after cooling down to room temperature is shown in panel (c') of **Figure 4**. The refinement of the diffraction pattern corresponds to β-UO_3 with a monoclinic structure (Space Group P12/11, No.4), see inset of **Figure 4**. Thus, during cooling from 600°C, U_3O_8 (panel c) absorbs oxygen and gives rise to crystalline β-UO_3 (panel c'). This result is in agreement with literature reporting that for the formation of U_3O_8 from UO_3, a heat treatment at 650°C for 1h is necessary, whereas at 600°C, the process would be extremely slow[19]. The structure of U_3O_8 appeared stable at 800°C as observed in both patterns (d) and (d'). The only difference induced by temperature is related to the lattice expansion. Thus, at this temperature, the loss of water

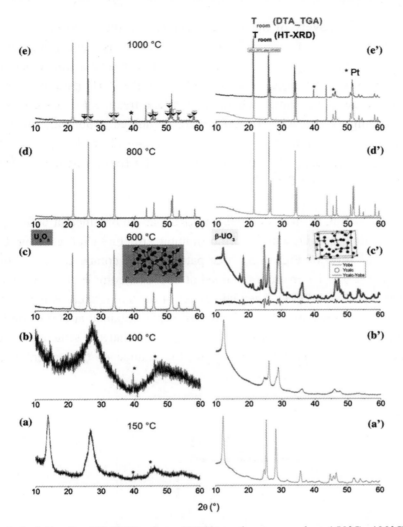

Figure 4. Left In situ HT-XRD of an ADU sample measured at 150°C, 400°C, 600°C, 800°C and 1000°C. Right Ex situ XRD of ADU samples measured at room temperature after heat treatment at 150°C, 400°C, 600°C, 800°C and 1000°C.

and nitrate is complete, U_3O_8 is stoichiometric and is the main phase even after heat treatment at 1000°C (patterns e and e').

However, at 1000°C (pattern e), extra peaks appear (indicated by arrows) that are not observed at 800°C. This could be assigned to a symmetry lower than orthorhombic or to the formation of a multiphase mixture. Therefore, the orthorhombic U_3O_8 structure is stable up to 800°C. However, the cooling of the sample in air shows the recovery of the stable structure and the presence of one single phase. Oxygen lost upon heating U_3O_8 above 800°C is rapidly replaced upon cooling[27]. In order to demonstrate that ex situ XRD measurements made at room temperature can be correlated to in situ HT-XRD, the XRD patterns recorded at room temperature on the sample obtained after heat treatment of 1000°C in in situ HT-XRD and in ex situ with TG/DTA were compared. Both patterns are completely identical, indicating that the processes taking place through ex situ heat treatment can be correlated to the in situ process. Le Bail refinement of HT-XRD pattern of ADU from 600°C to 1000°C

In this section, we focus on the HT-XRD pattern obtained at 600°C (**Figure 5**), corresponding to U_3O_8 with an ortho-rhombic structure (Space Group C2 mm, No. 38). In the inset of **Figure 5**, three patterns were compared on a limited 2θ range (50.4°–52.6°). The pattern obtained at 600°C is completely similar to those obtained at 700°C and 800°C. The main effect of the temperature is the shift of the peaks, due to the lattice expansion. However, peaks obtained at 600°C are broader than those obtained at higher temperature, showing that the structure is less ordered and less stable, as demonstrated by the change of the pattern when the

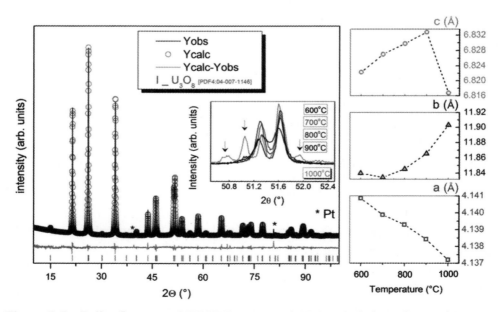

Figure 5. Le Bail refinement of HT-XRD pattern of ADU sample heated at 600°C. Inset Effect of heat treatment on ADU at 600, 700, 800, 900 and 1000, as shown on a 2θ range (50.4°–52.6°). Right Lattice parameters evolution from 600°C to 1000°C in air.

cooling of the sample is taking place [**Figure 4(c')**]. At 900°C, the pattern is also in a first glance similar to those obtained at 600°C, however a small extra peak appears, as shown in the inset at about 51° in 2θ. This demonstrates that the sample structure of U_3O_8 at this temperature is unstable in air. This structural change is emphasised in the heat treatment done at 1000°C, where additional and higher intensity peaks appear (indicated by arrows in the inset of **Figure 5**). This change clearly shows that the optimal temperature for obtaining U_3O_8 is around 700°C and that its stability in air decreases at higher temperatures.

The evolution of the lattice parameters of U_3O_8 as a function of temperature is reported on the right hand side of **Figure 5**. We still consider the same structure for the pattern obtained at 1000°C in order to refine the lattice parameters for comparisons with the patterns obtained at 600°C. The lattice expansion takes place mainly along the b and c axes, whereas along a, the lattice expansion is weakly negative. The trend along the lattice c is broken for the sample treated at 1000°C, thus emphasising the instability and/or the change of the U_3O_8 structure at this temperature. This result is in agreement with literature indicating that heating U_3O_8 during 1h in oxidising atmosphere at 750°C will achieve thermal stabilisation[19]. The thermal expansion of U_3O_8 is anisotropic as reported by Ackermann *et al.*[21].

4. Discussion

Figure 6 illustrates the evolution of the position of the Bragg diffraction peaks observed by HT-XRD together with the derivative curve dTGA/dT versus the temperature. To discuss the different steps taking place during the calcination of ammonium diuranate, the mechanism is summarised in **Table 1**.

Below 600°C, very few and broad peaks can be observed by HT-XRD, due to the disorder in the material. Indeed, the ADU sample obtained by precipitation still contains water and nitrate molecules giving rise to a weak crystallinity. Thus, in this range of temperature, XRD is not a very suitable characterisation tool.

Following the evolution of the dTGA/dT curve, the mass loss can be correlated with water molecules (peak I and II) and nitrate (peak III). The crystallisation

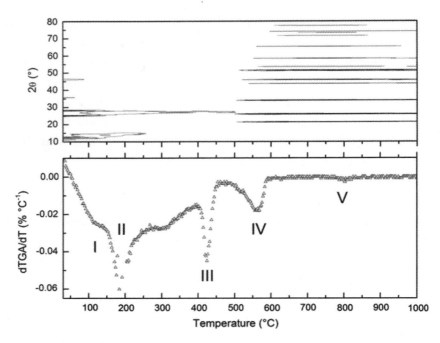

Figure 6. Thermal decomposition of ADU sample in air. Top HTXRD peaks position versus temperature with 2θ angle (10°–80°). Bottom dTGA/dT curve.

Table 1. Mechanisms of the process taking place in the calcination of ADU.

Step	I	II	III	IV	V
Reaction	$2UO_3 \cdot NH_3 \cdot xH_2O$ → $2UO_3 \cdot NH_3 \cdot xOH$ (− H_2O)	$2UO_3 \cdot NH_3 \cdot xOH$ → $2UO_3 \cdot NH_3$ (− H_2O)	$2UO_3 \cdot NHO_3$ → $2UO_3$ (− NO_x)	$3UO_3$ → U_3O_8 (− $1/2 O_2$)	U_3O_8 → U_3O_{8-x}

of the material observed at about 600°C by HT-XRD is preceded by a mass loss appearing at 550°C (peak IV) which can be explained by the self-reduction and phase transition of amorphous material UO_3 to highly crystalline material U_3O_8. Until the formation of crystalline UO_3 and U_3O_8, X-ray diffraction does not provide any information about the decomposition products.

The formation of U_3O_8 is stable in air in the temperature range of

700°C–800°C. Above this range, a mass loss takes place, as shown by peak V in TG/DTA curve, very likely related to oxygen loss. Also, this is emphasised by in situ HT-XRD by a change of structure, as demonstrated by extra peaks from 900°C on. Slow formation of another phase has been reported in literature[23] in α-U_3O_8 which can take place at the temperature range of 1000°C–1300°C. This has been linked to a possible U_8O_{21} phase with a homogeneity range extending between the compositions $UO_{2.60}$ and $UO_{2.65}$[23]. A slightly different composition range, $UO_{2.617}$–$UO_{2.655}$ has also been reported[28]. It is possible that the proper stoichiometry of β-U_3O_8 is U_8O_{21}, since β-U_3O_8 has been prepared by heating α-U_3O_8 to 1350°C followed by slow cooling to room temperature[29]. However it has often been reported that in this region, the phase is considered hypostoichiometric (U_3O_{8-z})[23]. Two modifications of U_3O_8, α and β-U_3O_8 crystallise both in orthorhombic system and their crystal structures are very similar. These compounds have a layer structure related to the hypostoichiometric "ideal" UO_3 structure which has uranyl bonds perpendicular to the layer. The difficulty to rearrange the oxygen atoms in these infinite layer structures is probably the reason[23] for the slow equilibration between U_3O_8 and the gas phase at different temperatures and the oxygen partial pressures. The O/U ratio of the "U_3O_8" phase varies with the experimental method used to determine it. Above 727°C, the upper phase boundary was observed to have O/U = 2.667 (stoichiometry of U3O8) up to 1127°C. At an ambient pressure of 0.21 atm O_2, the compound however becomes hypostoichiometric above 600°C[23].

The present study shows the potential use of HT-XRD together with TG/DTA as a potential tool for nuclear forensic investigations. Main results obtained in this study are in agreement with those reported in literature. The present material was prepared in our laboratory and analysed to collect the so-called 'fingerprints' in a data-base. This latter can then be used for comparison to identify any material relevant to nuclear security. For the future, different compositions of yellow cake can be prepared and analysed in the same way as the present study. This will provide hints on the processing history of any material found in the framework of nuclear forensics investigations. Based on the TG/DTA analyses, characteristic temperatures can be chosen and used to retrace the corresponding initiated product through the in situ analyses of HT-XRD. In this study, we did report for the first time, in situ HT-XRD analyses of ADU decomposition. This confirms the crystallisation of the material after 500°C and the stability of U_3O_8

under air in a temperature range between 700°C and 800°C. The transition of β-UO$_3$ to U$_3$O$_8$ could also be observed and confirmed only by using both HT and room temperature XRD, showing that U$_3$O$_8$ is metastable at 600°C.

5. Conclusion

The thermal decomposition of ammonium diuranate in air up to 1000°C was investigated, using in situ HT-XRD, TG/DTA. The results show that these two techniques are necessary and complementary to retrace the full process. Also in the framework of nuclear forensic investigations, XRD appears a useful technique for re-establishing the history of the heat treatment of uranium oxide of unknown origin. Such HT-XRD studies could be extended to other types of yellow cake material to complete the information basis on uranium oxide produced from different starting materials.

Acknowledgements

We thank G. Pagliosa and D. Bouëxière for their technical support in the X-ray diffraction. H. Hein and Ernstberger are also acknowledged for help with the thermo-gravimetry/differential thermal analyses.

Open Access: This article is distributed under the terms of the Creative Commons Attribution License which permits any use, distribution, and reproduction in any medium, provided the original author(s) and the source are credited.

Source: Eloirdi R, Lin D H M, Mayer K, *et al*. Investigation of ammonium diuranate calcination with high-temperature X-ray diffraction[J]. Journal of Materials Science, 2014, 49(24):8436–8443.

References

[1] Mayer K, Wallenius M, Ray I (2005) Nuclear forensics: a methodology providing clues on the origin of illicitly trafficked nuclear materials. Analyst 130:433–441.

[2] Wallenius M, Mayer K, Ray I (2006) Nuclear forensic investigations: two case studies. Forensic Sci Int 156(1):55–62.

[3] Mayer K, Wallenius M, Fanghänel T (2007) Nuclear forensic science: from cradle to maturity. J Alloy Compd 444–445:50–56.

[4] Wallenius M, Lützenkirchen K, Mayer K, Ray I et al (2007) Nuclear forensic investigations with a focus on plutonium. J Alloy Compd 444–445:57–62.

[5] Badaut V, Wallenius M, Mayer K (2009) Anion analysis in uranium ore concentrates by ion chromatography. J Radioanal Nucl Ch 280(1):57–61.

[6] Keegan E, Wallenius M, Mayer K, Varga Z, Rasmussen G (2012) Attribution of uranium ore concentrates using elemental and anionic data. Appl Geochem 27(8):1600–1609.

[7] Cordfunke EHP (1970) Composition and structure of ammonium uranates. J Inorg Nucl Chem 32:3129–3131.

[8] Cordfunke EHP (1962) On the uranates of ammonium-I. The ternary system NH_3–UO_3–H_2O. J Inorg Nucl Chem 24:303–307.

[9] Debets PC, Loopstra BO (1963) On the uranates of ammonium-II X-ray investigation of the compounds in the system NH_3–UO_3–H_2O. J Inorg Nucl Chem 25: 945–953.

[10] Mea Hermans, Markestein T (1963) Ammonium uranates and UO_3-hydrates-ammoniates. J Inorg Nucl Chem 25:461–462.

[11] Stuart WI, Whateley TL (1969) Composition and structures of ammonium uranates. J Inorg Nucl Chem 31:1639–1647.

[12] Stuart WI, Miller DJ (1973) The nature of ammonium uranates. J Inorg Nucl Chem 35(6):2109–2111.

[13] Urbanek V, Sara V, Moravec J (1979) Study of formation and composition of ammonium uranates. J Inorg Nucl Chem 41:537–540.

[14] Woolfrey JL (1978) The preparation of UO_2 powder: effect of ammonium uranate properties. J Nucl Mater 74:123–131.

[15] Manna S, Roy SB, Joshi JB (2012) Study of crystallization and morphology of ammonium diuranate and uranium oxide. J Nucl Mater 424:94–100.

[16] J.L.Woolfrey (1968) The preparation and calcination of ammonium uranates: A literature survey. Australian Atomic Energy Commission, AAEC/TM476.

[17] Price GH, Stuart WI (1973) Thermal decomposition of ammonium uranates. Australian Atomic Energy Commission, AAEC/ E276.

[18] Manna S, Karthik P, Mukherjee A, Banerjee J, Roy SB, Joshi JB (2012) Study of calcinations of ammonium diuranate at different temperatures. J Nucl Mater 426: 229–232.

[19] Thein SM and Bereolos PJ (2000) Thermal Stabilization of $233UO_2$, $233UO_3$, $233U_3O_8$. Oak Ridge National Laboratory, ORNL/TM-2000/82.

[20] Litz JE and Coleman RB (1980) Production of yellow cake and uranium fluorides In: Advisory Group Meeting, Paris, 1979. International Atomic Energy Agency, Vienna, STI/PUB/553, ISBN 92-0-041080-4.

[21] Ackermann RJ, Chang AT, Sorrell CA (1977) Thermal expansion and phase transformation of the U_3O_8-z phase in air. J Inorg Nucl Chem 39(1):75–85.

[22] Hoekstra HR, Siegel S (1961) The uranium-oxygen system: U_3O_8–UO_3. J Inorg Nucl Chem 18:154–165.

[23] Morss LR, Edelstein NM, Fuger J, Katz JJ (2006) The chemistry of the actinide and transactinide elements, vol 1, 3rd edn. Springer, Dordrecht.

[24] Lynch ED (1965) Studies of stoichiometric and hyperstoichiometric solid solutions in the thoria-urania system. Argonne National Laboratory, ANL-6894.

[25] Malinin GV, Tolmachev YM (1968) Differential thermal analysis of the decomposition of U_3O_8 in air. Radiokhimiya 10(3):362–366.

[26] Herak R (1969) The crystal structure of the high temperature modification of U_3O_8. Acta Crystallogr B25:2505–2508.

[27] Katz JJ, Seaborg GT, Morss LR (1986) The chemistry of the actinide elements, vol 1, 2nd edn. Chapman and Hall, New York.

[28] Caneiro A, Abriata JP (1984) Equilibrium oxygen partial pressure and phase diagram of the uranium-oxygen system in the composition range $2.61 < U/O < 2.67$ between 844 and 1371K. J Nucl Mater 126(3):255–267.

[29] Loopstra B (1970) The structure of [beta]-U_3O_8. Acta Crystallogr Sect B 26(5): 656–657.

Chapter 23
Metallic Muscles and Beyond: Nanofoams at Work

Eric Detsi[1,2], Sarah H. Tolbert[2], S. Punzhin[1], Jeff Th. M. De Hosson[1]

[1]Department of Applied Physics, Zernike Institute for Advanced Materials, University of Groningen, Nijenborgh 4, 9747AG Groningen, The Netherlands
[2]Department of Chemistry, UCLA, 607 Charles E. Young Drive East, Los Angeles, CA 90095-1569, USA

Abstract: In this contribution for the Golden Jubilee issue commemorating the 50th anniversary of the Journal of Materials Science, we will discuss the challenges and opportunities of nanoporous metals and their composites as novel energy conversion materials. In particular, we will concentrate on electrical-to-mechanical energy conversion using nanoporous metal-polymer composite materials. A materials system that mimic the properties of human skeletal muscles upon an outside stimulus is coined an 'artificial muscle.' In contrast to piezoceramics, nanoporous metallic materials offer a unique combination of low operating voltages, relatively large strain amplitudes, high stiffness, and strength. Here we will discuss smart materials where large macroscopic strain amplitudes up to 10% and strain-rates up to 10^{-2} s^{-1} can be achieved in nanoporous metal/polymer composite. These strain amplitudes and strain-rates are roughly 2 and 5 orders of magnitude larger than those achieved in common actuator materials, respectively. Continuing on the theme of energy-related applications, in the summary and outlook, we discuss two recent developments toward the integration of nanoporous metals into energy conversion and storage systems. We specifically focus on the exciting potential of nanoporous metals as anodes for high-performance water electrolyzers and in next-generation lithium-ion batteries.

1. Introduction

With the emphasis on miniaturization stemming from the electronics industry[1], the same push has been seen in submicron- and nanosized mechanical systems. In medicine and biology, for example, there is a need for high-precision actuators and manipulators for work on fluid filtration and living cell manipulation[2]. The increasingly popular lab-on-a-chip technology takes advantage of highly miniaturized mechanical systems—Micro-Electronic Mechanical Systems or MEMS—to fit efficient analysis systems in a very small space. For progress in these fields, there is a necessity for the continuous development of both materials with micro- and nanoscale functions and of tools that can facilitate the production and characterization of these materials.

Mechanical displacement that comes as a result of an electric signal passing through a material is called actuation. In materials that produce an actuation response, the reverse is often possible as well—an electric current can be induced to flow if the material is deformed. The most common type of material that shows such properties is described as piezoelectric, and of this class of materials, quartz is the most well-known. Indeed, it is the piezo-electric property of quartz that allows it to be used as an oscillating pace mechanism in the common wristwatch[3]. The typical piezoactuator delivers a ~0.2% strain at a high potential of 150V[4]. Considering that it is desirable to see the use of actuation in low-voltage devices, such as MEMS, much lower operational parameters are required for the modern actuating material. Polymer-based actuation materials have been developed, which offer extraordinary capacity for induced deformation, but have the drawback of being weak and compliant. In recent years, we have been exploring metallic nanofoams and have demonstrated the potential of nanostructured metals to act as actuators, creating so-called "metallic muscles," with the ability to demonstrate the properties required of the modern actuator: low throughput voltage requirements, high extension yield, strength, and stiffness[5]-[12]. In this respect, the beautiful work by Weißmüller et al. and the pioneering work by Herbert Gleiter and collaborators have to be mentioned[7][8].

In this contribution, we will review some aspects of this fast growing field and highlight a couple of ideas of applications in the outlook section that may also have a great impact onto the field of energy-related materials. While the production of nanoporous metallic structures is well-documented, up until recently very little was known about their mechanical properties—at submicron scales, sample size has the possibility to produce a large effect on mechanical properties, where in macroporous foams cell size specifically does not have an influence on material strength[13][14]. Indeed, it is highly uncertain that the behaviors of macroscopic and microscopic foams will be at all similar in principle and nature. Li and Sieradzki reported that porous Au undergoes a ductile-brittle transition that seemed to be influenced by the microstructural length scale of the material[15]. Biener *et al.* have continued this investigation into the mechanical properties of nanoporous Au through nanoindentation[10][16]. They reported the main deformation mechanism during nanoindentation as a ductile, plastic densification. Strong long-range stress fields, brittle fracture, and crack emission were not observed. They note that the scaling laws that are typically applied to macroporous foams apply poorly to nanoporous metals, as they observe experimental yield strength of 145MPa instead of the expected 16MPa.

Volkert *et al.* performed microcompression experiments on FIB-milled micronsized pillars of nanoporous Au with 15-nm diameter ligaments[17]. They find that, while Young's modulus values as determined experimentally and as predicted by scaling laws do not show significant difference, there is a major increase in yield strength as sample size decreases below 50mm length scales. A yield strength of 1.5GPa is predicted, which is several orders of magnitude above that of typical bulk Au. They interpret this effect as influenced by the increased required stress to activate dislocation sources as ligament size decreases, until theoretical shear strength is reached.

Further work by Biener *et al.*[10][16] investigated this elevated yield strength whether its origin was the microstructure of disordered nanoporous Au ligaments or the specific size-dependent mechanical properties of Au. This was performed by preparing multiple samples with varying ligament sizes. It is established that in the production of a nanoporous material it is possible to tune ligament and pore size through varying dealloying conditions. They observed a clear influence of ligament size such that the strength of nanoporous Au increases with decreasing ligament diameter, and thus propose that the Gibson and Ashby scaling model of foam plasticity[18] needs to be adjusted to take into account ligament size for nanoporous systems. Recently, we have applied a novel approach to the investigation of deformation of nanoporous metals at the nanoscale by exposing nanoporous nanopillars to a Ga^+ ion beam[19]. It will be not the main topic of this review but it is interesting to note that the results we have obtained with Au nanopillars have also been observed in Cu, Al, and Ni nano(porous) pillars, *i.e.*, a gradual massive deformation effect of the pillar during Ga ion beam exposure, where the pillar bends toward the ion beam. A relationship between the formation of defects due to ion collisions in the nanopillar and the pillar's deformation was derived, and we find that the deflection is linearly related to ion fluence. The high degree of control over deflection and the variables that influence it open an opportunity for use of ion-beam-induced bending as a characterization technique of nano(porous) materials.

Porous systems come in two types—interconnected and noninterconnected (alternatively, open-cell and closed-cell, respectively), describing the relationship of the material's pores: in the former, there exists a continuous pathway between every single pore in the material, and in the latter, the pores exist independently as separated islands. A porous system is typically characterized by a high surface area-to-volume ratio due to the high amount of air-to-solid interface area as well as by a lower density and, by connection, by a lower weight compared to its solid bulk counterpart. To briefly mention terminology, a porous material is made up of pores, struts, and nodes. Pores are the encompassing term for the volume of air within foam and struts are solid material that merge at nodes and connect nodes together.

The popularity of macrofoams, *i.e.*, porous materials where pore size is above the scale of tens of microns, stems from the intersection of a variety of desirable properties in industry. For example, aluminum macrofoam boasts of a high stiffness-to-density ratio, high capacity for energy absorption during compression, high-temperature resistance, electrical and thermal conductivities, good machinability, and cheap production costs[20]-[25], see **Figure 1**. Such properties make macrofoams attractive in the construction and automotive industries, for example. While macrofoams are common materials with known applications in industry, the class of materials known as nanofoams is more exotic. Operating under the same

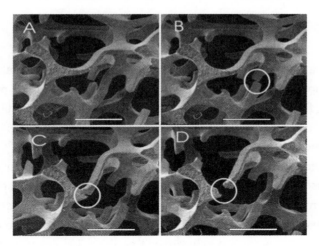

Figure 1. In-situ deformation in a Philips XL30-FEG-ESEM of Duocel 40 PPI macro foam with a relative density of approximately 7% (scale bar 1mm).

concepts as macrofoams in principle, nanofoams are characterized by pore and strut sizes being at the nanoscale, in other words the greatest diameter of a pore or strut must be considered a nanoscale dimension. For a more precise definition, the International Union of Pure and Applied Chemistry has categorized nanoporous metals into three groups, depending on the pore size: microporous metals have pore size under 2nm, mesoporous metals have pore sizes between 2 and 50nm, and macroporous metals have pore size above 50nm.

Nanofoams share many properties with their macro-foam counterparts, such as the high surface-area-to-volume ratio, but also including the capacity for cheap production and easy machinability. In addition, however, nanoporous foams have seen usage in many applications beyond those of macrofoams, including nanofiltration systems, drug delivery platforms, catalysis, sensing, and actuation[26]-[35]. A major advantage that nanoporous metals have is the ability to hold a lattice of nanoscale features while being able to be easily handled and transported, something metallic nanoparticles, for example, cannot provide.

The contribution for this Golden Jubilee issue of the Journal of Materials Science will be focused on the actuating properties of metallic foams, although these materials turned out be also sensors[36][37]. In the Discussion section, we also present an outlook onto two promising energy-related fields, involving oxygen evolution and Li-ion battery, in which porous metals may have a large impact.

2. Metallic Muscles: Why Does It Work?

The actuation mechanism in common artificial muscles makes use of microscopic phenomena in the bulk of the material such as dipoles polarization. In piezoceramics, for instance, asymmetry in the crystal lattice structure gives rise to domains with electric dipole moments. These dipole moments are randomly oriented, and they can be aligned with an external electric field[38]. Dipole alignment through the bulk of the material results in macroscopic dimensional changes. In metallic muscles, not only the bulk volume but also the interface surface area plays a central role during actuation. As a starting point, the fact that surface atoms in crystalline materials have a lower coordination than those in the bulk results in an excess bonds' charge at a newly created surface. This excess charge redistributes at the surface to strengthen the inter-atomic bonds and shorten the distance between surface atoms[39]. This results in a positive surface stress (tensile stress) in the material at mechanical equilibrium of the material system (*i.e.*, a tensile displacement to bring the atoms back to an equilibrium distances as in the bulk). That situation is schematized in **Figure 2** in the case of metallic bonding. The red circles represent positive metal ions (cations) consisting of nuclei and inner-shell electrons; the ovals illustrate delocalized free electrons in the bulk (light blue ovals) and at the surface (dark blue ovals) of the metal. It can be seen in **Figure 2(a)** that at mechanical equilibrium, the interatomic distance is reduced due to the tensile surface stress[40] with respect to the bulk.

The charge distribution at the metal surface (dark blue ovals in **Figure 2**) can be controlled in different ways, for instance, by bringing a layer of adsorbates at the interface[41]. A positive charge of the adsorbate will result in an electronic charge redistribution at the metal surface. It generates a lower tensile surface stress and results in a relaxation of the surface atom positions by increasing the interatomic spacing. In order to preserve the mechanical equilibrium[42], bulk atoms experience less compressive stress and move in a positive direction outward, as illustrated in **Figure 2(b)**. Therefore, since detection is due to the bulk atoms, a positive displacement is measured experimentally. Although these surface stress-induced bulk deformations are not detectable in macroscopic metals, they become significant in nanostructured metals where the properties are governed by the large surface area, rather than by the bulk volume as highlighted above. This phenomenon represents the basic operating principle of metallic muscles which consist of

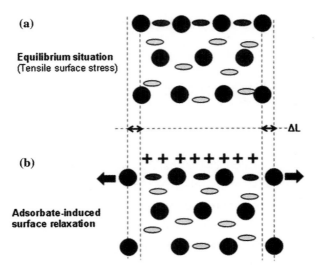

Figure 2. Actuation mechanism in metallic muscles. a A clean metal surface exhibits a tensile surface stress at equilibrium, resulting in the reduction of the interatomic distance. b Adsorbates with excess positive charges at the metal surface remove the bonds charge between surface atoms, inducing surface relaxation.

nanoporous metals with high surface-area-to-volume ratios.

The electronic charge distribution at a nanoporous metal interface can also be easily controlled in an aqueous electrolyte where an electrical voltage is used to bring positive or negative charge carriers (*i.e.*, ions) from the electrolyte to the nanoporous metal interface. Typically, referring to the above-mentioned actuation mechanism in metals, injection of negative charge in the space-charge region at a nanoporous metal/electrolyte interface during electroadsorption of charge compensating positive ions enhance the tensile surface stress in the metal[8][43], resulting in an increase in compressive stress in the bulk of the ligaments and in an overall macroscopic volume shrinkage of the nanoporous metal specimen[8][44], *i.e.*, a negative displacement is measured experimentally.

As highlighted above, nanoporous metal actuators offer a unique combination of low operating voltages, relatively large strain amplitudes, high stiffness, and strength. Despite these remarkable features, the emergence of nanoporous metal actuators in practicable applications is still delayed, a decade after their discovery[7][8]. The challenge to their further development in viable applications can be considered threefold, but principally concerns the aqueous electrolyte that is needed to inject electronic charge in the space-charge region at the metal/ electrolyte interface[45]:

- At first, an aqueous electrolyte limits the usage of metallic muscles to wet environments whereas most of the practical applications require artificial muscles that can operate in dry environments.

- A second major concern is that the aqueous electrolyte limits the actuation rate because of its relatively low ionic conductivity. Simply replacing the aqueous electrolyte by a solid one is an obvious solution but the actuation rate of all-solid-state electrochemical actuators is more severely hampered by the low room-temperature ionic conductivity of solid-state electrolytes.

- Completing the trio of challenges is the fact that the ligaments in nanoporous metals suffer from severe coarsening (undesired growth) during electrochemical processes[46] including actuation via redox reactions. Coarsening causes metallic muscles to lose performance efficiency after many cycles, because the charge-induced strain is ligament size-dependent as shown in **Figure 3** where strain amplitudes are plotted as a function of the ligament size. This shows that the growth of the ligaments during performance of electrochemical actuation is undesirable.

In addition to these technical boundaries, it is emphasized that metallic muscles operating as electrochemical actuators require a three-component configuration to function; explicitly these are a working electrode, an electrolyte, and a counter electrode. In such a configuration, the working and counter electrodes, which may separately actuate as demonstrated by Kramer et al.[9], are placed at a relatively large distance from each other. This fixed distance represents a limitation for the integration of metallic muscles into miniaturized devices[47]. In view of these various restrictions caused by the electrolyte, an electrolyte-free approach is desirable for actuation in nanoporous metals. In fact, the following features are a prerequisite for a breakthrough in the field of artificial muscles: (i) no usage of aqueous or solid electrolyte, (ii) a fast actuation rate, and (iii) a single actuating component as in piezoelectric materials.

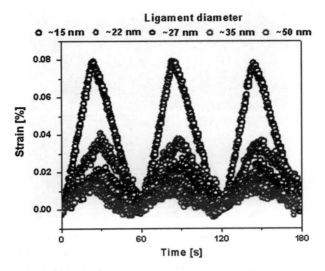

Figure 3. Ligament size-dependence of the charge-induced strain in nanoporous metals. The strain amplitude recorded on five NPG samples with different ligament sizes decreases with the increasing ligament size. This shows that ligaments' growth during electrochemical actuation is undesirable (Color figure online).

3. Processing

In the majority of cases, metal nanofoams are produced through dealloying. In the literature, several processing conditions and results are highlighted and reference can be made to[48]–[60]. For this process to work, an alloy must be produced between the required metal and another that can be etched away in some manner. It is also required that the two metals be able to form a solid solution, as any other morphology will not allow for the eventual formation of an isotropic structure of pores and ligaments. This is often the limiting factor for the dealloying process of nanofoam manufacture, as many metals do not easily form solid solutions, and if they do, it may not be possible to selectively etch one of the components of the alloy system while leaving the other intact. A typical example of a dealloyed Au–Ag system is displayed in **Figure 4**. Without going into details in[61], we present a Metropolis Monte Carlo study of the dealloying mechanism leading to the formation of nanoporous materials. A simple lattice-gas model, for gold, silver, and acid particles, vacancies, and products of chemical reactions, is adopted. The influences of temperature, concentration, and lattice defects on the dealloying process are investigated, and the morphological properties are characterized

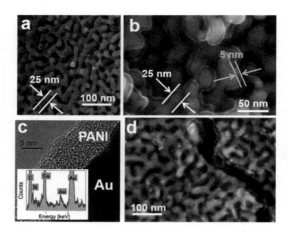

Figure 4. Microstructural characterization of NPG/PANI. a Scanning electron micrograph showing the bicontinuous morphology of NPG. b, c Scanning and transmission electron micrographs showing a ~5-nm-thick PANI skin covering the ligaments of NPG. The inset of c displays the EDX spectrum of PANI. C and N come from aniline (C_6H_7N), Cu and Au come, respectively, from the Cu grid used as sample holder and the NPG. d Fracture cross section of NPG/PANI; It can be seen that the polymer envelope covering the ligaments is present in the bulk of the composite material.

in terms of the Euler characteristic, volume, surface area, and the specific surface area[21]. It is shown that a minimal three-parameter model suffices to yield nanoporous gold structures which have morphological properties akin to those found in the experiment. The recent technique of Transmission Electron Back-Scatter Diffraction was used by us[62] to investigate the effect of dealloying on the microstructure of 140-nm-thin gold foils. Transmission electron backscatter diffraction (t-EBSD) combines the detection of Kikuchi bands of forward-scattered electrons with the scanning possibilities of the scanning electron microscope (SEM). This technique was first suggested by Keller and Geiss[63]. The terms SEM transmission Kikuchi diffraction (SEM-TKD)[64] and transmission electron forward-scattered diffraction (t-EFSD)[65][66] are coined as more appropriate alternatives. Although the acronym t-EBSD may sound peculiar, it describes the technique very well: transmission microscopy in an EBSD system. A global (statistical) and a local comparison of the microstructure between the nonetched and nanoporous gold foils were made. Characterizations of crystallographic texture, misorientation distribution, and grain structure clearly prove[67] that during formation of nanoporous materials by dealloying, the crystallographic texture is significantly enhanced with a clear decrease of internal strain while retaining the grain structure.

Silver is alloyed with gold to form the so-called "white gold," an alloy commonly used in jewelry. The versatility of the alloy stems from its ability to form a solid solution at any ratio of gold to silver allowing for fine control of porosity of a resultant pure gold system. While copper does not have alloy systems as simple as gold's with silver, copper-manganese forms a reliable solid solution across a wide range of compositions[68][69]. Unfortunately, except at very low percentages, Mn tends to segregate out of the solid solution and form a phase of pure Mn at low temperatures. This issue is solved through rapid quenching from solid solution temperature. This process prevents the formation of the pure Mn phase and, provided that the quenching had occurred successfully, yields an ingot with a microstructure comparable to that of Ag–Au alloy.

As is often the case for many novel materials, interest in producing a nanoporous structure with an ordered, aniso-tropic pore structure came from observing nature. Specifically, the opalescence effect—variation in color based on direction of observation—seen in butterfly wings and mother-of-pearl stems from the ordered chitinous scales for the former or calcium plates for the latter serving as photonic diffraction gratings. The brilliant color variation is a light effect described as opalescence and is a result of an ordered nanoporous lattice acting as a series of waveguides, only permitting through light of a particular frequency depending on viewing direction. The use of waveguide materials such as this has been proposed for use in optical circuitry[70][71].

With clear applications across many industries[72]–[75], three-dimensionally ordered materials have garnered much attention. As with all nanostructured materials, two methods of approach are viable for their production: top-down and bottom-up. Top-down production focuses on reducing a bulk source sample down to the correct size, shape, and morphology through a variety of destructive production methods. While the top-down method boasts a very logical application with easy potential for iterative improvement, its greatest limitation is the scale down to which its production methods can reach—detail at the nanometer scale is beyond the capabilities of the typical top-down method [76]–[78]. The alternative is the bottom-up approach, where nanoscale features are assembled piece by piece into a full structure. This method allows for far finer detail control and overall quality of the assembled structure, but is typically challenging to implement due to it often being a multistage process.

The top-down process of production of nanostructured materials typically involves a starting bulk macroscale solid and then the use of one of several techniques to achieve a nanostructure through size reduction. In medical applications, a top-down method for the production of nanoparticle suspensions is high-pressure homogenization, consisting of the repeated forcing of a suspension through a very thin gap at high velocity or media milling, which is the mechanical attrition of suspended particles using glass or zirconium oxide[79]. Mechanical attrition in general is a common procedure for the production of nanostructured materials, and one example of such a procedure is the method of ball milling, where powders are sealed in a strengthened container with a set of hard metal spheres and treated in a vibratory mill to elicit potential phase changes and the formation of nanostructured grains in the processed particles[80]. More nonstandard procedures are also known: Yan *et al.* have proposed that, considering its simplicity and capacity for high-resolution spatial imaging, an atomic force microscopy (AFM) apparatus can be used to machine nanoscale features, and have demonstrated this capability in aluminum[81].

As a whole, lithography is considered a top-down method, and is common in the electronics industry for the production of microchips. However, it can be converted into a bottom-up method by reducing the size of the initial template. This idea can be applied for 3D nanostructures: a porous 3D template is constructed that allows for the introduction of a particular material into the pore spaces within the template. After the template is removed, the result is a 3D "image" of the template's pore network. This is called inverse templating, and is the principal bottom-up method that allows for repeatable batch production in moderately large quantities, provided that the template is easily constructed. For this purpose, self-assembling templates are highly valued—templates whose component parts can, over time, arrange themselves into a desired 3D pattern with no further intervention aside from the initial process setup.

Nanosphere templating uses the natural ability of silica or polymer (PMMA, polystyrene, latex) nanospheres to reliably self-assemble into a template on a large scale. The nanospheres are suspended in a solution and are allowed to naturally settle over time, although centrifugation can be used to accelerate the process at the expense of quality of ordering. An alternative method is to allow the spheres to settle across a meniscus to improve ordering with the downside of the resultant template film being very thin. The spheres will preferentially settle into their most low-energy configuration, which is a crystalline face-centered cubic arrangement. As the nanospheres settle, the solution evaporates and eventually a dry nanosphere template remains.

A settled nanosphere template is typically a weak, brittle solid, as the spheres are held together only through van der Waals forces. Thermal processing can be applied to strengthen a finished template by furnace-treating the template to allow necks to form between adjacent spheres. The appropriate temperature and time vary significantly by material, e.g., several minutes at 120°C is enough for polymers like polystyrene, and processing for several hours at over 900°C is required for silica. This results in a much more robust template that can withstand mechanical rigor, but it may be more challenging to remove. See **Figure 5**.

In principle, any sort of material with submicron-scale features, which exhibits ordering can be used as a template, provided that interconnected voids exist between the features. One such material is the so-called "block copolymer." In such a material, two covalently linked polymers, or blocks, undergo a process called microphase separation, *i.e.*, attempt to separate as oil and water would, but are limited in their capacity to do so due to crosslinking. An enthalpy-entropy balance governs the specific manner in which this separation occurs. By controlling the compositions of the component block, it is possible to influence the phase behavior of the copolymer and thus form a variety of phases: lamellar, cylindrical,

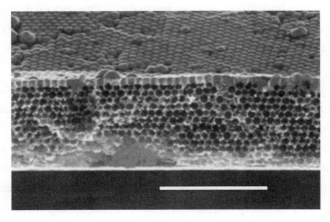

Figure 5. SEM micrograph of electroless plated nanosphere template on gold-plated silicon wafer after template removal (scale bar 5mm).

and spherical phases are common, while classical phases, with the perforated lamellar, gyroidal, and ordered bicontinuous double diamond phases are seen more rarely. In all cases, depending on the composition, one block forms the specific feature that the morphology is named after. and the other forms a matrix around the feature. Remarkably, all resultant structures show a high degree of ordering of their characteristic unit cell. As such, if, once a particular phase has been formed, one of the blocks of the copolymer could be removed while leaving the other preserved; the resultant polymer matrix could be used as an organic template. Of the possible phases formed by diblock copolymers, the gyroid phase is the most interesting due to its repeating, long-ranged ordering of a three-dimensional feature. In other words, the gyroid morphology offers the strongest potential for use as a template to produce a three-dimensionally ordered nanofoam. Recently, we have investigated this route via supramolecular PS-b-P4VP(PDP) (diblock copolymers of polystyrene and poly(4-vinylpyridine)) complexes with a bicontinuous gyroid morphology that were used as templates to produce metallic nickel nanofoams[82]. The complete dissolution of PDP from the complex with the major P4VP(PDP) component forming the matrix results in an open network structure with struts consisting of a PS core and a P4VP corona. The high specific surface area and the narrow pore size distribution of the structure formed are evidenced by nitrogen adsorption and mercury porosimetry. The open nature of the pores allows for electroless deposition of metal. During the processing, the symmetry and the size of the nanopattern are conserved. The subsequent removal of the polymer template by pyrolysis leads to the formation of an inverse gyroid nickel nanofoam with porosity exceeding 50% v/v (see **Figure 6**). The use of polymer templates with different compositions and sizes of the domains enables further tuning of the porosity characteristics.

It should be realized that the nanostructured metallic foams can show rather brittle behavior[83][84] (although we found way to circumvent that problem by making layered stacks[44]). Recently, Weißmüller and co-authors[85] investigated the anomalous compliance and early yielding of nanoporous gold by molecular dynamics simulation pointing out that nanoporous gold can be deformed to large strain in compression, but with very high strain-hardening coefficient. As regards brittleness, we have investigated size effects of these nanoporous materials. It is not difficult to understand that upon decreasing diameter (aspect ratio 3 because of stability) a lower stiffness is expected since a larger fraction of loosely bound

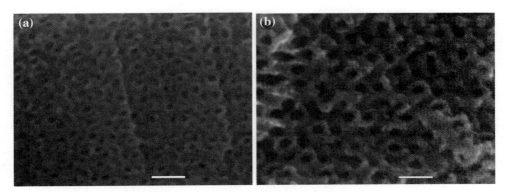

Figure 6. SEM images of the inverse gyroid nickel replicas obtained after the polymer template removal by pyrolysis (pyrolysis time = a 1 day, b 4 days). A 3D network structure composed of the interconnected nickel struts is clearly visible (scale bar 100nm).

struts appears whereas under pure shear the opposite will be observed (because of increasing fraction of confined material). Focused Ion Beam was employed to fabricate pillars with a controlled taper free shape. The use of non-tapered pillars prevents localization effects due to variation in diameter across the compression axis. Indentation experiments were carried out on the freestanding nontapered nanoporous metal pillars. Pillars were produced from both disordered nanoporous Au as well as ordered nano-porous Ni. Pillars produced in such a manner fully retained their porosity, as well as quality of ordering.

Two modes of deformation were observed during indentation (**Figure 7**): brittle fracture and ductile deformation with limited recovery. In the former, the pillar would spontaneously fracture at a particular load without developing any significant deformation prior. In the latter, the pillar would be observed to undergo a gradual bend without fracture until load release, where the pillar would partially regain its original shape. Full recovery was not observed due to the development of minor fracture regions during initial loading. Subsequent loads on the same pillar after recovery would increase the size of the fracture, leading to stress localization in subsequent loads and eventual spontaneous fracture nucleating from the affected area. It was observed that the tendency toward gradual deformation over spontaneous fracture was dependent on pillar diameter such that thinner pillars would deform gradually. Typically, pillars under 300–400-nm diameter would be thin enough for such behavior, and thicker pillars would be more likely to undergo spontaneous fracture. It is worth emphasizing that in all indentation experiments

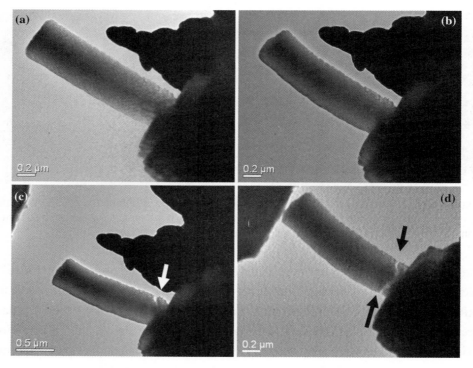

Figure 7. TEM micrograph of initial pillar (a), during loading (b), after loading with recovery and after subsequent loads (d). Minor fracture indicated by arrows after initial load is visible and the expansion thereof after subsequent loading.

where gradual deformation was the mode of deformation the nature of deformation was a bending, as opposed to a pure compression as would be expected from an indentation test This tendency made for challenging prediction on the true behavior of nanoporous pillars under compression, as bending adds additional deformation components to the otherwise simple cell wall buckling model of the effects of the compression of porous media. This bending stemmed from the specific design of the Hysitron Picoindenter, where the indentation tip is aligned such that, if a sample is sufficiently small, there is only a very specific set of sample positions and orientations that would allow for pure compression.

4. Electrolyte-Free Actuation in Metallic Muscles

As already aforementioned in Sect. 1, an important objective is to overcome the limitations that arise in metallic muscles when an aqueous electrolyte is used for actuation: existing metallic muscles made of nanoporous metals with high surface-area-to-volume ratios can exert work due to changes in their interface electronic charge density. However, they suffer from serious drawbacks caused by the usage of an aqueous electrolyte needed to modulate the interface electronic charge density by injection of electronic charge at the nanoporous metal/electrolyte interface. An aqueous electrolyte prohibits metallic muscles to operate in dry environments and hampers a high actuation rate due to the low ionic conductivity of electrolytes. Simply replacing the aqueous electrolyte by a solid is an obvious solution, but the actuation rate of all-solid-state electrochemical actuators is more severely hampered by the low room-temperature ionic conductivity of solid-state electrolytes. In addition, redox reactions involved in electrochemical actuation severely coarsen the ligaments of nanoporous metals, leading to a substantial loss in performance of the actuator. We have developed a new electrolyte-free concept to put metallic muscles to work via a metal/polymer interface[12]: A nanocoating of polyaniline (PANI) doped with sulfuric acid was grown onto the ligaments of nanoporous gold (NPG). Dopant sulfate anions co-adsorbed in the polymer coating matrix were exploited to tune the electronic charge density at the NPG interface and subsequently generate macroscopic dimensional changes in NPG. Strain rates achieved in the single-component NPG/PANI bulk heterojunction actuator are three orders of magnitude higher than that of the standard three-component nanoporous metal/electrolyte hybrid actuator. Also the combination of a nanoporous metal and a polymer has been further exploited to add a new functionality to metallic muscles operating as electrochemical actuators: in addition to the reversible dimensional changes in the nanoporous metal, the polymer undergoes reversible changes in color when it is electro-oxidized/reduced. This results in a smarter hybrid actuation material.

To show the proof of principle here: NPG was synthesized by the standard dealloying method, and the typical bicontinuous morphology of the synthesized NPG is shown on the scanning electron micrograph is shown in **Figure 4**. The corresponding interface surface area per unit mass computed from the analytical expression for the specific surface area of nanoporous materials[86]–[89] was found to be $\sim 7.7 m^2\ g^{-1}$. The internal surface area of the three-dimensional bicontinuous network of the synthesized NPG was uniformly coated by electropolymerization procedure, and a continuous doped polymer envelope covers the ligaments[90][91]. The potentiodynamic procedure (*i.e.*, monomers deposition with a variable voltage via cyclic voltammetry) was used[92], and the polymerization of aniline was carried out from an aqueous solution containing 50mM of aniline monomer and 0.5M of H_2SO_4 used as solvent and dopant molecules, respectively[90]. The scanning and transmission electron micrographs of **Figure 4** display a uniform PANI skin of average thickness of ~5nm grown onto the ligaments of NPG. The presence of carbon (C) and nitrogen (N) as constituents of the organic coating monomer C_6H_7N (aniline) was confirmed by energy-dispersive X-ray spectroscopy (EDS) performed during characterization of the PANI film by transmission electron microscopy (TEM). The typical EDS spectrum of this PANI envelope is displayed in the inset of **Figure 4(c)**. In this energy spectrum, the Cu and Au peaks are attributed to the Cu grid of the sample holder and to the NPG, respectively. The scanning electron micrograph of **Figure 4(d)** represents a fracture cross section of the NPG/PANI hybrid material. It can be seen that the nanocoating of polymer covering the ligaments of NPG is present in the bulk of the material. Detailed knowledge of the nature of electronic charge transport in the PANI film is necessary for a better insight into the actuation mechanism in the NPG/PANI hybrid material. A schematic illustration of this NPG/PANI bulk heterojunction actuator is shown in **Figures 8(a)-(d)**, and additional technical details are found in[93]. The actuator is connected to the voltage supplier in one of the following two configurations: NPG/PANI/Au with NPG as anode and solid Au as cathode, or Au/PANI/NPG with solid Au as anode and NPG as cathode. Both configurations can be used because the two contact electrodes (NPG and solid Au) are made of the same material. Since the work functions of NPG and solid Au are comparable with the highest occupied molecular orbital (HOMO) of PANI on the one hand[94], and since PANI is a p-type semiconductor on the other hand[95], electronic charge transport in the NPG/ PANI hybrid actuator is only controlled by holes conduction, rather than by both electrons and holes transports. Such a single carrier system is commonly referred to as a "hole-only" device because the high offset between the Fermi level of the metal and the lowest unoccupied molecular orbital (LUMO) of PANI restricts electrons injection from the metal into the LUMO of the polymer[96]. The energy-level diagram of the NPG/PANI/Au system in the absence of external electrical potential is shown in **Figure 8(e)**; it is seen that only holes can be injected from the anode into the HOMO of PANI.

An external electric potential was used to inject holes from the metal anode (either NPG or solid Au) into the HOMO of the PANI envelope[97]. A particular-

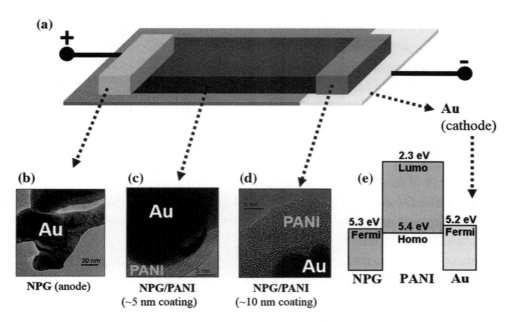

Figure 8. The NPG/PANI bulk heterojunction actuator. a, c The main part of the actuator consists of NPG whose ligaments are coated with a ~5-nm layer of PANI. b One edge connected to the positive terminal of the voltage supplier consists of NPG. d The other edge connected to the negative terminal consists of NPG having its ligaments covered with ~10-nm-thick layer of PANI. e Energy-level diagram of the system NPG/PANI/Au system in the absence of external electrical potential.

ity of the NPG/PANI/Au or Au/PANI/NPG configuration is that electronic charges injected from the anode flow through the PANI film before reaching the cathode. The typical current density-voltage (J-V) characteristic of the Au/ PANI/NPG system (holes injection from Au) is shown on the semi logarithmic plot of **Figure 9**. The inset of **Figure 9(a)** displays the corresponding forward J-V curve on a double logarithmic graph. Similarly, the J-V curve of the NPG/ PANI/Au system (holes injection from NPG) is shown on the semi logarithmic plot of **Figure 9(b)** for three successive forward–backward voltage sweeps in the same potential range between 0 and 2V. The inset of **Figure 9(b)** displays the corresponding forward J-V curve on a double logarithmic plot. The hysteresis[98] in the J-V curves are more pronounced when holes are injected from solid Au [**Figure 9(a)**] than from NPG [**Figure 9(b)**]; this can be caused by the difference in contact area for holes injection, which is larger for the NPG/PANI interface than the Au/PANI interface. Irrespective of this difference, it can be concluded from the slope values of 1 obtained on the double logarithmic graphs for both Au/PANI/NPG and NPG/PANI/

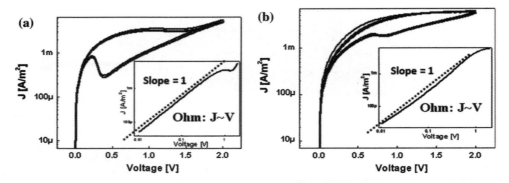

Figure 9. Nature of electronic charge transport in the PANI skin. Semi and double (insets)-logarithmic plots of the J-V curves for three successive forward-backward voltage cycles. It can be seen from the slope values of one that holes transport in the PANI coating is dominated by an Ohmic behavior in the potential range between 0 and 2V. a Holes injection from Au in the Au/PANI/NPG configuration. b Holes injection from NPG in the NPG/PANI/Au configuration. Reprinted from[12], with permission from the American Chemical Society.

Au configurations [see insets of **Figure 9(a), (b)**] that at low potentials, electronic charge transport through the polymer skin follows an Ohmic behavior at ambient temperatures[99]. A similar Ohmic behavior has been reported when the two contact electrodes are made of solid Au in the Au/PANI/Au configuration[12]. It is emphasized that the Ohmic nature of the current in the NPG/PANI hybrid actuator means that electronic charges do not accumulate at the metal/polymer interface during potential sweeps, in the contrary of a metal/electrolyte actuator where a (pseudo)capacitive double-layer formed at the metal/electrolyte interface is used for actuation. Therefore, the approach presented here for the modulation of the electronic charge density at the NPG/PANI interface is different from the building up of space-charge encountered in nanoporous metal/electrolyte hybrid materials.

Reversible dimensional changes were recorded in the NPG/PANI bulk heterojunction material during successive forward–reverse voltage cycles between 0 and 2V, and at various sweep rates ranging from 1 to 2000mV/s. A confocal displacement sensor (IFS2401-0.4 Micro-Epsilon) was used for this purpose. **Figure 10** displays typical dimensional changes as a function of the time recorded during 50 successive forward–backward potential sweeps at the sweep rate of 10mV/s, for the Au/PANI/NPG configuration. The NPG/PANI hybrid material expands during the forward sweep and contracts during the reverse process. The sign of the displacement is reversed when the actuator is connected in the NPG/PANI/Au

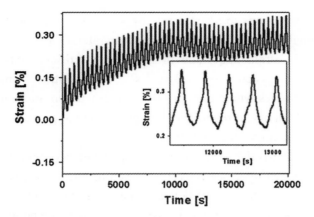

Figure 10. Electrolyte-free actuation in NPG/PANI. 50 well-reproducible expansion-contraction cycles recorded in response to 50 successive forward-reverse J-V cycles. The NPG/PANI hybrid material expands during the forward sweep and contracts during the reverse process. Reprinted from[12], with permission from the American Chemical Society.

configuration (*i.e.*, contraction during forward sweep and expansion during backward sweep). 50 well-reproducible expansion-contraction cycles are recorded in response to 50 forward-reverse voltage cycles. The strain amplitude at the sweep rate of 10mV/s is on the order of ~0.15%, and this is comparable to the one reported at lower sweep rates (between 0.2 and 1mV/s) in nanoporous metal/electrolyte composites[93].

It is emphasized that low sweep rates are required during actuation in nanoporous metals via an electrolyte; in fact dimensional changes in nanoporous metal/electrolyte composite actuators vanish at sweep rates beyond a few tens of mV/s. That behavior has two origins: (i) the low room-temperature ionic conductivity of electrolytes[100] does not favor a rapid transport of ions to the nanoporous metal/electrolyte interface; (ii) the equilibration of redox reactions involved in charged transferred at the nanoporous metal/electrolyte interface is not satisfied during fast sweep rates as highlighted in Ref.[93].

The impact of the voltage sweep rate on the performance of metallic muscles is clearly illustrated in the work of Viswanath *et al.*[93], who reported on a decrease of the strain amplitude in the nanoporous metal/electrolyte composite actuator from ~0.14% down to 0.05% when the sweep rate is increased from 30µV/s up to 1mV/s. That strain amplitude of ~0.05% is achieved in 1400s with the electrolyte, which corresponds to a strain rate of 3.6×10^{-7} s^{-1}. For the sake of comparison, a mammalian skeletal muscle has a strain rate of $\sim 10^{-1}$ s^{-1}[45] and is able to achieve the 0.05% strain amplitude in 5ms; most artificial muscles have their strain rate ranging between $\sim 10^{-3}$ and 10^{-1} s^{-1}[101]. In contrast to metallic muscles operating in electrolytes where dimensional changes are not present at sweep rates beyond a few tens of mV/s, reversible dimensional changes were still observed in our NPG/PANI electrolyte-free actuator at sweep rates far beyond 1mV/s as illustrated in **Figure 11(a), (b)**, where the strain amplitudes are plotted as a function of the time and sweep rate, respectively. By setting the sweep rate at 2000mV/s, the aforementioned strain of ~0.05% was achieved in our electrolyte-free actuator in 1s, rather than 1400s as with the electrolyte[93]. This corresponds to a strain rate of 5×10^{-4} s^{-1}, which is thus about 1400 times higher than that achieved in metallic muscles via an electrolyte. These results demonstrate that by virtue of the novel electrolyte-free actuation approach, metallic muscles can operate in dry environments at high strain rates, much higher than those of common electrochemical artificial muscles.

5. Discussion: Origin of the Dimensional Changes in Electrolyte-Free Actuation of Metallic Muscles

Referring to the nanoporous metal/electrolyte hybrid actuator, it is well-established that dimensional changes in this system are caused by changes in the

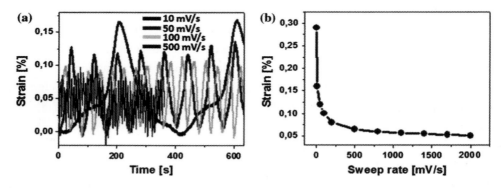

Figure 11. Fast actuation responses. Reversible dimensional changes are recorded at various sweep rates far beyond 1mV/s a as a function of the time and b as a function of the sweep rate. Reprinted from[12], with permission from the American Chemical Society (Color figure online).

nanoporous metal surface stress, when electronic charges are injected at the nanoporous metal/electrolyte interface during ions electroadsorption. In order to preserve the mechanical equilibrium, these changes in the nanoporous metal surface stress are compensated by opposite changes in the stress state in the bulk of the ligaments[102], resulting in an overall macroscopic dimensional changes in the nanoporous metal. For our NPG/PANI hybrid material, the situation is different because no electrolyte is used during actuation. In the absence of an electrolyte, changes in the NPG surface stress as a result of electronic charges accumulation in the space-charge region at the NPG interface are still possible, provided that an opposite space-charge builds up in the polymer coating during the voltage sweeps. However, as we have seen in the previous section, hole-transport in the PANI coating is governed by an Ohmic current (slope value of 1 in the double logarithmic plot), rather than a space-charge limited-current (slope value of 2 in the double logarithmic plot)[103]. This excludes the possibility of having changes in the surface stress of NPG as result of the build-up of a space-charge in the polymer coating.

Another possible origin of the measured dimensional changes in the NPG/PANI hybrid material points toward actuation in PANI. PANI can undergo reversible dimensional changes during electrochemical oxidation/reduction[104]. However, this option is not applicable to our electrolyte-free actuator because the electrochemical oxidation/reduction of PANI requires an electrolyte. Second, charge carriers in conducting polymers including PANI are susceptible to induce conformational changes in the polymer chains[105]-[108]. This later deformation mode does not necessarily require the oxidation or reduction of the polymer[105]. Conformational changes in PANI chains can therefore be responsible for the dimensional changes in our NPG/PANI composite material provided that stresses developed in the polymer chains during these conformational changes are fully transferred to the metal. This is not likely because mechanical adhesion at metal/polymer interfaces is commonly weak[109]. In addition, the relatively small PANI content in the NPG/PANI composite material (~Au95(PANI)5wt%) and the relatively low Young's modulus of PANI (~2GPa)[110] compared to that of the metallic ligaments (~79GPa) lead to the conclusion that the measured strains do not come from actuation in the thin polyaniline coating.

We propose the following physical picture: as schematized in **Figure 12(a)**,

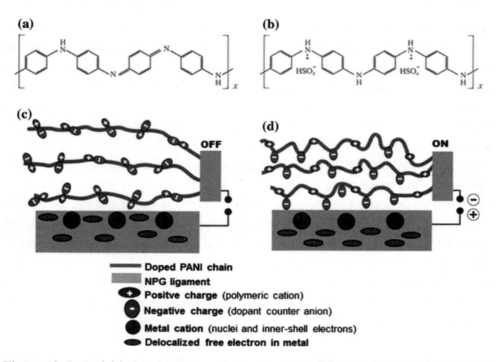

Figure 12. Potential-induced charge redistribution at the NPG surface. a Undoped PANI (blue insulating emeraldine base). b Doped PANI (green conducting emeraldine salt). c NPG/PANI interface in the absence of an electrical potential. Both positive and negative charge carriers along the polymer chains are held together by electrostatic interaction. An initial electronic charge distribution exists at the metal surface. d NPG/PANI interface in the presence of an electrical potential. Positive charge carriers along the polymer chains become involved in electrical conduction; localized negative charge carriers along the polymer chains electrostatically interact with the positive metal electrode, giving rise to electronic charge redistribution at the metal surface (Color figure online).

in its nonconducting state, the blue emeraldine base form of PANI consists of electrically neutral molecular chains[111]. PANI is made conducting (green emeraldine salt) by protonic acid doping or oxidative doping. During oxidative doping, an electron is removed from the pi-conjugated backbone, resulting in a free radical and a positive charge (polymeric cation) as schematized in **Figure 12(b)**. In the case of sulfuric acid doping[111], the charge neutrality in the doped PANI is maintained by negative sulfate counter ions co-adsorbed into the polymer matrix during the doping process[105][112]. The schematic structure of a doped PANI chain is shown in **Figure 12(b)**; both polymer cations and sulfate counter anions are held together by electrostatic interactions[113]. In the absence of an external electrical potential ["off" stand in **Figure 12(c)**], polymer chains adopt shapes that favor minimal intra- and inter-electrostatic interactions in the molecular chains. These shapes can be linear as reported by Lee et al. for a monolayer coating of PANI on a single Au crystal[105]. When a suitable electrical potential is applied on the hybrid actuator, e.g., in the NPG/PANI/Au configuration ["on" stand in **Figure 12(d)**], holes are injected from the positive NPG electrode into the PANI coating. The transport of these holes across the polymer film involves the cations and free radicals on the pi-conjugate backbone[95], the positive charges on the pi-conjugated backbone become mobile during electrical conduction, whereas the negative sulfate counter ions are localized along the chains.

The total amount of negative charges in the polymer matrix, arising from the co-adsorbed sulfate anions, was estimated for a 5-nm-thick PANI coating and was found to be ~3.2C per m^2 coating, assuming that each repeating unit of PANI contributes with two sulfate anions as illustrated in **Figure 12(b)**. This amount of charge is comparable to the quantity of electronic charge involved in dimensional changes in nanoporous metal/electrolyte hybrid actuators. During the potential sweeps, this relatively large amount of negative charge dispersed into the thin polymer matrix electrostatically interacts with the positive NPG electrode[105]. PANI molecular chains undergo conformational changes in order to bring the sulfate anions (i.e., negative charge carriers) in the proximity of the positive metal electrode; sulfate anions present in the first monolayer of PANI are eventually electroadsorbed onto the metal electrode as reported by Lee et al. and illustrated in **Figure 12(d)**[28]. The electrical potential-induced interactions between sulfate anions and the ligaments of NPG give rise to electronic charges redistribution at the ligaments interface[114]. Typically, the delocalized free electrons in the metal move from the interface toward the bulk, leaving the metal interface with positively charged metal ions [see **Figure 12(d)**]. These metal cations consist of nuclei and inner-shell electrons of metal atoms. The delocalization of negative charges from the metal surface toward the bulk weakens the interatomic bounds between metal surface atoms, resulting in relaxation of these metal surface atoms. This gives rise to an increase in tensile stress at the surface of the ligaments. The bulk of the ligaments opposes with an increase in compressive stress in order to preserve the mechanical equilibrium[102]. Due to the high surface-area-to-volume ratio of NPG, the dimensional changes in the ligaments result in an overall macroscopic volume change in the NPG electrode[6], which is experimentally measured during forward voltage sweeps in the NPG/ PANI/Au configuration.

During the reverse voltage sweep where the applied electrical potential is gradually removed, electrostatic interactions between the negative sulfate ions and the positive metal electrode gradually vanish, charge redistribution takes place again at the metal interface, and the initial charge distribution is restored. This causes lesser tensile surface stress in the ligaments and accordingly a reduction of the compressive bulk counter stress[102][114], resulting in ligaments expansion and in an overall macroscopic expansion of the NPG electrode back to its initial shape.

When the NPG/PANI hybrid actuator is connected in the configuration (Au/PANI/NPG), NPG is then used as negative electrode, and the sign of the strain is reversed (expansion during forward voltage sweeps), which suggests that in this later configuration conformational changes in the polymer chains take the negative sulfate anions away from the negative NPG electrode. Such a process will cause the delocalized electrons in the metal to move toward the metal surface, resulting in lesser tensile surface stress in NPG and, consequently, in a lesser compressive counter body stress in the bulk of the ligaments. In turn, the relaxation of the compressive stress in the bulk of the metal gives rise to a volume change.

Although the dimensional changes in the NPG/PANI hybrid actuator do not come from actuation in PANI as emphasized above, in the current understanding of the process it is believed that conformational changes in the polymer chains play an important role during actuation:

(i) Changes in molecular shapes of the polymer bring the sulfate anions in the proximity of the metal electrode, or take these counter anions away from the metal electrode depending on the sign of the potential applied at this electrode[28]. This process can be compared with the diffusion ions toward a metal/electrolyte interface in the case of actuation in an aqueous electrolyte. (ii) The high rate of which conducting polymers undergo conformational changes as highlighted by Yip and co-authors[107] might justify the high actuation rate recorded on the NPG/PANI composite material: rapid shape changes in polymer chains favor a fast exposure of sulfate anions to the positive NPG electrode and, consequently, rapid charge redistribution at the NPG interface. In contrast, when ions are transported through an electrolyte, a high actuation rate is hampered because of the low ionic conductivity of electrolytes[100].

The work density $W = 1/2Y\varepsilon^2$ of the NPG/PANI actuator ($\sim$113kJ/m^3), is comparable to the $\sim$130kJ/m^3 achieved in piezoceramics[4] and 90kJ/m^3 reported for the nanoporous metal/electrolyte actuator in Ref.[8]. In the above expression, W, Y, and e represent the volume work density, effective Young's modulus, and maximum strain amplitude, respectively[44][115]. Although the work density is the standard measure for the mechanical performance of artificial muscles, it is pointed out that a high value of W does not necessarily mean that the corresponding actuation material is suitable for every application. In fact, each actuation material satisfies only specific applications depending on how Y and e are combined: materials such as electroactive polymers can produce large actuation strokes ($\varepsilon\sim$4.5%), but they are weak (Y$\sim$1.1GPa); other like piezoceramics are strong (Y$\sim$64GPa) but their strain amplitudes are restricted to $\sim$0.2%. Metallic muscles are unique for a number of reasons, but none-more-so than in the sense that they can achieve a wide range of strengths as depicted by the effective Young's modulus of NPG, which is tunable from $\sim$5 to $\sim$45GPa through manipulation of the ligaments' size[116]. Additionally, they can also be designed to achieve a wide window of strain amplitudes ranging from the standard value of $\sim$0.1% up to large strains of $\sim$1.3% in binary nanoporous alloys[8].

As regards the small volumetric strain of about 0.1%, we have designed a new type of nanoporous gold architecture consisting of a two-microscopic length scale structure and details are presented in[44]. The nanoporous gold with the two-microscopic length scale structure consists of stacked gold layers with submicrometer thicknesses; in turn each of these layers displays nanoporosity through its entire bulk. This two-length scale structure strongly enhances the stain amplitude in metallic muscles up to 6%, compared to the standard strain of $\sim$0.1% achieved in nanoporous metals with one-microscopic length scale structure (*i.e.*, with uniform porous structure). The ratio between the work density of NPG actuator with a dual-microscopic length scale (volumetric strain 6%) and that of NPG actuator with a one-length scale porous morphology (volumetric strain 0.3%) was found to be $\sim$215. It is concluded that the relatively low effective Young's modulus of NPG with layered structure (compared to that of NPG with a uniform porous morphology) is largely compensated by the giant strain amplitudes. The large strains in NPG with layered structure give rise to an enhancement of the work density with at least two orders of magnitude.

Recently, various alternatives for achieving large displacements in actuation materials have been investigated[47], and several ideas were proposed for the displacement amplification including cantilever systems, hydraulic-piston devices, and piezoelectric motors. These techniques however are not always appropriate for microscale applications. Kramer *et al.* have achieved large relative displacements during cantilever bending experiments[9], up to ~3mm over a length of ~35mm–40 mm, by using a nanoporous metal strip to design a 40 mm-long bilayer strips. One advantage of the actuation mechanism associated to the layered structure we developed is the possibility to achieve comparable large relative displacements at smaller scales: displacements up to ~4μm can be achieved over a thickness of ~70μm. Furthermore, the multilength scale layered nanoporous systems operate at low voltages compared to common artificial muscles. Exploiting a polymer skin augmentation of the muscle for actuation, as we have demonstrated, is expected to stimulate the development of metallic muscles into a new class of actuation materials that operate at low voltages and combine large strain amplitudes with high stiffness and strength.

6. Summary and Outlook for Next Generation Applications of Nanoporous Metals

In conclusion, although metal nanofoams share many properties with their macrofoam counterparts, they have many applications beyond those of macrofoams. One promising application corresponds to metallic muscles based on nanoporous metals with high surface-area-to-volume ratios. For that specific application, we have demonstrated a new electrolyte-free approach to generate work from metallic muscles by exploiting a nanoporous metal/polymer interface rather than the common nanoporous metal/liquid electrolyte interface. In this actuation concept, a doped polymer coating is grown onto the ligaments of a nanoporous metal, and dopant counter ions present in the polymer coating matrix are exploited to modulate the electronic charge distribution at the nanoporous metal surface, resulting in surface stress changes and dimensional changes in the nanoporous metal. With this actuation approach, many of the drawbacks encountered in metallic muscles operating in aqueous electrolytes have been circumvented. In particular, the electrolyte-free actuator consists of a single-component hybrid material, in contrast to the three-component configuration required in nanoporous metal/electrolyte composite actuators; the nanoporous metal/polymer hybrid actuator is an all-solidstate device, like piezoceramic actuators, and its actuation rate is about three orders of magnitude higher than that of metallic muscles operating in aqueous electrolytes.

An interesting observation is that a thin polymer coating grown onto the metallic ligaments of nanoporous gold can be exploited to add a new functionality to nanoporous metals operating as electrochemical actuators. For example, a metallic muscle becomes a smart material because in addition to its reversible dimensional changes, it also undergoes a reversible change in color. This combination of electromechanical and optical changes could open the door to new applications in artificial muscles. A straightforward application includes a metallic muscle that can give feedback on the progress on its work simply by changing its color[117][118]. An interesting and rather new development in this field of actuation was recently published by Shih and co-authors[119] making an actuator made from botanic epidermal cells. This soft actuator changes its actuation direction by simply changing the magnitude of the applied voltage. In fact, the single-layered, latticed microstructure of onion epidermal cells after acid treatment became elastic and could simultaneously stretch and bend when an electric field was applied.

The exciting outlook for nanoporous metallic systems is not limited to the fields of sensors and actuators, however, where electrical or electrochemical energy is converted into mechanical work. The use of nanoporous metals and nanostructured materials in general[120] in other energy-related applications has recently been demonstrated. For example, there is currently much interest in the sustainable production of hydrogen fuel by the decomposition of water-based solutions into hydrogen and oxygen as the only products using alkaline electrolyzers. Interestingly, state-of-the-art oxygen evolution catalysts are increasingly grown on macroporous nickel foams, in order to improve the catalytic activity of these anode electrolyzers[121][122]. We have recently shown that catalytic activities toward oxygen evolution can be significantly enhanced if nanofoam catalysts are directly used as electrolyzer anodes[123], instead macroporous foams. In general, unsupported nonprecious metals oxygen-evolving catalysts require at least ~350mV overpotential to oxidize water with a current density of 10mA cm^{-2} in 1 M alkaline solution[124]. In our work, we have recently found a robust ultrafine mesoporous NiFe-based oxygen evolution catalyst made by partial removal of Fe and Mn from

Ni–Fe–Mn parent alloys[123]. The fine microstructure of our mesoporous NiFe-based oxygen evolution catalyst is shown at different magnifications in **Figure 13**[123]. Ligament and pore sizes are of the order of 10nm. In 0.5M KOH, only ~200 mV overpotential was required to oxidize water with a current density of 10mA cm^{-2}. In 1 M KOH, our material exhibits a catalytic activity toward water oxidation of 500mA cm^{-2} at 360mV overpotential and is stable for over 11 days. This exceptional performance is attributed to a combination of the small size of ligaments and pores in our mesoporous catalyst (~10nm), the high BET surface area that results from those sites (43m^2 g^{-1}), and therefore the high density of oxygen-evolving catalytic sites per unit mass. In addition, the open porosity facilitates effective mass transfer at the catalyst/electrolyte interface, and the high electrical conductivity of the mesoporous catalyst allows for effective current flow. Such a robust mesoporous catalyst is attractive for alkaline electrolyzers where water-based solutions are decomposed into hydrogen and oxygen as the only products.

Another promising application where nanoporous metallic system could have a significant impact is Li-ion and Na-ion batteries, where electrical energy is stored in the form of chemical energy. Porosity evolution in nano-porous tin (NP-Sn) during selective removal of Li from Li–Sn alloys was recently investigated by Chen and Sieradzki[125]. The ligaments in their NP-Sn exhibit an interpenetrating 'nanowire' type morphology, similar to that of common nanoporous metals. We have recently reported on the synthesis of NP-Sn powder with a unique

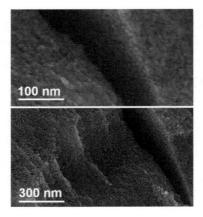

Figure 13. Mesoporous NiFe-based oxygen evolution catalyst at different magnifications. Ligament and pore sizes are of the order of 10nm.

granular-like ligament morphology using Sn–Mg binary alloys as pre-cursor and selectively removing the sacrificial Mg by free-corrosion dealloying. The typical microstructure of our NP-Sn powder is shown in **Figure 14(a)**[126]–[128]. In general, mechanically stable monolithic nanoporous metals are made by limiting the content of the sacrificial component in the parent alloys between 60at.% and 70at.%. However, for applications involving Li-ion or Na-ion battery electrode slurries, NP-Sn powder (instead of monolithic NP-Sn) is desirable. The high magnification SEM of typical powder particles indicate that they are porous [see **Figure 14(b)**]. While the porous architecture is disordered like in common nanoporous metals, the ligaments exhibit granular-like morphology [see **Figure 14(c)**], rather than an interpenetrating 'nanowire' type morphology. The ligament diameter exhibits a broad distribution, in the range between ~100 and ~200nm [**Figure 2(c)**]. Further magnification of these ligaments shows that they consist of clustered Sn nanoparticles with size in the sub-10-nm range (Not shown). These small feature sizes give rise to a relatively high BET surface area (~19m^2 g^{-1}) in our NP-Sn powder. Due to its unique granular-like ligament morphology, the NP-Sn powder

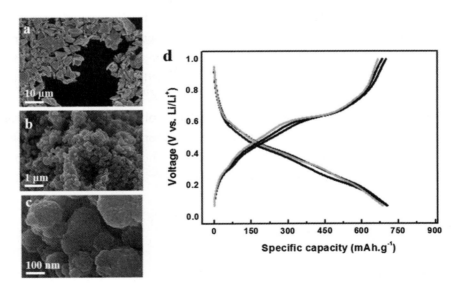

Figure 14. Nanoporous Sn (NPSn) as anode material for the next generation Li-ion battery. a As-synthesized fine NP-Sn powder dispersed on a carbon substrate. The powder particles have random shapes and random size in the sub-10-μm ranges. b High magnification SEM showing that the powder particles are porous. c The ligaments exhibit granular-like morphology. TEM analysis reveals that these granular ligaments are in turn porous (not shown). d Examples of galvanostatic curves obtained using Li$^+$. Good stability over a few hundred cycles can be achieved.

exhibits stable cycling as anodes in combination with both lithium and sodium. **Figure 14(d)** shows an example of a galvanostatic curve obtained using Li^+; good stability over a few hundred cycles can be achieved[126]–[128].

Overall we may conclude that metallic nanofoams sit at the centrepoint of a myriad of engineering disciplines, enabling a variety of applications because of their chemical and structural diversity. In this contribution, we have highlighted mainly the 'metallic muscle' performance, and as far as functional properties are concerned, modern actuating materials—*i.e.*, materials that have the capacity for controlled deformation under an applied electric current—nanoporous metals make ideal candidates for such roles. Actuation in nanoporous metals is enabled by the fact that injection of charge into a metal causes a change of surface charge, which is amplified due to the high surface-to-volume ratio present in a nanoporous metal; this ultimately leads to measurable deformation. Clearly as porous materials, these systems display excellent surface-area-to-volume ratios and attractive commercial properties such as reasonable low-cost production. As nanostructured materials, they also offer many properties often sought-out for potential applications: good electrical and thermal conductivity, strong capacity as catalyst and catalyst carrier, and potentially exploitable optical properties, to name just a few. As metals, they boast the robust mechanical properties required of a structural material: strength, impact resistance, and resistance to aging[129][130]. In particular, the exploitation of nanoporous metals in energy-related applications such as hydrogen fuel production and batteries opens novel avenues for fundamental and applied materials research that may result in many new publications during the next 50 years in the Journal of Materials Science.

Acknowledgements

The authors are thankful to the Netherlands Organization for Scientific Research (NWO-the Hague, Mozaıek Grant 2008 BOO Dossiernr: 017.005.026 and the Rubicon Grant Dossiernr: 680-50-1214) and the Zernike Institute for Advanced Materials, University of Groningen, the Netherlands. Fruitful discussions with Patrick Onck and the testing of NP-Sn by John Cook are gratefully acknowledged.

Chapter 23

Compliance with Ethical Standards

Conflict of Interest: The authors declare that they have no conflict of interest.

Open Access: This article is distributed under the terms of the Creative Commons Attribution 4.0 International License (http://creativecommons.org/licenses/by/4.0/), which permits unrestricted use, distribution, and reproduction in any medium, provided you give appropriate credit to the original author(s) and the source, provide a link to the Creative Commons license, and indicate if changes were made.

Source: Detsi E, Tolbert S H, Punzhin S, *et al*. Metallic muscles and beyond: nanofoams at work[J]. Journal of Materials Science, 2016, 51(1):615–634.

References

[1] Schaller RR (1997) Moore's law: past, present, and future. Spectr IEEE 34(52):59.

[2] Hunter IW, Lafontaine SA (1992) Comparison of muscle with artificial actuators. In: Solid-state sensor and actuator workshop, 5th technical digest., IEEE, pp 178–185.

[3] Momosaki E, Kogure S (1982) The application of piezoelectricity to watches. Ferroelectrics 40:203–216.

[4] Jin HJ, Wang XL, Parida S, Wang K, Seo M, Weissmüller J (2010) Nanoporous Au-Pt alloys as large strain electrochemical actuators. Nano Lett 10:187–194. doi:10.1021/nl903262b.

[5] Mirfakhrai T, Madden JDW, Baughman RH (2007) Polymer artificial muscles. Mater Today 10:30–38. doi:10.1016/S1369-7021(07)70048-2.

[6] Li DB, Paxton WF, Baughman RH, Huang TJ, Stoddart JF, Weiss PS (2009) Molecular, supramolecular, and macromolecular motors and artificial muscles. MRS Bull 34:671–681. doi:10.1557/mrs2009.179.

[7] Gleiter H, Weissmüller J, Wollersheim O, Würshum R (2009) Nanocrystalline materials: a way to solids with tunable electronic structures and properties? Acta Mater 49:737–745. doi:10.1016/S1359-6454(00)00221-4.

[8] Weissmüller J, Viswanath RN, Kramer D, Zimmer P, Wür-schum R, Gleiter H (2003) Charge-induced reversible strain in a metal. Science 300:312–315. doi:10.1126/science.1081024.

[9] Kramer D, Viswanath RN, Weissmuller J (2004) Surface-stress induced macroscopic

bending of nanoporous gold cantilevers. Nano Lett 4:793–796. doi:10.1021/nl 049927d.

[10] Biener J, Hodge AM, Hamza AV, Hsiung LM, Satcher JH (2005) Nanoporous Au: a high yield strength material. J Appl Phys 97:024301-1–024301-4. doi:10.1063/1.1832742.

[11] Balk TJ, Eberl C, Sun Y, Hemker KJ, Gianola DS (2009) Tensile and compressive microspecimen testing of bulk nanoporous gold. JOM 61:26–31. doi:10.1007/s11837-009-0176-6.

[12] Detsi E, Onck P, De Hosson JTM (2013) Metallic muscles at work: High rate actuation in nanoporous gold/polyaniline composites. ACS Nano 7:4299–4306. doi:10. 1021/nn400803x.

[13] Biener J, Hodge AM, Hayes JR, Volkert CA, Zepeda-Ruiz LA, Hamza AV, Abraham FF (2006) Size effects on the mechanical behavior of nanoporous Au. Nano Lett 6:2379–2382. doi:10. 1021/nl061978i.

[14] Xia R, Xu C, Wu W, Li X, Feng XQ, Ding Y (2009). Microtensile tests of mechanical properties of nanoporous Au thin Films. J Mater Sci 44:4728–4733. doi:10.1007/s10853-009-3731-1.

[15] Li R, Sieradzki K (1992) Ductile–brittle transition in random porous Au. Phys Rev Lett 68:1168–1171. doi:10.1103/Phys RevLett.68.1168.

[16] Biener J, Hodge AM, Hamza AV (2005) Microscopic failure behavior of nanoporous gold. Appl Phys Lett 87:121908. doi:10. 1063/1.2051791.

[17] Volkert CA, Lilleodden ET, Kramer D, Weissmüller J (2006) Approaching the theoretical strength in nanoporous Au. Appl Phys Lett 89:061920. doi:10.1063/1.1832742.

[18] Gibson LJ, Ashby MF (1997) Cellular solids: structure and properties, 2nd edn. Cambridge University Press, Cambridge.

[19] Punzhin S, Detsi E, Kuzmin A, De Hosson JTM (2014) Deformation of nanoporous nanopillars by ion beam-induced bending. J Mater Sci 49:5598–5605. doi:10.1007/s10853-014-8269-1.

[20] Amsterdam E, De Vries JHB, De Hosson JTM, Onck PR (2008) The influence of strain-induced damage on the mechanical response of open-cell aluminum foam. Acta Mater 56:609–618. doi:10.1016/j.actamat.2007.10.034.

[21] Michielsen K, De Raedt H, De Hosson JTM (2002) Aspects of integral-geometry. Adv Imaging Electron Phys 125:119–195.

[22] Amsterdam E, Onck PR, De Hosson JTM (2005) Fracture and microstructure of open cell aluminum foam. J Mater Sci 40:5813–5819. doi:10.1007/s10853-005-4995-8.

[23] Amsterdam E, Goodall R, Mortensen A, Onck PR, De Hosson JTM (2008) Fracture behavior of low-density replicated aluminum alloy foams. Mater Sci Eng A 496:

376–382. doi:10. 1016/j.msea.2008.05.036.

[24] Amsterdam E, De Hosson JTM, Onck PR (2008) On the plastic collapse stress of open-cell aluminum foam. Scr Mater 59:653–656. doi:10.1016/j.scriptamat.2008.05.025.

[25] Amsterdam E, Van Hoorn H, De Hosson JTM, Onck PR (2008) The influence of cell shape anisotropy on the tensile behavior of open cell aluminum foam. Adv Eng Mater 10:877–881. doi:10. 1002/adem.200800128.

[26] Wittstock A, Zielasek V, Biener J, Friend CM, Baumer M (2010) Nanoporous gold catalysts for selective gas-phase oxidative coupling of methanol at low temperature. Science 327:319–322. doi:10.1126/science.1183591.

[27] Nagle LC, Rohan JF (2011) Nanoporous gold anode catalyst for direct borohydride fuel cell. Int J Hydrog Energy 36:10319–10326. doi:10.1016/j.ijhydene.2010.09.077.

[28] Lang XY, Yuan HT, Iwasa Y, Chen MW (2011) Three-dimensional nanoporous gold for electrochemical supercapacitors. Scr Mater 64:923–926. doi:10.1016/j.scriptamat.2011.01.038.

[29] Gittard SD, Pierson BE, Ha CM, Wu C-AM, Narayan RJ, Robinson DB (2010) Supercapacitive transport of pharmacologic agents using nanoporous gold electrodes. Biotechnol J 5:192–200. doi:10.1002/biot.200900250.

[30] Yavuz MS, Cheng Y, Chen J, Cobley CM, Zhang Q, Rycenga M, Xie J, Kim C, Song KH, Schwartz AG, Wang LV, Xia Y (2009) Gold nanocages covered by smart polymers for controlled release with near-infrared light. Nat Mater 8:935–939. doi:10.1038/nmat2564.

[31] Au L, Zheng D, Zhou F, Li Z-Y, Li X, Xia Y (2008) A quantitative study on the photothermal effect of immuno gold nanocages targeted to breast cancer cells. ACS Nano 2:1645–1652. doi:10.1021/nn800370j.

[32] Chuan C; Ngan AHW (2015) Reversible electrochemical actuation of metallic nanohoneycombs induced by pseudocapacitive redox processes. ACS Nano 9: 3984–3995. doi:10.1021/nn507466n.

[33] Biener J, Wittstock A, Baumann TF, Weissmuller J, Baumer M, Hamza AV (2009) Surface chemistry in nanoscale materials. Materials 2:2404–2428. doi:10.3390/ma2042404.

[34] Kwan KW, Gao P, Martin CR, Ngan AHW (2015) Electrical bending actuation of gold-films with nanotextured surfaces. Appl Phys Lett 106:023701. doi:10.1063/1.4905676.

[35] Cheng C, Ngan AHW (2013) Charge-induced reversible bending in nanoporous alumina-aluminum composite. Appl Phys Lett 102:213119–2131122. doi:10.1063/1.4808212.

[36] Detsi E, Chen ZG, Vellinga WP, Onck PR, De Hosson JTM (2012) Actuating and sensing properties of nanoporous gold. J Nanosc Nanotechnol 11:4951–4955.

doi: 10.1166/jnn.2012. 4882.

[37] Detsi E, Chen ZG, Vellinga WP, Onck PR, De Hosson JTM (2011) Reversible strain by physisorption in nanoporous gold. Appl Phys Lett 99:083104. doi:10.1063/1.3625926.

[38] Jayachandran KP, Guedes JM, Rodrigues HC (2011) Ferro-electric materials for piezoelectric actuators by optimal design. Acta Mater 59:3770–3778. doi:10.1016/j.actamat.2011.02.005.

[39] Ibach H (1997) The role of surface stress in reconstruction, epitaxial growth and stabilization of mesoscopic structures. Surf Sci Rep 29:193–263. doi:10.1016/S0167-5729(97)00010-1.

[40] Needs RJ, Godfrey MJ, Mansfield M (1991) Theory of surface stress and surface reconstruction. Surf Sci 242:215–221. doi:10. 1016/0039-6028(91)90269-X.

[41] Salomons E, Griessen R, De Groot DG, Magerl A (1988) Surface-tension and subsurface sites of metallic nanocrystals determined by H-absorption. Europhys Lett 5:449–454.

[42] Jin HJ, Weissmüller J (2010) Bulk nanoporous metal for actuation. Adv Eng Mater 12:714–723. doi:10.1002/adem. 200900329.

[43] Saane SSR, Mangipudi KR, Loos KU, DecHosson JTM, Onck PR (2014) Multiscale modeling of charge-induced deformation of nanoporous gold structures. J Mech Phys Sol 66:1–15. doi:10. 1016/j.jmps.2014.01.007.

[44] Detsi E, Punzhin S, Rao J, Onck PR, De Hosson JTM (2012) Enhanced strain in functional nanoporous gold with a dual microscopic length scale structure. ACS Nano 6:3734–3744. doi:10.1021/nn300179n.

[45] Baughman RH (2003) Muscles made from metal. Science 300:268–269. doi:10.1126/science.1082270.

[46] Zhang J, Liu P, Ma H, Ding Y (2007) Nanostructured porous gold for methanol electro-oxidation. J Phys Chem C 111:10382. doi:10.1021/jp072333p.

[47] Conway NJ, Traina ZJ, Kim SG (2007) A strain amplifying piezoelectric MEMS actuator. J Micromech Microeng 17:781–787. doi:10.1088/0960-1317/17/4/015.

[48] Erlebacher Aziz MJ, Karma A, Dimitrov N, Sieradzki K (2001) Evolution of nanoporosity in dealloying. Nature 410:450. doi:10.1038/35068529.

[49] Kimling J, Maier M, Okenve B, Kotaidis V, Ballot H, Plech A (2006) Turkevich method for gold nanoparticle synthesis revisited. J Phys Chem B 110:15700–15707. doi:10.1021/jp061667w.

[50] Detsi Van, de Schootbrugge M, Punzhin S, Onck PR, De Hosson JTM (2011) On tuning the morphology of nanoporous gold. Scr Mater 64:319–322. doi:10.1016/j.scriptamat.2010.10.023.

[51] Hodge A, Biener J, Hayes J, Bythrow P, Volkert C, Hamza A (2007) Scaling equation for yield strength of nanoporous open-cell foams. Acta Mater 55:1343. doi:10.1016/j.actamat.2006.09. 038.

[52] Detsi E, Selles MS, Onck PR, De Hosson JTM (2013) Nanoporous silver as electrochemical actuator. Scr Mater 69:195–198. doi:10.1016/j.scriptamat.2013.04.003.

[53] Sun L, Chien CL, Searson PC (2004) Fabrication of nanoporous nickel by electrochemical dealloying. Chem Mater 16:3125–3129. doi:10.1021/cm0497881.

[54] Wang XG, Qi Z, Zhao CC, Wang WM, Zhang ZH (2009). Influence of alloy composition and dealloying solution on the formation and microstructure of monolithic nanoporous silver through chemical dealloying of Al–Ag alloys. J Phys Chem C113: 13139–13150. doi:10.1021/jp902490u.

[55] Li W-C, Balk TJ (2010) Achieving finer pores and ligaments in nanoporous palladium-nickel thin films. Scr Mater 62:167–169. doi:10.1016/j.scriptamat.2009.10.009.

[56] Sun Y, Burger SA, Balk TJ (2014) Controlled ligament coarsening in nanoporous gold by annealing in vacuum versus nitrogen. Philos Mag 94:1001–1011. doi:10. 1080/14786435.2013.876113.

[57] Wang L, Balk TJ (2014) Synthesis of nanoporous nickel thin films from various precursors. Philos Mag Let 94:573–581. doi:10.1080/09500839.2014.944600.

[58] Kim MS, Nishikawa H (2013) Fabrication of nanoporous silver and microstructural change during dealloying of melt-spun Al-20at.%Ag in hydrochloric acid. J Mater Sci 48:5645–5652. doi:10.1007/s10853-013-7360-3.

[59] Wang J, Xia R, Zhu J, Ding Y, Zhang X, Chen Y (2012) Effect of thermal coarsening on the thermal conductivity of nanoporous gold. J Mater Sci 47:5013–5018. doi:10.1007/s10853-012-6377-3.

[60] Li ZQ, Li BQ, Qin ZX, Lu X (2010) Fabrication of porous Ag by dealloying of Ag–Zn alloys in H2SO4 solution. J Mater Sci 45:6494–6497. doi:10.1007/s10853-010-4737-4.

[61] Zinchenko O, De Raedt HA, Detsi E, Onck PR, De Hosson JTM (2013) Nanoporous gold formation by dealloying: a metropolis Monte Carlo study. Comput Phys Commun 184:1562–1569. doi:10.1016/j.cpc.2013.02.004.

[62] Van Bremen R, Ribas Gomes D, de Jeer LTH, Ocelík, De Hosson JTM (2015). On the optimum resolution of transmission—electron backscattered diffraction (t-EBSD). Ultramicroscopy (submitted).

[63] Keller R, Geiss RH (2012) Transmission EBSD from 10nm domains in a scanning electron microscope. J Microsc 245:245–251. doi:10.1111/j.1365-2818.2011.03566.x.

[64] Trimby PW (2012) Orientation mapping of nanostructured materials using transmission Kikuchi diffraction in the scanning electron microscope. Ultramicroscopy 120: 16–24. doi:10.1016/ j.ultramic.2012.06.004.

[65] Brodusch N, Demers H, Trudeau M, Gauvin R (2013) Acquisition parameters optimization of a transmission electron forward scatter diffraction system in a cold-field emission scanning electron microscope for nanomaterials characterization. Scanning 35:375–386. doi:10.1002/sca.21078.

[66] Geiss R, Keller R, Sitzman S, Rice P (2011) New method of transmission electron diffraction to characterize nanomaterials in the SEM. Microsc Microanal 17:386–387. doi:10.1017/ S1431927611002807.

[67] de Jeer LTH, Ribas Gomes D, Nijholt J, Ocelik V, De Hosson JTM (2015) Formation of nanoporous gold studied by transmission kikuchi diffraction. J Microsc Microanal, submitted 2015.

[68] Hayes JR, Hodge AM, Biener J, Hamza AV, Sieradzki K (2006) Monolithic nanoporous copper by dealloying Mn–Cu. J Mater Res 21:2611–2616. doi:10.1557/jmr.2006.0322.

[69] Min US, Li JCM (1994) The microstructure and dealloying kinetics of a Cu-Mn alloy. J Mater Res 9:2878–2883. doi:10. 1557/JMR.1994.2878.

[70] Foresi JS, Villeneuve PR, Ferrera J, Thoen ER, Steinmeyer G, Fan S, Joannopoulos JD, Kimerling LC, Smith HI, Ippen EP (1997) Photonic-bandgap microcavities in optical waveguides. Nature 390:143–145. doi:10.1038/36514.

[71] Palasantzas G, De Hosson JTM, Michielsen KFL, Stavenga DG (2005) Optical properties and wettability of nanostructured biomaterials: moth eyes, lotus leaves, and insect wings. Hand-book of nanostructured bio-materials and their applications in nanobiotechnology (Chap. 7). American Science Publications, Valencia, pp 274–301.

[72] Li X, Tao F, Jiang Y, Xu Z (2007) 3-D Ordered macroporous cuprous oxide: fabrication, optical, and photoelectrochemical properties. J Colloid Interface Sci 308:460–465. doi:10.1016/j. jcis.2006.12.044.

[73] Soni K, Rana BS, Sinha AK, Bhaumik A, Nandi M, Kumar M, Dhar GM (2009) 3-D ordered mesoporous KIT-6 support for effective hydrodesulfurization catalysts. Appl Catal B 90:55–63. doi:10.1016/j.apcatb.2009.02.010.

[74] Meng Y, Gu D, Zhang F, Shi Y, Yang H, Li Z, Yu C, Tu B, Zhao D (2005) Ordered mesoporous polymers and homologous carbon frameworks: amphiphilic surfactant templating and direct transformation. Angew Chem 117:7215–7221. doi:10. 1002/anie.200501561.

[75] Fang TH, Wang TH, Kang SH (2019) Nanomechanical and surface behavior of polydimethylsiloxane filled nanoporous anodic alumina. J Mater Sci 44:1588–1593. doi: 10.1007/ s10853-008-3232-7.

[76] Yu H, Wang D, Han MY (2007) Top-down solid-phase fabrication of nanoporous cadmium oxide architectures. J Am Chem Soc 2007:2333–2337. doi:10.1021/ja06 6884p.

[77] Garrigue P, Delville MH, Labrugere C, Cloutet E, Kulesza PJ, Morand JP, Kuhn A (2004) Top-down approach for the preparation of colloidal carbon nanoparticles. Chem Mater 16:2984–2986. doi:10.1021/cm049685i.

[78] Jeon HJ, Kim KH, Baek YK, Kim DW, Jung HT (2010) New top-down approach for fabricating high-aspect-ratio complex nanostructures with 10 nm scale features. Nano Lett 10:3604–3610. doi:10.1021/nl1025776.

[79] Van Eerdenbrugh B, Van den Mooter G, Augustijns P (2008) Top-down production of drug nanocrystals: nanosuspension stabilization, miniaturization and transformation into solid products. Int J Pharm 364:64–75. doi:10.1016/j.ijpharm.2008.07.023.

[80] Koch CC (2003) Top-down synthesis of nanostructured materials: mechanical and thermal processing methods. Rev Adv Mater Sci 5:91–99.

[81] Yan Y, Hu Z, Zhao Z, Sun T, Dong S, Li X (2010) Top-down nanomechanical machining of three-dimensional nanostructures by atomic force microscopy. Small 6:724–728. doi:10.1002/smll.200901947.

[82] Vukovic I, Punzhin S, Vukovic Z, Onck P, De Hosson JTM, ten Brinke G, Loos K (2011) Supramolecular route to well-ordered metal nanofoams. ACS Nano 5:6339–6348. doi:10.1021/nn201421y.

[83] Hakamada M, Mabuchi M (2007) Mechanical strength of nanoporous gold fabricated by dealloying. Scr Mater 56:1003–1006.

[84] Hodge AM, Hayes JR, Caro JA, Biener J, Hamza AV (2006) Characterization and mechanical behavior of nanoporous gold. Adv Eng Mater 8:853–857.

[85] Dinh NgôBao-Nam, Stukowski A, Mameka N, Markmann J, Albe K, Weissmüller J (2015) Anomalous compliance and early yielding of nanoporous gold. Acta Mater 93:144–155.

[86] Detsi E, De Jong E, Zinchenko A, Vukovic Z, Vukovic I, Punzhin S, Loos K, Ten Brinke G, De Raedt HA, Onck PR, De Hosson JTM (2011) On the specific surface area of nanoporous materials. Acta Mater 59:7488–7497. doi:10.1016/j.actamat.2011.08.025.

[87] Tan YH, Davis JA, Fujikawa K, Ganesh NV, Demchenko AV, Stine KJ (2012) Surface area and pore size characteristics of nanoporous gold subjected to thermal, mechanical, or surface modification studied using gas adsorption isotherms, cyclic voltammetry, thermogravimetric analysis, and scanning electron microscopy. J Mater Chem 22:6733–6745. doi:10.1039/C2JM16633J.

[88] Detsi E, Vuković Z, Punzhin S, Bronsveld PM, Onck PR, De Hosson JTM (2012) Fine tuning the feature size of nanoporous silver. Cryst Eng Comm 14:5402–5406. doi:10.1039/C2CE25313E.

[89] Detsi E, Punzhin S, Onck PR, De Hosson JTM (2012) Direct synthesis of metal nanoparticles with tunable porosity. J Mater Chem 22:4588–4591. doi:10.1039/C2JM15801A.

[90] Lang X, Zhang L, Fujita T, Ding Y, Chen MW (2012) Three-dimensional bicontinuous nanoporous Au/polyaniline hybrid films for high-performance electrochemical supercapacitors. J Power Sour 197:325–329. doi:10.1016/j.jpowsour.2011.09.006.

[91] Meng F, Ding Y (2011) Sub-micrometer-thick all-solid-state supercapacitors with high power and energy densities. Adv Mater 23:4098–4102. doi:10.1002/adma.201101678.

[92] Heinze J, Frontana-Uribe BA, Ludwigs S (2010) Electrochemistry of conducting polymers-persistent models and new concepts. Chem Rev 110:4724–4771. doi:10.1 021/cr900226k.

[93] Viswanath RN, Kramer D, Weissmuller J (2008) Adsorbate effects on the surface stress-charge response of platinum electrodes. Electrochim Acta 53:2757–2767. doi:10.1016/j.electacta. 2007.10.049.

[94] Reis FT, Santos LF, Faria RM, Mencaraglia D (2006) Temperature dependent impedance spectroscopy on polyaniline based devices. IEEE Trans Dielectr Electr Insul 13:1074–1081. doi:10.1109/TDEI.2006.247834.

[95] Bhadra S, Khastgir D, Singha NK, Lee JH (2009) Progress in preparation, processing and applications of polyaniline. Prog Poly Sci 34:783–810. doi:10.1016/j.progpolymsci.2009.04.003.

[96] Parker ID (1994) Carrier tunneling and device characteristics in polymer light-emitting diodes. J Appl Phys 75:1656–1666. doi:10.1063/1.356350.

[97] Shen Y, Hosseini AR, Wong MH, Malliaras GG (2004) How to make ohmic contacts to organic semiconductors. Chem Phys Chem 5:16–25. doi:10.1002/cphc.200300942.

[98] Craciun NI *et al* (2010) Hysteresis-free electron currents in poly(p-phenylene vinylene) derivatives. J Appl Phys 107:124504–124509. doi:10.1063/1.3432744.

[99] Kronemeijer AJ, Huisman EH, Katsouras I, van Hal PA, Geuns TCT, Blom PWM (2010) Universal scaling in highly doped conducting polymer films. Phys Rev Lett 105:156604–156608. doi:10.1103/PhysRevLett.105.156604.

[100] Allebrod F, Chatzichristodoulou C, Mollerup PL, Mogensen MB (2012) Electrical conductivity measurements of aqueous and immobilized potassium hydroxide. Int J Hydrog Energy 37:16505–16514. doi:10.1016/j.ijhydene.2012.02.088.

[101] Madden JD, Cush RA, Kanigan TS, Hunter IW (2000) Fast contracting polypyrrole actuators. Synth Met 113:185–192. doi:10.1016/S0379-6779(00)00195-8.

[102] Jin HJ, Weissmüller J (2011) A material with electrically tunable strength and flow stress. Science 332:1179–1182. doi:10. 1126/science.1202190.

[103] Blom PWM, de Jong MJM, Vleggaar JJM (1996) Electron and hole transport in poly(p-phenylene vinylene) devices. Appl Phys Lett 68:3308. doi:10.1063/1.116583.

[104] Yan H, Tomizawa K, Ohno H, Toshima N (2003) All-solid actuator consisting of polyaniline film and solid polymer electrolyte. Macromol Mater Eng 288:578–584.

doi:10.1002/mame. 200200007.

[105] Lee YH, Chang CZ, Yau SL, Fan LJ, Yang YW, Yang LY, Itaya K (2009) Conformations of polyaniline molecules adsorbed on Au(111) probed by in situ STM and ex Situ XPS and NEXAFS. J Am Chem Soc 131:6468–6474. doi:10.1021/ja809263y.

[106] Botelho AL, Lin X (2009) Am Phys Soc Meeting 5 4 (http://meetings.aps.org/link/BAPS.2009.MAR.H20.2).

[107] Lin X, Li J, Yip S (2005) Controlling bending and twisting of conjugated polymers via solitons. Phys Rev Lett 95:198303–198307. doi:10.1103/PhysRevLett.95.198303.

[108] Lin X, Li J, Smela E, Yip S (2005) Polaron-induced conformation change in single polypyrrole chain: an intrinsic actuation mechanism. Int J Quant Chem 102:980–985. doi:10.1002/qua.20433.

[109] Vellinga WP, Detsi E, De Hosson JTM (2008) MRS fall meeting—symposium I—reliability and properties of electronic devices on flexible substrates editors: J.R. Greer, J. Vlassak, J. Daniel, T. Tsui. Mater Res Soc Sym Proc 1116:37–42.

[110] Wan M, Liu LJ, Wang J (1998) Electrical and mechanical properties of polyaniline films—effect of neutral salts added during polymerization. Chin J Poly Sci 16:1–8.

[111] Krinichnyi VI, Roth H-K, Hinrichsen G, Lux F, Lüders K (2002) EPR and charge transfer in H2SO4-doped polyaniline. Phys Rev B 65:155205–155219. doi:10.1103/PhysRevB.65.155205.

[112] Nahar MS, Zhang J (2011) Int Conf Sign Im Proc Appl IPCSIT 21. IACSIT Press, Singapore.

[113] Choi M-R et al (2011) Polyaniline-based conducting polymer compositions with a high work function for hole-injection layers in organic light-emitting diodes: formation of ohmic contacts. ChemSusChem 4:363–368. doi:10.1002/cssc.201000338.

[114] Ibach H (1997) The role of surface stress in reconstruction, epitaxial growth and stabilization of mesoscopic structures. Surf Sci Rep 29:195–263. doi:10.1016/S0167-5729(97)00010-1.

[115] Baughman RH, Cui C, Zakhidov AA, Lqbal Z, Barisci JN, Spinks GM, Wallace GG, Mazzoldi A, Rossi DD, Rinzler AG et al (1999) Carbon nanotube actuators. Science 284:1340–1344. doi:10.1126/science.284.5418.1340.

[116] Mathur A, Erlebacher J (2007) Size dependence of effective Young's modulus of nanoporous gold. Appl Phys Lett 90:061910–061913. doi:10.1063/1.2436718.

[117] Detsi E, Onck PR, De Hosson JTM (2013) Electrochromic artificial muscles based on nanoporous metal-polymer composites. Appl Phys Lett 103:193101–193104. doi:10.1063/1.4827089.

[118] Detsi E, Salverda M, Onck PR, De Hosson JTM (2014) On the localized surface plasmon resonance modes in nanoporous gold films. J Appl Phys 115:044308–0443016. doi:10.1063/1.4862440.

[119] Chen CC, Shih WP, Chang PZ, Lai HM, Chang SY, Huang PC, Jeng HA (2015) Onion artificial muscles. Appl Phys Lett 106:183702–183711.

[120] Sahaym U, Norton EMG (2008) Advances in the application of nanotechnology in enabling a 'hydrogen economy'. J Mater Sci 43:5395–5429. doi:10.1007/s10853-008-2749-0.

[121] Esswein AJ, Surendranath Y, Reecea SY, Nocera DG (2011) Highly active cobalt phosphate and borate based oxygen evolving catalysts operating in neutral and natural waters. Energy Environ Sci 4:499–504. doi:10.1039/c0ee00518e.

[122] Luo J, Im JH, Mayer MT, Schreier M, Nazeeruddin MK, Park N-G, Tilley SD, Fan HJ, Grätzel M (2014) Water photolysis at 12.3% efficiency via perovskite photovoltaics and earth-abundant catalysts. Science 345:1593–1596. doi:10.1126/science.1258307.

[123] Detsi E, Lesel B, Turner C, Liang Y-L, Cook JB, Tolbert SH (2015) Unpublished results.

[124] McCrory CCL, Jung S, Peters JC, Jaramillo TF (2013) Bench-marking heterogeneous electrocatalysts for the oxygen evolution reaction. J Am Chem Soc 135:16977–16987. doi:10.1021/ja407115p.

[125] Chen Q, Sieradzki K (2013) Spontaneous evolution of bicontinuous nanostructures in dealloyed Li-based systems. Nat Mater 12:1102–1106. doi:10.1038/nmat3741.

[126] Detsi E, Cook JB, Tolbert SH (2015) UCLA Provisional patent application.

[127] Detsi E, Petrissans X, Cook JB, Dunn B, Tolbert SH (2015) Unpublished results.

[128] Cook JB, Detsi E, Petrissans X, Liang Y-L, Dunn B, Tolbert SH (2015) Unpublished results.

[129] Wang J, Lam DCC (2009) Model and analysis of size-stiffening in nanoporous cellular solids. J Mater Sci 44:985–991. doi:10.1007/s10853-008-3219-4.

[130] Askes H, Aifantis EC (2011) Comments on "Model and analysis of size-stiffening in nanoporous cellular solids" by Wang and Lam [J. Mater. Sci. 44, 985–991 (2009)]. J Mater Sci 46:6158–6161. doi:10.1007/s10853-011-5637-y.

Chapter 24

Microstructural Evolution in 316LN Austenitic Stainless Steel during Solidification Process under Different Cooling Rates

Congfeng Wu[1], Shilei Li[1], Changhua Zhang[1], Xitao Wang[2]

[1]State Key Laboratory for Advanced Metals and Materials, University of Science and Technology Beijing, Beijing 100083, China
[2]Collaborative Innovation Center of Steel Technology, University of Science and Technology Beijing, Beijing 100083, China

Abstract: The solidification sequence and microstructure evolution during solidification process of two 316LN stainless steels with different compositions under different cooling rates were in situ observed with confocal scanning laser microscope. The results show that 316LN solidifies with primary austenite or primary δ ferrite when the cooling rate is small in the range of conventional casting process, depending on the value of C_{req}/N_{ieq} which are calculated by Hammar and Svensson equations. As the cooling rate increases in the range of 0–100°C s^{-1}, the solidification sequences do not change, but both the dendrite arm spacing and the mean free path between δ ferrite decrease. In addition, concomitant with the variations of chemical composition in δ ferrite and austenite are the shape transformation of interdendritic δ ferrite from island-like to lacy-like and the coarsening of dendrite δ ferrite with cooling rate increasing. The mechanism of three-phase reaction in

316LN with different compositions, *i.e.*, eutectic reaction or peritectic reaction, was analyzed. The bigger diffusivities of Cr and Ni in primary δ ferrite than that in primary austenite and the positions of alloys in phase diagram were thought to be the main reasons for the difference in type of the reaction.

1. Introduction

The solidification microstructure of austenitic stainless steel has always been the interest of researches in academia and industry because it determines the castability, weldability, hot workability, mechanical properties, and corrosion resistance[1]-[4]. In austenitic stainless steels, a three-phase reaction region (L + δ + γ), which can be either eutectic or peritectic, exists for compositions of over 15wt% Cr and 10wt% Ni according to the Fe–Cr–Ni ternary phase diagram[5][6]. Therefore, the solidification microstructure, which mainly depends on both composition and cooling rate, is complex as a result of the complicated three-phase reaction[5][7]. Suutala[8] investigated the solidification conditions on solidifying sequence of a range of AISI 300 series steels by autogenous gas tungsten arc (GTA) welding and concluded that the composition was of primary importance while the cooling rate was only of secondary importance. However, a given austenitic stainless steel with composition passing through the Cr-rich part of the three-phase region can solidify with primary δ ferrite or primar γ phase under different cooling rates[8]-[10]. The different solidifying sequences would result in change of elements redistribution path, and thus may alter the type of three-phase reaction and the solidification microstructure. Ma *et al.*[11] and Fu *et al.*[12][13] investigated the detailed microstructural evolution process in directional solidified 304 austenitic stainless steel under estimated cooling rates of 3.3°C s^{-1}, 1°C s^{-1}, and 4°C s^{-1}, respectively. They concluded that eutectic reaction (L → γ + δ) which resulted in the formation of coupled structure occurred among the dendrite arms after primary δ ferrite precipitated from liquid. Liang *et al.*[14] observed a different phenomenon in 301 austenitic stainless steel at cooling rates of 4°C s^{-1}–25°C s^{-1} under non-directional solidification condition using differential thermal analysis (DTA). They found that peritectic reaction and eutectic reaction coexisted in the microstructure of the sample cooled at 25°C s^{-1}. In addition, many other researchers[15]-[17] investigated the solidification microstructure of various austenitic stainless steels by sorts of welding methods which have much higher cooling rate than casting and directional solidification. However, the three-phase reaction mechanism is still unclear and it was not directly observed in the above-mentioned literatures owing to the limitations of test method. Confocal scanning laser microscope (CSLM) enables the in situ observation of phase transformation at high temperature, as shown in references[18]–[20]. Huang et al.[19] and McDonald et al.[20] observed the δ/γ interface and microstructure evolution during the peritectic reaction at a cooling rate of 0.05°C s^{-1} and constant undercooling degree, respectively. Nevertheless, the effect of larger cooling rate, which may be confronted in many types of conventional casting processes, on the microstructure evolution has not been studied, nor has been the in situ observation.

AISI 316LN steel, a type of nitrogen-alloyed ultralow carbon (<0.02wt%) stainless steel, is used as the material of main pipelines in AP1000 pressurized water reactor (PWR) which is about to commercially serve in China. The pipes are manufactured by integral hot forging and the ingots for hot forging are obtained by mold casting or electro slag remelting (ESR) which has typical cooling rates of 0°C s^{-1}–100°C s^{-1}[17]. Therefore, in this regard, the studies of phase transformation in 316LN during solidification under different cooling rates would be of interest.

In this paper, two 316LN stainless steels with different compositions were used to investigate their solidification sequence and microstructure evolution. The three-phase reaction was in situ observed in CSLM under different cooling rates and the variation of chemical composition in different phases was discussed.

2. Experimental

Two ingots of AISI 316LN stainless steel with different values of Cr_{eq}/Ni_{eq} were produced by vacuum induction furnace (VIF) and the casting was carried out at a constant superheat of 50°C in a cast-iron ingot mold. The ingots were forged at 1200°C and then solution treated at 1100°C for 4 h. The chemical compositions are given in **Table 1**.

The cooling rate of this type of mold casting was estimated to be ~2.00°C s^{-1} according to the relationship of secondary dendrite arm spacing (SDAS, λ_2) to cooling rate (ε): $\lambda_2 = 68 \times (\varepsilon)^{-0.45}$[17][21][22]. In order to directly observe the

Table 1. Chemical compositions of AISI 316LN stainless steel.

Alloy	C	N	Si	Mn	Cr	Ni	Mo	P	S	Fe
316LN-1	0.0076	0.13	0.25	1.30	17.86	12.72	2.80	0.0052	0.0046	Bal.
316LN-2	0.0043	0.11	0.17	1.31	17.00	11.24	2.32	0.0053	0.0046	Bal.

microstructural evolution under different cooling rates, a CSLM (VL2000DX, Lasertec, Japan) was used. The principle of this equipment is described in many literatures[19][20]. The disc samples with dimensions of φ7mm × 3mm for observation were machined from the solution-treated materials, and then they were mechanically polished with diamond paste. Subsequently, they were put in an alumina crucible which was installed into a furnace at the focal point of the sample chamber. After the chamber was filled with argon, the samples were heated to the melting points and kept for 60 s, and next they were cooled down to 1100°C with cooling rates of $2°C\ s^{-1}$, $10°C\ s^{-1}$, and $100°C\ s^{-1}$, respectively (see **Figure 1**).

After the samples were cooled down to room temperature, they were prepared by conventional process for metallographic observation. The etchant solution contains 0.5g $K_2S_2O_5$, 20.0g NH_4FHF, and 100 ml distilled water (Beraha's etchant modified by Lichtenegger[16]), whose pH value was maintained at about 2.5 by addition of NH4OH or HNO_3. Electron probe microanalyzer (EPMA) (JEOL JXA-8230, Japan) was employed to characterize the variation of chemical composition. The volume fractions of primary δ ferrite and primary austenite during in situ observation were measured by image analysis using software of Image Tool Version 3.0.

3. Results

3.1. As-Cast Microstructure of Mold Casting Ingots

The as-cast microstructures of mold casting 316LN-1 and 316LN-2 are shown in **Figure 2**. Beraha's etchant makes Cr-rich areas show light color contrast and Cr-depleted areas darker (blue or green) color contrast[16]. In other words, ferrite remains unaffected showing as whiteness, while austenite exhibits a range of colors depending on its Cr concentration. Therefore, the white island-like phase which lies in interdendritic areas in **Figure 2(a)** is ferrite. This indicates that

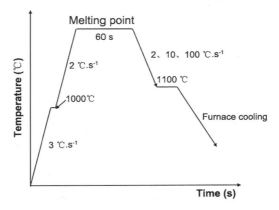

Figure 1. Schematic of heating and cooling curves of specimens during in situ observation.

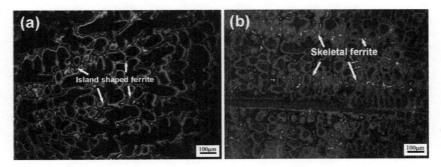

Figure 2. As-cast microstructure of 316LN stainless steel: a 316LN-1, b 316LN-2.

austenite is primary phase and δ ferrite forms at the final stage of solidification in alloy 316LN-1. While in 316LN-2, skeletal δ ferrite exists in dendrite branches [see **Figure 2(b)**], which means that δ ferrite precipitates from liquid firstly and austenite forms later. The solidification behaviors during mold casting process may approximate the equilibrium process because the cooling rate of ~2°C s^{-1} is relatively small in the range of conventional casting processes. When the cooling rate becomes high enough, the solidification behavior will deviate from the equilibrium process, and different solidification microstructures may appear.

3.2. Solidified Microstructures of 316LN under Different Cooling Rates

Figure 3 presents the solidified microstructures of alloy 316LN-1 under

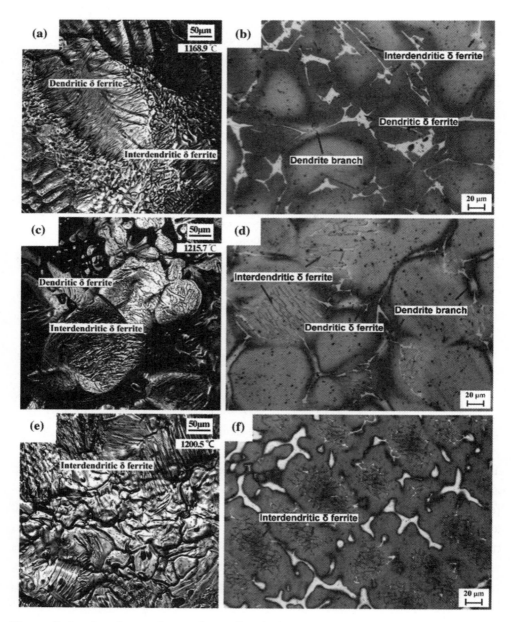

Figure 3. In situ observation and metallurgical micrographs of the solidified microstructure in alloy 316LN-1 under different cooling rates. a, b 2°C s^{-1}; c, d 10°C s^{-1}; e, f 100°C s^{-1}.

different cooling rates. When the steel was cooled at 2°C s^{-1}, δ ferrite exists in both interdendritic areas and dendrite branches, as shown in **Figure 3(a)** and **(b)**.

However, the interdendritic δ ferrite is predominant. The dendrite branch shows light color contrast which means that its Cr concentration is higher than adjacent area [see **Figure 3(b)**]. This characteristic is not consistent with the feature of elements segregation in **Figure 2(a)**, which may be associated with the fact that the steel was not fully melted owing to the limitation of the equipment as the temperature reached the peak point (see **Figure 4**). The unmelted parts, which are richer in Cr element as a result of redistribution of elements during the reaction of L + δ + γ → L + γ on heating, become dendrite branches as the solidification goes on. When the temperature decreases, the dendritic δ ferrite appears owing to the higher Cr concentration.

As the cooling rate increases, δ ferrite still mainly forms in interdendritic areas which can be seen in **Figure 3(c)-(f)**, but the amount of dendritic δ ferrite is reduced. The shape of ferrite changes from island-like to lacy-like with increasing of cooling rate due to the shortening of diffusion path[23]. Furthermore, both the dendrite arm spacing (DAS) and the mean free path between δ ferrite decrease owing to the inhibition of element redistribution and quicker heat rejection under higher undercooling[17]. Overall, in view of the larger variances of content changes at $10°C\ s^{-1}$, contents of Cr and Mo can be thought to decrease slightly in both δ ferrite and γ phase while Ni increases in γ phase and decreases in δ ferrite in a small scale with the increasing of cooling rate, as shown in **Figure 5**. The larger variances at $10°C\ s^{-1}$ may be caused by the bigger scattering

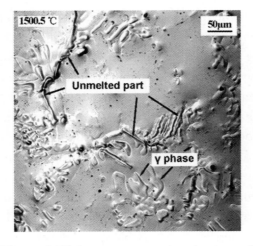

Figure 4. High temperature microstructure of alloy 316LN-1.

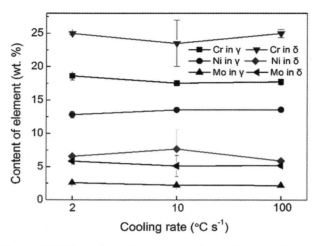

Figure 5. Effect of cooling rate on the variations of composition in 316LN-1.

of measurement points.

The solidified microstructures of alloy 316LN-2 under different cooling rates are shown in **Figure 6**. δ ferrite precipitates in dendrite branches under cooling rate of 2°C s^{-1}, and a coupled growth microstructure of δ + γ was observed when cooled at 10 and 100°C s^{-1}. In particular, δ ferrite becomes coarser due to the suppression of time for solid-state transformation from δ to γ when cooling rate increases. In pace with this coarsening behavior, the element contents in δ ferrite and γ phase change regularly, as shown in **Figure 7**. Cr and Mo increase and Ni decreases in δ ferrite, while in γ phase they change in a reverse direction with the increasing of cooling rate. This phenomenon is thought to be caused by the inhibition of element diffusion during δ → γ transformation as the cooling rate increases[11][17]. It can be seen that the contents change trend in 316LN-2 is more obvious than that in 316LN-1 and their variation trend of Cr and Mo in δ ferrite is opposite. This can be understood based on that the δ ferrite in 316LN-1 generates through the reaction of L → δ + γ at the final stage of solidification, as will be discussed in next section. As the cooling rate increases, the amount of remaining liquid between primary austenite decreases and therefore the amount of δ ferrite decreases. Thus, the contents of Cr and Mo in δ ferrite decrease, while they increase in 316LN-2 because its δ ferrite is primary phase and the outward diffusion rate of Cr and Mo from δ ferrite was suppressed by increasing cooling rate.

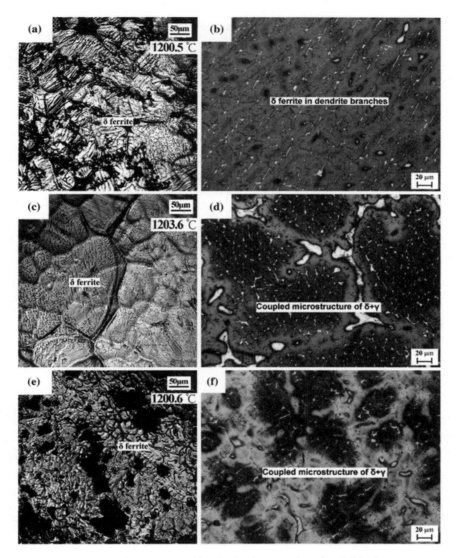

Figure 6. In situ observation and metallurgical micrographs of solidified microstructure in alloy 316LN-2 under different cooling rates. a, b 2°C s^{-1}; c, d 10°C s^{-1}; e, f 100°C s^{-1}.

4. Discussion

4.1. Solidification Mode

From the above results, the precipitation sequences and morphologies of ferrite can vary, depending on the composition and cooling rate. This can be illustrated by the concept of solidification mode. The solidification mode of austenitic stainless steel can be divided into four types according to the value of Cr_{eq}/Ni_{eq}[8][24]:

F mode : $L \rightarrow L + \delta \rightarrow \delta \rightarrow \delta + \gamma$ $Cr_{eq}/Ni_{eq} > 2.00$

FA mode : $L \rightarrow L + \delta \rightarrow L + \delta + \gamma \rightarrow \delta + \gamma \rightarrow \gamma$ $1.50 < Cr_{eq}/Ni_{eq} < 2.00$

AF mode : $L \rightarrow L + \gamma \rightarrow L + \delta + \gamma \rightarrow \gamma + \delta \rightarrow \gamma$ $1.37 < Cr_{eq}/Ni_{eq} < 1.50$

A mode : $L \rightarrow L + \gamma \rightarrow \gamma$ $Cr_{eq}/Ni_{eq} < 1.37$

where A and F refer to austenite and ferrite, respectively. The values of Cr_{eq} and Ni_{eq} can be calculated using the following Hammar and Svensson equations[24]:

$$Cr_{eq} = Cr + 1.37Mo + 1.5Si + 2Nb + 3Ti \quad (1)$$

$$Ni_{eq} = Ni + 22C + 14.2N + 0.31Mn + Cu \quad (2)$$

For the alloys 316LN-1 and 316LN-2 studied here, the ratios of Cr_{eq} to Ni_{eq} equal 1.47 and 1.55, respectively. Hence, alloy 316LN-1 falls into AF mode while alloy 316LN-2 follows FA mode. This indicates that austenite is the primary phase in 316LN-1 and the precipitation of δ ferrite occurs first in 316LN-2, then three-phase reaction (L + δ + γ) comes about at the terminal stage of solidification in both alloys. Subsequently, solid-state transformation of δ → γ continues below solidus lines[12][25]. Therefore, the results predicted by Hammar and Svensson equations work fairly concordant with the solidified microstructure of mold casting in **Figures 3** and **6**.

In respect of the effect of cooling rate on the solidification mode, the results in **Figures 3** and **6** show that this effect can be almost ignored in the range of 0–100°C s^{-1}, except that a small region of primary ferrite which co-existed with primary austenite was found in alloy 316LN-2 when it was cooled at 100°C s^{-1} (see **Figure 8**). That is to say, the solidification mode of FA type in 316LN may change to AF mode if the cooling rate is greater than 100°C s^{-1}. In this case,

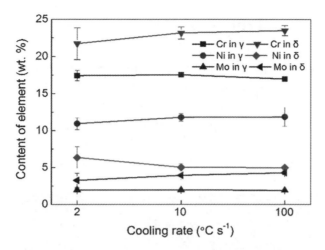

Figure 7. Effect of cooling rate on the variations of chemical composition in alloy 316LN-2.

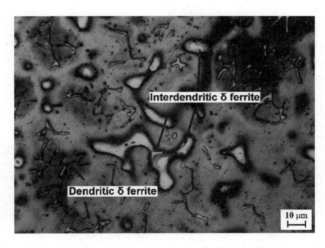

Figure 8. Solidified microstructure of alloy 316LN-2 cooled at rate of 100°C s^{-1}.

the available time for segregation in front of dendrite tip becomes less on account of the high cooling rate[17]. Consequently, the dendritic growth transforms into cellular growth, which results in the achievement of critical value of undercooling degree for metastable austenite formation[8][17][26]. While in alloy 316LN-1, its solidification mode may change to A mode under large enough undercooling conditions because the precipitation of interdendritic ferrite will be suppressed by high cooling rate, whereas it does not occur under current conditions.

4.2. Solidification Process of 316LN

Figure 9 presents some representative micrographs of phase formation during in situ observation in alloy 316LN-1 when it was cooled at 2°C s^{-1}. Based on AF solidification mode, the solidification process can be described as below: when the melt undercooling degree reached a critical value, primary γ phase appeared in liquid and grew in the form of dendrite with the temperature decreasing [see **Figure 9(a), (b)**]. During this process, Cr and Mo were rejected to liquid[16] and the

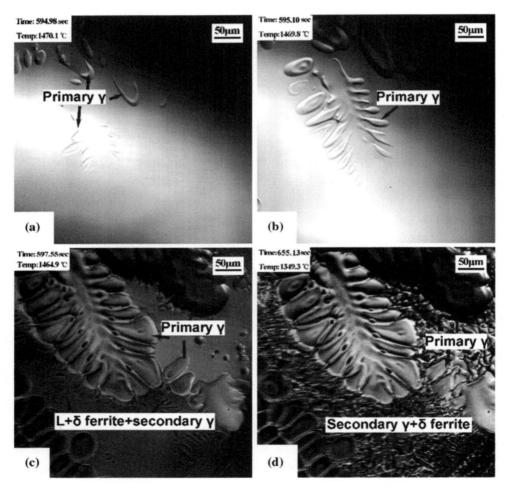

Figure 9. In situ observation of phase formation during solidification in alloy 316LN-1 at 2°C s^{-1}. a, b Primary austenite appeared and grew in liquid; c eutectic reaction occurred among the dendrite; d liquid disappeared at 1349.3°C.

austenite liquidus line was also lowered down, which both favored for the formation of δ ferrite. Subsequently, a eutectic reaction of L → γ + δ was observed among the primary austenite dendrite when the temperature dropped to 1464.9°C [see **Figure 9(c)**]. As temperature reached 1349.3°C, the liquid phase disappeared and the solid-state transformation of δ → γ occurred. As discussed in the above section, the solidification mode and δ ferrite's distribution in 316LN-1 did not change as the cooling rate increased, so did the solidification process. **Figure 10** shows the phases evolution during in situ observation in 316LN-1 at 100°C s^{-1}, and we can see that the characteristics of phases formation are similar to that in **Figure 9** except the corresponding time and temperature.

In alloy 316LN-2, different features of phase evolution were observed when it was also cooled at 2°C s^{-1}. **Figure 11(a)** and **(b)** show that primary δ ferrites precipitated along impurities and grew into liquid with the temperature decreasing. As this process went on, Ni was partitioned to the remaining liquid while Cr and Mo were absorbed into δ ferrites[13]. Subsequently, γ phase formed at the boundaries of δ ferrite and liquid through the peritectic reaction of L + δ → γ, as shown in **Figure 11(c), (d)**. At temperature of 1377.90°C, liquid phase disappeared and the solid-state transformation from δ ferrite to austenite occurred [see **Figure 11(e)**]. Concurrently, Cr was rejected and Ni was accepted by austenite, which resulted in the increasing of austenite's relative stability. However, as a result of the

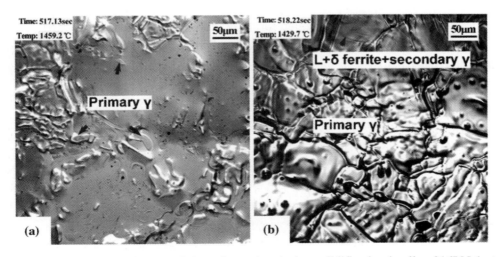

Figure 10. In situ observation of phase formation during solidification in alloy 316LN-1 at 100°C s^{-1}. a Primary austenite grew in liquid; b eutectic reaction occurred among the dendrite.

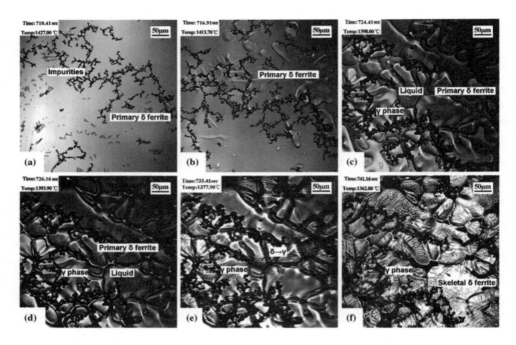

Figure 11. In situ observation of phase formation during solidification in alloy 316LN-2 at 2°C s^{-1}. a, b Primary ferrite appeared and grew in liquid; c, d peritectic reaction occurred at ferrite/liquid boundaries; e liquid disappeared at 1377.9°C and solid state transformation occurred; f skeletal ferrite was retained in core area of dendrites.

limitation of elements diffusion on cooling, the transformation from δ ferrite to austenite was uncomplete, and therefore, skeletal ferrite, which is enriched in Cr and Mo while depleted in Ni, was retained in the core area of dendrites [see **Figure 11(f)**]. Particularly, it should be mentioned that the impurities were mostly retained around the boundaries of δ ferrite and interdendritic austenite because the solidifying interfaces can push the impurities if the critical radius for pushing or engulfment transition of nonmetallic is larger than inclusion size[27][28]. Liang et al.[29] also observed this phenomenon in their work and thought that δ ferrite can absorb the inclusions moving near the δ-L interface.

The biggest difference between the solidification process of alloy 316LN-1 and 316LN-2 is the precipitation sequence of ferrite and austenite, which has been discussed by many authors[16][17][30][31]. In addition, the alloys behaved in different types of reaction in three-phase region, *i.e.*, eutectic reaction in 316LN-1 and peritectic reaction in 316LN-2, as shown in **Figures 9(c)** and **11(c)**. **Figure 12**

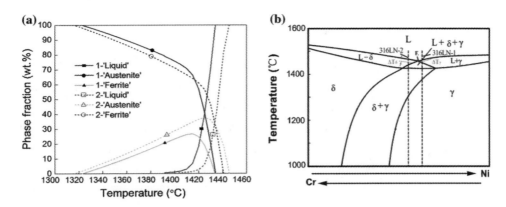

Figure 12. Equilibrium phase diagram calculated by JMatPro software (a) and vertical section of Fe–Cr–Ni ternary diagram at constant Fe (b).

shows the equilibrium phase diagrams calculated by JMatPro software (version 6.1) and illustrates the positions of 316LN-1 and 316LN-2 in vertical section of Fe–Cr–Ni ternary diagram at constant Fe content. Based on the above results of observation and analysis, it is certain that the melt undercooling degree before solidification process which started did not reach the difference between the liquidus temperature of austenite T_L^γ and the eutectic trough temperature [T_E, indicated by point E in **Figure 12(b)**] in 316LN-1 and that between T_L^δ and T_E in 316LN-2. Therefore, when the critical undercooling degree for austenite nucleation (ΔT_γ) in 316LN-1 and that for δ ferrite nucleation (ΔT_δ) in 316LN-2 were reached, austenite or δ ferrite firstly precipitated from liquid phase, respectively. However, the volume fraction of primary δ ferrite [φ_δ = 85.4% in **Figure 9(c)**] in 316LN-2 was more than that of primary austenite [φ_γ = 50.2% in **Figure 11(c)**] in 316LN1, because of the about 100 times larger diffusivities of Cr and Ni in ferrite than that in austenite[32] and the more nucleation sites in 316LN-2. Consequently, the remaining liquid phase in 316LN-1 when three-phase reaction began was more than that in 316LN-2. Moreover, the difference between T_L^δ and the temperature of three-phase reaction beginning (T_t) in 316LN-2 was calculated to be 29°C, which is much larger than that between T_L^γ and T_t, 5.1°C, in 316LN-1. Therefore, the bigger temperature drop and less remaining liquid phase in 316LN-2 facilitated the peritectic reaction owing to the more amount of primary phase and shorter diffusion distance between δ and L. While in alloy 316LN-1, the longer diffusion distance between primary γ and L, combined with the fact that the cooling line of 316LN-1 in the ternary phase diagram is nearer to eutectic trough than that of 316LN-2 [austenite and δ ferrite almost precipitate simultaneously in 316LN-1 according to **Figure 12(a)**], prompted the eutectic reaction.

While cooling rate increases, the type of three-phase reaction in 316LN-1 does not change. However, under the conditions of 10 and 100°C s^{-1}, the peritectic reaction transforms into eutectic reaction in 316LN-2, which results in the appearance of a coupled growth microstructure of γ + δ. The main reason for this transformation is that the melt undercooling degrees at the beginning of solidification become higher than the difference in values between T_L^δ and *TE*. Therefore, δ + γ microstructure formed directly without the occurrence of primary δ ferrite, as shown in **Figure 13**.

5. Conclusions

In situ observation with CSLM and color metallographic method were carried out in two 316LN stainless steels with different compositions in order to investigate their solidification sequence and microstructure evolution during solidification process. The effects of cooling rate on variations of microstructure and chemical composition were also studied. These experimental investigations allow the following conclusions to be reached:

(1) During mold casting process, alloy 316LN-1 follows the solidifying path of AF solidification mode, while 316LN-2 solidifies in FA mode. Both the results are consistent with the prediction of Hammar and Svensson equations.

Figure 13. In situ observation of coupled growth microstructure of δ + γ in 316LN-2 cooled at 10°C s^{-1}.

(2) As the cooling rate increases in the range of 0 °C s^{-1}–100 °C s^{-1}, the solidification modes of both alloys do not change on the whole. However, the ferrite in 316LN-1 changes from island-like to lacy-like and it becomes coarser in 316LN-2 with cooling rate increasing. In alloy 316LN-1, the contents of Cr and Mo decrease slightly in both δ ferrite and γ phase while Ni increases in γ phase and decreases in δ ferrite in a small scale with the increasing of cooling rate. On the other hand, Cr and Mo increase in δ ferrite and decrease in austenite, while Ni has a contrary variation trend in alloy 316LN-2.

(3) Eutectic reaction occurs in the three-phase region of 316LN-1, while peritectic reaction occurs in 316LN-2 when it is cooled at 2 °C s^{-1}. The bigger diffusivities of Cr and Ni in primary δ ferrite than that in primary austenite, as well as the positions of alloys in phase diagram were thought to be the main reasons accounting for the difference in the type of three-phase reaction. Peritectic reaction transforms into eutectic reaction in 316LN-2 when it is cooled at 10 and 100 °C s^{-1}.

Acknowledgements

The authors would like to thank the National High-Tech Research and Development Program of China (863 Program) for the financial support through Grant No. 2012AA03A507. We also acknowledge Dr. Yu Liu and Dr. Jianhua Wu in the Institute of New Materials in Shandong Academy of Science for their help in carrying out CSLM experiment.

Open Access: This article is distributed under the terms of the Creative Commons Attribution 4.0 International License (http://creativecommons.org/licenses/by/4.0/), which permits unrestricted use, distribution, and reproduction in any medium, provided you give appropriate credit to the original author(s) and the source, provide a link to the Creative Commons license, and indicate if changes were made.

Source: Wu C, Li S, Zhang C, *et al*. Microstructural evolution in 316LN austenitic stainless steel during solidification process under different cooling rates[J]. Journal of Materials Science, 2016, 51(5):2529–2539.

References

[1] Saha S, Mukherjee M, Kumar Pal T (2015) Microstructure, texture, and mechanical property analysis of gas metal arc welded AISI 304 austenitic stainless steel. J Mater Eng Perform 24:1125–1139. doi:10.1007/s11665-014-1374-0.

[2] de Lima MSF, SankaréS (2014) Microstructure and mechanical behavior of laser additive manufactured AISI 316 stainless steel stringers. Mater Des 55:526–532. doi:10.1016/j.matdes.2013.10.016.

[3] Amudarasan NV, Palanikumar K, Shanmugam K (2013) Mechanical properties of AISI 316L austenitic stainless steels welded by GTAW. Adv Mater Res 849:50–57. doi:10.4028/www.scientific.net/AMR.849.50.

[4] Plaut RL, Herrera C, Escriba M, Rios PR, Padilha AF (2007) A short review on wrought austenitic stainless steels at high temperatures: processing, microstructure, properties and performance. Mater Res 10:453–460. doi:10.1590/S1516-14392007000400021.

[5] Hunter A, Ferry M (2002) Phase formation during solidification of AISI 304 austenitic stainless steel. Scr Mater 46:253–258. doi:10.1016/S1359-6462(01)01215-5.

[6] Takalo T, Suutala N, Moisio T (1979) Austenitic solidification mode in austenitic stainless steel welds. Metall Trans A 10A:1173–1181. doi:10.1007/BF02811663.

[7] Brooks JA, Thompson AW (1991) Microstructural development and solidification cracking susceptibility of austenitic stainless steel welds. Int Mater Rev 36:16–44. doi:10.1179/imr.1991.36.1.16.

[8] Suutala N (1983) Effect of solidification conditions on the solidification mode in austenitic stainless steels. Metall Trans A 14A:191–197. doi:10.1007/BF02651615.

[9] Di Schino A, Mecozzi MG, Barteri M, Kenny JM (2000) Solidification mode and residual ferrite in low-Ni austenitic stainless steels. J Mater Sci 35:375–380. doi:10.1023/A:1004774130483.

[10] Huang FX, Wang XH, Wang WJ (2012) Microstructures of austenitic stainless steel produced by twin-roll strip caster. J Iron Steel Res Int 19:57–61. doi:10.1016/S1006-706X(12)60060-0.

[11] Ma JC, Yang YS, Tong WH, Fang Y, Yu Y, Hu ZQ (2007) Microstructural evolution in AISI304 stainless steel during directional solidification and subsequent solid-state transformation. Mater Sci Eng, A 444:64–68. doi:10.1016/j.msea.2006.08.039.

[12] Fu JW, Yang YS, Guo JJ, Ma JC, Tong WH (2009) Formation of two-phase coupled microstructure in AISI 304 stainless steel during directional solidification. J Mater Res 24:2385–2390. doi:10.1557/jmr.2009.0282.

[13] Fu JW, Yang YS, Guo JJ, Ma JC, Tong WH (2008) Formation of a two-phase microstructure in Fe–Cr–Ni alloy during directional solidification. J Cryst Growth

311:132–136. doi:10.1016/j.jcrys gro.2008.10.021.

[14] Bai L, Ma YL, Xing SQ, Liu CX, Zhang JY (2015) Phase morphology evolution in AISI301 austenite stainless steel under different cooling rates. J Wuhan Univ Technol-Mater Sci Ed 30:392–396. doi:10.1007/s11595-015-1158-x.

[15] Tate SB, Liu S (2014) Solidification behaviour of laser welded type 21Cr–6Ni–9Mn stainless steel. Sci Technol Weld Joi 19:310–317. doi:10.1179/1362171813Y.0000000189.

[16] Rajasekhar K, Harendranath CS, Raman R, Kulkarni SD (1997) Microstructural evolution during solidification of austenitic stainless steel weld metals: a color metallographic and electron microprobe analysis study. Mater Charact 38:53–65. doi:10.1016/ S1044-5803(97)80024-1.

[17] Elmer JW, Allen SM, Eagar TW (1989) Microstructural development during solidification of stainless steel alloys. Metall Mater Trans A 20:2117–2131. doi:10.1007/BF02650298.

[18] Zhu ZL, Ma GJ, Yu CF (2014) In situ observation of inclusions behavior in molten state and solidifying interface of tire cord steel. Adv Mater Res 881–883:1584–1587. doi:10.4028/www. scientific.net/AMR.881-883.1584.

[19] Huang FX, Wang XH, Zhang JM, Ji CX, Fang Y, Yu Y (2008). In situ observation of solidification process of AISI 304 austenitic stainless steel. J Iron Steel Res Int 15:78–82. doi:10.1016/S1006-706X(08)60271-X.

[20] McDonald NJ, Sridhar S (2005) Observations of the advancing δ-ferrite/γ-austenite/liquid interface during the peritectic reaction. J Mater Sci 40:2411–2416. doi:10.1007/s10853-005-1967-y.

[21] Katayama S, Matsunawa A (1984) Solidification microstructure of laser welded stainless steels. Proc. ICALEO, pp 60–67.

[22] Loser W, Thiem S, Jurish M (1993) Solidification modeling of microstructures in near-net-shape casting of steels. Mater Sci Eng, A 173:323–326. doi:10.1016/0921-5093(93)90237-9.

[23] Baldissin D, Baricco M, Battezzati L (2007) Microstructures in rapidly solidified AISI 304 interpreted according to phase selection theory. Mater Sci Eng, A 449–451: 999–1002. doi:10. 1016/j.msea.2006.02.248.

[24] Hammar O, Svensson U (1979) Solidification and casting of metals. The Metals Society, London.

[25] Li JY, Sugiyama S, Yanagimoto J (2005) Microstructural evolution and flow stress of semi-solid type 304 stainless steel. J Mater Process Technol 161:396–406. doi:10.1016/j.jmatprotec. 2004.07.063.

[26] Fu JW, Yang YS, Guo JJ, Tong WH (2008) Effect of cooling rate on solidification microstructures in AISI 304 stainless steel. Mater Sci Tech-Lond 24:941–944. doi:10.1179/174328408X 295962.

[27] Juretzko FR, Dhindaw BK, Stefanescu DM, Sen S, Curreri PA (1998) Particle engulfment and pushing by solidifying interfaces: part 1. Ground experiments. Metall Mater Trans A 29:1691–1696. doi:10.1007/s11661-998-0091-4.

[28] Stefanescu DM, Catalina A (1998) Calculation of critical velocity for pushing/engulfment transition of nonmetallic inclusions in Steel. ISIJ Int 38:503–505. doi:10.2355/isijinternational.38.503.

[29] Liang GF, Wang CQ, Fang Y (2006) In-situ observation on movement and agglomeration of inclusion in solid-liquid mush zone during melting of stainless steel AISI304. Acta Metall Sin 42:708–714. doi:10.3321/j.issn:0412-1961.2006.07.007.

[30] Tsuchiya S, Ohno M, Matsuura K (2012) Transition of solidification mode and the as-cast γ grain structure in hyperperitectic carbon steels. Acta Mater 60:2927–2938. doi:10.1016/j.actamat.2012.01.056.

[31] Valiente Bermejo MA (2012) Influence of the [Cr_{eq} + Ni_{eq}] alloy level on the transition between solidification modes in austenitic stainless steel weld metal. Weld World 56:2–14. doi:10.1007/ BF03321390.

[32] Moharil DB, Jin I, Purdy GR (1974) The effect of δ-ferrite formation on the post-solidification homogenization of alloy steels. Metall Trans 5:59–63. doi:10.1007/BF02642927.

Chapter 25
Orientation of Cellulose Nanocrystals in Electrospun Polymer Fibres

N. D. Wanasekara[1], R. P. O. Santos[1,2], C. Douch[1], E. Frollini[2], S. J. Eichhorn[1]

[1]College of Engineering, Mathematics and Physical Sciences, University of Exeter, Physics building, Stocker Road, Exeter EX4 4QL, UK
[2]Macromolecular Materials and Lignocellulosic Fibers Group, Center for Research on Science and Technology of BioResources, Institute of Chemistry of São Carlos, University of São Paulo, São Carlos, São Paulo 13560-970, Brazil

Abstract: Polystyrene and poly(vinyl alcohol) nanofibres containing cellulose nanocrystals (CNCs) were successfully produced by electrospinning. Knowledge of the local orientation of CNCs in electrospun fibres is critical to understand and exploit their mechanical properties. The orientation of CNCs in these electrospun fibres was investigated using transmission electron microscopy (TEM) and Raman spectroscopy. A Raman band located at ~1095cm^{-1}, associated with the C-O ring stretching of the cellulose backbone, was used to quantify the orientation of the CNCs within the fibres. Raman spectra were fitted using a theoretical model to characterize the extent of orientation. From these data, it is observed that the CNCs have little orientation along the direction parallel to the axis of the fibres. Evidences for both oriented and non-oriented regions of CNCs in the fibres are presented from TEM images of nanofibres. These results contradict previously published work in this area and micromechanical modelling calculations suggest a uniform orientation of CNCs in electrospun polymer fibres. It is demonstrated that this explains why the mechanical properties of electrospun fibre mats containing CNCs are not always the same as that would be expected for a fully oriented system.

1. Introduction

Cellulose is the most abundant biopolymer on earth and exhibits the potential to be reshaped into an array of forms ranging from non-woven paper to high performance fibres[1]. In addition, cellulose offers an attractive alternative to synthetic polymers due its biodegradability, renewability and sustainability. Cellulose in the native state consists of both amorphous and crystalline regions, which can be isolated to form cellulose nanocrystals (CNCs) by a controlled acid hydrolysis process[2]. CNCs can be derived from a myriad of source materials ranging from cotton to grape skin[3]. CNCs are a potential reinforcement for other polymers due to their high aspect ratio, high elastic modulus (120–150GPa) and excellent chemical and thermal properties[4]. CNCs have therefore attracted a lot of attention as a reinforcement for thin and functional polymer films[5] and nano and microfibres[6][7].

CNCs have been utilized as a reinforcement in composite nanofibres produced by electrospinning using an array of matrix polymers. Peresin *et al.*[8] have successfully electrospun CNC-reinforced poly(vinyl alcohol) (PVA) fibres with improved morphological and thermomechanical properties. It was shown that the increase in the CNC concentration in the fibres led to a significant enhancement of mechanical properties[9]. Electrospun PVA fibres have been extensively studied to understand the effects of molecular weight[10], concentration[9] and the presence of additives on mechanical properties and thermal stability[11]. Polystyrene (PS), another commodity polymer, has also been electrospun into fibres[12]. PS exhibits different chemical, physical and functional attributes to PVA because of the presence of aromatic groups. Rojas *et al.*[12] have produced electrospun PS fibres reinforced with CNCs, observing unique ribbon shapes and the presence of twists along the axes of the fibres. The incorporation of CNCs was found to increase the glassy modulus of the fibres due to their mechanical percolation, forming a stiff and continuous network held together by hydrogen bonding in a hydrophobic matrix environment.

Understanding CNC-polymer interfaces and the orientation of CNCs within electrospun fibres is critical to explain and exploit mechanical properties. Strong matrix-filler interactions leading to efficient stress transfer within composite fibres and oriented CNCs along the fibre axis are expected to enhance anisotropic mechanical properties such as axial stiffness. There have been several studies on directing orientation of different forms of cellulosic materials during processing, via electric fields[13], magnetic fields[14] and conventional wet spinning[15]. However, little work has been published on the orientation of CNCs in fibres produced by electrospinning. Electrospinning is a robust technique of producing networks of composite nanofibres that offer advantages of higher aspect ratios and surface area to suit an array of end applications ranging from biomedical scaffolds to filtration membranes[16]. The possibility of orientating CNCs during electrospinning should be evaluated against the processing conditions and confinement effects. The nanoscale confinement effects resulting from electrospun fibre diameters less than the length of the nanocrystal may lead to orientation of nanocrystals along the fibre axis direction. In addition, large electrostatic fields, and the orientation of polymer chains during electrospinning, may influence the alignment of CNCs in the fibres.

Transmission electron microscopy (TEM) is the most widely used technique to assess the orientation of CNCs in fibres. Changsarn et al.[17] have utilized TEM to visualize the CNCs embedded within electrospun poly(ethylene oxide) (PEO) fibres of ~100nm diameter. The absence of any protruding segments from fibres seen in TEM images suggested that the CNCs were indeed oriented along the fibres' axes. Also, they suggested that CNCs were embedded in the core of the electrospun fibre with a high degree of uniaxial alignment. However, they were not able to investigate the orientation of CNC in fibres with larger diameters. Similarly, Raman spectroscopy has been reported as an essential tool in mapping the orientation of CNCs in polymeric matrices[14]; but to date not electrospun fibres. This approach utilizes the measurement of the intensity of a Raman peak related to the nanofiller as a function of rotation of the specimen with respect to the fixed axis of polarization. Using this technique, the polarization direction of the excitation laser is typically parallel to the main orientation direction of the polymer; for fibres this would be along their axes. The intensity of this band remains invariant with rotation angle for nanofillers with random in-plane orientations; however, this intensity changes dramatically for oriented nanofillers, e.g. CNCs in a composite[18].

The present work utilizes both TEM and Raman spectroscopy techniques to examine the orientation of CNCs in electrospun fibres of CNC-reinforced PVA and PS. Both TEM images and Raman orientation maps indicate a lack of orientation of CNCs along the fibre axis direction. However, smaller regions with local orientations are found to be present in the fibres. This result contradicts previous work that has suggested uniform orientation of CNCs in electrospun polymer fibres. Some calculations are given for the predicted mechanical properties of electrospun fibres published in the literature, which, based on composite theory applied to the data, are lower than what would be expected for fully oriented systems.

2. Materials and Methods

Freeze-dried CNCs were purchased from the Process Development Center at the University of Maine. High molecular weight 87%–89% hydrolysed PVA was purchased from Alfa Aesar. Atactic PS with an average molecular weight of ~280,000 determined by GPC was purchased from Sigma Aldrich (Dorset, UK). Dimethyl-formamide [DMF (purity grade ≥99%)] was purchased from Sigma Aldrich (Dorset, UK) and was used as received.

Aqueous solutions of PVA with varying concentrations of CNCs (10wt% and 20wt%) were created as follows. PVA in water was stirred at 80°C until it was fully dissolved. Once the PVA solution cooled to room temperature, the CNCs were added and stirred for 2 days until a homogenously mixed PVA/CNC solution was obtained. The total concentration of PVA in the final solution was 10(w/v)%. The CNC concentrations (10wt% and 20wt% of PVA) were calculated relative to the PVA weight in the solution. High wt% values of CNCs were used to ensure a strong signal from the Raman spectrometer, emanating from the cellulosic component. Similarly, PS was dissolved in DMF for 24h at room temperature to have a concentration of 25% (w/v). Once the PS was fully dissolved, CNCs were added to the solution and stirred for two more days at ambient temperature, until they were homogenously dispersed. These colloidal solutions were then drawn into a syringe, which was attached to a pump in the electrospinning apparatus. Fibres were then electrospun using the following conditions: flow rate of 1.0mL/h, 20kV, 15cm tip-to-rotating collector distance (TCD) for PVA/CNC solution and 1.5mL/h, 15kV, and 10cm (TCD) for PS/CNC. These conditions produced the best electrospun fibres, and they were chosen after trial experiments using a range of spinning conditions. PS/CNC composite fibres were produced using an electrospinning instrument (Electrospinz ES1a) equipped with a custom-built rotating mandrel collector. PVA/CNC composite fibres were produced at a custom-built electro-spinning set-up at the Materials Science Centre, University of Manchester. Similar rotational speeds were used to electro-spin both PS and PVA fibres although different instruments were used. Once an electrospun mat of fibres was formed, it was removed from the collector.

The PVA and the CNCs are thought to hydrogen bond to each other in the solid state via the primary alcohol groups present on their respective polymer backbones. In the aqueous state, CNCs (produced using sulphuric acid) are known to possess negative surface charge due to the presence of sulphate half-ester groups[2]. These groups aid in the CNCs being dispersed in PVA.

Raman spectra were recorded using a Renishaw 1000 Raman imaging microscope equipped with a thermoelectrically cooled CCD detector. A near-IR laser with a wave-length of 785nm was used to record spectra from oriented electrospun scaffold samples using an exposure time of 30s and two accumulations. The laser beam was focused using a 4× objective lens onto a few fibres contained within the electrospun mat. Care was taken to focus onto fibres in fully oriented regions of the mats. For orientation mapping experiments, one or a few fibres parallel to the polarization direction of the laser were chosen. The incident and scattered light radiations were polarized parallel to the principal axis of the Raman spectrometer. The polarization direction of the incident light was rotated using a half-wave plate, and a polarizer was used to maintain the polarization direction of scattered light parallel to the macroscopic orientation direction of the electrospun fibres. The intensities of a Raman band located at ~1095cm^{-1} were recorded as a function of the rotation angle of the incoming polarizer with respect to the axis of the fibres, and a Lorentzian function was used to fit this Raman peak.

TEM samples were prepared by embedding the electrospun mats in an epoxy resin (TAAB Low Viscosity Resin Hard from TAAB Laboratories Equipment LTD, UK.). These samples were microtomed into thin sections of ~100-nm thickness, both parallel and perpendicular to the fibre axis. Samples were imaged using a TEM (JEM-2100 LaB6) using a voltage of 100kV.

SEM samples were sputter coated with gold to a thickness of ~10Å. The samples were imaged using a SEM (HITACHI S3200N SEM–EDS) at a voltage of 20kV.

3. Results and Discussion

3.1. Orientation and Morphology of Nanofibres

To achieve macroscopic alignment of fibres, a rotating mandrel was utilized as the grounded collector during electrospinning. The orientations of both PVA and PS fibres were found to increase with the rotational speed of the mandrel. We utilized a rotational speed of ~1500rpm for all the experiments to produce aligned fibres. As shown in **Figure 1**, the PS/CNC and PVA/CNC fibres were observed to

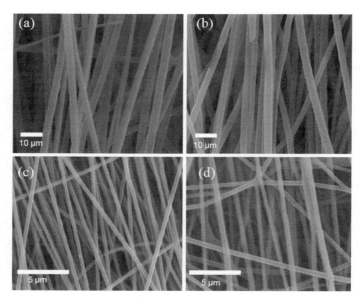

Figure 1. SEM images of oriented fibres of PS/CNC a 10%, b 20%; and PVA/CNC c 10%, d 20%.

be oriented parallel to the rotation direction. The fibre surface appeared to be smooth, and very few 'beads' were observed. Fibre diameter distributions were produced (**Figure 2**), and they show that the diameters of fibres in the mats have normal distributions. The mean fibre diameters of PS/CNC mats (3.00μm ± 0.12μm for 10% CNCs and 2.83μm ± 0.09μm for 20% CNCs) were significantly higher than those of the PVA/CNC fibre mats (0.29μm ± 0.01μm for 10% CNCs and 0.36μm ± 0.01μm for 20% CNCs). These differences in fibre diameters can be attributed to the differences in electrospinning conditions, solvents, chemical structures and tacticity of the polymers used. Either single fibres or oriented fibres were deemed necessary for Raman analysis since filament axes were needed to be placed parallel to the polarization direction of the laser. Attempts to obtain isolated single fibres were unsuccessful, and thus oriented fibres were utilized for analysis under the Raman spectrometer.

3.2. Raman Spectroscopic Characterization of CNC Orientation

Raman spectra of CNC-reinforced PVA and PS fibres are shown in **Figure 3**.

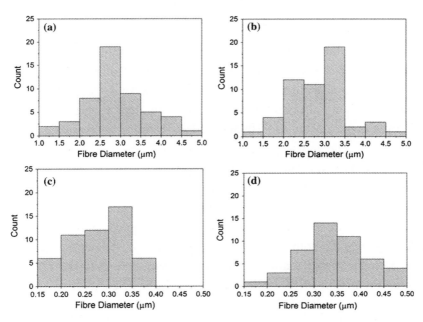

Figure 2. Fibre diameter histograms for PS/CNC a 10%, b 20%; and PVA/CNC c 10%, d 20%.

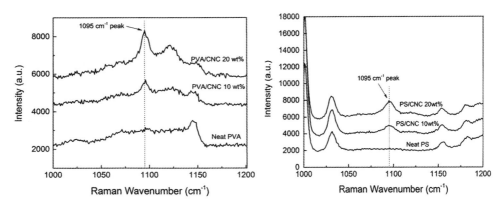

Figure 3. Typical Raman spectra of (left) PVA/CNC (right) PS/CNC.

A Raman peak characteristic of cellulose (and not present for either pure PS or PVA) is located at ~1095cm^{-1}. This peak can be clearly observed in nanocomposite fibres with 10wt% and 20wt% of CNCs; below this CNC concentration, we were not able to observe CNCs. The Raman laser spot size of 1–2µm was focused on one or two fibres in a PS/CNC mat and a few fibres in PVA/CNC electrospun mats. There may be a small error involved in the measurements on the PVA/CNC sample due to a few fibres possibly lying in directions [**Figure 1(c), (d)**] other than the polarization direction of the laser. The intensity of the Raman peak located at ~1095cm^{-1} is found to increase with increasing CNC concentration. For orientation mapping experiments, the oriented fibre mat was placed under the lens of the Raman microscope in such a way that the polarization of the incident laser was parallel to the longitudinal axis of either a single fibre or a group of aligned fibres. The scattered radiation was sent through a parallel polarization filter. The intensity of the Raman band located at ~1095cm^{-1} was recorded as a function of rotation angle (0°–360°). The most intense peak recorded was fitted with a Lorentzian function to determine the maximum intensity from which a normalization of other intensities was performed. As shown in **Figure 4**, polar plots of the intensity as a function of the rotational angle are presented for the evaluation of the orientation of CNCs in the fibres. The Raman band intensity remained approximately constant with rotation angle for both PS/CNC- and PVA/CNC-based composite fibres. Previous research has shown that more ellipsoidal shapes are obtained for these polar plots when orientation of the cellulose occurs, be it in chain form or emanating from CNCs themselves[18][19]. This result suggests a lack of orientation of CNCs in both PS and PVA fibres. However, at a higher concentration of CNCs of 20%, both PS/CNC and PVA/CNC fibres, some orientation of the reinforcing

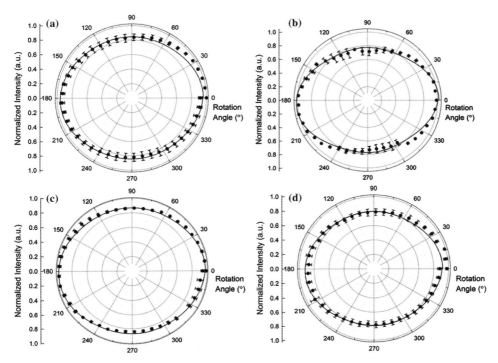

Figure 4. Polar plots of the normalized intensity of the Raman band located at ~1095cm^{-1} as a function of rotation angle of the polarization with respect to the fibre axis showing the orientation of CNCs in composite nanofibres of PS/CNC a 10 %, b 20%; and PVA/CNCs c 10%, d 20%. The solid line represents a fit of $I = r + t \times \cos^4\theta$.

phase is observed, as evidenced by the slightly ellipsoidal shape of the data points in the polar plot. The ratio of the intensity of the band at 90° to the band at 0° ($I_{90°}/I_{0°}$) can be used to characterize the degree of alignment of CNCs along the fibre axis. The values of $I_{90°}/I_{0°}$ are 0.84 and 0.71 for 10% and 20% of CNCs in PS fibres, and those of $I_{90°}/I_{0°}$ are 0.86 and 0.75 for 10% and 20% CNCs in PVA fibres. These values suggest that both PS/CNC and PVA/CNC fibres with 20% CNCs exhibit a slight orientation of CNCs along the fibre axis direction. These data were fitted using the equation:

$$I = r + t \times \cos^4\theta$$

where r and t are fitting parameters, θ is the rotation angle and I is the intensity.

The values of these fitting parameters for these polar plots are given in

Table 1. Equation 1 has been previously used to fit orientation data for graphene oxide nanocomposites[20].

We acknowledge that misorientation of the fibres could affect the values obtained via Raman spectroscopy. Image analysis was therefore used to calculate the fibre orientation parameter f based on the following equation:

$$f = \frac{3\langle \cos^2 \theta \rangle - 1}{2} \qquad (2)$$

where θ represents the angle between the fibres' axes and the orientation direction; and f = 1 corresponds to perfect alignment of the fibres, f = 0 to a lack of preferred orientation and f = −1/2 to perpendicular orientation. Equation 2 has been previously used to assess the orientation of electrospun fibres[21]. We obtained values of 0.94 and 0.95 for PS/CNC 10wt% and 20wt% mats, and 0.94 and 0.67 for PVA/CNC 10wt% and 20wt% mats, respectively. These values suggest that there is a high degree of orientation of fibres in our fibre mats. We do note, however, that a value of 0.67 for the 20wt% sample would suggest some degree of misalignment.

To further elucidate the orientation of CNCs in the fibres, TEM characterization was performed on thin sections of fibres embedded in a supporting matrix. The fibres were sectioned to have a thickness of ~100nm suitable for viewing in the TEM. Sections were taken along and plane perpendicular to the electrospun fibre axis. CNCs were stained using uranyl acetate before imaging. This staining agent applies a passive outline for CNCs. Several attempts were made to image the PS/CNC fibres using this approach (**Figure 5**). However, CNCs were not clearly visible in the fibres (darker regions in the fibres suggest their presence), possibly due to lack of contrast between CNCs and PS.

Table 1. Parameter values for fits of equation to orientation data in **Figure 4**.

Composite fibre	r	t	R^2
PS/CNC 10 %	0.84	0.11	0.71
PS/CNC 20 %	0.77	0.24	0.88
PVA/CNC 10 %	0.85	0.09	0.82
PVA/CNC 20 %	0.76	0.15	0.74

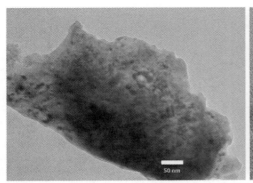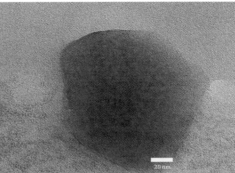

Figure 5. Typical TEM images showing the orientation of CNCs in PS/CNC 10% fibres. Left image showing CNCs present (darker regions) along the axis of the electrospun fibre, in the plane of the page; right image showing CNCs (darker regions) with the axis of the electrospun fibre out of the plane of the page.

Figure 6 shows a cross-sectional image of a PVA/CNC 10% nanofibre exhibiting clearly randomly oriented CNCs perpendicular to the direction of the fibre axis (lighter rod-like entities seen in the plane of the image of **Figure 6**); a few CNCs were found to be oriented along the fibres' axis (also **Figure 6**). The aspect ratios of our CNCs have been measured and found to be ~18 (lengths 109nm ± 34.5nm and widths 6.0nm ± 1.9nm) with some variability. It is noted that our electrospun fibres are much wider than the average length of a CNC, and so little protrusion of the CNCs is observed. It is acknowledged that microtoming using a glass knife may have assisted the orientation of CNCs in the cutting direction due to the application of shear forces. If this were the case, however, we might have expected to see dominant orientation perpendicular to the fibre axis in one particular direction. This was not observed in the images we obtained. Microtoming was also carried out at room temperature, well below the glass transition of PVA (~88°C)[22]. Therefore, the PVA was thought to be in the glassy state, which, with the supporting epoxy resin, should offer significant resistance to the movement of CNCs within the PVA during the microtoming process. Attempts to microtome fibres at sub-zero temperatures were not successful due to the complete embrittlement of the resin.

Overall, these results combined with Raman spectroscopic results give evidence for a lack of orientation of CNCs along the axis of the fibres. This has been true for both PS/ CNC and PVA/CNC fibres regardless of the differences in electrospinning conditions and fibre diameters. Previous work has shown that single-

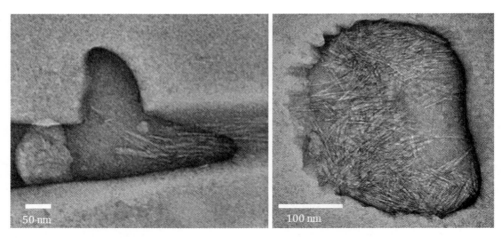

Figure 6. Typical TEM images showing the orientation of CNCs in PVA/CNC 10% fibres. Left image showing CNCs orientation (lighter regions) with the axis of the electrospun fibre in the plane of the page; right image showing CNCs orientation (lighter regions) with the axis of the electrospun fibre out of the page.

walled carbon nanotubes (SWNTs) can be oriented in electrospun PVA fibres[23]-[25]. Given that the lateral sizes of CNCs and SWNTS are similar, it is surprising that orientation is not achieved with the present system. However, different hydrophobic/hydrophilic characteristics of SWNTS and CNCs and their interfacial interactions with the polymer matrix may play a role in dictating orientation. The lack of orientation of CNCs should result in lower axial mechanical properties than what would be expected for the electrospun fibres.

3.3. Micromechanical Modelling of Tensile Modulus of Composite Fibres

Tensile moduli of CNC-reinforced composite fibres can be predicted using the rule of mixtures[26] and the Cox model[27]. This model was developed based on the classical shear-lag theory and has been commonly used to predict the tensile moduli of randomly oriented short fibre composites[27]. Several assumptions are required to apply this model to composite fibres: (i) both fibre and matrix undergo elastic deformation, (ii) fibre ends bear no axial loads and (iii) the presence of a fully bonded fibre-matrix interface[26][27]. In this case, CNCs are assumed to behave as nanoscale reinforcing fibres, which are embedded in a polymer matrix to form the larger composite fibres.

The rule of mixtures model, with added efficiency factors (according to Cox), is governed by the equation:

$$E_{fibre} = \eta_L \times \eta_O^{CNC} \times E_{CNC} \times V_{CNC} + E_{polymer}(1 - V_{CNC})$$

where E_{fibre}, η_L, η_O^{CNC}, E_{CNC}, V_{CNC} and $E_{polymer}$ represent composite fibre modulus, fibre length efficiency factor (according to Cox theory), orientation factor for CNCs (again according to Cox), the modulus of the CNCs, CNC volume fraction and the elastic modulus of the polymer matrix, respectively. Equation 3 predicts the modulus of one electrospun fibre with a volume fraction of misoriented CNCs. Electrospun fibre mats that are considered in this study are mostly themselves randomly oriented, and therefore, modulus of the fibre mat can be calculated by using a fibre orientation factor of 1/3 according to Cox's theory[1]

$$E_c = \eta_o^{fibre} \times E_{fibre} \qquad (4)$$

where E_c and η_o^{fibre} are the moduli of the mat and fibre orientation factors, respectively. The orientation factors (η_o^{fibre} or η_o^{CNC}) are equal to 1 for highly oriented fibres or CNCs and is taken as 1/3 if either is randomly oriented. This is true for in-plane orientation. However, if the CNCs are also oriented out of the plane then the latter of these factors could be equal to 1/5. This would mean some reduction in the estimation of the modulus of our reinforced fibres. It is worth noting that Cox derived his theory[27] for fibrous networks (in this case paper). We note that electrospun networks can be very porous, and perhaps there are fewer interactions between fibres than for a network of hydrogen bonded fibres (for paper). Therefore, there could be some influence of the porosity of the networks. Evidence for this comes from the data presented in **Table 2** for electrospun fibre networks. Data for single electrospun fibres is scarce in the literature, but one study has reported tensile moduli of ~120MPa for single PCL fibres[30]. A value of 6MPa for the PCL networks[6] (it is noted that this value is with fillers) is much lower than 1/3 of 120MPa (40MPa), and so it is reasonable to assume that Eq. 4 may make some overprediction of the mat modulus. We have considered two possibilities of fibre arrangement electrospun fibres: (i) fibres are randomly oriented: $\left(\eta_o^{fibre} = 1/3\right)$ in the mats, and CNCs are perfectly oriented $\left(\eta_o^{CNC} = 1\right)$

---
[1]Krenchel[28] uses a factor of 3/8—which is close to 1/3—for randomly oriented fibres.

Table 2. Summary of data utilized in the calculation of theoretical moduli.

Matrix polymer	CNC (%)	Young's modulus (MPa)	Average fibre diameter (nm)	Aspect ratio of CNCs	Reference
PVA	10	89	235	27	[8]
PAA	10	224	141	30	[28]
PCL	7.5	6	200	30*	[6]
PEO	10	267	448	100	[17]

* The reported values of CNC dimensions are 3-10 nm for diameter and 100-250 nm for length

along the electrospun fibres' axes [**Figure 7(a)**]; and (ii) both electrospun fibres and CNCs (within the fibres) are randomly oriented ($\eta_o^{fibre} = 1/3$ and $\eta_o^{CNC} = 1/3$) [**Figure 7(b)**]. Upper and lower limits of η_L are considered in the calculations for (i) high aspect ratios nanofibres ($\eta_L = 1$) and (ii) low aspect ratio nanofibres as $\eta_L = 0.1$ (an estimate for short fibres[31]). **Figure 7(a), (b)** compares experimental values and theoretical values of Young's moduli for four different CNC-reinforced electrospun fibre systems. All experimental Young's moduli values have been obtained from mechanical data reported for electro-spun fibre webs (**Table 2**)[6][8][17][31]. The aspect ratios of the CNCs used for the studies used in this analysis were ~30 for PVA, PAA, PCL; and ~100 for PEO fibres. Further, all electrospun fibres in the studies we focused on were reported to be randomly oriented, except for PEO electrospun fibres; however, the SEM images presented in this paper[17] show a few unoriented fibres as well. A Young's modulus of the CNCs is taken as 57GPa[32].

When modelling, we have considered the possibility that the electrospun fibres and the CNCs can either both be oriented or non-oriented and variants of both. Non-oriented electrospun fibres in a mat, containing oriented low aspect ratio CNCs, would give an efficiency factor of 1/3 (and 0.1 for the fibre length efficiency). This is expected to give good agreement between theoretical and experimental data. Indeed, this is evidenced by the results shown in **Figure 5(a)**. This suggests that CNCs could be oriented within the fibres for all the electrospun fibre webs considered in the studies in the literature. Further, we have examined the experimental Young's moduli and theoretical values when both fibres and CNCs are randomly oriented. As expected, for higher aspect ratio fibres ($\eta_L = 1$), the theoretical values were always higher [**Figure 5(a)**] than experimental moduli due to the fact that CNCs typically have low aspect ratios (<100). This was also observed for PEO/CNC fibres where the CNC aspect ratio was higher (~100). We do note again that Eq. 4 could overpredict the modulus of a mat of fibres, due to the influence of significant porosity in the networks. An over-prediction may also occur

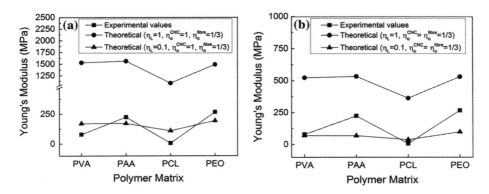

Figure 7. A comparison between experimental and theoretical values of Young's moduli electrospun polymer/CNC fibre mats with different polymeric matrices; a only CNCs are highly oriented ($\eta_o = 1/3$), b both fibres and CNCs are randomly oriented. Theoretical values were calculated using the Eqs. (3) and (4). Experimental values were taken from references[6][8][17][29].

due to the use of 1/3 for the orientation factor for the CNCs. If they occupy an out-of plane random orientation then this factor may be 1/5. If either of these scenarios this the case then better agreement with the upper set of predicted data in **Figure 7** (represented in each case by solid circles) and experimental data is expected.

Further, if we assume that the PEO/CNC fibres are highly oriented (with $\eta_o^{fibre} = 1$), then the theoretical moduli will be even higher than experimental values. However, when a lower aspect ratio of CNCs ($\eta_L = 0.1$) was used, with random orientation of both the CNCs and fibres, there is again agreement with the experimental moduli [**Figure 5(b)**]. It may therefore be true that low aspect ratio CNCs (such as cotton) remain unoriented in the fibres, in agreement with our own experimental evidence. Modelling suggests that CNCs could indeed be fully oriented along the fibre axis direction. It is likely that the real situation will be some way between these two extreme cases for our fibres, with some misorientation remaining. Indeed, we have shown that orientation of the CNCs can occur, although it appears that random orientation remains. It may be that thinner fibres induce orientation of CNCs more readily. Further work is required to carefully control the spinning procedure and thus to ensure a more uniform orientation.

4. Conclusions

PS and PVA nanofibres containing CNCs were successfully produced by electrospinning. PS/CNC fibres were observed to have larger fibre diameters in the range of ~3μm when compared to PVA/CNC fibres (~0.3μm). Raman spectroscopic analysis of oriented fibres within mats showed some evidence for a lack of orientation of CNCs in the polymer fibres. Slightly ellipsoidal shapes to the polar intensity maps and a lower value for the intensity ratio $I_{90°}/I_{0°}$ indicates that CNCs exhibited some orientation in polymer fibres with a 20% volume fraction. TEM images of CNC-reinforced polymer fibres confirmed the random orientation of CNCs in PVA fibres with local regions of oriented CNCs in the cross section and a few CNCs oriented along the axis of the fibre. These data are expected to provide a vital insight into the orientation of CNCs in polymer fibres produced by electrospinning. The strong electric fields and confinement effect associated with electrospinning may not necessarily induce full orientation of CNCs. A comparison of mechanical data from previously published work and theoretical calculations using the rule of mixtures also seems to suggest that this is possibly the case.

Acknowledgements

The authors would like to thank Peter Splatt, the College of Life and Environmental Sciences, the University of Exeter for assistance in microtoming and staining samples. The authors also thank Dr. Hong Chang for assistance in TEM imaging. The Engineering and Physical Sciences Research Council (EPSRC) is acknowledged for the funding provided under Grant No. EP/L017679/ The authors would also like to thank Dr. Fenglei Zhou, the Materials Science Centre at the University of Manchester, the UK for assistance with electrospinning. The authors would like to thank CNPq (National Council of Scientific Research) for the fellowship to R.P.O.S.

Compliance with Ethical Standards

Conflict of Interest: The authors declare that they have no conflict of interest.

Open Access: This article is distributed under the terms of the Creative Commons Attribution 4.0 International License (http://creativecommons.org/licenses/by/4.0/), which permits unrestricted use, distribution, and reproduction in any medium, provided you give appropriate credit to the original author(s) and the source, provide a link to the Creative Commons license, and indicate if changes were made.

Source: Wanasekara N D, Santos R P O, Douch C, *et al*. Orientation of cellulose nanocrystals in electrospun polymer fibres[J]. Journal of Materials Science, 2016, 51(1):1–10.

References

[1] Qiu X, Hu S (2013) "Smart" materials based on cellulose: a review of the preparations, properties, and applications. Materials 6:738–781.

[2] Samir MASA, Dufresne A (2005) Review of recent research into cellulosic whisker, their properties and their application in nanocomposites field. Biomacromolecules 6:612–626.

[3] Hsieh Y-L (2013) Cellulose nanocrystals and self-assembled nanostructures from cotton, rice straw and grape skin: a source perspective. J Mater Sci 48:7837–7846. doi:10.1007/s10853-013-7512-5.

[4] Vallejos ME, Peresin MS, Rojas OJ (2012) All-cellulose composite fibers obtained by electrospinning dispersions of cellulose acetate and cellulose nanocrystals. J Polym Environ 20:1075–1083.

[5] Valentini L, Bittolo Bon S, Fortunati E, Kenny JM (2014) Preparation of transparent and conductive cellulose nanocrystals/graphene nanoplatelets films. J Mater Sci 49: 1009–1013. doi:10.1007/s10853-013-7776-9.

[6] Zoppe JO, Peresin MS, Habibi Y, Venditti RA, Rojas OJ (2009) Reinforcing poly(e-caprolactone) nanofibers with cellulose nanocrystals. ACS Appl Mater Inter 1:1996–2004.

[7] Martínez-Sanz M, Olsson RT, Lopez-Rubio A, Lagaron JM (2011) Development of electrospun EVOH fibres reinforced with bacterial cellulose nanowhiskers. Part I: characterization and method optimization. Cellulose 18:335–347.

[8] Peresin M, Habibi Y, Zoppe JO, Pawlak JJ, Rojas OJ (2010) Nanofiber composites of polyvinyl alcohol and cellulose nanocrystals: manufacture and characterisation. Biomacro-molecules 11:674–681.

[9] Ding B, Kim H-Y, Lee S-C, Lee DR, Choi KJ (2002) Preparation and characterization of nanoscaled poly(vinyl alcohol) fibers via electrospinning. Fiber Polym 3: 73–79.

[10] Koski A, Yim K, Shivkumar S (2004) Effect of molecular weight on fibrous PVA produced by electrospinning. Mater Lett 58:493–497.

[11] Lee HW, Karim MR, Ji HM, Choi JH, Ghim HD, Park SM, Oh W, Yeum JH (2009)

Electrospinning fabrication and characterization of poly (vinyl alcohol)/montmorillonite nanofiber mats. J Appl Polym Sci 113:1860–1867.

[12] Rojas OJ, Montero GA, Habibi Y (2009) Electrospun nanocomposites from polystyrene loaded with cellulose nanowhiskers. J Appl Polym Sci 113:927–935.

[13] Bordel D, Putaux J-L, Heux L (2006) Orientation of native cellulose in an electric field. Langmuir 22:4899–4901.

[14] Pullawan T, Wilkinson AN, Eichhorn SJ (2012) Influence of magnetic field alignment of cellulose whiskers on the mechanics of all-cellulose nanocomposites. Biomacromolecules 13:2528–2536.

[15] Iwamoto S, Isogai A, Iwata T (2011) Structure and mechanical properties of wet-spun fibers made from natural cellulose nano-fibers. Biomacromolecules 12:831–836.

[16] Huang ZM, Zhang YZ, Kotaki M, Ramakrishna S (2003) A review on polymer nanofibers by electrospinning and their applications in nanocomposites. Compos Sci Technol 63:2223–2253.

[17] Changsarn S, Mendez JD, Shanmuganathan K, Foster EJ, Weder C, Supaphol P (2011) Biologically inspired hierarchical design of nanocomposites based on poly-(ethylene oxide) and cellulose nanofibers. Macromol Rapid Commun 32:1367–1372.

[18] Pullawan T, Wilkinson AN, Eichhorn SJ (2013) Orientation and deformation of wet-stretched all-cellulose nanocomposites. J Mater Sci 48:7847–7855. doi:10.1007/s10853-013-7404-8.

[19] Bakri B, Eichhorn SJ (2010) Elastic coils: deformation micromechanics of coir and celery fibres. Cellulose 17:1–11.

[20] Li Z, Young RJ, Kinloch IA (2013) Interfacial stress transfer in graphene oxide nanocomposites. ACS Appl Mater Inter 5:456–463.

[21] Wang M, Yu JH, Kaplan DL, Rutledge GC (2006) Production of submicron diameter silk fibres under benign processing conditions by two-fluid electrospinning. Macromolecules 39:1102–1107.

[22] Wanasekara ND, Stone DA, Wnek GE, Korley LTJ (2012) Stimuli-responsive and mechanically-switchable electrospun composites. Macromolecules 45:9092–9099.

[23] Deng L, Young RJ, Sun R, Zhang GP, Lu DQD, Li H, Eichhorn SJ (2014) Unique identification of single-walled carbon nanotubes in electrospun fibers. J Phys Chem C 118:24025–24033.

[24] Kannan P, Eichhorn SJ, Young RJ (2007) Deformation of isolated single-wall carbon nanotubes in electrospun polymer nanofibres. Nanotechnology 18:235707.

[25] Kannan P, Young RJ, Eichhorn SJ (2008) Debundling, isolation, and identification of carbon nanotubes in electrospun nanofibers. Small 4:930–933.

[26] Harris B (1999) Engineering composite materials. IOM Communications Ltd, London.

[27] Cox HL (1952) The elasticity and strength of paper and other fibrous materials. Br J Appl Phys 3:72–79.

[28] Krenchel H (1964) Fibre reinforcement. Akademisk Forlag, Copenhagen.

[29] Lu P, Hsieh Y-L (2009) Cellulose nanocrystal-filled poly(acrylic acid) nanocomposite fibrous membranes. Nanotechnology 20: 415604.

[30] Tan EPS, Ng SY, Lim CT (2005) Tensile testing of a single ultrafine polymeric fiber. Biomaterials 26:1453–1456.

[31] Lee K-Y, Aitoma"ki Y, Berglund LA, Oksman K, Bismarck A (2014) On the use of nanocellulose as reinforcement in polymer matrix composites. Compos Sci Technol 105:15–27.

[32] Rusli R, Eichhorn SJ (2008) Determination of the stiffness of cellulose nanowhiskers and the fiber-matrix interface in a nanocomposite using Raman spectroscopy. Appl Phys Lett 93:1–3.

Chapter 26

Oxide Dispersion-Strengthened Steel PM2000 after Dynamic Plastic Deformation: Nanostructure and Annealing Behaviour

Z. B. Zhang[1,2,5*], N. R. Tao[3,5], O. V. Mishin[1,5], W. Pantleon[4,5]

[1]Section for Materials Science and Advanced Characterization, Department of Wind Energy, Technical University of Denmark, Risø Campus, 4000 Roskilde, Denmark
[2]School of Materials, University of Manchester, Manchester M13 9PL, UK
[3]Institute of Metal Research, Chinese Academy of Science, Shenyang 110016, China
[4]Section for Materials and Surface Engineering, Department of Mechanical Engineering, Technical University of Denmark, 2800 Kgs. Lyngby, Denmark
[5]Sino-Danish Center for Education and Research, Aarhus, Denmark

Abstract: The microstructure, texture and mechanical properties have been studied in PM2000 compressed via dynamic plastic deformation to a strain of 2.1. It is found that dynamic plastic deformation results in a duplex $\langle 111 \rangle + \langle 100 \rangle$ fibre texture and refines the initial microstructure by nanoscale lamellae, which substantially increases the strength of the material, but decreases its thermal stability. In the as-deformed microstructure, the stored energy density is found to be higher in $\langle 111 \rangle$-oriented regions than in $\langle 100 \rangle$-oriented regions. Recovery during annealing at 715 °C reduces the energy stored in the deformed microstructure. This reduction is more pronounced in the ⟨111⟩-oriented regions. Orientation-dependent recrystallisation takes place in the recovered microstructure, leading to strengthening of the ⟨111⟩ fibre texture component at the expense of the ⟨100⟩ fibre texture component.

1. Introduction

Oxide dispersion-strengthened (ODS) steels are promising structural materials for the next-generation fission and fusion reactors because of their excellent resistance to both irradiation damage and high-temperature creep[1][2]. It has been reported that irradiation tolerance in steels and other materials may further be improved if their microstructures are refined to the submicrometre or nanometre scale[3][4]. One well-known way to refine the microstructure is via plastic deformation when original grains are subdivided by deformation-induced dislocation boundaries[5][6]. For example, structural refinement in the range of 0.3–0.6μm was produced by equal channel angular extrusion (ECAE) in a ferritic/martensitic steel T91 and a 12Cr ODS steel[3][7]. The swelling rate during the irradiation of the submicrometre-grained T91 steel after warm ECAE to a strain of 2.3 was found to be three times lower than that of a coarse-grained sample[3].

Compared to the structural refinement achieved by ECAE in[3][7], a much finer boundary spacing can be obtained if the deformation is performed at high strain rates. For instance, a lamellar structure with a boundary spacing of only 0.1 μm developed in a modified 9Cr-1Mo steel (T91)[8] due to compression via dynamic plastic deformation (DPD)[9] to a strain of 2.3 at a strain rate of 10^2–10^3 s^{-1}. A more effective structural refinement by DPD as compared to lowstrain-rate deformation has also been documented for other metals such as aluminium and nickel[10][11]. It is reasonable to expect that DPD can also be very effective in refining the microstructures of ODS steels. The primary aim of this work is therefore to investigate the effect of compression by DPD on the microstructure of an ODS material. For this purpose, we chose to study PM2000[12], an iron-based ODS alloy developed for application in power plants. Since nanostructured materials produced by deformation are in general less thermally stable than their less-refined counterparts[13], the annealing behaviour of DPD-processed PM2000 is also investigated in the present work. Both transmission electron microscopy (TEM) and electron backscatter diffraction (EBSD) are used here to enable a detailed microstructural analysis of changes taking place during DPD and subsequent annealing. The microstructural analysis is complemented by hardness measurements and tensile tests.

2. Experimental

The nominal chemical composition of PM2000[12] is shown in **Table 1**. The material was received in the form of a hot-extruded rod with a diameter of 13mm. Two cylindrical specimens with a diameter of 6mm and height of 9mm were then machined with their cylinder axis along the extrusion direction (ED) see **Figure 1(a)**. The specimens were compressed at room temperature by DPD in five steps to a final thickness of 1.1mm, which corresponds to an equivalent strain, $\varepsilon_{vM} = \ln(h_0/h)$ of 2.1, where h_0 is the initial height, and h is the final sample thickness, respectively. The diameter of the compressed samples was about 17.2mm [**Figure 1(b)**]. Parts of the deformed samples were then annealed for 1h at different temperatures between 500°C and 800°C and for different time intervals at 715°C.

Table 1. Nominal chemical composition (wt%) of PM2000[12].

Cr	Al	Ti	Y$_2$O$_3$	Fe
20	5.5	0.5	0.5	Balance

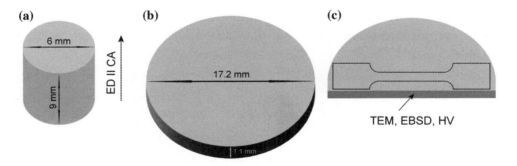

Figure 1. Schematic illustration of samples before and after DPD: a initial sample; b sample after DPD to a strain of 2.1; c locations of specimens used for microstructural examinations and mechanical tests after DPD.

TEM foils from the longitudinal section in the as received sample and from the longitudinal section near the centre of the compressed disc [see **Figure 1(c)**] were prepared by twin jet electropolishing in a solution of ethanol (70vol%), water (12vol%), 2-butoxy-ethanol (10vol%), and perchloric acid (8vol%). TEM images were obtained using a JEOL 2000FX transmission electron microscope operating at 200kV.

EBSD analysis was performed using a Zeiss Supra 35 field emission gun scanning electron microscope equipped with a Channel 5 system. In the deformed sample, several regions with a total area of 750µm² were investigated by the EBSD technique. Larger areas, at least 2500µm², were investigated in each annealed sample. Due to the limited angular resolution of this technique[14][15], misorientation angles less than 2° were not considered in the analysis. Boundaries with misorientation angles θ between 2° and 15° were defined as low-angle boundaries (LABs), while boundaries with misorientation angles larger than 15° were classified as high-angle boundaries (HABs). Recrystallised grains in the orientation maps were defined as regions at least partly surrounded by HABs, having an equivalent circular diameter (ECD) above 3µm and an internal point-to-point misorientation below 2°[16]. Area fractions of different fibre texture components were calculated allowing a 10° orientation deviation from the exact ⟨uvw⟩ fibres. These area fractions are representative of volume fractions.

The boundary area density S_V was also determined from the EBSD data[17][18], and the energy density stored in the form of boundaries in the deformed or recovered microstructure was estimated as $u = \gamma\, S_V$. The specific boundary energy of HABs was assumed to be $\gamma_{HAB} = 617 mJ/m^2$[19]. The specific boundary energy of LABs with misorientation angles $\theta < \theta_{cr} = 15°$ was calculated using the Read-Shockley equation[20][21]:

$$\gamma_{LAB} = \gamma_{HAB} \frac{\theta}{\theta_{cr}}\left(1 - \ln\left(\frac{\theta}{\theta_{cr}}\right)\right)$$

Vickers hardness was determined using a load of 1kg with a 10s dwell time. Dog-bone shaped specimens with a gauge section of $5 \times 1 \times 0.5 mm^3$ prepared from the as-received sample and the compressed sample were tensile tested at room temperature using an Instron 5848 MicroTester with an initial strain rate of

5×10^{-3} s^{-1}. The tensile axis was aligned along the ED for the as-received condition and perpendicular to the compression axis (CA) for the compressed sample. A MTS LX300 laser extensometer was used for measuring strain during the test.

3. Results

3.1. As-Received PM2000

A bright field TEM image in **Figure 2(a)** shows that the as-received material is characterised by a high dislocation density and contains well-dispersed oxide nanoparticles identified in our previous work[22] as orthorhombic yttrium-aluminium oxide YAlO$_3$. The orientation map obtained by EBSD in **Figure 2(b)** demonstrates both almost equiaxed and highly elongated grains. As is evident from the colour code in the inset, most grains have either $\langle 100 \rangle$ or $\langle 110 \rangle$ directions aligned along the extrusion direction. Approximately 50% of all boundaries in this material are HABs. The average spacing $d_{\theta > 2°}$, defined as the mean distance between boundaries with misorientation angles greater than 2°, is 580nm, and the HAB spacing is 940nm as measured along the ED.

The hardness of this material is 332 HV1. The tensile test reveals a combination of moderate ultimate tensile strength (UTS) of 932MPa and reasonably high ductility (see **Figure 3**) with an elongation to failure of about 15%.

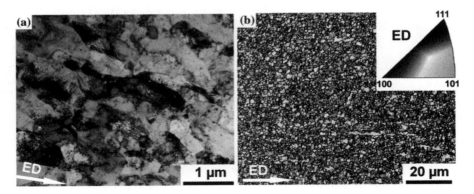

Figure 2. Microstructure of the as-received PM2000 sample: a TEM image and b orientation map showing crystallographic directions along the ED. The colour code is given in the inset. In b, white and black lines represent LABs and HABs, respectively.

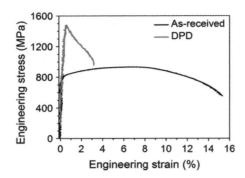

Figure 3. Stress-strain curves for PM2000 in the as-received condition and after compression by DPD. Minor horizontal oscillations are an artefact of the strain measurement.

3.2. PM2000 after Compression by DPD

A characteristic deformation structure with lamellar boundaries almost perpendicular to the CA is observed after DPD (see **Figure 4**). The mean thickness of the lamellae measured in TEM images [see an example in **Figure 4(a)**] is 72nm, and the $d_{\theta>2°}$ measured along the CA in the orientation map in **Figure 4(b)** is 76nm, *i.e.* after compression by DPD the sample is truly nanostructured. The fraction of HABs in this material is 46%.

The orientation map in **Figure 4(b)** indicates that lamellae are arranged in bands with either $\langle 100 \rangle$ directions [red in **Figure 4(b)**] or $\langle 111 \rangle$ directions [blue in **Figure 4(b)**] along the CA. The texture of the DPD-processed sample can therefore be described as a duplex $\langle 111 \rangle + \langle 100 \rangle$ fibre texture.

Compared to the as-received condition, the hardness and the UTS of the DPD-processed sample are increased to 443 HV1 and 1480MPa, respectively. The elongation to failure of PM2000 after DPD is only 3% (see **Figure 3**).

3.3. PM2000 Annealed after Compression by DPD

3.3.1. Softening

The effect of annealing was first evaluated by conducting Vickers hardness

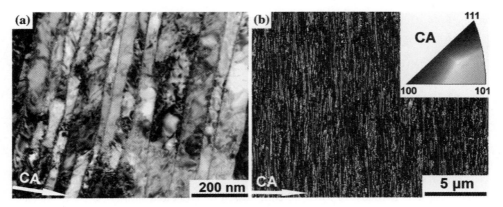

Figure 4. Microstructure of PM2000 after compression by DPD: a TEM image and b orientation map showing crystallographic directions along the CA. The colour code is given in the inset. In b, white and black lines represent LABs and HABs, respectively.

measurements after annealing at different temperatures for 1h. Results of these measurements (see **Figure 5**) demonstrate that annealing for 1h at temperatures up to 500°C does not appreciably affect the hardness reached by DPD. Within the temperature range, 500°C < T ≤ 700°C, the material softens gradually to 383 HV1. A sharp drop in hardness is observed when the annealing temperature exceeds 700°C (**Figure 5**). For all the samples annealed in the temperature range, 715°C ≤ T ≤ 800°C, the hardness is below 300 HV1.

As the transition between slight and large reductions in hardness occurs at 715°C, the microstructural evolution during annealing was investigated at this temperature. Samples annealed for 10, 20 and 80min at 715°C were chosen for a detailed microstructural analysis.

3.3.2. Evolution of Microstructure and Texture

TEM images and orientation maps for the annealed samples selected for microstructural investigations are presented in **Figures 6** and **7**, respectively. The TEM images provide evidence that the oxide nanoparticles have a certain pinning effect on dislocations [**Figure 6(a)**] and migrating boundaries [**Figure 6(b)**]. The microstructure after 10min of annealing at 715°C is considerably coarser ($d_{\theta>2°}$ = 176nm) than that after DPD [see **Figure 8(a)**]. This coarsening, however, does not change the lamellar morphology induced by DPD [see **Figure 7(a)**]. The

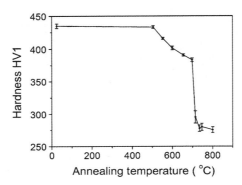

Figure 5. Vickers hardness of PM2000 compressed by DPD to a strain of 2.1 and annealed at different temperatures for 1h. Error bars represent the standard deviation of the distribution.

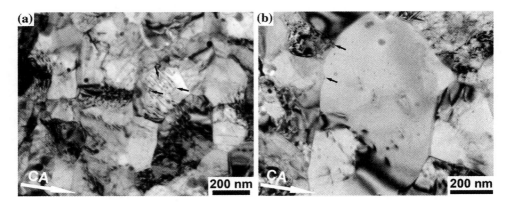

Figure 6. TEM images of PM2000 after compression by DPD and subsequent annealing at 715°C for 10min. Small black arrows indicate dislocations in (a) and a migrating boundary in (b) pinned by oxide nanoparticles.

fraction of HABs after 10min is similar to that in the as-deformed condition [see **Figure 8(b)**]. Recrystallisation nuclei are very rare after 10min [one nucleus is marked by an arrow in **Figure 7(a)**], and the area fraction of recrystallised material f_{RX} is only 1%.

After 20 min of annealing, f_{RX} is much larger, 68%, and the average boundary spacing $d_{\theta>2°}$ is increased to 750nm [**Figures 7(b), 8**]. After annealing for 80min, the material is almost fully recrystallised ($f_{RX} = 97\%$) with $d_{\theta>2°} = 3750$nm [**Figures 7(c), 8(a), (c)**]. The fraction of LABs in this almost fully recrystallised microstructure is only 19% [**Figure 8(b)**]. The orientation maps demonstrate

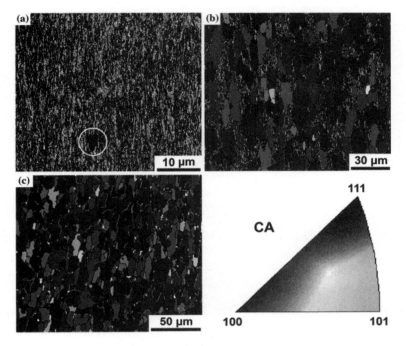

Figure 7. Orientation maps for PM2000 after compression by DPD and subsequent annealing at 715°C for a 10min, b 20min and c 80min. The colours correspond to crystallographic directions along the CA according to the inverse pole figure in the lower right corner. One recrystallised grain found after 10min of annealing is encircled in a. White and black lines represent LABs and HABs, respectively. In each map, the CA is horizontal.

that the majority of recrystallised grains have a ⟨111⟩ direction along the CA (blue in **Figure 7**). In the sample annealed for 80min [**Figure 7(c)**], such ⟨111⟩-oriented grains occupy 67% of the area.

Figure 9 presents an example of the non-recrystallised regions still present in the microstructure after 80min of annealing. A very fine lamellar structure with an average lamellae thickness of 165nm and a high density of dislocations are preserved in such non-recrystallised regions. Analysis of selected area diffraction patterns from several lamellae in **Figure 9(b)** indicates that their orientations belong to the ⟨100⟩ fibre texture.

The evolution of crystallographic texture measured by the EBSD technique in large sample areas is illustrated by inverse pole figures (IPFs) in **Figure 10**. It is seen that the duplex ⟨111⟩ + ⟨100⟩ fibre texture formed by DPD evolves during

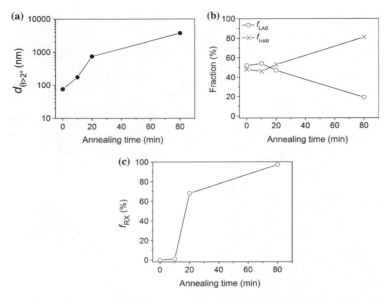

Figure 8. Microstructural parameters (from EBSD data) of PM2000 after compression by DPD and subsequent annealing at 715°C: a boundary spacing $d_{\theta>2°}$ measured along the CA; b fractions of HABs and LABs; c fraction of recrystallised material.

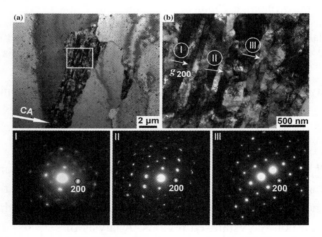

Figure 9. Microstructures of PM2000 after compression by DPD and subsequent annealing at 715°C for 80min: a TEM images showing recrystallised grains and a small non-recrystallised (recovered) region; b enlarged image of the non-recrystallised region marked by a frame in a. Circles I, II, and III in b indicate areas from which selected area diffraction patterns were obtained (shown in the lower part of the figure). The g_{200} diffraction vectors are aligned almost parallel to the CA, signifying that crystallographic 200 poles are closely aligned with the CA. The corresponding regions thus belong to the ⟨100⟩ fibre texture.

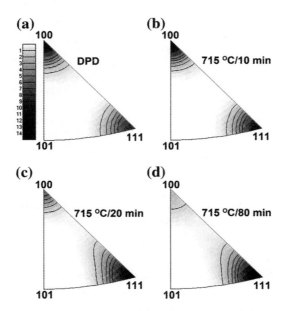

Figure 10. Inverse pole figures showing probability densities for crystallographic directions along the CA in PM2000 after compression by DPD (a), and subsequent annealing at 715°C for 10min (b), 20min (c) and 80min (d). Contour lines are 1, 2, 3, 4 and 5 times random.

annealing into a well-defined ⟨111⟩ fibre texture, whereas the strength of the ⟨100⟩ fibre texture component decreases. Considering the area fractions of the individual fibre texture components in **Figure 11**, it is apparent that during recrystallisation the ⟨111⟩ fibre strengthens at the expense of the ⟨100⟩ fibre, *i.e.* recrystallisation in PM2000 after compression by DPD is orientation-dependent.

4. Discussion

4.1. Nanostructure and Strength

Compression of PM2000 by DPD to a strain of 2.1 results in nanoscale lamellar structures with a boundary spacing of 72nm as measured along the CA using TEM. This spacing is considerably smaller than the value of 98nm measured in a modified 9Cr-1Mo steel compressed by DPD to a similar strain[8]. The smaller spacing in PM2000 as compared to the modified 9Cr-1Mo steel may be attributed to a finer initial microstructure in PM2000 before DPD as well as to the

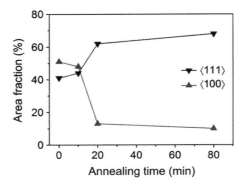

Figure 11. Area fractions of the ⟨111⟩ and ⟨100⟩ fibre texture components in PM2000 after compression by DPD and subsequent annealing at 715°C.

presence of oxide nanodispersoids, which could make structural refinement of this ODS alloy more efficient. Such accelerated structural refinement has previously been reported for rolled aluminium containing small alumina particles[23]. Similar to other compressed body-centred cubic materials, PM2000 develops a strong duplex ⟨111⟩ + ⟨100⟩ fibre texture during compression by DPD[7][8][24][25].

The structural refinement by DPD makes PM2000 very hard and increases the UTS to 1480MPa, which is significantly higher than the strength in the as-received condition and the UTS previously reported for PM2000 and other ODS steels such as mechanically alloyed MA956 and MA957[26]. The strength of PM2000 after DPD is also higher than that of the modified 9Cr-1Mo steel processed by DPD to a similar strain, in which the UTS is only 1240MPa[8]. The higher strength in the DPD-processed PM2000 is consistent with the finer microstructure in this material. The increase in strength of PM2000 during DPD is accompanied by a significant reduction in ductility—a phenomenon which has previously been observed in many cold-deformed materials due to their limited capacity of work hardening[8][27][28].

4.2. Recovery and Recrystallisation

Our investigation demonstrates that the nanostructure formed in PM2000 after compression by DPD is fairly stable during annealing for 1h at temperatures as high as 500°C. However, annealing at temperatures higher than 500°C makes the nanostructure unstable, as follows from the significant softening observed in **Figure 5** after annealing above 500°C. The shape of the hardness curve in **Figure 5** implies that the deformed microstructure only recovers within the temperature range from 500°C to 700°C, whereas at temperatures higher than 700°C pronounced recrystallisation takes place. It is therefore apparent that compared to the as-received condition, in which no evidence of recrystallisation was observed even after annealing 1100°C for 1h[29], the thermal stability of the nanostructured condition is considerably reduced.

The microstructural observations made in PM2000 after DPD and annealing at 715°C indicate that during recovery (within the first 10 min of annealing) the microstructure coarsens despite a certain pinning effect imposed by nanoparticles on boundary migration. The coarsened microstructure largely retains the lamellar morphology and maintains a similar proportion of LABs and HABs to that observed in the as-deformed microstructure [see **Figure 8(b)**]. Although the density of interior dislocations was not determined quantitatively in the present experiment, it is apparent that the dislocation density in the TEM images taken after 10 min at 715°C is lower than that directly after DPD [cf. **Figures 4(a), 6**]. This suggests that at 715°C dislocations annihilate despite the pining effect of nanoparticles.

Very few recrystallised grains appear within the first 10min of annealing, but within the next 10min (a total annealing time of 20min) recrystallisation progresses very quickly and is almost complete after 80min [see **Figures 7, 8(c)**]. As is evident from **Figure 7**, recrystallisation proceeds in such a way that the frequency of $\langle 111 \rangle$-oriented grains growing into the recovered environment is greater than the frequency of growing grains of any other orientation. This orientation-dependent recrystallisation leads to a clear predominance of the $\langle 111 \rangle$ fibre when the material is almost fully recrystallised [see **Figures 10, 11**].

To understand the reason for the observed differences in the annealing behaviour of regions having different orientations, structural parameters of the deformed and recovered material are shown in **Figure 12** separately for the two different texture components. The data in **Figure 12(a)** indicate that after DPD the average boundary spacing within the $\langle 100 \rangle$-oriented lamellae is larger (80nm) than within the $\langle 111 \rangle$-oriented lamellae (68 nm). After 10min of annealing, how-

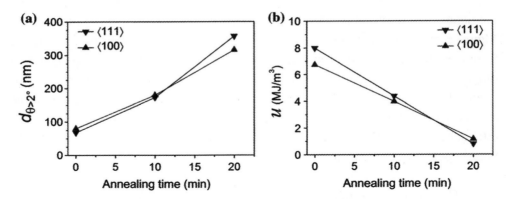

Figure 12. Changes in deformed and recovered regions of either ⟨111⟩ or ⟨100⟩ fibre texture components in PM2000 after compression by DPD and subsequent annealing at 715°C: a average boundary spacing $d_{\theta>2°}$ along the CA; b stored energy density u calculated from the EBSD data.

ever, the difference between the average boundary spacings $d_{\theta>2°}$ in the recovered ⟨111⟩ and ⟨100⟩-oriented regions is reduced, while after 20 min of annealing, $d_{\theta>2°}$ for the ⟨111⟩-oriented regions is even larger than that for the ⟨100⟩-oriented regions.

Figure 12(b) shows the stored energy density u estimated from the EBSD data for the ⟨111⟩- and ⟨100⟩-oriented regions in the same deformed and recovered samples. It is seen that in the deformed microstructure u ⟨111⟩ is considerably higher than u ⟨100⟩. This difference can be attributed to the difference in the Taylor factor M between the ⟨111⟩ and ⟨100⟩-oriented regions. According to the pencil-glide model proposed by Taylor, in body-centred cubic metals for grains compressed along ⟨100⟩ M is 2.1, while for compression along ⟨111⟩ the value of M is 3.2[30]. Considering that the Taylor factor describes the ratio between the magnitude of accumulated plastic slip and the magnitude of applied strain, the higher Taylor factor for regions of the ⟨111⟩ fibre texture component suggests an increased slip activity in these regions. Consequently, a larger dislocation density and a higher stored energy are expected for the ⟨111⟩-oriented regions compared to the ⟨100⟩-oriented regions, which is fully consistent with the stored energy values obtained for the as-deformed microstructure [see **Figure 12(b)**]. Such a higher stored energy in the ⟨111⟩-oriented regions provides a higher driving force for nucleation of recrystallisation, and many more nuclei are formed within the ⟨111⟩

-oriented regions. These nuclei have ⟨111⟩ orientations, thus leading to the dominance of the ⟨111⟩-oriented recrystallized grains observed after annealing for 20min [see **Figure 7(b)**]. The growth of abundant ⟨111⟩-oriented nuclei results in the dominant ⟨111⟩ fibre texture and explains the orientation dependent recrystallisation behaviour. Such a correlation between the Taylor factor, stored energy and preferential nucleation is consistent with previous reports describing the annealing behaviour of body-centred cubic steels after rolling[31]-[34].

Nevertheless, it should be emphasised that the initially large difference in the stored energy density between ⟨111⟩- and ⟨100⟩-oriented regions in the deformed condition becomes rather small after annealing for 10min at 715°C. Moreover, after 20min the stored energy density within the ⟨111⟩-oriented regions is even lower than of the ⟨100⟩-oriented regions. Obviously, the stored energy density within the ⟨111⟩-oriented regions is reduced during recovery much faster than within regions of the ⟨100⟩ fibre texture component [see **Figure 12(b)**]. For nucleation of recrystallisation beyond 10min, when the stored energy density in both texture components is similar, there might be other reasons for preferred formation of ⟨111⟩-oriented nuclei. One possible reason is that the morphological characteristics of recovered subgrains may be different for the ⟨111⟩ component and the ⟨100⟩ component[29]. Considering that rather straight lamellar boundaries are still observed within recovered ⟨100⟩-oriented regions even after 80min at 715°C (**Figure 9**), it is reasonable to suggest that the recovered lamellar structure impedes the formation of nuclei in these regions.

5. Concluding Remarks

Compression of PM2000 by DPD to a strain of 2.1 significantly refines the initial microstructure by nanoscale lamellae having a duplex ⟨111⟩ + ⟨100⟩ fibre texture. The strength of material increases substantially, whereas the thermal stability of the nanostructured material is reduced. This reduced thermal stability is attributed to a high density of dislocations and dislocation boundaries induced by DPD. In the as-deformed microstructure, the stored energy density estimated from the EBSD data is found to be higher in the ⟨111⟩-oriented regions than in the ⟨100⟩-oriented regions. Recovery during annealing at 715°C reduces the energy stored in the deformed microstructure, and this reduction is more pronounced in the ⟨111⟩

-oriented regions. Orientation-dependent recrystallisation takes place in the recovered microstructure, leading to the strengthening of the ⟨111⟩ fibre texture component at the expense of the ⟨100⟩ fibre texture component. As a result, a strong ⟨111⟩ fibre texture is observed when recrystallisation is almost complete.

Acknowledgements

Financial support from the Sino-Danish Center for Education and Research is gratefully acknowledged. The authors are grateful to Prof. M. Heilmaier for providing the initial PM2000 sample.

Open Access: This article is distributed under the terms of the Creative Commons Attribution 4.0 International License (http://creativecommons.org/licenses/by/4.0/), which permits unrestricted use, distribution, and reproduction in any medium, provided you give appropriate credit to the original author(s) and the source, provide a link to the Creative Commons license, and indicate if changes were made.

Source: Zhang Z B, Tao N R, Mishin O V, *et al.* Oxide dispersion-strengthened steel PM2000 after dynamic plastic deformation: nanostructure and annealing behaviour[J]. Journal of Materials Science, 2016, 51(11):5545–5555.

References

[1] Odette GR, Alinger MJ, Wirth BD (2008) Recent developments in irradiation-resistant steels. Annu Rev Mater Res 38:471–503.

[2] Charit I, Murty KL (2010) Structural materials issues for the next generation fission reactors. JOM 62:67–74.

[3] Song M, Wu YD, Chen D, Wang XM, Sun C, Yu KY, Chen Y, Shao L, Yang Y, Hartwig KT, Zhang X (2014) Response of equal channel angular extrusion processed ultrafinegrained T91 steel subjected to high temperature heavy ion irradiation. Acta Mater 74:285–295.

[4] Shen TD, Feng S, Tang M, Valdez JA, Wang Y, Sickafus KE (2007) Enhanced radiation tolerance in nanocrystalline $MgGa_2O_4$. Appl Phys Lett 90:263115.

[5] Hansen N (1990) Cold deformation microstructures. Mater Sci Technol 6:1039–1047.

[6] Hansen N, Juul Jensen D (2011) Deformed metals—structure, recrystallisation and strength. Mater Sci Technol 27:1229–1240.

[7] Song M, Sun C, Jang J, Han CH, Kim TK, Hartwig KT, Zhang X (2013) Microstructure refinement and strengthening mechanisms of a 12Cr ODS steel processed by equal channel angular extrusion. J Alloys Compd 577:247–256.

[8] Zhang ZB, Mishin OV, Tao NR, Pantleon W (2015) Microstructure and annealing behavior of a modified 9Cr-1Mo steel after dynamic plastic deformation to different strains. J Nucl Mater 458:64–69.

[9] Li YS, Tao NR, Lu K (2008) Microstructural evolution and nanostructure formation in copper during dynamic plastic deformation at cryogenic temperatures. Acta Mater 56:230–241.

[10] Huang F, Tao NR, Lu K (2011) Effects of strain rate and deformation temperature on microstructures and hardness in plastically deformed pure aluminum. J Mater Sci Technol 27:1–7.

[11] Luo ZP, Mishin OV, Zhang YB, Zhang HW, Lu K (2012) Microstructural characterization of nickel subjected to dynamic plastic deformation. Scr Mater 66:335–338.

[12] Schneibel JH, Heilmaier M, Blum W, Hasemann G, Shanmugasundaram T (2011) Temperature dependence of the strength of fineand ultrafine-grained materials. Acta Mater 59:1300–1308.

[13] Zhang ZB, Mishin OV, Tao NR, Pantleon W (2015) Effect of dynamic plastic deformation on microstructure and annealing behaviour of modified 9Cr-1Mo steel. Mater Sci Technol 31:715–721.

[14] Humphreys FJ (2001) Review grain and subgrain characterisation by electron backscatter diffraction. J Mater Sci 36:3833–3854. doi:10.1023/A:1017973432592.

[15] Godfrey A, Mishin OV, Liu Q (2006) Processing and interpretation of EBSD data gathered from plastically deformed metals. Mater Sci Technol 22:1263–1270.

[16] Wu GL, Juul Jensen D (2008) Automatic determination of recrystallization parameters based on EBSD mapping. Mater Charact 59:794–800.

[17] Godfrey A, Cao WQ, Hansen N, Liu Q (2005) Stored energy, microstructure, and flow stress of deformed metals. Metall Mater Trans A 36A:2371–2378.

[18] Godfrey A, Hansen N, Juul Jensen D (2007) Microstructural-based measurement of local stored energy variations in deformed metals. Metall Mater Trans A 38A:2329–2339.

[19] Humphreys FJ, Hatherly M (2004) Recrystallization and related annealing Phenomena. Elsevier, Oxford.

[20] Read WT, Shockley W (1950) Dislocation models of crystal grain boundaries. Phys Rev 78:275–289.

[21] Read WT (1953) Dislocations in crystals. McGraw-Hill, New York, pp 155–172.

[22] Zhang ZB, Mishin OV, Tao NR, Pantleon W (2014) Evolution of oxide nanoparticles during dynamic plastic deformation of ODS steel. In: Fæster S et al (Eds.), Proc 35th Risø Int Symp Mater Sci, p. 423–430.

[23] Barlow CY, Hansen N (1989) Deformation structures in aluminum containing small particles. Acta Metall 37:1313–1320.

[24] Dillamore IL, Katoh H, Haslam K (1974) The nucleation of recrystallisation and the development of textures in heavily compressed iron-carbon alloys. Texture 1:151–156.

[25] Hu H (1974) Texture of metals. Texture 1:233–258.

[26] Klueh RL, Shingledecker JP, Swindeman RW, Hoelzer DT (2005) Oxide dispersion-strengthened steels: a comparison of some commercial and experimental alloys. J Nucl Mater 341:103–114.

[27] Song R, Ponge D, Raabe D, Speer JG, Madock DK (2006) Overview of processing, microstructure and mechanical properties of ultrafine grained bcc steels. Mater Sci Eng A 441:1–17.

[28] Zhang YB, Mishin OV, Kamikawa N, Godfrey A, Liu W, Liu Q (2013) Microstructure and mechanical properties of nickel processed by accumulative roll bonding. Mater Sci Eng A 576:160–166.

[29] Zhang ZB (2015) Nanostructures in a ferritic and an oxide dispersion strengthened steel induced by dynamic plastic deformation. PhD dissertation. Technical University of Denmark.

[30] Rosenberg JM, Piehler HR (1971) Calculation of the Taylor factor and lattice rotations for bcc metals deforming by pencil glide. Metall Trans 2:257–259.

[31] Every RL, Hatherly M (1974) Oriented nucleation in low-carbon steels. Texture 1:183–194.

[32] Hutchinson WB (1984) Development and control of annealing textures in low-carbon steels. Int Mater Rev 29:25–42.

[33] Samajdar I, Verlinden B, Van Houtte P, Vanderschueren D (1997) γ-fibre recrystallization texture in IF-steel: an investigation on the recrystallization mechanisms. Mater Sci Eng A 238:343–350.

[34] Gazder AA, Sanchez-Araiza M, Jonas JJ, Pereloma EV (2011) Evolution of recrystallization texture in a 0.78wt% Cr extra-low-carbon steel after warm and cold rolling. Acta Mater 59:4847–4865.

Chapter 27

Early Root Development of Field-Grown Poplar: Effects of Planting Material and Genotype

Grant B. Douglas[1], **Ian R. McIvor**[2], **Catherine M. Lloyd-West**[1]

[1]AgResearch, Private Bag 11008, Palmerston North, New Zealand
[2]Plant and Food Research, Private Bag 11600, Palmerston North, New Zealand

Abstract: Background: Poplar trees (Populus spp.) are used widely for soil conservation. A key advantage is their ability to establish from unrooted stem sections of varying dimensions, ranging from small cuttings to large poles. This study determined root length and biomass of young trees from three different-sized stem sections and quantified clonal variation. Methods: Two concurrent field trials were conducted: trial 1 compared root attributes of trees from cuttings, stakes, and poles of a single poplar clone, while trial 2 compared those of trees from cuttings of six poplar clones. Excavations of entire trees were conducted in autumn for three (trial 1) or two (trial 2) years after planting. Results: Total root mass averaged over 3 years was in the order poles (364g) > stakes (70g) > cuttings (17g), and total root length was in the order poles (73m) > stakes (21m) > cuttings (7m). Maximum lateral root extension was approximately 2.6m from poles, 1.7m from stakes, and 0.8m from cuttings. Clonal variation in trees from cuttings was found for both mean total root mass (10.4–45.9g) and total root length (3.5–11.8m). In both trials, root mass and length increased, decreased, or were unchanged with increasing 0.5-m increments of the distance from stem and soil depth, depending on year, planting material, root diameter, and their interaction. Conclusions: Early root development from poles was greater than from cuttings, with development from stakes being intermediate. Different poplar clones exhibited large variation in root biomass development within 2 years of planting. The results provide an understanding of the differences in early root development of poplar planting materials and clones used for soil conservation and other purposes and guidance on appropriate tree spacings of different planting materials to achieve root interlock.

Keywords: Pasture-Tree Systems, Erosion Control, Root Systems, Root Diameter, Tree Spacing

1. Background

Soil erosion in diverse ecosystems and landscapes modified by human activity is a significant global problem (Toy *et al.* 2002; Blanco and Lal 2008; Liu *et al.* 2011). Methods to control various erosion types and processes comprise built structures (Gebrernichael *et al.* 2005; Posthumus and De Graaff 2005; Yang *et al.* 2009; Wang *et al.* 2012) or the use of biological solutions, principally the establishment of a live vegetation cover (Zuazo and Pleguezuelo 2008; Stokes *et al.* 2014). The choice of approach depends on factors such as the extent and severity of the erosion problem, initial and ongoing costs of implementation, anticipated maintenance requirements, and expected useful life of works. Vegetation reduces erosion through above- and below-ground mechanical and hydrological effects that are governed by factors including species (e.g. growth form, morphology, root depth, and distribution), age, and spatial and temporal configurations (Bischetti *et al.* 2005; de Baets *et al.* 2009; Stokes *et al.* 2009; Genet *et al.* 2010). Vegetation used in erosion control programmes ranges from indigenous or introduced herbaceous grasses, legumes, and herbs through to woody vegetation comprising assorted shrub and tree species (Blanco and Lal 2008; Evette *et al.* 2009; Kuzovkina and Volk 2009; Wu *et al.* 2010).

Species and hybrids of the genus Populus (poplar) are important examples of trees used in erosion control and remediation programmes in many regions of the world (Wu *et al.* 1994; Lammeranner *et al.* 2005; Licht and Isebrands 2005; Reisner *et al.* 2007; Blanco and Lal 2008). The advantages of using poplar for erosion control include rapid growth, high evapotranspiration rates, extensive lateral root systems, abundant fine root production, and rapid tree establishment from vegetative material. There continue to be many above-ground attributes of poplars assessed, including in Europe (Pulkkinen *et al.* 2013; Toillon *et al.* 2013; Verlinden *et al.* 2013), China (Fang *et al.* 2013), and North America (Hart *et al.* 2013; Kaczmarek *et al.* 2013), mainly with respect to the rapidly developing use of short-rotation coppice systems for biofuels and other products. In contrast, fewer belowground investigations have been conducted recently (Benomar *et al.* 2013; Berhongaray *et al.* 2013; Hajek *et al.* 2014; Phillips *et al.* 2014), and this imbalance needs addressing because root characteristics and their spatial and temporal changes are particularly important in determining efficacy for soil stabilisation (Stokes *et al.* 2009). Furthermore, in view of the many natural and bred poplar genotypes available (Eckenwalder 1996), variation in root characteristics between species/hybrids of poplar in the field has received scant attention (Al Afas *et al.* 2008; Benomar *et al.* 2013; Berhongaray *et al.* 2013; Phillips *et al.* 2014). Consequently, there is limited knowledge of the genetic variation in root attributes of populations of poplar which hinders identifying the most appropriate clones for stabilising soil.

Poplars have been planted in significant numbers in New Zealand since the 1960s to reduce the extent and severity of mass-movement erosion processes such as shallow landslides, earth flows, and gully erosion (Thompson and Luckman 1993). Plantings are particularly widespread on hill country supporting a pastoral cover for livestock (predominantly sheep and beef cattle) grazing enterprises, where established trees at densities of less than 100 stems per hectare (sph) provide effective slope stabilisation and enable understorey pasture production (Wilkinson 1999; Guevara-Escobar *et al.* 2007; Benavides *et al.* 2009). For example, the area of shallow landsliding was reduced by 50%–80% by trees of Populus spp. and species of other genera at 70sph or greater (Hicks 1995), and trees of mostly Populus spp. at 32–65sph reduced the occurrence of shallow landslides by an average of 95% (Douglas *et al.* 2013). Annual pasture production beneath widely spaced poplars is reduced by 20%–50% depending mainly on tree spacing and tree size/age (Douglas *et al.* 2006; Wall *et al.* 2006; Guevara-Escobar *et al.* 2007).

In pasture-poplar systems, the growth and distribution of 'Veronese' poplar (Populus deltoides Marshall × nigra L.) roots on erodible slopes have been measured for entire trees aged 5–11.5 years (McIvor *et al.* 2008; McIvor *et al.* 2009), but there appear to be scant data for other clones, regardless of age. Nine months after establishing poplar clones from various planting materials (PMs) on flat, cultivated terrace soils, the mean root depth of Veronese was almost twice that of P. deltoides × yunnanensis clone 'Kawa' (Phillips *et al.* 2014). There were no significant differences between the clones in below-ground biomass, maximum lateral root diameter, and total root length. In a greenhouse study of up to 10 weeks, genetic variation was detected for number of root nodes, root mass, and root length among ten poplar genotypes established from 0.2-m-long stem sections (McIvor *et al.* 2014). Additionally, root development was reduced significantly with increasing soil bulk density, which has implications for likely field performance of the genotypes in different substrates.

Poplars can be established in the presence of grazing livestock by planting stem sections protected with a plastic sleeve (Wilkinson 1999), rather than using nursery-grown seedlings and ceasing grazing within at least the first few years following transplanting. In New Zealand, unrooted stems 3m long, 50–70mm diameter, and aged 2 years, henceforth referred to as poles, are typically planted vertically in grazed pastures. Smaller and younger vegetative material such as 1-m-long stems of 20–30mm diameter and aged 1 year, henceforth referred to as stakes, may be planted where livestock are excluded (Wilkinson 1999). Stakes also have potential use in non-farm applications such as protecting roadside and railway embankments. Similarly, shorter and thinner stems, e.g. 0.4m long and 15–20mm diameter, henceforth referred to as cuttings, are used in agricultural and non-agricultural situations. The effect of the size of PM on early tree development, particularly in relation to root length and biomass, is poorly understood (Sidhu and Dhillon 2007; DesRochers and Tremblay 2009). It is hypothesised that over the first few years of tree establishment, poles will have significantly greater rates of above- and below-ground biomass development than stakes and cuttings.

The first objective of this study was to determine the effect of PM (poles, stakes, and cuttings) on above- and below-ground biomass development of Veronese poplar within the first 3 years after planting in the field. The second objective aimed to quantify the variation in root development of selected experimental and released poplar clones in the first 2 years after planting as cuttings. The results from two concurrent trials are reported.

2. Methods

2.1. Study Sites

Two trials were conducted in a cultivated block on flat land at a commercial plant propagation nursery (175°39'E, 40°21'S), approximately 10km from Palmerston North in the south-west of the North Island. A flat site was deemed likely to maximise tree growth and facilitate whole-tree excavation. The ground was cultivated 2 months before planting when soil water levels were high, so it did not develop a fine tilth. Data obtained from the National Institute of Water and Atmospheric Research (NIWA) showed that the site has a temperate climate, with mean monthly air temperature over the period 1981–2010 ranging from 8.6°C in July (winter) to 18.3°C in February (summer), and mean annual rainfall of 917 mm (NIWA 2013). The soil was a Manawatu fine sandy loam (Fluvial Recent soil, Dystric Fluventic Eutrochrept derived from greywacke alluvium) (Hewitt 1998). Soil samples from $0 < 25$, $25 < 50$, and 50–75cm depth at the site were analysed for pH (1:2.1 v/v water slurry), Olsen phosphate (Olsen extraction), and sulphate-sulphur content (potassium phosphate extraction).

2.2. Trial 1: Planting Material Comparison

2.2.1. Plant Material and Experimental Design

Hybrid poplar clone Veronese was established from vegetative PM of three dimensions and ages: poles, stakes, and cuttings, as described previously. Veronese has been available commercially for 15+ years (Wilkinson 1999) and is used widely in erosion control programmes throughout New Zealand. Experience has found that the unrooted stems of Veronese are able to establish roots readily in a range of field and glasshouse conditions. Cuttings and stakes were removed from the middle of vertical stems of parent trees, and poles were obtained from lower-middle parts of stems. All PMs were dormant, obtained from a local nursery, and soaked in a water bath for 1–3 days before planting.

In August 2009, three units of each PM were planted in each of five randomised complete blocks that provided five replicates for destructive sampling in each of three consecutive years. Each PM unit was treated as an experimental unit, and in total, there were 45 experimental units in the trial. The volume (V) of each of the three PMs was estimated using the formula $V = \pi d^2 L 4^{-1}$ where d = diameter of PM and L = length of PM. A single value was calculated for each of the three PMs assuming all poles were 3m long and 60mm diameter (midpoint of 50–70mm range), all stakes were 1m long and 25mm diameter (20–30mm), and all cuttings were 0.4m long and 17.5mm diameter (15–20mm). Units were at 2-m spacings within two adjacent rows 3m apart, with two or three buffer rows of stakes planted either side to reduce any border effects. The PMs were planted vertically at depths of 0.3m for cuttings, 0.5m for stakes, and 0.8m for poles to facilitate anchorage of the developing trees in the substrate. Soil surrounding the PM was compacted to reduce the likelihood of air pockets (van Kraayenoord *et al.* 1986). Weeds, principally grasses (mostly Holcus lanatus L.), but also legumes (e.g. Lotus uliginosus Schkuhr) and other broadleaf species (e.g. Ranunculus repens L.), were controlled by hand-weeding and application of herbicide (glyphosate at 1.8kg a.i. ha^{-1}). For example, in the first year, weeds were sprayed once after planting and once in summer.

2.2.2. Measurements

Above- and below-ground attributes were measured in autumn 2010 [year (Y)1], 2011 (Y2), and 2012 (Y3). Five replicates of establishing trees from stakes and poles were measured each year whereas the death of several trees established from planted cuttings resulted in the measurement of four replicates of this PM in Y1 and Y2 and three replicates in Y3. Above-ground attributes measured were as follows: longest shoot length, henceforth referred to as shoot length (m, measured in Y1–Y3); diameter of the base of the longest shoot, henceforth referred to as shoot diameter (mm, Y1 and Y3); distance from ground level to the base of the longest shoot, henceforth referred to as shoot position (m, Y1–Y3); root collar diameter (mm, Y1); and shoot mass (g), comprising leaf and stem (Y1–Y3). All leaves and stems were severed from the PM units, and their total mass determined by oven-drying at 70°C for 24h. Mass of the destructively sampled PMs (g) was determined in Y1–Y3 by drying at 70°C to a constant weight (up to 4 days).

Whole-tree root excavations were conducted using handheld implements within a series of 0.5-m radial widths radiating outwards from the establishing trees and at 0.5-m soil-depth increments (**Figure 1**). Width and depth each comprised three classes of 0 < 0.5, 0.5 < 1.0, and ≥1.0m. Roots extending beyond 1.0m laterally or vertically were followed to their end point, often characterised by a venation of very thin roots. Roots were removed from the original PMs and classified into diameters of <1 (fibrous), 1 < 2, 2 < 5, 5 < 10, 10 < 20, and ≥20mm. The total length of roots in each diameter class, except fibrous, was measured, and all material was oven-dried (70°C to a constant weight) to determine root mass.

2.3. Trial 2: Clonal Variation Comparison

2.3.1. Plant Material and Experimental Design

Cuttings from six different poplar clones ('San Rosa', 'Fraser', 'Veronese', 'Kawa', 'Geyles' and 'PN471') from various poplar species/hybrids (**Table 1**) (Van Kraayenoord and Hathaway 1986; MAF 2011) were planted in a randomised design with five replicate cuttings of each clone in each of 2 years. Spacings between cuttings were 2m within rows and 3m between rows and, as in trial 1, were

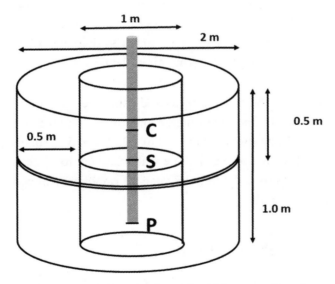

Figure 1. Arrangement of sampling depths and widths around poplar trees established from cuttings (C), stakes (S), and poles (P). Horizontal bars on central axis represent depth of planting of cuttings (C, 0.3m), stakes (S, 0.5m) and poles (P, 0.8m). Planting materials have different diameters (not shown). Excavations exceeded 1.0-m radius and depth when roots extended beyond shown dimensions.

Table 1. Clones of Populus spp. established from cuttings.

Clone	Description
Populus deltoides Marshall × *P. ciliata* (Wall. Ex Royle) clone San Rosa	New Zealand (NZ)-bred vigorous clone, best suited to drier regions because of rust susceptibility; copes well with wind.
Populus deltoides Marshall × *P. nigra* L. clone Fraser	NZ-bred, very narrow tree with a light open canopy that casts minimal shade; light stems are prone to breakage at windy sites.
Populus deltoides Marshall × *P. nigra* L. clone Veronese	European-bred, straight-stemmed, narrow-crowned tree that has good drought and wind tolerance.
Populus deltoides Marshall × *P. yunnanensis* (Dode) clone Kawa	NZ-bred, narrow-crowned tree, resistant to browsing by possums; well suited to moist slopes but not to strong winds.
Populus maximowiczii (A. Henry) × *P. nigra* L. clone Geyles	NZ-bred, straight-stemmed tree with a narrow crown, high rust resistance; grows well on moist sites.
Populus trichocarpa (Torrey and Gray) clone PN471	European-selected, American balsam poplar, narrow-crowned tree for slope stabilisation and gully plantings; not recommended for planting on windy and exposed sites.

planted in August 2009. Experimental units were as defined previously, and the total number of units in the experiment was 60 (6 clones × 5 reps × 2 years). Planting method and depth, and plot management, were as in trial 1.

2.3.2. Measurements

Above-ground measurements were conducted in autumn 2010 (Y1) and 2011 (Y2) for shoot length (m), shoot diameter (mm, Y1 only), and shoot mass (g), as in trial 1. Mass of cuttings (g) was determined as previously. The number of replicates measured per clone varied in Y1 (three to five) and Y2 (one to four) because of inadvertent tree deaths occurring after planting.

Below-ground attributes were measured using the same protocol used in trial 1, including soil width and depth increments of $0 < 0.5$, $0.5 < 1.0$, and ≥ 1.0m. Roots were classified into diameter classes of <2, $2 < 5$, $5 < 10$, and ≥ 10mm. The total soil volume excavated for each clone was relatively low because of the small root systems developed over the initial 2 years of establishment compared to those for stakes and poles in trial 1.

2.4. Statistical Analyses of Data

All analyses were conducted using GenStat© 16th edition (VSN International 2014). Data for above-ground attributes were analysed using mixed effects models with a restricted maximum likelihood (REML) approach. In trial 1, data for root collar diameter, shoot mass, and PM mass were log-transformed to achieve variance homogeneity. Similarly in trial 2, shoot mass data were log-transformed. In trial 1 for above-ground attributes measured in each of 3 years, sources of variation in the model were block [4 degrees of freedom (df)], year (2 df), PM (2 df), year × PM interaction (4 df), and residual (28 df), giving a total of 40 df. The total df in models was reduced for attributes measured in one (13 df) or two (26 df) years. In trial 2, key sources of variation in the model were year (1 df), clone (5 df), and year × clone interaction (5 df). Significant terms in the fitted models were examined further using Fisher's protected least significant differences (LSDs) at the 5% level. For transformed data, back-transformed predicted means and approximate standard errors are presented.

Root data in both trials were analysed using a two-step method, similar to that used previously (Douglas *et al.* 2010), comprising (1) analysis of the proportion of width × depth × year combinations (cells) with roots and (2) analysis of data for cells where roots were present. This method was used because of the large number of zero values in the datasets. Logistic regression was used to analyse presence/absence data for step 1. Data for step 2 were log-transformed and analysed using REML, and back-transformed predicted means and approximate standard errors are presented.

In trial 1, two separate analyses of root data (A, B) were conducted for step 1. Analysis A was for all PMs and comprised data for 3 years, two depths (0 < 0.5, 0.5 < 1.0m) and two widths (0 < 0.5, 0.5 < 1.0m). The data were restricted to a maximum of 0.5 < 1.0m width because trees from cuttings did not produce roots beyond 1.0m within the study time frame. Terms in the model comprised year, depth, width, PM, and depth × width, depth × year, and depth × PM interactions, with a regression df of 11 and total df of 35. Analysis B was for pole data only and was conducted because the roots of trees from poles extended frequently beyond 1m. This enabled three (0 < 0.5, 0.5 < 1.0, ≥1.0m) rather than two widths to be included in the analysis to obtain a greater understanding of variation between widths. The model used comprised terms of depth, width, year, and depth × year interaction with a regression df of 7 and total df of 89. For step 2, total values for root attributes were calculated in two ways. Firstly, data were summed over all diameter classes, soil depths, and widths to calculate total root mass (TRM) and total root length (TRL) per tree. Secondly, data were summed over all diameter (D) classes only, providing attributes of TRMD and TRLD for all other factor combinations (width and depth). As an example of models used, TRM and TRL data were analysed using a model with terms block (4 df), year (2 df), PM (2 df), PM × year interaction (4 df), and residual (28 df). There was considerable data imbalance from the absence of roots, for example, in the larger diameter classes for trees from all PMs in Y1 and from many trees from cuttings in Y2 and Y3. The widest distribution of diameter classes (<2 to ≥20mm) occurred for roots from trees from poles in Y3. For these data, root mass and root length in each diameter class were analysed separately to determine the effect of depth (0 < 0.5, ≥ 0.5m) and width (0 < 0.5, 0.5 < 1.0, ≥1.0m). The model comprised block (4 df), depth (1 df), width (2 df), depth × width interaction (2 df), and residual (17 df).

In trial 2, root presence/absence was analysed (step 1). Roots were absent or detected rarely in the larger diameter classes (5 < 10, ≥10mm) and width zones (0.5 < 1.0, ≥1.0m), and therefore, diameter was reduced to three classes for root mass (<2, 2 < 5, ≥5mm) and root length (1 < 2, 2 < 5, ≥5mm), and width from the trees was reduced to two classes (0 < 0.5, ≥0.5m). Classification by depth was discarded because only 3 of 25 trees in Y1 and 4 of 23 trees in Y2 had root development exceeding 0.5-m soil depth. A ratio of TRL to TRM was calculated.

3. Results

3.1. Soil Properties

Soil had a pH of 6.0 at all depths: 0 < 25, 25 < 50, and 50–75cm and corresponding values for Olsen phosphate of 16, 11, and 4µg mL^{-1}, respectively. Sulphatesulphur levels were 1ppm or less at the three soil depths.

3.2. Trial 1: Planting Material Comparison

3.2.1. Above-Ground Attributes

There was a large variation in stem volume between PMs—approximately 8490cm^3 for poles, 490cm^3 for stakes, and 95cm^3 for cuttings. Tree growth over the first 3 years of establishment varied markedly between PMs with the largest growth achieved from poles. Mean shoot length over 3 years was in the order poles (3.4m) > stakes (2.9m) > cuttings (1.7m), and differences among PMs increased over time [PM × year interaction P < 0.001; **Figure 2(a)**]. In Y3, shoot length from cuttings averaged 2.3m which was less than half that of the longest shoots from stakes (4.8m) and poles (5.6m).

Shoot diameter averaged over years was in the order poles (41mm) > stakes (34mm) > cuttings (20mm) (P < 0.001) and increased from an average across PMs of 15mm in Y1 to 56mm in Y3 (P < 0.001). Shoot diameter of PMs was similar in Y1 (range 13–18mm) whereas, in Y3, the diameter of poles (66mm) was 16% greater than from stakes (57mm) and 113% greater than from cuttings

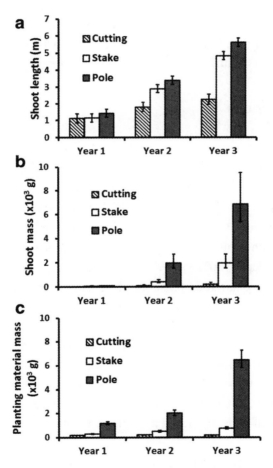

Figure 2. Above-ground attributes of (a) shoot length, (b) shoot mass, and (c) planting material mass of Veronese poplar established from three planting materials. Values are means and capped vertical lines are standard errors.

(31mm). The longest shoots from poles originated at an average height above the ground of 1.77m which was significantly greater than those from cuttings (0.05m above ground) and stakes (0.34m). There were more and thicker shoots from PMs in the order poles > stakes > cuttings. In Y1, root collar diameter averaged 58mm for poles, 43mm for stakes, and 35mm for cuttings, which all differed from each other ($P < 0.001$).

Shoot mass of trees from the three PMs in Y1 ranged from 30 to 70g whereas, in Y2 and Y3, shoot mass was in the order poles > stakes > cuttings [PM × year interaction $P < 0.01$; **Figure 2(b)**]. Stakes and poles produced significant increases in shoot mass over the 3 years. Similar trends occurred over time for the total mass of pole PM, with a marked increase occurring between Y2 and Y3 [**Figure 2(c)**]. In contrast, between Y1 and Y3, there was only a small increase in mass of stakes and no significant change in cutting mass.

3.2.2. Root Attributes

Root presence increments decreased with increasing width from stem and soil depth ($P < 0.001$). Roots were present within the top 0.5m of soil in 72% of cells for cuttings, 90% for stakes, and all cells for poles whereas, at 0.5 < 1.0m depth, root presence was 33% for cuttings, 47% for stakes, and 60% for poles (PM × depth interaction $P = 0.014$). A decrease in root presence at 0 < 0.5m depth with increasing width (100% at 0 < 0.5m width, 78% at 0.5 < 1.0m) was smaller than at 0.5 < 1.0m depth (66% vs. 29%) (depth × width interaction $P = 0.024$). In the analysis of pole data, root presence decreased with increasing width ($P < 0.001$), being 87%, 77%, and 50% for 0 < 0.5, 0.5 < 1.0, and ≥1.0m, respectively, and this trend was not influenced by other factors. There were large increases in root presence over time at 0.5 < 1.0m depth compared with at 0 < 0.5m depth (depth × year interaction) in the analyses comprising data for all PMs ($P = 0.001$) and poles only ($P = 0.049$). For example, root presence of poles at 0.5 < 1.0m depth increased from 27% in Y1 to 87% in Y3, whereas, at 0 < 0.5 m depth, roots were found in 73% of cells in Y1 and all cells in Y2 and Y3.

Each year, roots from cuttings did not extend beyond 0.5-m width, except for one instance where a root extended between 0.5 and 1.0m, in contrast to those from poles where extension was >1.0m, but no intersection between roots from different trees was observed. An exception was the intersection of the roots of two trees from poles in Y3. Root extension from stakes was intermediate between that from cuttings and poles, with roots found at both $0 < 0.5$m and $0.5 < 1.0$m widths in Y1 and Y2 and at ≥ 1.0m in Y3. Roots extended between 0.5- and 1.0-m soil depth at all widths from poles in Y2 and Y3, and Y3 roots were also found at ≥ 1.0-m depth within 0.5m of poles. No roots were found at this depth for the other PMs. The greatest radial distance attained by the individual roots of PMs was about 0.8, 1.7, and 2.6m for those from cuttings, stakes, and poles, respectively. A root 4.6m long was found in Y3 from a pole, but it was not in a straight line. Roots <5mm diameter were produced by all PMs each year whereas roots $5 < 10$mm diameter were only found in Y2 and Y3. Roots ≥ 10mm diameter were only excavated from poles in Y2 and Y3 and from stakes in Y3.

The value of TRM averaged over 3 years was in the order poles (364g) > stakes (70g) > cuttings (17g) ($P < 0.001$). There was much greater variation between years for TRM of trees from poles than from cuttings and stakes [PM × year interaction $P < 0.05$; **Figure 3(a)**]. In Y3, the TRM of poles was 8-fold greater than stakes and 80-fold greater than cuttings.

The value of TRMD decreased dramatically with increasing width from trees and with increasing soil depth, and a greater increase in TRMD occurred at $0 < 0.5$m depth over time than at ≥ 0.5m [depth × year interaction $P = 0.033$; **Figure 4(a)**]. In Y1 and Y3, TRMD at the shallower depth was 5-fold greater than at ≥ 0.5-m depth, whereas, in Y2, it was 11-fold greater. TRMD in Y1 within 0.5m of trees was significantly greater than beyond 0.5m in contrast to other years, where no difference was found between widths [width × year interaction $P < 0.001$; **Figure 4(b)**].

For trees established from poles, in Y3, mass of roots of <2, $2 < 5$, and $5 < 10$mm diameter was 3–6-fold greater in $0 < 0.5$m soil depth than deeper in the profile (**Table 2**). Within 0.5m of trees, mass of the thinnest roots was 3-fold greater than at $0.5 < 1.0$ and ≥ 1.0m widths, which were not significantly different from each other (**Table 2**). Trends for depth and width were inconsistent for these smaller-diameter classes (depth × width interaction $P < 0.05$).

The relative differences in TRL between trees established from the three PMs were less than those found for TRM but still followed a similar order of poles

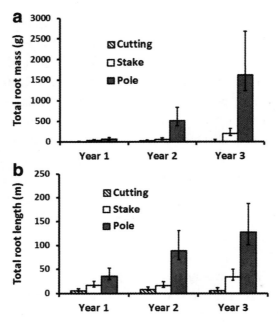

Figure 3. Total root mass (a) and length (b) of Veronese poplar established from three planting materials. Values are means and capped vertical lines are standard errors.

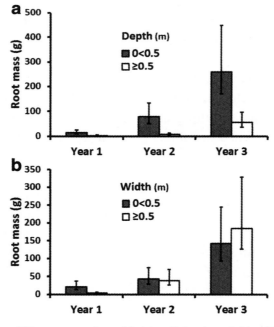

Figure 4. Root mass of Veronese poplar with (a) soil depth and (b) width over time. Values are means and capped vertical lines are standard errors.

Table 2. Mean root mass and length of Veronese poplar trees 3 years after establishment from poles.

Factor	Root mass (g)					Root length (m)				
	Root diameter (mm)					Root diameter (mm)				
	<2	2 < 5	5 < 10	10 < 20	≥20	1 < 2	2 < 5	5 < 10	10 < 20	≥20
Soil depth (m)										
0 < 0.5	11.1^a (2.0)	29.8^a (6.4)	96.1^a (19.3)	114.1 (32.6)	120.4 (67.3)	7.3^a (1.6)	10.3^a (1.8)	6.7^a (1.4)	2.6 (0.6)	2.2^a (0.4)
≥0.5	2.0^b (0.4)	10.1^b (2.4)	28.4^b (6.4)	66.2 (25.4)	11.8 (9.1)	1.2^b (0.3)	2.6^b (0.5)	1.7^b (0.4)	1.0 (0.3)	1.1^b (0.2)
P value	<0.001	0.004	<0.001	0.556	0.087	<0.001	<0.001	<0.001	0.102	0.040
Width (m)										
0 < 0.5	10.7^a (2.4)	27.0 (7.1)	77.3 (19.1)	199.5 (74.7)	93.2 (54.7)	5.8^a (1.6)	8.8 (1.9)	4.2 (1.1)	3.3 (0.9)	2.6 (0.6)
0.5 < 1.0	3.5^b (0.8)	13.9 (3.9)	37.2 (9.7)	45.9 (18.1)	27.4 (21.5)	2.4^b (0.7)	4.2 (1.0)	3.4 (0.9)	1.2 (0.4)	1.4 (0.4)
≥1.0	3.7^b (0.9)	17.9 (5.3)	70.0 (19.3)	88.7 (40.0)	103.9 (180.1)	2.6^b (0.8)	5.2 (1.3)	4.0 (1.1)	2.1 (0.7)	1.1 (0.3)
P value	0.002	0.190	0.184	0.084	0.689	0.047	0.054	0.794	0.141	0.067
Depth × width										
P value	0.029	0.049	0.010	0.250	0.261	0.072	0.145	0.044	0.187	0.067

Within factors within columns, means with different letters differ significantly at $P = 0.05$. P values that were significant ($P < 0.05$) are shown in italics. Figures in brackets are standard errors

(73m) > stakes (21m) > cuttings (7m) ($P < 0.001$), and there was also less variation between years [**Figure 3(b)**]. The values of TRLD decreased with increasing distance from trees and increasing soil depth (not presented), but responses varied with year. Over time, the difference between root lengths at 0 < 0.5m and ≥0.5m depths increased [$P = 0.017$; **Figure 5(a)**]. Root length at 0 < 0.5m width exceeded that beyond 0.5m from trees in Y1, whereas, in subsequent years, length did not vary significantly between widths [width × year interaction $P < 0.001$; **Figure 5(b)**].

The length of roots of trees established from poles of <2, 2 < 5, and 5 < 10mm diameter was 4–6-fold greater in 0 < 0.5m soil depth than deeper in the profile ($P < 0.001$; **Table 2**). Within 0.5m of trees, the length of 1 < 2mm roots was over 2-fold greater than at 0.5 < 1.0m and ≥1.0m widths. For 5 < 10mm roots, length did not vary significantly between widths at 0 < 0.5m depth whereas, at ≥0.5-m depth, length beyond 1.0m from trees (0.8m) was significantly greater than at 0 < 0.5m width (0.6m) (depth × width interaction $P = 0.044$).

The partitioning of biomass between the root and shoot changed over time for the three sizes of PM, with the root as a proportion of total biomass (shoot + root, excluding PM mass) in trees from poles reducing from 48% in Y1 to 21% in Y2 and 19% in Y3. Corresponding values for stakes were 39%, 13%, and 9% and for cuttings were 25%, 19%, and 9%, respectively.

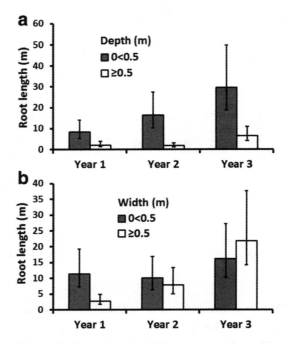

Figure 5. Root length of Veronese poplar with (a) soil depth and (b) width over time. Values are means and capped vertical lines are standard errors.

3.3. Trial 2: Clonal Variation Comparison

3.3.1. Above-Ground Attributes

Clones varied in shoot length, shoot mass, and cutting mass (**Table 3**) but not in shoot diameter (P = 0.078). Mean shoot length of Kawa was greater than for San Rosa, Fraser, Veronese, and PN471, and it had greater shoot mass than all other clones, except for the mass of Geyles, which did not vary significantly from that of Kawa. The cutting mass of Veronese was twice that of San Rosa, and it was also significantly greater than for PN471 and Geyles (**Table 3**). Across clones, shoot length increased (P < 0.001) 77% between Y1 (1.19m) and Y2 (2.09m), shoot mass increased 4.8-fold between Y1 (29.1g) and Y2 (168.0g), and cutting mass increased only slightly over the same period (133 to 144 g). Shoot diameter of clones averaged 13 (s.e. = 0.7)mm.

Table 3. Mean above-ground attributes and total root mass and length of clones of poplar established from cuttings.

Clone	Shoot length(m)	Shoot mass(g)	Cutting mass(g)	Total root mass(g)	Total root length(m)
P. deltoides × P. ciliata San Rosa	1.52[bc] (0.34)	48.8[b] (23.6)	93.6[c] (26.9)	15.2[b] (5.0)	6.4[abc] (1.7)
P. deltoides × P. nigra Fraser	1.47[bc] (0.24)	58.7[b] (19.5)	141.4[abc] (19.5)	13.7[b] (4.4)	4.1[bc] (1.1)
P. deltoides × P. nigra Veronese	1.51[bc] (0.21)	60.1[b] (16.6)	190.0[a] (16.5)	16.4[b] (5.3)	7.3[ab] (2.0)
P. deltoides × P. yunnanensis Kawa	2.30[a] (0.17)	156.2[a] (27.1)	164.7[ab] (13.3)	39.1[a] (12.7)	11.8[a] (3.2)
P. maximowiczii × P. nigra Geyles	1.84[ab] (0.20)	85.9[ab] (22.4)	108.6[c] (15.6)	45.9[a] (14.9)	10.3[a] (2.8)
P. trichocarpa PN471	1.20[c] (0.22)	51.0[b] (15.3)	132.1[bc] (17.8)	10.4[b] (3.4)	3.5[c] (0.9)
P value	*0.004*	*0.021*	*0.009*	*0.001*	*0.006*

Means are averaged over 2 years. Means for shoot mass and root attributes are back-transformed from logarithms. Within columns, means with different letters differ significantly at $P = 0.05$. P values that were significant ($P < 0.05$) are shown in italics. Figures in brackets are standard errors

3.3.2. Root Attributes

The presence of roots varied between clones, years, and widths from trees. The root presence of all clones was more than 80% in both years except for San Rosa and PN471 in their first year (50%) (clone × year interaction $P = 0.043$), where they only produced roots within 0.5m of trees. In Y1, all roots of clones were <5mm diameter except for those of San Rosa and Geyles where roots ≥5mm were also found.

Mean TRM per tree varied between clones ($P = 0.001$) and ranged from 10.4g for PN471 to 45.9g for Geyles (**Table 3**). Clones Kawa and Geyles had similar TRM that was 2.5–4.5-fold greater than for the other clones, all of which did not vary significantly from each other. Across all clones, mean TRM in Y2 (43.5g) was almost 5-fold greater than in Y1 (9.2g) ($P < 0.001$).

Clones varied in mass of roots <2 and 2 < 5mm diameter, but there were no significant differences between them for ≥5-mm roots (**Table 4**). The root mass of Kawa and Geyles exceeded that of most other clones. For all root diameters, mass of roots beyond 0.5m was at least 75% less than within 0.5m of trees (**Table 4**). There was a trend of increasing root mass between Y1 and Y2, with mass of 2 < 5mm roots increasing 5-fold ($P < 0.001$). The mass of this root class beyond 0.5-m width was 12% of that at 0 < 0.5m in Y1, increasing to 48% in Y2 (width × year interaction $P < 0.001$).

The TRL of Kawa and Geyles was greater than for Fraser and PN471, and the mean for San Rosa was not significantly different from that for any other clone

Table 4. Mean root mass and length of six clones of poplar established from cuttings.

Factor	Root mass (g)			Root length (m)		
	Root diameter (mm)			Root diameter (mm)		
	<2	2 < 5	≥5	1 < 2	2 < 5	≥5
Clone						
P. deltoides × P. ciliata San Rosa	1.7^{ab} (0.5)	2.6^{bc} (0.9)	2.0 (1.3)	1.5 (0.4)	1.2^{abc} (0.3)	0.2 (0.1)
P. deltoides × P. nigra Fraser	1.0^{b} (0.3)	2.8^{bc} (0.9)	4.5 (3.0)	1.2 (0.3)	1.1^{bc} (0.3)	0.3 (0.2)
P. deltoides × P. nigra Veronese	1.1^{b} (0.3)	3.8^{b} (1.2)	2.0 (1.3)	1.3 (0.4)	1.5^{abc} (0.4)	0.2 (0.1)
P. deltoides × P. yunnanensis Kawa	2.7^{a} (0.8)	5.9^{ab} (1.9)	12.9 (8.6)	2.1 (0.6)	2.1^{ab} (0.6)	0.8 (0.6)
P. maximowiczii × P. nigra Geyles	2.5^{a} (0.8)	9.4^{a} (3.0)	8.6 (5.7)	1.6 (0.4)	2.8^{a} (0.8)	0.6 (0.4)
P. trichocarpa PN471	0.8^{b} (0.2)	1.5^{c} (0.5)	6.7 (4.5)	1.1 (0.3)	0.7^{c} (0.2)	0.5 (0.4)
P value	*0.027*	*0.005*	0.118	0.333	*0.013*	0.322
Width (m)						
0 < 0.5	4.2^{a} (0.7)	7.2^{a} (1.2)	9.8^{a} (3.6)	3.2^{a} (0.5)	2.1^{a} (0.3)	0.7^{a} (0.3)
≥0.5	0.5^{b} (0.1)	1.8^{b} (0.3)	2.4^{b} (0.9)	0.7^{b} (0.1)	1.0^{b} (0.1)	0.2^{b} (0.1)
P value	*<0.001*	*<0.001*	*<0.001*	*<0.001*	*0.002*	*0.006*
Year after planting						
1	1.2 (0.2)	1.5^{b} (0.3)	2.9 (1.3)	1.2 (0.2)	0.9^{b} (0.1)	0.3 (0.1)
2	1.9 (0.4)	8.9^{a} (1.7)	8.2 (3.7)	1.7 (0.3)	2.2^{a} (0.4)	0.7 (0.3)
P value	0.085	*<0.001*	0.137	0.081	*<0.001*	0.158

Within factors within columns, means with different letters differ significantly at P = 0.05. P values that were significant (P < 0.05) are shown in italics. Figures in brackets are standard errors

(Table 3). Clone Kawa had approximately 3-fold greater TRL than PN471. The length of 2 < 5mm diameter roots of Geyles was greater than for Fraser and PN471 but was not significantly different from that for San Rosa, Veronese, and Kawa (Table 4). There was no significant variation between clones for the length of the other root diameter classes. For each diameter, the length decreased 52%–78% with increasing width from trees. For all clones and in both years, roots ≥5mm diameter made negligible contribution to mean root length. The ratio of TRL to TRM was 0.22 for Geyles, 0.30 for Kawa and Fraser, 0.34 for PN471, 0.42 for San Rosa, and 0.45 for Veronese.

4. Discussion

The ability to establish trees of Populus spp. rapidly and easily using unrooted stems is a major reason for their widespread use for stabilising erodible pastoral hill country (Wilkinson 1999). This study showed the significant benefits in early establishment from planting poles compared with shorter and thinner material in terms of above-ground (shoot length, shoot diameter, shoot mass) and below-ground (root length, root mass, lateral and vertical root distribution) attributes. The different responses were predominantly due to the large variation in stem volume between PMs, ranging from 95cm^3 for cuttings to 8490cm^3 for poles, which suggested significant differences in stored energy reserves available to initiate growth under suitable conditions. The differences in surface area between the PMs would have influenced the number of root initials and potential root production. Variation in planting depth of the PMs would have ensured root initiation to varying depths and subsequent access to water. These factors favour larger PM over smaller such as cuttings and aid survival. Poles would have also been able to hold more water than the two other PMs, and proportionally less of their surface area was cut and exposed to air (DesRochers and Thomas 2003), which would have slowed water loss and facilitated growth. However, larger PMs will also be at greater risk of desiccation because they produce more foliage rapidly after planting, when roots are limited (DesRochers and Tremblay 2009).

In a range of species, the carbohydrate content used for initiating and supporting adventitious root growth and shoot meristem development is related directly to the dimensions of the PM (diameter and length), although it is affected by many other factors including environment (e.g. temperature, soil moisture, soil texture), physiology (e.g. phytohormonal balances, particularly of auxin and cytokinins), condition of the parent trees from which the PM was obtained, and sampling position within parent trees (Zalesny Jr *et al.* 2003; Leakey 2004; da Costa *et al.* 2013; Zhao *et al.* 2014). In this study, the sampling of specific parts of parent trees to obtain the different PMs suggests that stem units of each PM had similar physiological states and propensities for root initiation and development.

Comparisons between the PMs were confounded because of factors including different depths of burial and variation in the ratio of above- to below-ground stem length. These were considered minor issues because the study aimed to compare PMs established using best practice techniques. For example, depth of pole planting in this study aligned with recommendations of planting at depths of 0.6–0.8m or even up to one third of their length depending on site conditions such as dryness and exposure to wind (Hathaway 1986; National Poplar and Willow Users Group 2007).

The development of fewer and thinner shoots from the smaller PMs was probably because of relatively low stored energy reserves and fewer epicormic buds (Meier *et al.* 2012) along the shorter stem sections. Greater shoot initiation and growth from the upper compared with lower parts of all PMs may have been in response to slightly elevated irradiation higher up the stems; variation in bark thickness along the above-ground stem sections, particularly for the longer poles; and differential distribution of soluble sugars and carbohydrate reserves (Morisset *et al.* 2012). Field observations in Y3 indicated that the significantly greater shoot mass produced by trees from poles than from cuttings and stakes was likely because of more and larger shoots and the greater total photosynthetic leaf area. No carbohydrates were assayed, but the contribution to growth of reserves and current photosynthate (Tromp 1983; Magel *et al.* 2000) in trees from poles was likely considerably greater than in those from the other PMs.

The lower variation among PMs and among years within each PM for root length compared with root mass was considered due to the predominance of roots <5mm contributing to the length in all treatments and the investment in root diameter of trees established from poles in Y3. The ≥10-mm roots contributed 55% to root mass but only 10% to root length. Proliferation of fine roots is essential for enhancing water and nutrient uptake, but there is variation within and among seasons in their longevity and mass depending on soil factors (e.g. water content), tree species, age, and root diameter (Block *et al.* 2006). Conducting root excavations at the end of the growing season each year, as was done in this study, would have minimised any confounding effects when comparing responses between years. The increased partitioning of biomass to shoots and reduced allocation to roots with increasing age, found in trial 1, has been reported for young hybrid poplars grown from cuttings from plantation stock in North America (Wullschleger *et al.* 2005). Such results have implications for growth and survival of young trees in dry environments.

Root extension is important in aiding tree survival and growth. The larger PM produced greater root extension and accessed a greater soil volume for water and nutrients than smaller PMs over the same time period. The results reflected the greater resources available in the larger PM. Greater abundance of roots in shallow soil layers than deeper in the soil profile is a common occurrence because of factors such as enhanced soil organic matter content and higher soil nutrient status (e.g. trends for Olsen phosphate in this study) (Block *et al.* 2006; Johnston *et al.* 2009; McIvor *et al.* 2009). The spacings used were appropriate to allow development of young root systems unaffected by those of neighbouring trees because, for all trees except for two from poles in Y3, no intersection between roots from different trees was observed.

The variation among clones grown from cuttings in trial 2 supported findings for field-grown hybrid poplar clones evaluated elsewhere (Pregitzer *et al.* 1990; Dickmann *et al.* 1996; Wullschleger *et al.* 2005; Phillips *et al.* 2014). Clones Kawa and Geyles had relatively high above- and below-ground growth. Kawa is one of the main poplar clones grown in New Zealand for soil stabilisation and other purposes (National Poplar and Willow Users Group 2007), and this study demonstrated its ability to rapidly establish an extensive root system. Clone Geyles would also seem worthy of further appraisal for root proliferation in soil and was released for commercialisation in 2011 (McIvor *et al.* 2011). The role of cutting mass in contributing to the shoot and root growth of Kawa and Geyles was uncertain because Geyles had 30% lower cutting mass than Kawa whereas the two clones did not differ significantly with respect to shoot length, shoot mass, and root attributes. Veronese is grown for a range of applications (National Poplar and Willow Users Group 2007), but this study found that its early growth from cuttings was less than Kawa for shoot length and shoot mass and less than Kawa and Geyles for total root mass. The results suggest that where cuttings are used in applications, clones such as Kawa and Geyles could be used as alternatives to plantings of Veronese or be included in mixtures with Veronese. Phillips *et al.* (2014) found that 9-month-old Veronese and Kawa trees established from 0.5-m-long stem sections were similar for above- and below-ground attributes, except for maximum root depth where Veronese was about 40% greater than Kawa (1.0 vs. 0.7m). They did not recommend a particular clone for planting. Based on its overall low above- and below-ground growth, PN471 could not be recommended for providing rapid and extensive early root development when grown from cuttings and could be difficult to establish from cuttings in regions where summer rainfall is low and irrigation is not available.

The development of relatively thick roots (≥5mm diameter) in Y1, as shown by trees of San Rosa and Geyles, could be advantageous for storing reserves for maintenance and growth, enhancing anchorage of the developing tree to the soil, and providing a foundation for further development and position within the soil of thinner roots essential for water and nutrient uptake (Stokes *et al.* 2009). The differences between all clones in TRL:TRM ratio indicated clonal variation in root development and morphology. Higher TRL:TRM ratio for clones such as Veronese and San Rosa suggested that they had more numerous roots, particularly fine roots, than clones such as Geyles and Kawa. It is possible that the focus of some clones is root extension into new nutrient sources, whereas other clones tend to exploit smaller nutrient sources but to a greater degree. This has possible implications for risk from drought in the first few establishment years.

The death of trees established from cuttings could have been due to several factors. Cuttings had potentially fewer reserves to initiate growth above and below ground. They also had shorter stem lengths buried than stakes and poles, presenting lower surface areas for potential root development and extraction of soil water and nutrients. Inspection of cuttings and the other PMs within 3 months of planting found that all were alive and therefore that post-planting factors were responsible for any tree mortality. There was no evidence of pest or disease attack, the area of the trials was not grazed by livestock, and browsing damage by small mammals was not observed.

Although soil pH at the trial sites was within the range of 5.5 to 7.0 recommended for poplar-nursery production (van Kraayenoord *et al.* 1986), levels of Olsen phosphate were low, particularly at 50–75-cm soil depth, and levels of sulphate-sulphur were very low. Raising the levels of phosphate and sulphur through strategic fertiliser application may have accentuated the differences found between the PMs in root and shoot responses (van den Driessche 1999; Guillemette and DesRochers 2008), particularly by favouring trees from poles with their larger and more extensive root systems and consequent exploration of greater soil volumes.

Poles are more costly to produce than cuttings and stakes and are heavier and bulkier to transport. However, these disadvantages are more than compensated for in field applications with their potentially higher survival and lower maintenance requirements. Also, the planting of poles (with sleeve protection) is the only practicable option for establishment in pastoral areas continuing to be grazed by livestock (Wilkinson 1999).

In a comparison of poplar PMs at Gisborne on the east coast of the North Island, Phillips *et al.* (2014) found that 9 months after planting, the above- and below-ground growth of trees from poles exceeded that from stakes and cuttings, as found presently. Both studies also found no significant differences in root length between clones Veronese and Kawa. However, growth at Gisborne was considerably greater than found in this study with, for example, root collar diameter of Veronese PMs at Gisborne (77–104mm) being almost 2-fold greater than in Y1 (35–58mm). At Gisborne, TRL ranged from 155m for trees established from stakes to 255m for trees established from poles, compared with less than 40m for trees from any PM after one growing season in this study. At Gisborne, roots extended 3 to 5m from the stem after 9 months, whereas, currently, maximum radial extension was 2.6m. Finally, the lowest estimates of TRM of Veronese at Gisborne ranged from 1.50 to 3.05 × 10^3g, which were considerably greater than those found in Y1. By Y3, pole TRM was at the lower end of the earlier estimates (Phillips *et al.* 2014). The much greater tree growth at Gisborne than Palmerston North may have been because of higher soil nutrient status, higher average air temperatures (long-term values of 12°C to 23°C at Gisborne; 8.6°C to 18.3°C at Palmerston North), irrigation during summer (Gisborne only), and loose structured soil at Gisborne compared with more compacted upper soil layers at Palmerston North, which would have facilitated root extension.

The rapid interlocking of roots of neighbouring trees on man-made or natural slopes is an important attribute to minimise the risk of mass movement erosion of the ground between adjacent trees (Reubens *et al.* 2007; Stokes *et al.* 2009). The lower lateral root extension from stakes and particularly from cuttings, compared with that from poles, indicated that these smaller PMs need to be planted at much closer spacings than poles to achieve, within a given time period, a similar amount of root interlocking in a defined volume of shallow soil. The results indicated that for roots of trees established from each PM to touch those of neighbours by Y3, spacings would need to be about 5.2m for poles (370sph), 3.4m for stakes (865sph), and 1.6m for cuttings (3900sph). Spacings of established poplar trees on erodible pastoral hill country are often 10–15m (100–45sph), which are effective in reducing erosion processes such as shallow landslides (Cairns *et al.* 2001; Douglas *et al.* 2013). These spacings are achieved by planting at closer spacings with thinning later or the more common (but higher risk) approach of planting at final spacings and relying on near 100% survival to achieve desired root coverage of the slope. There would also need to be negligible mass movement erosion in the meantime. Another important element in slope stabilisation by woody vegetation is vertical root development and its potential penetration and anchoring into underlying bedrock (Reubens *et al.* 2007; Stokes *et al.* 2009). In trial 1, the early soil conservation advantage of poles was shown further by the development in Y3 of vertical roots from the bottom or near bottom of poles that extended deeper than 1m within 0.5-m radius of poles. The ability of these roots to penetrate stable strata would depend on factors such as their depth, pedological structure, and the presence of cracks in the strata.

5. Conclusions

Poles of Veronese hybrid poplar produced trees that had considerably greater above and below-ground growth and development than those from cuttings within the first 3 years after planting. Trees from stakes were intermediate between poles and cuttings in their growth and development. After 3 years, maximum lateral root extension was about 0.8, 1.7, and 2.6m for trees grown from cuttings, stakes, and poles, respectively. This indicated that large differences in spacing would be required to achieve similar levels of contact between roots of neighbouring trees established from the different PMs over the same period of time. Survival of trees from cuttings was <100% whereas all trees from stakes and poles survived. The proportion of biomass allocated to roots decreased with age for trees from the three PMs. Among six different clones from cuttings, there was a 4.4-fold variation in TRM and a 3.4-fold variation in TRL. Hybrid poplar clones Kawa and Geyles produced the greatest amount of above- and below-ground biomass. This study indicates that there are poplar clones that root more readily and produce greater root mass in the first year than Veronese; they should be considered for greater commercial production than occurs presently.

The findings on early root development of poplar add to the very limited information available. Different-sized stem sections are used in a range of applications, and it is important to know how they influence root development spatially and temporally. This impacts directly on decisions regarding initial spacing of stem sections to achieve interlock of roots within a specific time frame. Practitioners will have to accommodate the delay in time of root interlock of trees established from stakes, and particularly cuttings, in their applications. The data for Veronese complement those for trees aged 5+ years collected for trees established from poles on slopes and will enable an improved understanding of root development over time. Results are for clones used widely in New Zealand, such as Veronese and Kawa, and therefore have immediate applicability.

Abbreviations

LSD: least significant difference; PM: planting material; REML: restricted maximum likelihood approach; sph: stems per hectare; TRL: total root length; TRLD: total root length diameter; TRM: total root mass; TRMD: Total root mass diameter; Y1: year 1; Y2: year 2; Y3: year 3.

Competing Interests

The authors declare that they have no competing interests.

Authors' Contributions

GD' participated in the design of the study, data acquisition, and drafted the manuscript. IM conceived of the study, participated in its design, coordination and data acquisition, and helped to draft the manuscript. CL-W conducted the statistical analyses and helped draft the section on analyses. All authors read and approved the final manuscript.

Acknowledgements

Funding from New Zealand Foundation for Research, Science and Technology's 'Sustainable Land Use Research Initiative' (SLURI) programme (Contract No. C02X0813, Objective 2); large teams of post-graduate students from Massey University, Palmerston North, for assisting with tree excavations and sample processing; Drs. Mike Dodd and Andrew Wall, AgResearch, and two anonymous reviewers for constructive comments and helpful suggestions.

Source: Douglas G B, Mcivor I R, Lloyd-West C M. Early root development of field-grown poplar: effects of planting material and genotype[J]. New Zealand

References

[1] Al Afas, N., Marron, N., Zavalloni, C., & Ceulemans, R. (2008). Growth and production of a short-rotation coppice culture of poplar—IV: Fine root characteristics of five poplar clones. Biomass and Bioenergy, 32(6), 494–502.

[2] Benavides, R., Douglas, G. B., & Osoro, K. (2009). Silvopastoralism in New Zealand: Review of effects of evergreen and deciduous trees on pasture dynamics. Agroforestry Systems, 76, 327–350.

[3] Benomar, L., DesRochers, A., & Larocque, G. (2013). Comparing growth and fine root distribution in monocultures and mixed plantations of hybrid poplar and spruce. Journal of Forestry Research, 24(2), 247–254.

[4] Berhongaray, G., Janssens, I. A., King, J. S., & Ceulemans, R. (2013). Fine root biomass and turnover of two fast-growing poplar genotypes in a short-rotation coppice culture. Plant and Soil, 373(1–2), 269–283.

[5] Bischetti, G., Chiaradia, E., Simonato, T., Speziali, B., Vitali, B., Vullo, P., et al. (2005). Root strength and root area ratio of forest species in Lombardy (Northern Italy). Plant and Soil, 278, 11–22.

[6] Blanco, H., & Lal, R. (2008). Principles of soil conservation and management. Dordrecht, Netherlands: Springer Science + Business Media B.V.

[7] Block, R., Van Rees, K., & Knight, J. (2006). A review of fine root dynamics in Populus plantations. Agroforestry Systems, 67(1), 73–84.

[8] Cairns, I., Handyside, B., Harris, M., Lambrechtsen, N., & Ngapo, N. (2001). Soil conservation technical handbook (New Zealand Ministry for the Environment Series; no. 404). Wellington: New Zealand Association of Resource Management.

[9] da Costa, C. T., de Almeida, M. R., Ruedell, C. M., Schwambach, J., Maraschin, F. S.& Fett-Neto, A. G. (2013). When stress and development go hand in hand: main hormonal controls of adventitious rooting in cuttings. Frontiers in Plant Science, 4, 1–19.

[10] 10. de Baets, S., Poesen, J., Reubens, B., Muys, B., de Baerdemaeker, J., & Meersmans, J. (2009). Methodological framework to select plant species for controlling rill and gully erosion: Application to a Mediterranean ecosystem. Earth Surface Processes and Landforms, 34(10), 1374–1392.

[11] DesRochers, A., & Thomas, B. (2003). A comparison of pre-planting treatments on hardwood cuttings of four hybrid poplar clones. New Forests, 26(1), 17–32.

[12] DesRochers, A., & Tremblay, F. (2009). The effect of root and shoot pruning on early

growth of hybrid poplars. Forest Ecology and Management, 258(9), 2062–2067.

[13] Dickmann, D. I., Nguyen, P. V., & Pregitzer, K. S. (1996). Effects of irrigation and coppicing on above-ground growth, physiology, and fine-root dynamics of two field-grown hybrid poplar clones. Forest Ecology and Management, 80(1–3), 163.

[14] Douglas, G. B., Walcroft, A. S., Hurst, S. E., Potter, J. F., Foote, A. G., Fung, L. E., et al. (2006). Interactions between widely spaced young poplars (Populus spp.) and introduced pasture mixtures. Agroforestry Systems, 66, 165–178.

[15] Douglas, G. B., McIvor, I. R., Potter, J. F., & Foote, L. G. (2010). Root distribution of poplar at varying densities on pastoral hill country. Plant and Soil, 333, 147–161.

[16] Douglas, G. B., McIvor, I. R., Manderson, A. K., Koolaard, J. P., Todd, M., Braaksma, S., et al. (2013). Reducing shallow landslide occurrence in pastoral hill country using wide-spaced trees. Land Degradation and Development, 24, 103–114.

[17] Eckenwalder, J. E. (1996). Systematics and evolution of Populus. In R. F. Stettler, H. D. Bradshaw Jr., P. E. Heilman, & T. M. Hinckey (Eds.), Biology of Populus and its implications for management and conservation (pp. 7–32). Ottawa: NRC Research Press.

[18] Evette, A., Labonne, S., Rey, F., Liebault, F., Jancke, O., & Girel, J. (2009). History of bioengineering techniques for erosion control in rivers in Western Europe. Environmental Management, 43(6), 972–984.

[19] Fang, S., Zhai, X., Wan, J., & Tang, L. (2013). Clonal variation in growth, chemistry and calorific value of new poplar hybrids at nursery stage. Biomass and Bioenergy, 54, 303–311.

[20] Gebrernichael, D., Nyssen, J., Poesen, J., Deckers, J., Haile, M., Govers, G., et al. (2005). Effectiveness of stone bunds in controlling soil erosion on cropland in the Tigray Highlands, northern Ethiopia. Soil Use and Management, 21(3), 287–297.

[21] Genet, M., Stokes, A., Fourcaud, T., & Norris, J. E. (2010). The influence of plant diversity on slope stability in a moist evergreen deciduous forest. Ecological Engineering, 36(3), 265–275.

[22] Guevara-Escobar, A., Kemp, P. D., Mackay, A. D., & Hodgson, J. (2007). Pasture production and composition under poplar in a hill environment in New Zealand. Agroforestry Systems, 69, 199–213.

[23] Guillemette, T., & DesRochers, A. (2008). Early growth and nutrition of hybrid poplars fertilized at planting in the boreal forest of western Quebec. Forest Ecology and Management, 255(7), 2981–2989.

[24] Hajek, P., Hertel, D., & Leuschner, C. (2014). Root order- and root age-dependent response of two poplar species to belowground competition. Plant and Soil, 377(1–2), 337–355.

[25] Hart, J., de Araujo, F., Thomas, B., & Mansfield, S. (2013). Wood quality and growth characterization across intra- and inter-specific hybrid aspen clones. Forests, 4(4),

786–807.

[26] Hathaway, R. L. (1986). Plant establishment. In C. W. S. Van Kraayenoord & R. L. Hathaway (Eds.), Water and soil miscellaneous publication no. 93 (pp. 21–37). Wellington, New Zealand: Water and Soil Directorate, Ministry of Works and Development.

[27] Hewitt, A. E. (1998). New Zealand soil classification. (2nd ed., Landcare Research Science Series No. 1.). Lincoln, Canterbury, New Zealand: Manaaki Whenua Press.

[28] Hicks, D. L. (1995). Control of soil erosion on farmland: a summary of erosion's impact on New Zealand agriculture, and farm management practices which counteract it (MAF Policy Technical Paper 95/4). Wellington, New Zealand: MAF Policy, Ministry of Agriculture.

[29] Johnston, A., Poulton, P., & Coleman, K. (2009). Soil organic matter: its importance in sustainable agriculture and carbon dioxide fluxes. Advances in Agronomy, 101, 1–57.

[30] Kaczmarek, D. J., Coyle, D. R., & Coleman, M. D. (2013). Survival and growth of a range of Populus clones in central South Carolina USA through age ten: Do early assessments reflect longer-term survival and growth trends? Biomass and Bioenergy, 49, 260–272.

[31] Kuzovkina, Y. A., & Volk, T. A. (2009). The characterization of willow (Salix L.) varieties for use in ecological engineering applications: Co-ordination of structure, function and autecology. Ecological Engineering, 35, 1178–1189.

[32] Lammeranner, W., Rauch, H., & Laaha, G. (2005). Implementation and monitoring of soil bioengineering measures at a landslide in the middle mountains of Nepal. Plant and Soil, 278(1–2), 159–170.

[33] Leakey, R. R. B. (2004). Physiology of vegetative reproduction. In J Burley, J Evans,& J Youngquist (Eds.), Encyclopaedia of Forest Sciences (pp. 1655–1668). Oxford, UK: Elsevier Academic Press.

[34] Licht, L. A., & Isebrands, J. G. (2005). Linking phytoremediated pollutant removal to biomass economic opportunities. Biomass and Bioenergy, 28(2), 203–218.

[35] Liu, X., Zhang, S., Zhang, X., Ding, G., & Cruse, R. M. (2011). Soil erosion control practices in Northeast China: A mini-review. Soil and Tillage Research, 117, 44–48.

[36] MAF (2011). Poplars for the farm. Specially bred clones for New Zealand conditions. Poplar [Brochure No. 5]. http://www.poplarandwillow.org.nz/documents/brochure-5-poplars-for-the-farm.pdf. Accessed 27 June 2014.

[37] Magel, E., Einig, W., & Hampp, R. (2000). Carbohydrates in trees. In G. Anil Kumar & K. Narinder (Eds.), Developments in crop science (Vol. 26, pp. 317–336). Amsterdam, Oxford: Elsevier.

[38] McIvor, I. R., Douglas, G. B., Hurst, S. E., Hussain, Z., & Foote, A. G. (2008). Structural root growth of young Veronese poplars on erodible slopes in the southern North Island, New Zealand. Agroforestry Systems, 72, 75–86.

[39] McIvor, I. R., Douglas, G. B., & Benavides, R. (2009). Coarse root growth of Veronese poplar trees varies with position on an erodible slope in New Zealand. Agroforestry Systems, 76, 251–264.

[40] McIvor, I. R., Hedderley, D. I., Hurst, S. E., & Fung, L. E. (2011). Survival and growth to age 8 of four Populus maximowiczii × P. nigra clones in field trials on pastoral hill slopes in six climatic zones of New Zealand. New Zealand Journal of Forestry Science, 41, 151–163.

[41] McIvor, I. R., Sloan, S., & Pigem, L. R. (2014). Genetic and environmental influences on root development in cuttings of selected Salix and Populus clones—a greenhouse experiment. Plant and Soil, 377(1–2), 25–42.

[42] Meier, A. R., Saunders, M. R., & Michler, C. H. (2012). Epicormic buds in trees: a review of bud establishment, development and dormancy release. Tree Physiology, 32(5), 565–584.

[43] Morisset, J. B., Mothe, F., Bock, J., Bréda, N., & Colin, F. (2012). Epicormic ontogeny in Quercus petraea constrains the highly plausible control of epicormic sprouting by water and carbohydrates. Annals of Botany, 109(2), 365–377.

[44] NIWA. (2013). Climate data and activities. http://www.niwa.co.nz/education-and-training/ schools/resources/climate. Accessed 21 May 2014.

[45] Phillips, C., Marden, M., & Suzanne, L. (2014). Observations of root growth of young poplar and willow planting types. New Zealand Journal of Forestry Science, 44, 15.

[46] National Poplar and Willow Users Group. (2007). Growing poplar and willow trees on farms. Guidelines for establishing and managing poplar and willow trees on farms. Palmerston North, New Zealand: National Poplar and Willow Users Group.

[47] Posthumus, H., & De Graaff, J. (2005). Cost-benefit analysis of bench terraces, a case study in Peru. Land Degradation & Development, 16(1), 1–11.

[48] Pregitzer, K. S., Dickmann, D. I., Hendrick, R., & Nguyen, P. V. (1990). Whole-tree carbon and nitrogen partitioning in young hybrid poplars. Tree Physiology, 7, 79–93.

[49] Pulkkinen, P., Vaario, L.-M., Koivuranta, L., & Stenvall, N. (2013). Elevated temperature effects on germination and early growth of European aspen (Populus tremula), hybrid aspen (P. tremula × P. tremuloides) and their F2-hybrids. European Journal of Forest Research, 132(5–6), 791–800.

[50] Reisner, Y., de Filippi, R., Herzog, F., & Palma, J. (2007). Target regions for silvoarable agroforestry in Europe. Ecological Engineering, 29(4), 401–418.

[51] Reubens, B., Poesen, J., Danjon, F., Geudens, G., & Muys, B. (2007). The role of fine and coarse roots in shallow slope stability and soil erosion control with a focus on root system architecture: A review. Trees-Structure and Function, 21(4), 385–402.

[52] Sidhu, D. S., & Dhillon, G. P. S. (2007). Field performance of ten clones and two

sizes of planting stock of Populus deltoides on the Indo-gangetic plains of India. New Forests, 34(2), 115–122.

[53] Stokes, A., Atger, C., Bengough, A., Fourcaud, T., & Sidle, R. (2009). Desirable plant root traits for protecting natural and engineered slopes against landslides. Plant and Soil, 324, 1–30.

[54] Stokes, A., Douglas, G., Fourcaud, T., Giadrossich, F., Gillies, C., Hubble, T., et al. (2014). Ecological mitigation of hillslope instability: Ten key issues facing researchers and practitioners. Plant and Soil, 377(1–2), 1–23.

[55] Thompson, R. C., & Luckman, P. G. (1993). Performance of biological erosion control in New Zealand soft rock hill terrain. Agroforestry Systems, 21, 191–211.

[56] Toillon, J., Fichot, R., Dallé, E., Berthelot, A., Brignolas, F., & Marron, N. (2013). Planting density affects growth and water-use efficiency depending on site in Populus deltoides × P. nigra. Forest Ecology and Management, 304, 345–354.

[57] Toy, T. J., Foster, G. R., & Renard, K. G. (2002). Soil erosion: Processes, prediction, measurement and control. New York: John Wiley & Sons Inc.

[58] Tromp, J. (1983). Nutrient reserves in roots of fruit trees, in particular carbohydrates and nitrogen. Plant and Soil, 71(1–3), 401–413.

[59] van den Driessche, R. (1999). First-year growth response of four Populus trichocarpa × Populus deltoides clones to fertilizer placement and level. Canadian Journal of Forest Research, 29(5), 554–562.

[60] Van Kraayenoord, C. W. S., & Hathaway, R. L. (Eds.). (1986). Plant materials handbook for soil conservation. Volume 2, introduced plants. (Water and Soil Miscellaneous Publication No. 94). Wellington, New Zealand: Water and Soil Directorate, Ministry of Works and Development.

[61] Van Kraayenoord, C. W. S., Wilkinson, A. G., & Hathaway, R. L. (1986). Nursery production of soil conservation plants. In CWS Van Kraayenoord, & RL Hathaway (Eds.), Plant materials handbook for soil conservation. Volume 1, Principles and practices. (pp. 149–160, Water and Soil Miscellaneous Publication No. 93.). Wellington, New Zealand: Water and Soil Directorate, Ministry of Works and Development.

[62] Verlinden, M. S., Broeckx, L. S., Van den Bulcke, J., Van Acker, J., & Ceulemans, R. (2013). Comparative study of biomass determinants of 12 poplar (Populus) genotypes in a high-density short-rotation culture. Forest Ecology and Management, 307, 101–111.

[63] VSN International. (2014). GenStat for Windows (17th ed.). United Kingdom: Hemel Hempstead. Retrieved from: http://www.vsni.co.uk/software/genstat/.

[64] Wall, A. J., Kemp, P. D., & MacKay, A. D. (2006). Predicting pasture production under poplars using canopy closure images. Proceedings of the New Zealand Grassland Association, 68, 325–330.

[65] Wang, Z., Qi, L., & Wang, X. (2012). A prototype experiment of debris flow control with energy dissipation structures. Natural Hazards, 60(3), 971–989.

[66] Wilkinson, A. G. (1999). Poplars and willows for soil erosion control in New Zealand. Biomass and Bioenergy, 16, 263–274.

[67] Wu, Q. X., Liu, X. D., & Zhao, H. Y. (1994). Soil water characteristics in mountain poplar stand and its benefits to soil and water conservation in loess hilly region. Journal of Environmental Sciences, 6(3), 347–354.

[68] Wu, J. Y., Huang, D., Teng, W. J., & Sardo, V. I. (2010). Grass hedges to reduce overland flow and soil erosion. Agronomy for Sustainable Development, 30(2), 481–485.

[69] Wullschleger, S. D., Yin, T. M., DiFazio, S. P., Tschaplinski, T. J., Gunter, L. E., Davis, M. F., *et al.* (2005). Phenotypic variation in growth and biomass distribution for two advanced-generation pedigrees of hybrid poplar. Canadian Journal of Forest Research, 35(8), 1779–1789.

[70] Yang, Q., Zhao, Z., Chow, T. L., Rees, H. W., Bourque, C. P. A., & Meng, F.-R. (2009). Using GIS and a digital elevation model to assess the effectiveness of variable grade flow diversion terraces in reducing soil erosion in northwestern New Brunswick, Canada. Hydrological Processes, 23(23), 3271–3280.

[71] Zalesny, R. S., Jr., Hall, R. B., Bauer, E. O., & Riemenschneider, D. E. (2003). Shoot position affects root initiation and growth of dormant unrooted cuttings of Populus. Silvae Genetica, 52(5–6), 273–279.

[72] Zhao, X., Zheng, H., Li, S., Yang, C., Jiang, J., & Liu, G. (2014). The rooting of poplar cuttings: A review. New Forests, 45(1), 21–34.

[73] Zuazo, V. H. D., & Pleguezuelo, C. R. R. (2008). Soil-erosion and runoff prevention by plant covers. A review. Agronomy for Sustainable Development, 28(1), 65–86.

Chapter 28

Preparation of (001) Preferentially Oriented Titanium Thin Films by Ion-Beam Sputtering Deposition on Thermal Silicon Dioxide

Imrich Gablech[1,2], Vojtěch Svatoš[1,2], Ondřej Caha[3,4], Miloš Hrabovský[1,5], Jan Prášek[1,2], Jaromír Hubálek[1,2], Tomáš Šikola[1,5]

[1]Central European Institute of Technology, Brno University of Technology, Technická 3058/10, 61600 Brno, Czech Republic
[2]Department of Microelectronics, Faculty of Electrical Engineering and Communication, Brno University of Technology, Technická 3058/10, 61600 Brno, Czech Republic
[3]Central European Institute of Technology, Masaryk University, Kamenice 753/5, 62500 Brno, Czech Republic
[4]Department of Condensed Matter Physics, Faculty of Science, Masaryk University, Kotlářská 2, 60200 Brno, Czech Republic
[5]Institute of Physical Engineering, Brno University of Technology, Technická 2896/2, 61669 Brno, Czech Republic

Abstract: We propose the ion-beam sputtering deposition providing Ti thin films of desired crystallographic orientation and smooth surface morphology not obtainable with conventional deposition techniques such as magnetron sputtering and vacuum evaporation. The sputtering was provided by argon broad ion beams generated by a Kaufman ion-beam source. In order to achieve the optimal properties of thin film, we investigated the Ti thin films deposited on an amorphous thermal silicon dioxide using X-ray diffraction, and atomic force microscopy. We have optimized deposition conditions for growing of thin films with the only (001) preferential orientation of film crystallites, and achieved ultra-low surface roughness of 0.55nm. The deposited films have been found to be stable upon annealing up to 300 °C which is often essential for envisaging subsequent deposition of piezoelectric AlN thin films.

1. Introduction

Titanium has been a frequently used material in microelectronics and MEMS technology. Titanium thin films have been used as sensing electrodes, buffer, or adhesive layers. The advantages of titanium thin films are good electric conductivity, extraordinary chemical resistivity, thermal stability, high hardness, high melting point, and lower number of crystallographic imperfections[1]-[5]. Crystallographic orientation of titanium thin films has to be controlled during the deposition process to obtain specific properties (e.g., mechanical, chemical) suitable for an eventually required application[6][7]. Some of recent MEMS devices use the piezoelectric effect for energy harvesting or sensing purposes[8]. Titanium has been often utilized in MEMS technology as a compatible material for fabrication of thin conductive underlying electrodes on which the piezoelectric layers are deposited[6].

The crystallographic orientation of titanium thin films is crucial for properties of consequently deposited piezoelectric layers[9]. Obviously, the properties of these layers (AlN, ZnO), namely electromechanical properties, are significantly affected by the crystallographic orientation, surface morphology, and the roughness layers beneath[10]. There have been many papers published investigating the degree of orientation which is strongly dependent on the texture and roughness of underlying material[11][12]. Titanium thin films have been usually deposited by the physical vapor deposition (PVD) particularly magnetron sputtering[13] or e-beam evaporation[4]. Obviously, different deposition parameters result in various film properties. Considering the surface roughness, low process pressure during deposition (*i.e.*, $< 5 \times 10^{-3}$ mbar) usually results in a smooth film surface (*i.e.*, R_{rms}

<10nm). At higher deposition pressure (*i.e.*, $< 14 \times 10^{-3}$ mbar), the titanium RMS surface roughness of 45nm was achieved. Consequently, a negative effect on the piezoelectric coefficients of AlN layers deposited on Ti thin film due to a porous surface was observed[10]. The magnetron sputtering is the most frequent method for titanium thin films deposition[14]. Titanium thin films deposited by the magnetron sputtering generally possess (100), (001), and (101) crystallographic orientations of crystallites parallel to the surface[1]. Without a special modification of the magnetron sputtering process (closed-field unbalanced magnetron sputtering, pulsed magnetron sputtering), the number of various and independent deposition parameters capable of optimizing deposited layers is limited[15][16]. Due to the lack of the process control parameters, it is often an issue to achieve thin films of desired properties for specific application.

In this paper, we report on a deposition of Ti thin film with the (001) preferential crystallite orientation growth on amorphous thermal silicon dioxide using a 3-grid radio frequency inductive-coupled plasma (RFICP) Kaufman ion-beam source. As generally known, the Kaufman ion-beam source provides more efficient control and optimization of the deposition process compared to the conventional deposition method[17][18]. Here, ion-beam flux, energy, and ion-beam geometry, can be controlled independently. We have studied the influence of deposition parameters on properties of Ti thin films especially on their preferential crystallographic orientation and topography.

2. Experimental Details

2.1. Deposition Process

The deposition of titanium layers was done on substrates (20 × 20)mm diced from 4-inch P-type silicon wafer with the (100) crystallographic orientation and the resistivity of 6–12Ω cm covered with thermal silicon dioxide (ON Semiconductor). The thickness of silicon dioxide was approximately 1μm. Before the sputtering process, we cleaned all substrates in the standard piranha solution (96% H_2SO_4 + 30% H_2O_2 in the 3:1 ratio) for 5min, rinsed in deionized water (18.7MΩ cm), and dried them by compressed nitrogen.

The deposition process was done using the ion-beam sputtering apparatus (Bestec) equipped with RFICP Kauf-man ion-beam source (Kaufman & Robinson—KRI®) with the molybdenum 3-grid dished focused ion optics (4cm in diameter and with 45° ellipse pattern) providing an Ar ion beam bombarding the Ti target under an incidence angle of 45°. The 99.9996% pure argon gas was used during the deposition process. Titanium of 99.995% purity (Porexi) was sputtered from a (100 × 100)mm target. Reduction of the ion-beam space charge was provided by KRI LFN 2000 charge neutralizer (KRI®). Before each deposition process, the sputtering chamber was evacuated to pressure of 5×10^{-9} mbar using a turbomolecular HiPace 1200 turbopump (Pfeiffer Vacuum) with the pumping speed of 1200L/s for argon backed by TriScroll® 300 series dry scroll vacuum pump (Agilent Technologies). The schematic illustration of the apparatus setup including all significant dimensions and angles is shown in **Figure 1**.

The operation parameters of the deposition process are beam voltage (BV) determining energy of the ion beam at the target, acceleration voltage (AV) controlling the extraction and the optical parameters of the beam, beam current (BC) equal to the ion-beam current at the target, radio frequency power (RFP) supplied to the discharge, argon flow rate, and substrate temperature. During all deposition experiments, the substrate was rotating with the speed of 5rpm. A particular setup

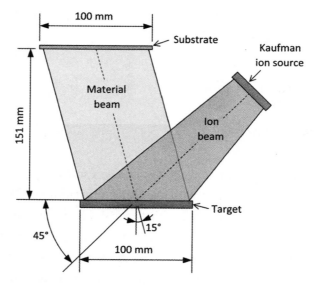

Figure 1. Schematic of sputtering apparatus geometry with Kaufman ion-beam source.

of the major parameters for individual deposition experiments is listed in **Tables 1 and 2**. In all deposition experiments, the substrate was not heated up and its temperature was affected only by the energy of sputtered material. The substrate temperature did not exceed the value of 100°C. The Inficon SQM-242 card with a quartz crystal sensor was used to monitor the deposition parameters as thickness and rate of deposition.

Table 1. Ion-beam source experimental setup (BV, AV, BC, RFP, Ar flow) and deposition pressure and rate.

Sample no.	BV (V)	AV (V)	BC (mA)	RFP (W)	Ar flow (sccm)	Process pressure (mbar)	Deposition rate (Å/s)
1–1	200	−100	6.0	74	2.2	2.1×10^{-4}	0.04–0.06
1–2	400	−80	13.0	73	2.4	2.2×10^{-4}	0.15–0.18
1–3	500	−100	18.0	86	2.6	2.3×10^{-4}	0.24–0.26
1–4	600	−120	23.0	94	2.8	2.4×10^{-4}	0.32–0.34
1–5	700	−140	29.0	107	3.0	2.5×10^{-4}	0.42–0.44
1–6	800	−160	36.0	118	3.3	2.6×10^{-4}	0.56–0.58
1–7	900	−180	43.0	130	3.6	2.7×10^{-4}	0.70–0.72
1–8	1000	−200	50.0	140	3.9	2.9×10^{-4}	0.84–0.86
1–9	1200	−240	65.0	168	4.4	3.1×10^{-4}	1.18–1.20
1–10	1200	−600	82.0	188	5.3	3.5×10^{-4}	1.49–1.51

Table 2. Ion-beam source setup deposition pressure and rate in experiments on optimization of the film structure with respect to the (001) preferential crystallographic orientation.

Sample no.	BV (V)	AV (V)	BC (mA)	RFP (W)	Ar flow (sccm)	Deposition pressure (mbar)	Deposition rate (Å/s)
2–1	200	−100	6.0	73	2.1	1.9×10^{-4}	0.04–0.06
2–2	200	−120	6.0	74	2.1	1.9×10^{-4}	0.04–0.07
2–3	200	−140	7.0	73	2.1	1.9×10^{-4}	0.05–0.07
2–4	200	−160	7.0	74	2.1	1.9×10^{-4}	0.05–0.07
2–5	200	−180	8.0	74	2.2	2.0×10^{-4}	0.06–0.07
2–6	200	−200	8.0	73	2.2	2.0×10^{-4}	0.07–0.08
2–7	200	−220	9.0	74	2.2	2.0×10^{-4}	0.07–0.09
2–8	200	−240	9.0	74	2.2	2.0×10^{-4}	0.07–0.09

2.2. Diagnostic Methods

X-ray diffraction (XRD) technique was used for crystallography analysis. These analysis were done with X-ray diffractometer (SmartLab, Rigaku) containing a linear D/teX Ultra detector and working in the Bragg-Brentano (BB) focusing geometry. Pole figures were measured using parallel beam setup with multilayer parabolic mirror as a collimator and a scintillation detector. The surface roughness (RRMS) of deposited layers was investigated by Atomic Force Microscopy (AFM, Dimension Icon, Bruker) in the ScanAsyst®-Air mode using the corresponding probe (ScanAsyst-Air) with the cantilever spring constant of 0.4N m^{-1} and tip radius of 2nm.

2.3. Annealing Procedure

The deposited samples were exposed to an annealing procedure using the annealing oven (Vakuum Praha). The annealing procedure was done for three temperatures 100°C, 200°C, and 300°C at the pressure of 5×10^{-7} mbar. The annealing protocol was set with following parameters: heating rate 5°C per minute; peak temperature was held for 60min; cooling rate 5°C per minute.

3. Results and Discussion

In the first series of experiments, we set the absolute values of BV and AV in the ratio 5:1 to reduce the flux of electrons according to the rule of electron-back-streaming limit experimentally determined by the manufacturer[19]. The higher ratios of BV and AV cause flowing of electrons (secondary electrons, neutralizing electrons) through ion optics into the ion source affecting the discharge. Utilizing the feedback control, we set BC appropriately to BV and AV according to a recommendation in the KRI® datasheet. These values of the ion-beam current were optimum ones at which a direct impingement of beam ions into the accelerator and decelerator grids were suppressed.

The first sample (Sample 1–1) was deposited using higher AV because it was not possible to keep stable plasma discharge, and the last one (Sample 1–10) was deposited at maximum possible voltage settings of ion-beam source. In **Table 1**, different settings of deposition parameters together with the resultant deposition pressure and deposition rate are listed.

We used XRD in the BB setup with the 2θ angle ranging from 20° to 90° to perform the phase analysis of all deposited layers. We detected diffraction peaks belonging to (100), (101), and (001) crystallographic planes. For (001) crystallographic plane, second-order diffraction 002 was measured since the first-order diffraction is forbidden. However, in the following text we will note it as (001). The corresponding X-ray diffractogram is depicted in **Figure 2**. The diffractograms

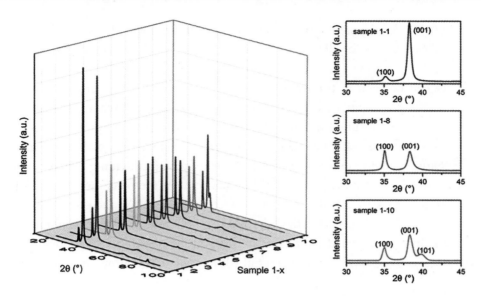

Figure 2. X-ray diffractograms of deposited thin films obtained with different BV and AV; 3D plot of diffractograms of all samples prepared at deposition parameters listed in **Table 1** (left), three detailed diffractograms showing the samples with the most distinguished crystallographic orientations of the films (right).

show also the slight peak at 2θ of 68° to 70° which originates from the silicon substrate and sometimes a very weak peak corresponding to the fourth-order diffraction on the Ti (001) plane at 83°. The obtained results show that crystallographic orientation of sputtered layer depends mainly on ion energy which is given by BV. In case of low ion-beam energies (200eV and 400eV, samples 1–1 and 1–2, respectively), the required (001) preferential crystallographic orientation of the crystallites with a small contribution of (100) planes parallel to the surface was obtained. On the other hand, the (100) plane orientation was much more represented at higher ion-beam energies, namely in the range of BV from 500 to 1000V (Samples 1–3 to 1–8). At the highest ion-beam energy (BV of 1200V, sample 1–9 and 1–10), the (101) plane orientation was observed along with the (001) and (100) ones.

The second series of experiments was aimed at finding the optimal deposition conditions in order to achieve the only (001) preferential orientation of crystallites in the Ti films parallel to the surface. We were changing AV (affecting ion extraction and ion-beam formation, and thus the ion-beam space charge as well) from −100 to −240V at the fixed BV of 200V (*i.e.*, at the constant ion-beam energy of 200eV). All deposition parameters are summarized in **Table 2**. The obtained X-ray diffractograms for all prepared samples are shown in **Figure 3**.

As can be seen in **Figure 3**, at an AV value of −220V (sample 2–7), it is clear that the preferential (001) orientation of thin film was achieved. The other settings of the AV resulted in a minor peak in the diffractogram proving the presence of (100) plane. This behavior was probably attributed to distinct values of space charge potential depending on the AV setting, and providing different energies of charge-exchange argon ions which are leaving the beam and bombarding the substrate surface[17]. In this way, these factors can assist in the growths of thin films and modify their properties similarly to ion-beam-assisted techniques[20]. These results of performed optimizations showed the possibility of producing the titanium layer with only one preferential (001) plane orientation. We have determined the lattice parameters of the optimized thin films; the c lattice parameter was ascertained c = 4.72Å while the lattice parameter was determined by grazing incidence diffraction a = 2.95Å. The tabulated values of these parameters are a = 2.951Å and c = 4.695Å[21].

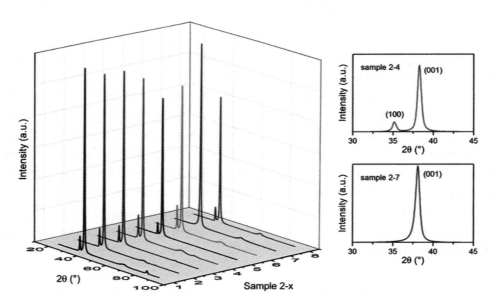

Figure 3. X-ray diffractograms of thin films deposited at the fixed BV of 200V and different AVs; 3D plot of diffractograms of all samples prepared at deposition parameters listed in Table 2 (left), two detailed diffractograms showing the samples with most presented (100) crystallographic orientation (sample 2–4) and the layer with only (001) orientation (sample 2–7) (right).

Carrying out identical experiments for four times, we proved a good repeatability of the results. In all four cases, we obtained the identical diffractograms of thin films prepared in the independent experiments with the same deposition parameters leading to the (001) preferential orientation as sample 2–7.

Full width at half maximum (FWHM) of the diffraction peak belonging to (001) planes was 0.4° for all samples with good reproducibility. This width corresponds to the average coherently diffracting domain of 22nm calculated using Scherrer formula[22]. This value is underestimating the real average crystallite size since the internal strain is neglected. The higher order diffractions were too weak to be detected with sufficient statistics and further analysis was not possible.

Further, we performed the pole figure analysis of the optimized sample (*i.e.*, the sample 2–7) for the diffraction angles belonging to (100), (001), and (101) diffraction planes to determine their preferential orientation. The pole figures shown in **Figure 4** are the stereographic projections of the diffracted intensity plotted with respect to the sample coordinates, *i.e.*, pole figure center corresponds to the crystallographic planes parallel to the surface while the edge of the circle corresponds to the planes being perpendicular to the surface. In the previously shown Bragg-Brentano scans, one can detect only crystallographic planes parallel to the surface. The experimental results show the (001) planes are predominantly oriented parallel to the sample surface, while (100) planes are predominantly perpendicular to the surface. The average misorientation of individual crystallites determined from FWHM of the pole figure peak is 5° [see **Figure 4(b), (d)**]. Accordingly, the (100) planes being perpendicular to (001) planes reveal their maximum intensity at the edge of the pole figure as shown in **Figure 4(a), (d)**. The weak maximum at the pole figure center in **Figure 4(a)** is caused by the fact that the diffraction peaks belonging to (100) and (001) planes are partially overlapping in the angle 2θ and they are not completely separated with used experimental resolution [see **Figure 4(d)**]. The angle between the (101) and (001) crystallographic planes in the Ti lattice is 61.3° which perfectly corresponds to the observed maximum in the pole **Figure 4(c)**. All the three pole figures have a perfect rotational symmetry which means the orientation of the individual crystallites in the azimuthal direction is random with no indication of any preferred azimuthal direction. However, we expected such behavior since the substrate was rotating during the Ti deposition and therefore no unique azimuthal axis is present.

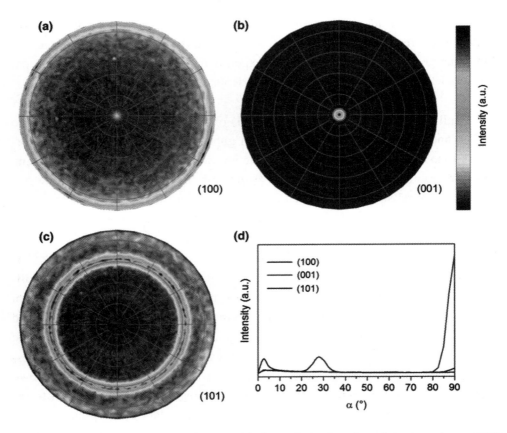

Figure 4. Pole figures of the sample 2–7 with the optimized preferential orientation; a (100) plane angular (*i.e.*, polar and azimuthal) distribution, b (001) plane angular distribution, c (101) plane angular distribution d azimuthally averaged intensity profiles extracted from the preceding pole figures plotted as a function of the crystallographic plane inclination with respect to the sample normal.

Consequent deposition of some layers such as AlN or ZnO over a titanium film often requires process temperatures up to 300°C[8]. Considering this fact, we exposed several samples (sample 2–7) with the optimized (001) preferential crystallites orientation to an annealing process in vacuum of 5×10^{-7} mbar. The annealing process was supposed to simulate conditions similar to those needed for deposition of these binary thin film compounds on a titanium layer and thus to learn its possible thermal instability. The consequent XRD analysis (see **Figure 5**) shows no obvious changes in the crystal lattice after three different annealing processes which were carried out for three maximum temperatures 100°C, 200°C, and 300°C.

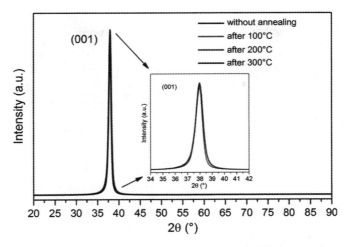

Figure 5. X-ray diffractograms obtained after each annealing step up to 300°C for 1 h.

We used the AFM in the ScanAsyst® mode to determine the surface topography and roughness of deposited thin films. **Figure 6(a)** shows topography of the sample 2–7 containing titanium crystallites with the preferential orientation (001) with surface roughness of only 0.55nm. Topography of the sample 1–8 containing crystallites with both (100) and (001) crystallographic planes (see **Figure 2**) parallel to the surface is shown in **Figure 6(b)**. This sample has one of the highest surface roughnesses with a value of 0.75nm. We attribute this fact to the equal distribution of these two crystallographic orientations recognizable as sharp peaks (001) combined with the rounded elongated islands (100). **Figure 6(c)** shows the topography of the sample 1–10 where the major contribution to the crystallographic structure comes from the crystallographic orientation 001 along with (100) and (101) planes. We presume this fact is reflected in a surface roughness of the value 0.67nm which is lower compared to the previous sample. The individual AFM profiles of these three samples (sample 2–7, 1–8, 1–10) are depicted in **Figure 6(d)**.

The results of the AFM measurement are listed in **Table 3**. The table contains the values of RMS surface roughness including its standard deviations for each sample with corresponding BV and AV. It is obvious that the sample 2–7 possessing the (001) preferential orientation has the lowest surface roughness.

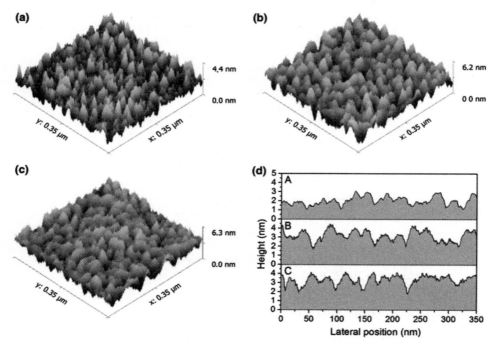

Figure 6. Surface topography obtained by AFM; a the sample 2–7 with the (001) preferential crystallographic orientation; b the sample 1–8 with the (100) and (001) crystallographic orientations; c the sample 1–10 with (100), (001), and (101) crystallographic planes; d profiles corresponding to the three shown AFM surface topography.

Table 3. Measured RRMS surface roughness of samples prepared with different BV and AV settings.

Sample no.	BV (V)	AV (V)	R_{RMS} (nm)
1–1	200	−100	0.72 ± 0.08
2–7	200	−220	0.55 ± 0.07
1–2	400	−80	0.73 ± 0.09
1–4	600	−120	0.76 ± 0.09
1–6	800	−160	0.76 ± 0.12
1–8	1000	−200	0.75 ± 0.09
1–9	1200	−240	0.66 ± 0.08
1–10	1200	−600	0.67 ± 0.08

4. Conclusion

We have presented the optimized deposition process of the preferentially oriented titanium thin film using the Kaufman ion-beam source. The performed experiments have shown that both the low energy and the low ion-beam current are necessary for deposition of highly oriented Ti thin films with the only (001) preferential orientation of films crystallites parallel to the surface.

The crystallites with the (100) orientation parallel to the surface were present in the thin films deposited at higher ion-beam energies (BV of 400V and higher), and those with the (101) plane parallel to the surface in the films deposited at the highest value of ion-beam energy (BV of 1200V). The RRMS roughness of all deposited films was less than 1 nm according to the AFM measurements which confirms an ultra-smooth character of surface. The lowest value of surface roughness is only 0.55nm. We have shown that the surface roughness of thin films depends on the preferential crystallographic orientation.

The observed behavior and properties of the Ti thin films are attributed to specific deposition conditions provided by the Kaufman ion-beam source which generally are not achievable with the conventional magnetron sputtering.

Acknowledgement

We acknowledge the support by the European Regional Development Fund (project No. CZ.1.05/1.1.00/02.0068). The work was also carried out with the support of the CEITEC Nano Core Facility under the CEITEC—open access project, ID number LM2011020, funded by the Ministry of Education, Youth and Sports of the Czech Republic under the activity "Projects of major infrastructures for research, development and innovations."

Open Access: This article is distributed under the terms of the Creative Commons Attribution 4.0 International License (http://creativecommons.org/licenses/by/4.0/), which permits unrestricted use, distribution, and reproduction in any medium, provided you give appropriate credit to the original author(s) and the source, provide a link to the Creative Commons license, and indicate if changes were made.

Source: Gablech I, Svatoš V, Caha O. Preparation of (001) preferentially oriented titanium thin films by ion-beam sputtering deposition on thermal silicon dioxide[J].

Journal of Materials Science, 2016, 51(7):3329–3336.

References

[1] Chawla V, Jayaganthan R, Chawla AK, Chandra R (2009) Microstructural characterizations of magnetron sputtered Ti films on glass substrate. J Mater Process Technol 209(7):3444–3451. doi:10.1016/j.jmatprotec.2008.08.004.

[2] Chen C-N (2012) Fully quantitative characterization of CMOS-MEMS polysilicon/titanium thermopile infrared sensors. Sens Actuators B 161(1):892–900. doi:10.1016/j.snb.2011.11.058.

[3] Doll JC, Petzold BC, Ninan B, Mullapudi R, Pruitt BL (2010) Aluminum nitride on titanium for CMOS compatible piezoelectric transducers. J Micromech Microeng 20(2):25008. doi:10. 1088/0960-1317/20/2/025008.

[4] López JM, Gordillo-Vázquez FJ, Fernández M, Albella JM (2001) Influence of oxygen on the morphological and structural properties of Ti thin films grown by ion beam-assisted deposition. Thin Solid Films 384(1):69–75. doi:10.1016/S0040-6090(00)01804-6.

[5] Tsuchiya T, Hirata M, Chiba N (2005) Young's modulus, fracture strain, and tensile strength of sputtered titanium thin films. Thin Solid Films 484(1–2):245–250. doi:10.1016/j.tsf.2005.02.024.

[6] Tran AT, Wunnicke O, Pandraud G, Nguyen MD, Schellevis H, Sarro PM (2013) Slender piezoelectric cantilevers of high quality AlN layers sputtered on Ti thin film for MEMS actuators. Sens Actuators A 202:118–123. doi:10.1016/j.sna.2013.01.047.

[7] Jackson N, O'Keeffe R, Waldron F, O'Neill M, Mathewson A (2013) Influence of aluminum nitride crystal orientation on MEMS energy harvesting device performance. J Micromech Microeng. doi:10.1088/0960-1317/23/7/075014.

[8] Jackson N, Keeney L, Mathewson A (2013) Flexible-CMOS and biocompatible piezoelectric AlN material for MEMS applications. Smart Mater Struct. doi:10.1088/0964-1726/22/11/115033.

[9] Xiong J, Gu H-s HuK, M-z Hu (2010) Influence of substrate metals on the crystal growth of AlN films. Int J Miner Metall Mater 17(1):98–103. doi:10.1007/s12613-010-0117-y.

[10] Ababneh A, Alsumady M, Seidel H, Manzaneque T, Hernando-García J, Sánchez-Rojas JL, Bittner A, Schmid U (2012) c-axis orientation and piezoelectric coefficients of AlN thin films sputter-deposited on titanium bottom electrodes. Appl Surf Sci 259:59–65. doi:10.1016/j.apsusc.2012.06.086.

[11] Iriarte GF, Bjurstrom J, Westlinder J, Engelmark F, Katardjiev IV (2005) Synthesis of c-axis-oriented AlN thin films on high-conducting layers: Al, Mo, Ti, TiN, and Ni.

IEEE Trans Ultrason Ferroelectr Freq Control 52(7):1170–1174. doi:10.1109/tuffc.2005.1504003.

[12] Boeshore SE, Parker ER, Lughi V, MacDonald NC, Bingert M (2005) Aluminum nitride thin films on titanium for piezoelectric microelectromechanical systems. In 2005 IEEE ultrasonics symposium, vol 1–4. New York, pp. 1641–1643.

[13] Jung MJ, Nam KH, Shaginyan LR, Han JG (2003) Deposition of Ti thin film using the magnetron sputtering method. Thin Solid Films 435(1–2):145–149. doi:10.1016/S0040-6090(03)00344-4.

[14] PalDey S, Deevi SC (2003) Single layer and multilayer wear resistant coatings of (Ti, Al)N: a review. Mater Sci Eng A 342(1–2):58–79. doi:10.1016/S0921-5093(02)00259-9.

[15] Jing FJ, Yin TL, Yukimura K, Sun H, Leng YX, Huang N (2012) Titanium film deposition by high-power impulse magnetron sputtering: influence of pulse duration. Vacuum 86(12):2114–2119. doi:10.1016/j.vacuum.2012.06.003.

[16] Henderson PS, Kelly PJ, Arnell RD, Bäcker H, Bradley JW (2003) Investigation into the properties of titanium based films deposited using pulsed magnetron sputtering. Surf Coat Technol 174–175:779–783. doi:10.1016/S0257-8972(03)00397-9.

[17] Harper JME, Cuomo JJ, Kaufman HR (1982) Technology and applications of broad-beam ion sources used in sputtering. 2. Applications. J Vac Sci Technol 21(3):737–756. doi:10.1116/1.571820.

[18] Kaufman HR, Cuomo JJ, Harper JME (1982) Technology and applications of broad-beam ion sources used in sputtering.1. Ion-source technology. J Vac Sci Technol 21(3):725–736. doi:10.1116/1.571819.

[19] Catalog: The Kaufman & Robinson Inc. (2013) RFICP 40 ion optics supplement: molybdenum three-grid dished focused 4-cm diameter 45° ellipse.

[20] Sikola T, Spousta J, Dittrichova L, Nebojsa A, Perina V, Ceska R, Dub P (1996) Dual ion-beam deposition of metallic thin films. Surf Coat Technol 84(1–3):485–490. doi:10.1016/S0257-8972 (95)02823-4.

[21] PDF-2 database entry 00-044-1294, ICDD-JCPDS.

[22] Patterson AL (1939) The scherrer formula for X-ray particle size determination. Phys Rev 56(10):978–982.

Chapter 29

Probing Trace Levels of Prometryn Solutions: From Test Samples in the Lab toward Real Samples with Tap Water

Rafael J. G. Rubira[1], Sabrina A. Camacho[1], Pedro H. B. Aoki[1,2], Fernando V. Paulovich[3], Osvaldo N. Oliveira Jr.[2], Carlos J. L. Constantino[1]

[1]Faculdade de Ciências e Tecnologia, UNESP Univ Estadual Paulista, Presidente Prudente, SP 19060-900, Brazil
[2]São Carlos Institute of Physics, University of São Paulo, CP 369, São Carlos, SP 13560-970, Brazil
[3]Institute of Mathematical Sciences and Computing, University of São Paulo, CP 668, São Carlos, SP 13560-970, Brazil

Abstract: Growing food demand has been addressed by protecting crops from insects, weeds, and other organisms by increasing the application of pesticides, thus increasing the risk of environmental contamination. Many pesticides, such as the triazines, are poorly soluble in water and require trace detection methods, which are normally achieved with high-cost sophisticated chromatography techniques. Here, we combine surface-enhanced Raman scattering (SERS) with multidimensional projection techniques to detect the toxic herbicide prometryn in ultrapure, deionized, and tap waters. The SERS spectra for prometryn were recorded with good signal-to-noise ratio down to 5×10^{-12} mol/L in ultrapure water, approaching single-molecule levels, and 5×10^{-9} mol/L in tap water. The latter is one order of magnitude below the threshold allowed for drinking water. In addition to providing a fingerprint of prometryn molecules at low concentrations, SERS is advantageous compared to other methods since it does not require pretreatment or chemical separation. The multidimensional projection methods and the detection procedure with SERS are entirely generic, and may be extended to any other pesticide or water contaminants, thus allowing environmental control to be potentially low cost if portable Raman spectrophotometers are used.

1. Introduction

The growing demand for food in conjunction with competition from international markets forces the agribusiness to increasingly resort to agrochemicals, as is the case of Brazil where the sales of pesticides have increased dramatically to position the country as the largest consumer in the world, ahead of the USA[1]. The lack of criteria for controlling the use of pesticides represents a severe environmental issue due to contamination of plants, soil, and groundwater. Furthermore, the easy dispersion of pesticides in the atmosphere and oceans can rapidly reach global scale contamination[2]. According to the World Health Organization (WHO), millions of people are poisoned by these substances every year, leading to approximately 220,000 deaths[3]. From the various types of herbicides, triazine derivatives are prominent for their use in the pre- and post-emergent control of weed seeds in a variety of crops, including corn, sugar cane, sorghum, pineapple, banana, coffee, and grapes[4]. The triazine derivatives and their degradation products are toxic and nonbiodegradable, which can lead to environmental contamination. Particularly relevant is contamination in groundwater and drinking water[5]. Prometryn is one of the most used herbicides of the triazine family, whose maximum concentration allowed in drinking water is 23.2μg/L (9.62×10^{-8} mol/L or 23.2ppb), according to regulations from the United States Environmental Protection Agency (USEPA)[6].

Only a few analytical methods are available that can reach such level of dilution, most of which involve chromatography experiments[7][8]. Electrochemical techniques are also employed in detecting prometryn, although electrochemical sensors based on mercury electrodes[9]-[11] have negative environmental implications. Motivated by the need to develop sensitive and selective methods to monitor residual amounts of prometryn, in this study we combine the highly sensitive and selective surface-enhanced Raman scattering (SERS)[12] with advanced computational data analysis in order to detect prometryn below the threshold allowed for drinking water. SERS is a vibrational technique in which signal enhancement of target molecules is achieved due to localized surface plasmon resonance (LSPR) excited using metallic nanostructures[13]–[15]. Basically, the incident electric field (E_{inc}), from the probing laser, interacts with the metallic nanoparticles and induces a collective oscillation of the electronic metal cloud, which is able to sustain LSPR. In terms of electric field, the nanoparticle excitation induces an electric field (E_p) in its surroundings, leading to a local electric field (E_{local}) given by the sum $E_{local} = E_{inc} + E_p$, which is much larger than E_{inc} itself and oscillates with the same frequency of the probing laser. When the target molecule is close enough to the nanoparticle, E_{local} polarizes the target molecule, inducing a molecular dipole (p_1). The electric field irradiated by p_1 can also polarize the metallic nanoparticle, inducing a dipole in the nanoparticles (p_2), which oscillates with the same frequency of p_1 (Raman frequency). The intense local electric field (E_{local}) is proportional to E^4 and can achieve an enhancement factor from 10^3 to 10^6 in average and up to 10^{10} in the so-called "hot spots" (interstices of AgNPs aggregates where extremely high density of electric field is found)[13]–[16].

The samples used in the experiments included prometryn solutions in ultrapure, deionized, and tap waters, in which the latter mimics the conditions found in real samples.

2. Materials and Methods

2.1. Reagents

The reagents silver nitrate ($AgNO_3$, MM = 169.88g/mol), hydroxylamine hydrochloride ($NH_2OH\cdot HCl$, MM = 69.49g/mol), and sodium hydroxide (NaOH, MM = 40.00g/mol) were acquired from Sigma-Aldrich. Prometryn (C10H19N5S, MM = 241.36g/mol), purity = 98.8%, was purchased from Fluka Analytical. All the chemicals were used without further purification. Ultrapure water with resistivity of 18.2MΩ cm and pH 5.6, acquired from a Milli-Q system (model Simplicity), was used to prepare the Ag colloid and prometryn stock solution. Deionized tap water with resistivity of 0.3MΩ cm, acquired from a deionization system (model SP-050C), was also used to prepare a prometryn stock solution. The neutralization process removes nitrates, calcium, magnesium, cadmium, barium, lead, and some forms of radium from the water[17].

2.2. Synthesis of Ag Colloid by Hydroxylamine Reduction

The Ag colloid (AgNPs) obtained by hydroxylamine reduction was synthesized according to the methodology described by Leopold and Lendl[18]. The synthesis consisted in adding 4.5mL aqueous solution of NaOH 0.1mol/L, at room temperature, into 5mL of $NH_2OH\text{-}HCl$ 43.3mmol/L solution. This solution was added to 90mL of $AgNO_3$ at 1.2mmol/L, under stirring, thus yielding a AgNPs colloidal dispersion with a final concentration of ca. 1.0mmol/L[18]. The UV-Vis spectrum and SEM image of the Ag colloid are shown in Figure S1 in the Supplementary Material, in which results from zeta potential, size and shape of the AgNPs are discussed.

2.3. UV-Vis Absorption, SEM, Zeta Potential, and Raman Scattering

The UV-Vis absorption spectrum of the Ag colloid was recorded using a Varian spectrophotometer, model Cary 50, from 190 to 1100nm. Scanning electron microscope (SEM) images were taken with a Carls Zeiss equipment, EVO LS15 model (Laboratory LabMMEV at FCT/UNESP). The zeta potential was determined using a ZetaSizer3000 HS. The Micro-Raman analysis was performed using a micro-Raman Renishaw spectrograph, model in-Via, equipped with a Leica microscope, where a 950 objective lens allows collecting spectra with ca. $1\mu m^2$ spatial resolution. The spectrograph contains a charge-coupled device (CCD) detector, laser line at 633nm, 1800 grooves/mm grating with additional edge filters. In all measurements, the Raman scattering and SERS spectra were acquired with spectral acquisition times of 10s and laser power within the µW range at the sample.

2.4. Solutions for SERS Measurements

Three prometryn stock solutions at 1×10^{-4} mol/L were prepared by dissolving 6.0mg of prometryn in 250mL of ultrapure, deionized, and tap waters, under sonication. These stock solutions were diluted in Ag colloid to achieve SERS. The final solution concentrations of prometryn were 5×10^{-5}, 5×10^{-6}, 5×10^{-7}, 5×10^{-8}, 5×10^{-9}, 5×10^{-10}, 5×10^{-11}, and 5×10^{-12} mol/L. In order to acquire SERS spectra, a small droplet of prometryn solution (diluted in Ag colloid) was placed in a holder under the microscope and the laser focus was adjusted onto the air/water interface. Usually, several spectra were recorded from 2 different drops until getting at least 8 spectra with suitable signal/noise ratio for each drop. For each concentration (2 drops), the measurements were performed in a total of approximately 2h, for which the solvent evaporation (water) was minimal. **Figure 1** illustrates the procedure adopted in the SERS measurements.

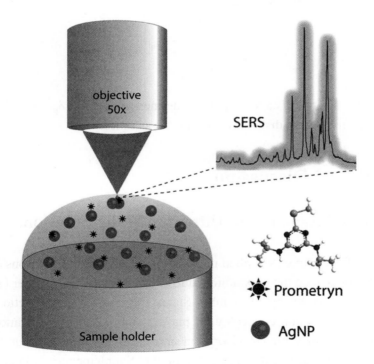

Figure 1. Schematic procedure for the SERS measurements. The prometryn molecular structure is represented in 3D where the atoms are distinguished by the colors: H (white), C (gray), N (blue), and S (yellow) (Color figure online).

2.5. Data Analysis

The SERS spectra were analyzed using a multidimensional projection technique, in which data from a multidimensional space can be projected onto a 2D space creating a plot with maximum preservation of similarity relationships. Formally, the data in the original space are represented by $X = \{x_1, x_2, ..., x_n\}$, and $\delta(x_i, x_j)$ is defined as the distance between two data instances i and j. The 2D plot is created by projecting them onto the plane with graphical markers represented by $Y = \{y_1, y_2, ..., y_n\}$, with the positions on the 2D plot being determined in an optimization procedure using an injective function $f: X \rightarrow Y$ that minimizes $|\delta(x_i, x_j) - d(f(x_i), f(y_j))| \approx 0$, $x_i, x_j \in X^{[19]}$, where $d(y_i, y_j)$ is the distance function on the projected plane. The flexibility of this optimization approach arises from the availability of several cost (or error) functions used for placing the graphical markers on the 2D plot. Here we used the so-called Interactive Document Map (IDMAP)[20], whose function is defined as follows:

$$S_{IDMAP} = \frac{\delta(x_i, x_j) - \delta_{min}}{\delta_{max} - \delta_{min}} - d(y_i, y_j)$$

where δ and d are the distance functions defined above and δ_{min} and δ_{max} are the minimum and maximum distances between the samples.

3. Results and Discussion

3.1. Probing Prometryn in Ultrapure Water Solutions

The SERS effect was applied in detecting highly diluted solutions of prometryn in order to include the maximum value allowed for drinking water (10^{-8} mol/L) according to regulation by the United States Environmental Protection Agency (USEPA)[6]. The solution concentration plays a role because aggregation of prometryn molecules may hinder adsorption onto the Ag colloid[21], thus leading to smaller SERS enhancement. Figure S2 in the Supplementary Material shows SERS spectra collected at relatively high concentrations of 5×10^{-5} and 5×10^{-6} mol/L. Despite the low signal/noise ratio, these SERS spectra were found in all the measurements for these high concentrations. In contrast, intense, well-defined vibrational bands are seen in **Figure 2** for solutions from 5×10^{-7} mol/L down to 10^{-12} mol/L. Two main points should be highlighted here: (i) the spectral acquisition from 5×10^{-7} down to 10^{-12} mol/L is not straightforward because only the molecules adsorbed at the interstices of AgNPs aggregates ("hot spots") will give rise to measurable signals; (ii) the SERS spectra in **Figure 2** are normalized and a linear increase of the SERS signal with concentration was not observed at this range of concentration. Obtaining analytical curves using SERS (e.g., band intensity vs. prometryn concentration) is not straightforward because band intensity depends on several factors, in addition to the analyte concentration. For example, the SERS intensity may fluctuate owing to the movement of metallic nanoparticles-target molecules in and out of the volume probed by the laser. It also depends on the size, shape, and aggregation of the metallic nanoparticles, which is the reason why there are only a few works in the literature with attempts to correlate band intensity with analyte concentration[22].

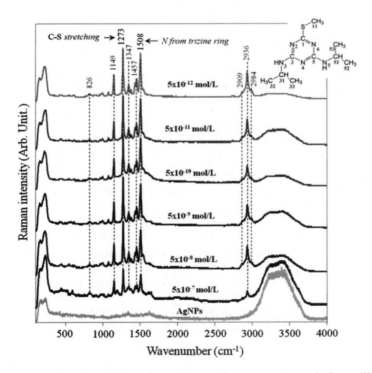

Figure 2. SERS spectra (raw data) of prometryn ultrapure water solutions diluted in Ag colloid at 5×10^{-7}, 5×10^{-8}, 5×10^{-9}, 5×10^{-10}, 5×10^{-11}, and 5×10^{-12} mol/L. The Raman intensity (Y-axis) is normalized. An Ag colloid spectrum is given as reference (control) at the bottom.

The spectra taken from neat Ag colloid, shown in Figure S3 in the Supplementary Material, are used as reference (control). Since only the signal expected from the Ag colloid was observed in these samples, one may be sure of the low level of impurities. The main vibrational bands in the SERS spectra for the solutions in **Figure 2** are highlighted by dotted lines, with the assignments given in **Table 1**.

According to the selection rules for SERS[13]–[16][23], vibrational modes with dipole moment perpendicular to Ag surface (consequently, parallel to the electric field) result in a high intensity of the SERS signal. Larger enhancements are found for bands assigned to the vibrational modes involving the ring, within the 1500–1100 cm^{-1} region[24], suggesting that the aromatic N atoms play a role in the prometryn adsorption onto the AgNPs. Furthermore, the largest enhancement of the band at 1273 cm^{-1} (in-plane deformation of the ring mixed with the stretching of C–S[24][25]) indicates that prometryn molecules are preferentially adsorbed onto

Table 1. Vibrational assignments of prometryn characteristic bands[24][25].

Prometryn (cm^{-1})	Assignments
703	(C1–S stretching) [24, 25]
826	(Ring breathing) [24]
902	(Ring breathing) [24]
970	(Ring breathing) [24]
1149	(C–S stretching); (isopropyl group deformation) [24, 25]
1176	(C–S stretching) [24, 25]
1273	(Plane deformation of the ring); (C–S stretching); (N–C–H deformation); (C–C–H deformation) [24, 25]
1310	(Plane deformation of the ring); (N–C–H deformation); (C–S stretching); (C–C–H deformation) [24, 25]
1347	(S-CH$_3$ symmetric bending + lateral chains C–C–H deformation) [24, 25]
1457	(CH$_3$ deformation on the sulfanyl group) [24, 25]
2760	(C–H stretching); (H–C–C bending) [24]
2873	(C–H stretching of the CH$_3$ groups) [24]
2909	(C51-H + C52-H + C53-H stretching) [24]
2936	(C31-H + C32-H + C33-H stretching) [24]
2984	(C31-32-H symmetric and antisymmetric stretchings) [24]

AgNPs through the C–S groups, as depicted in **Figure 3**. Besides, we highlight the band at 970cm^{-1} (ring breathing), which is the most intense in the powder spectrum but decreased drastically in the SERS spectrum. This suggests that the prometryn molecules have their triazinic ring positioned parallel to the AgNPs surface (selection rules[13]). This conclusion is supported by the SERS experimental results from Bonora et al.[24] working with atrazine, simetryn, and prometryn herbicides, and by theoretical calculations of Benassi et al.[25].

A visual inspection of **Figure 2** allows one to infer whether it is possible to detect the different prometryn concentrations. In order to demonstrate this distinction ability when many samples are compared, one has to resort to statistical or computational methods for analyzing the data. Here we confirmed such distinction ability by treating the SERS spectra shown in **Figure 2** with the IDMAP multidimensional projection technique, whose results are given in **Figure 4**. Details of multidimensional projection techniques can be found in Paulovich et al.[26] and Oliveira et al.[27]. Basically, each circle in the plot represents a whole spectrum, and the closer the circles, the more similar the SERS spectra. Samples with distinct prometryn concentrations are clustered apart from each other, indicating that SERS can easily distinguish samples even down to 10^{-12}mol/L. A larger dispersion of the data is seen for prometryn solutions at 10^{-10}, 10^{-11}, and 10^{-12}mol/L, which

Figure 3. Schematic representation of the prometryn adsorption onto AgNPs, which is believed to occur via interactions of their N and C–S groups, with the ring lying parallel to the AgNPs.

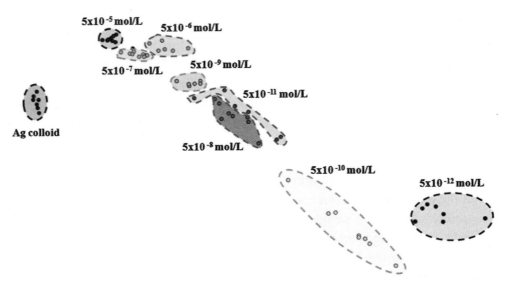

Figure 4. IDMAP multidimensional projection grouping the results by different concentrations of prometryn in ultrapure water solutions. Each circle represents a whole SERS spectrum. The closer the circles, the more similar the SERS spectra.

is a direct consequence of fluctuations on the SERS spectra affecting bandwidth, band shape, Raman shift, and absolute and relative intensities for highly diluted solutions. This is typical of the unique behavior observed as single-molecule limits are approached, revealing the breakdown of ensemble averaging SERS spectra and the local changes of molecular environment[13][28]. Indeed, the number of prometryn molecules per picoliter (10^{-12}L) at 5×10^{-12}mo/L is estimated to be three. The picoliter scale is the order of magnitude of the volume probed by a laser in single-molecule experiments[29][30].

The smallest concentration tested here is the lowest ever detected for prometryn solutions, approaching single-molecule levels. In fact, the use of SERS for prometryn in the literature has never gone below 10^{-4}mol/L[24]. Prometryn detection with chromatography techniques has reached limits from 10^{-6} to 10^{-10}mol/L[31][32] while the differential pulse polarographic method led to detection of atrazine, prometryn, and simazine herbicides down to 8×10^{-8}mol/L[7]. Oliveira-Brett et al.[33] reported an electrochemical biosensor to investigate the interactions between DNA and herbicides from the s-triazine group, achieving a detection limit of 5×10^{-4}mol/L.

3.2. Probing Trace Levels of Prometryn in Deionized and Tap Water Solutions

One important challenge in sensing experiments is to deal with real samples, as is the case of probing trace levels of prometryn in deionized and tap water solutions, which mimic the conditions prevailing in real samples. **Figure 5(a)** shows the SERS spectra collected for deionized water solutions of prometryn diluted in Ag colloid at 5×10^{-7}, 5×10^{-8}, 5×10^{-9}, 5×10^{-10}, 5×10^{-11}, and 5×10^{-12} mol/L. The SERS spectra collected for tap water solutions of prometryn diluted in Ag colloid at 5×10^{-5}, 5×10^{-6}, 5×10^{-7}, 5×10^{-8}, and 5×10^{-9} mol/L are displayed in **Figure 5(b)**. Several Ag colloid spectra are given in the Supplementary Material (Figure S4). The signal arising from the Ag colloid containing either deionized or tap water indicates the presence of impurities in the medium that might interfere in the SERS signal of the target molecule. Indeed, there is an overlap of bands between 1100 and 1621 cm^{-1} for prometryn and Ag colloid spectra, as shown in **Figure 5(a), (b)**.

The suitability of SERS to identify prometryn in deionized tap water solutions is more clearly visualized in the 2D plot in **Figure 6** where the SERS spectra were treated using the IDMAP multidimensional technique. The results are grouped according to the similarity of the analyzed data. Prometryn concentrations at 5×10^{-5}, 5×10^{-6}, 5×10^{-7}, 5×10^{-8}, and 5×10^{-9} mol/L are clustered apart from each other. However, the clusters associated with more diluted prometryn solutions (5×10^{-10}, 5×10^{-11}, and 5×10^{-12} mol/L) are lumped together and some overlap occurs. Hence, though the samples containing prometryn can be distinguished from deionized water—even down to 5×10^{-12} mol/L—they cannot be clearly separated among themselves.

The distinguishing ability using SERS spectra is the poorest for prometryn solutions obtained from tap water, which should be expected due to its larger amount of impurities. **Figure 7** shows the projection map of the data acquired for tap water solutions of prometryn only for concentrations starting at 10^{-9} mol/L. We omitted the data for lower concentrations, since their clusters would collapse into the cluster for the Ag colloid reference (control).

The SERS spectra of prometryn solutions prepared with ultrapure, deionized,

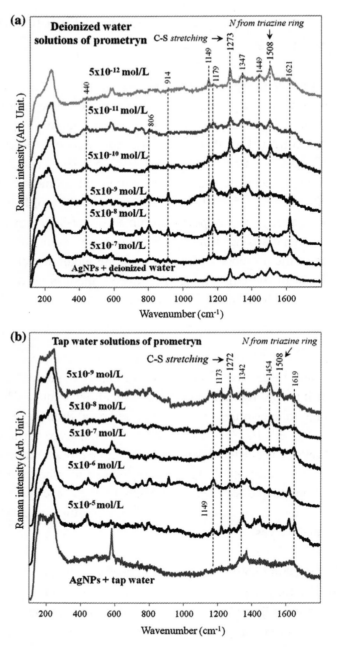

Figure 5. SERS spectra (raw data) collected for: a deionized water solutions of prometryn diluted in Ag colloid at 5×10^{-7}, 5×10^{-8}, 5×10^{-9}, 5×10^{-10}, 5×10^{-11}, 5×10^{-12} mol/L and b tap water solutions of prometryn at 5×10^{-5}, 5×10^{-6}, 5×10^{-7}, 5×10^{-8}, and 5×10^{-9} mol/L. The Raman intensity (Y-axis) is normalized. Ag colloid spectra containing deionized and tap water are given as reference (control) at the bottom of each figure.

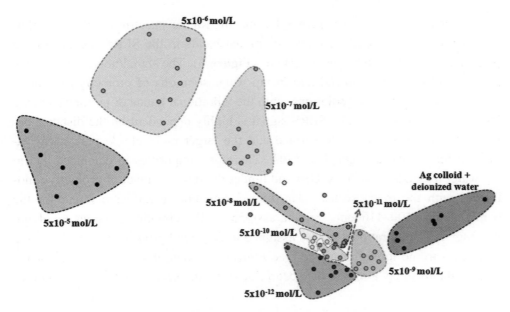

Figure 6. IDMAP multidimensional projection grouping the results by different concentrations of prometryn in deionized water solutions. Each circle represents a whole SERS spectrum. The closer the circles, the more similar the SERS spectra.

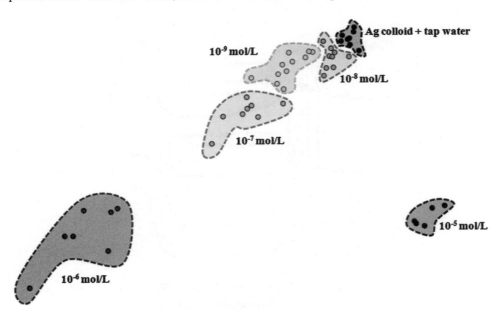

Figure 7. IDMAP multidimensional projection grouping the results by different concentrations of prometryn in tap water solutions. Each circle represents a whole SERS spectrum. The closer the circles, the more similar the SERS spectra.

or tap waters are shown in **Figure 8**, besides the Raman spectrum of the Ag colloid containing tap water. The main vibrational bands in the SERS spectra for the solutions are highlighted by dotted lines (**Figure 8**). The signal/noise ratio for the spectra collected in deionized and in tap water solutions of prometryn is not as high as in ultrapure water solutions, with the vibrational bands of prometryn being less defined (**Figure 8**). The SERS signal is highly dependent on the distance between the metal nanoparticle surface and the target molecule[13], practically vanishing for distances larger than 150Å. Therefore, impurities in deionized and in tap water may adsorb onto AgNPs and hinder the direct contact of prometryn molecules with the metal surface. The higher signal/noise ratio are found for the bands within 1500–1100cm^{-1} region, assigned to the vibrational modes involving the ring[24]. The vibrational modes assigned to isopropyl groups between 2909 and 2984cm^{-1} are no longer observed (see **Figure 8**). Then, it seems that the triazine ring might also play a role in the adsorption onto the AgNPs surface for deionized and tap water.

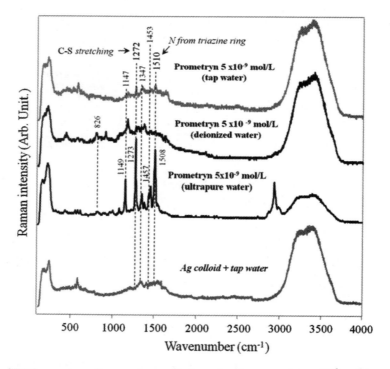

Figure 8. SERS spectra of prometryn solutions in ultrapure (5 × 10^{-9}mol/L), deionized (5 × 10^{-9}mol/L), and tap water (5 × 10^{-9}mol/L) compared with the reference spectrum (control) of the Ag colloid containing tap water.

The role of impurities is highlighted by the IDMAP projection in **Figure 9** in which the data are plotted from different Ag colloid references (ultrapure, deionized, and tap waters) and prometryn solutions at 10^{-9} mol/L (ultra-pure, deionized, and tap waters). It is clear that the projection technique (IDMAP) results show different patterns depending on the system analyzed. For instance, in general it is seen that the circles (within a cluster) are more spread for lower concentrations when prometryn is diluted in ultrapure water. On the other hand, considering the prometryn diluted in deionized or tap water, the circles (within a cluster) are more spread for higher concentrations. These opposite patterns could be understood considering that for ultrapure water (high purity medium), the addition of small amounts of impurity (prometryn) would be enough to lead to dramatic changes in the medium. For deionized or tap water (high impurity medium), the addition of small amounts of impurity (prometryn) would not be enough to interfere substantially in the medium because it is already impure. It is likely that a larger number of spectra will be required to understand the dispersion in the data for real samples, *i.e.*, containing considerable amounts of impurities.

Therefore, even in the worst scenario represented by solutions made with tap water, treating the SERS data with IDMAP allows one to detect prometryn to concentrations one order of magnitude below the limit allowed for drinking water. This highlights the usefulness of projection techniques not only to handle the data

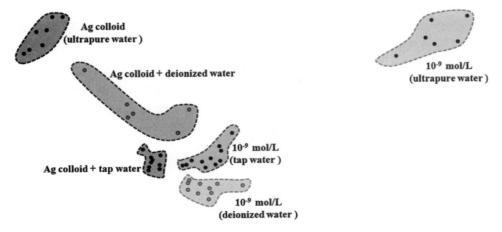

Figure 9. IDMAP multidimensional projection grouping the results from the different Ag colloid references and prometryn solutions at 10^{-9} mol/L in ultrapure, deionized, and tap water. Each circle represents a whole SERS spectrum. The closer the circles, the more similar the SERS spectra.

but also to optimize the sensing performance. The performance of SERS-based detection is competitive with a few works in the literature for detection of pesticides in real samples, most of them based on extraction and size exclusion methods in chromatography techniques[8][34]–[36]. Koeber et al. were able to detect triazine herbicides in river samples with a detection limit of 3.3×10^{-10} mol/L for simazine by combining column size exclusion and adsorption chromatography[31]. Djozan et al. reported detection of s-triazine herbicides in tap water with detection limit of 8.29×10^{-8} mol/L using gas chromatography[37]. Electrochemical techniques have also been applied to detect herbicides in solutions, foods, and real samples. Some limits of detection (LOD) reported were 1.7×10^{-8} mol/L for methidathion[38], 1.9×10^{-11} mol/L for paraquat in buffer or in potato extracts[39], and 8.9×10^{-8} mol/L for lindane in ultrapure water[40].

In conclusion, the combination of SERS with information visualization methods was successfully applied to detect prometryn not only in ultrapure water but also in tap water, which resembles real samples. The concentration detected is the lowest ever reported for prometryn in ultrapure water (5×10^{-12} mol/L) and among the lowest for real samples of nonpurified water (10^{-9} mol/L). The prometryn detection in ultrapure water approached single-molecule levels, which is promising since the experimental procedure for the measurement is carried out with the herbicide being directly detected in the sample, with no pretreatment or chemical separation. The lower SERS signal/noise ratio for prometryn in tap water is related to impurities in the samples that may hinder adsorption of the analyte molecules onto the colloidal AgNPs. The lowest prometryn concentration detected (10^{-9} mol/L) in tap water is still one order of magnitude below the threshold for drinking water. Since both the detection principle based on SERS and the information visualization methods are entirely generic, the approach may be extended to other pesticides and contaminants, including real samples.

Acknowledgements

This work was supported by FAPESP, CNPq, CAPES, INEO, and nBioNet (Brazil).

Open Access: This article is distributed under the terms of the Creative Commons

Attribution 4.0 International License (http://creativecommons.org/licenses/by/4.0/), which permits unrestricted use, distribution, and reproduction in any medium, provided you give appropriate credit to the original author(s) and the source, provide a link to the Creative Commons license, and indicate if changes were made.

Source: Rubira R J G, Camacho S A, Aoki P H B. Probing trace levels of prometryn solutions: from test samples in the lab toward real samples with tap water[J]. Journal of Materials Science, 2016, 51(6):3182–3190.

References

[1] Sanitária ANDV (2005–2010) Avaliação dos Projeto da cooperação Técnica Anvisa, OPAS/OMS. Brasilia-DF.

[2] Alloway BJ, Ayres DC (1997) Chemical principles of environmental pollution, 2nd edn. Blackie Academic Professional, Glasgow.

[3] Li R-H, Liu D-H, Yang Z-H, Zhou Z-Q, Wang P (2012) Vortex-assisted surfactant-enhanced-emulsification liquid-liquid microextraction for the determination of triazine herbicides in water samples by microemulsion electrokinetic chromatography. Electrophoresis 33(14):2176–2183. doi:10.1002/elps.201200104.

[4] Coutinho CFB, Tanimoto ST, Galli A, Garbellini GS, Takayama M, Amaral RB, Mazo LH, Avaca LA, Machado SAS (2005) Pesticidas: mecanismo de ação, degradação e toxidez. Pesticidas: Revista de Ecotoxicologia e Meio Ambiente 15:65–72.

[5] Pinto GMF, Jardim I (2000) Use of solid-phase extraction and high-performance liquid chromatography for the determination of triazine residues in water: validation of the method. J Chromatogr A 869(1–2):463–469. doi:10.1016/s0021-9673(99)01242-x.

[6] USEPA (1996) United States Environmental Protection Agency (USEPA). vol 2014.

[7] Lippolis MT, Concialini V (1988) Differential pulse polarographic-determination of the herbicides atrazine, prometrine and simazine. Talanta 35(3):235–236. doi:10.1016/0039-9140(88) 80072-9.

[8] Tolcha T, Merdassa Y, Megersa N (2013) Low-density extraction solvent based solvent-terminated dispersive liquid-liquid microextraction for quantitative determination of ionizable pesticides in environmental waters. J Sep Sci 36(6):1119–1127. doi:10.1002/jssc.201200849.

[9] Nouws HPA, Delerue-Matos C, Barros AA, Rodrigues JA, San-tos-Silva A (2005) Electroanalytical study of fluvoxamine. Anal Bioanal Chem 382(7):1662–1668. doi:10.1007/s00216-005-3310-5.

[10] dos Santos LBO, Silva MSP, Masini JC (2005) Developing a sequential injection-square wave voltammetry (SI-SWV) method for determination of atrazine using a hanging mercury drop electrode. Anal Chim Acta 528(1):21–27. doi:10.1016/j.aca.2004. 10.008.

[11] Guse D, Bruzek MJ, DeVos P, Brown JH (2009) Electrochemical reduction of atrazine: NMR evidence for reduction of the triazine ring. J Electroanal Chem 626(1–2): 171–173. doi:10.1016/j.jele chem.2008.12.006.

[12] Liu J, Hu Y, Zhu G, Zhou X, Jia L, Zhang T (2014) Highly sensitive detection of zearalenone in feed samples using competitive surface-enhanced Raman scattering immunoassay. J Agric Food Chem 62(33):8325–8332. doi:10.1021/jf503191e.

[13] Aroca R (2006) Surface-enhanced vibrational spectroscopy. Wiley, Chichester.

[14] Kneipp K (2007) Surface-enhanced Raman scattering. Phys Today 60:40–47.

[15] Willets KA, Van Duyne RP (2007) Localized surface plasmon resonance spectroscopy and sensing. Ann Rev Phys Chem 58:267–297. doi:10.1146/annurev.physchem.58.032806.104607.

[16] Le Ru EC, Etchegoin PG (2009) Principles of surface enhanced raman spectroscopy (and related plasmonic effects). Elsevier, Amsterdam.

[17] Lopes HJJ (2003) Garantia e Controle da Qualidade no Laboratório Clínico. Belo Horizonte-MG.

[18] Leopold N, Lendl B (2003) A new method for fast preparation of highly surface-enhanced Raman scattering (SERS) active silver colloids at room temperature by reduction of silver nitrate with hydroxylamine hydrochloride. J Phys Chem B 107(24): 5723–5727. doi:10.1021/jp027460u.

[19] Tejada E, Minghim R, Nonato LG (2003) On improved projection techniques to support visual exploration of multi-dimensional data sets. Inf Vis 2:218–231.

[20] Minghim R, Paulovich FV, Lopes ADA (2006) Content-based text mapping using multi-dimensional projections for exploration of document collections. In: Erbacher RF, Roberts JC, Grohn MT, Borner K (eds) Visualization and data analysis 2006, vol 6060. Proceedings of SPIE. doi:10.1117/12.650880.

[21] Cai J, Elrassi Z (1992) Micellar electrokinetic capillary chromatography of natural solutes with micelles of adjustable surface-charge density. J Chromatogr 608(1–2):31–45. doi:10.1016/0021-9673(92)87103-f.

[22] Furini LN, Sanchez-Cortes S, López-Tocón I, Otero JC, Aroca RF, Constantino CJL (2015) Detection and quantitative analysis of carbendazim herbicide on Ag nanoparticles via surface-enhanced Raman scattering. J Raman Spectrosc 46(11):1095–1101. doi:10.1002/jrs.4737.

[23] Moskovits M (1985) Surface-enhanced spectroscopy. Rev Mod Phys 57(3):783–826. doi:10.1103/RevModPhys.57.783.

[24] Bonora S, Benassi E, Maris A, Tugnoli V, Ottani S, Di Foggia M (2013) Raman and SERS study on atrazine, prometryn and simetryn triazine herbicides. J Mol Struct 1040:139–148. doi:10. 1016/j.molstruc.2013.02.025.

[25] Benassi E, Di Foggia M, Bonora S (2013) Accurate computational prediction of the structural and vibrational properties of s-triazine derivatives in vacuo. A DFT approach. Comput Theor Chem 1013:85–91. doi:10.1016/j.comptc.2013.03.010.

[26] Paulovich FV, Moraes ML, Maki RM, Ferreira M, Oliveira ON Jr, de Oliveira MCF (2011) Information visualization techniques for sensing and biosensing. Analyst 136(7):1344–1350. doi:10. 1039/c0an00822b.

[27] Oliveira ON Jr, Pavinatto FJ, Constantino CJL, Paulovich FV, de Oliveira MCF (2012) Information visualization to enhance sensitivity and selectivity in biosensing. Biointerphases 7(1–4):53. doi:10.1007/s13758-012-0053-7.

[28] Aoki PHB, Carreon EGE, Volpati D, Shimabukuro MH, Constantino CJL, Aroca RF, Oliveira ON Jr, Paulovich FV (2013) SERS mapping in Langmuir-Blodgett films and single-molecule detection. Appl Spectrosc 67(5):563–569. doi:10.1366/12-06909.

[29] Rubira RJG, Camacho SA, Aoki PHB, Maximino MD, Alessio P, Martin CS, Oliveira ON Jr, Fatore FM, Paulovich FV, Constantino CJL (2014) Detection of trace levels of atrazine using surface-enhanced Raman scattering and information visualization. Colloid Polym Sci 292(11):2811–2820. doi:10.1007/ s00396-014-3332-7.

[30] Aoki PHB, Alessio P, Riul A Jr, De Saja Saez JA, Constantino CJL (2010) Coupling surface-enhanced resonance Raman scattering and electronic tongue as characterization tools to investigate biological membrane mimetic systems. Anal Chem 82(9): 3537–3546. doi:10.1021/ac902585a.

[31] Koeber R, Fleischer C, Lanza F, Boos KS, Sellergren B, Barcelo D (2001) Evaluation of a multidimensional solid-phase extraction platform for highly selective on-line cleanup and high-throughput LC-MS analysis of triazines in river water samples using molecularly imprinted polymers. Anal Chem 73(11):2437–2444. doi:10.1021/ac001483s.

[32] Djozan D, Farajzadeh MA, Sorouraddin SM, Baheri T, Norouzi J (2012) Inside-needle extraction method based on molecularly imprinted polymer for solid-phase dynamic extraction and preconcentration of triazine herbicides followed by GC-FID determination. Chromatographia 75(3–4):139–148. doi:10.1007/ s10337-011-2173-5.

[33] Oliveira Brett AM, Da Silva LA (2001) Validation of novel biosensors in real environment and food sample. Paper presented at the 1st Workshop on Evaluation, Athens, Greece, May.

[34] Djozan D, Ebrahimi B, Mahkam M, Farajzadeh MA (2010) Evaluation of a new method for chemical coating of aluminum wire with molecularly imprinted polymer layer. Application for the fabrication of triazines selective solid-phase microextraction fiber. Anal Chim Acta 674(1):40–48. doi:10.1016/j.aca.2010.06. 006.

[35] Tuzimski T (2012) Application of RP-HPLC-diode array detector after SPE to the determination of pesticides in pepper samples. J AOAC Int 95(5):1357–1361. doi:10.5740/jaoacint.SGE_Tuzimski.

[36] Wang Y, Chang Q, Zhou X, Zang X, Wang C, Wang Z (2012) Application of liquid phase microextraction based on solidification of floating organic drop for the determination of triazine herbicides in soil samples by gas chromatography with flame photometric detection. Int J Environ Anal Chem 92(14): 1563–1573. doi:10.1080/03067319.2011.564618.

[37] Djozan D, Ebrahimi B (2008) Preparation of new solid phase micro extraction fiber on the basis of atrazine-molecular imprinted polymer: Application for GC and GC/MS screening of triazine herbicides in water, rice and onion. Anal Chim Acta 616(2):152–159. doi:10.1016/j.aca.2008.04.037.

[38] Bakas I, Hayat A, Piletsky S, Piletska E, Chehimi MM, Noguer T, Rouillon R (2014) Electrochemical impedimetric sensor based on molecularly imprinted polymers/sol-gel chemistry for methidathion organophosphorous insecticide recognition. Talanta 130:294–298. doi:10.1016/j.talanta.2014.07.012.

[39] Valera E, Garcia-Febrero R, Isabel Pividori M, Sanchez-Baeza F, Marco MP (2014) Coulombimetric immunosensor for paraquat based on electrochemical nanoprobes. Sens Actuators B Chem 194:353–360. doi:10.1016/j.snb.2013.12.029.

[40] Lopez Rodriguez ML, Madrid RE, Giacomelli CE (2013) Evaluation of impedance spectroscopy as a transduction method for bacterial biosensors. IEEE Latin Am Trans 11(1):196–200. doi:10.1109/tla.2013.6502802.

Chapter 30
Effect of Addition of BaO on Sintering of Glass-Ceramic Materials from SiO_2–Al_2O_2–Na_2O–K_2O–CaO/MgO System

Janusz Partyka, Katarzyna Gasek, Katarzyna Pasiut, Marcin Gajek

Department of Ceramics and Refractory Materials, Faculty of Material Science and Ceramics, AGH University of Science and Technology, al. A. Mickiewicza 30, 30-059 Kraków, Poland

Abstract: Glass-ceramic materials due to the various physical and chemical parameters have a more and more wide application in various areas of engineering. This paper presents a glass-ceramic material originating from the Al_2O_2–SiO_2–Na_2O–K_2O–CaO/MgO modified by the addition of barium oxide. For the oxide composition which the molar ratio of SiO_2/Al_2O_3 is constant and equal to 6.42, and containing 0.375mol% of Na_2O + K_2O and 0.625mol% of MgO was introduced barium oxide in quantities: 4mass%, 9mass% and 14mass%. For research of the sintering process, the following thermal analysis techniques were used: differential scanning calorimetry, dilatometry measurement and hot-stage microscopy. Analysis of the observed thermal transitions, marked on the respective charts, allows for appropriate design of the sintering curves in order to obtain a material with a high degree of crystallisation. The results of thermal analysis were compared with the phase compositions determined by X-ray diffraction and images from electron scanning microscopy of polished samples of glass-crystalline materials. Phenomena related to the influence of addition of barium oxide and heating treatment are analytically discussed.

1. Introduction

Glass-ceramic materials are produced through controlled crystallisation of a completely or partially melted raw material batch in a manner that allows for obtaining a fine-grained crystalline structure in the crystalline phase matrix. The material obtained in this way should be characterised by lack of porosity and an appropriately selected degree of re-crystallisation[1]–[4]. Designing the input oxide composition and the selection of the firing curve determine the type and quantity of the resulting crystalline phase and the chemical composition of the vitreous phase bonding crystallite grains[1][2][4]. The final phase composition determines the performance of the material obtained.

The proposed multi-component oxide system, given as SiO_2–Al_2O_3–Na_2O–K_2O–CaO–MgO with an additive of BaO, is not commonly used for the production of technical glass-crystalline materials. Such compositions are more often used for the production of crystalline or glass-crystalline glazes. The most important components in the system being discussed include silicone oxides, which form the basic structural network and amphoteric aluminium oxide, which are built into the silico-oxide network by isomorphous substitutions of Al^{3+} ions instead of Si^{4+}[3][4]. The SiO_2/Al_2O_3 ratio has the main influence on characteristic temperatures of the glass-crystalline materials, their crystallisation tendencies and thus the scope of their application[3]–[5][8][9]. Modifications of the characteristic temperatures can be obtained by introducing alkaline metal oxides, Na_2O and K_2O (low-and medium-temperature fluxing agents), or alkaline earth oxides CaO, MgO, BaO, etc. (medium-and high-temperature fluxing agents)[1][2][4][6][7].

Barium oxide, as the only fluxing agent in the group of alkaline earth oxides, is active even in a small amount, but only at temperatures above 1250°C. However, in the presence of alkaline metal oxides, BaO begins to co-form the liquid phase at a considerably lower temperature than 1000°C, as in the SiO_2–BaO–K_2O system at a temperature of 907°C[6][7][9][11][12] and in the SiO_2–BaO–Na_2O system at a temperature of 785°C[11][12]. The BaO additive effectively improves mechanical parameters and chemical resistance[1]–[3]. Sources of barium oxide include BaO-containing frits, barium sulphate ($BaSO_4$) and barium carbonate ($BaCO_3$). Barium sulphate, due to sulphur dioxide emissions during thermal processing, is used relatively rarely. Barium carbonate is also applied as a raw material to a limited extent, mostly due to its harmfulness to humans. Both raw materials can also pose problems related to the quality of surface in low-melting materials or materials with a low content of alkaline metal oxides. During thermal processing in an oxidising atmosphere, they decompose in the liquid vitreous phase rich in alkaline oxides (Na_2O, K_2O or Li_2O) and produce a large number of gaseous products, which cause surface defects, such as holes, pinholes or blisters situated near the surface. However, if released, BaO quickly reacts with the oxygen-aluminium-silicon lattice of the vitreous phase; barium oxide reacts with other oxides causing crystallisation of new phases under favourable conditions[10][11].

2. Research Methodology

The research focused on two glass-crystalline materials with the composition selected from among the multi-component system of SiO_2–Al_2O_3–Na_2O–K_2O–CaO/MgO, into which BaO was introduced in the mass amounts of 4, 9 and 14%, over 100% of the base material. Molar base compositions of the materials, without the additive of barium oxide, are presented in **Table 1**. Naturally enriched natural raw materials were used in the study: Sibelco Norfolat K600 and Na600 feldspars, Sobótka MK40 silica powder, Ottavi wollastonite, Luzenac talc and Surmin KOC kaolin. The compositions of the raw materials were calculated using the SEGER software developed in the course of the project. Sample preparation consisted in grinding individual compositions in a planetary ball mill for 40min, resulting in residues of 0.8% ± 0.05% by mass on the 56-µm sieve. Once dried to obtain a constant mass at 150°C, some powder was burnt in a porcelain crucible at 1230°C. The raw powder was subject to thermal processing; in this case, the characteristic temperatures were measured and differential scanning calorimetry (DSC) was performed[9][12][13]. Using the thermally processed material, samples were cut out in the form of cuboids with dimensions of 4mm × 4mm × 10mm for

Table 1. Oxide compositions of PORC 01-4 Ca/Mg base glass-ceramic materials.

Description of glass-ceramic materials	SiO_2	Al_2O_3	CaO	MgO	$Na_2O + K_2O$
I PORC 01-4 Ca	70.45	10.75	11.69	0.36	6.75
I PORC 01-4 Mg	70.89	11.04	0.42	10.87	6.78

the purpose of dilatometric tests and the observation on an electron scanning microscope (SEM) with an EDS attachment. The remaining part underwent size reduction to below 56μm, which was used for determination of the XRF oxide composition and XRD phase composition. For examination of the characteristic temperatures and thermal transformation, the following equipment was used:

- HSM Misura 3 hot-stage temperature microscope, each measurement was carried out up to determine the melting temperature, increasing the temperature at a constant rate of $10°$ min^{-1};

- Netzsch DL 402C mechanical dilatometer, heating the samples at a constant rate of $10°$ min to a temperature above the softening temperature;

- Netzsch STA 449 F3 Jupiter thermal analyser to test thermal transformation during thermal processing, heating the samples at a constant rate of $10°$ min^{-1} to a maximum temperature 1230°C.

To determine other parameters, the following devices were applied:

- Spectrometer WDXRF Axios mAX with the RH lamp with a power 4 kW, analyser, to determine chemical composition;

- PANalytical X-ray diffractometer (X'Pert Pro), with scanning range was 5°–90° (2θ) at scanning rate of 0.05° 2θ/2s with a recording speed of 0.05° 2θ/2s, to determine the phase composition;

- NOVA NANO SEM 200 electron microscope with an EDS X-ray chemical microanalyser supplied by EDAX to observe the surface of microsections of the glass-ceramic materials.

3. Results and Discussion

The chemical composition was investigated to verify the theoretical assumptions on the molar compositions used for the conversion into raw material batches and comparison with the composition of materials after grinding and burning. The results presented in **Table 1**, mostly the SiO_2/Al_2O_3 molar ratio and the CaO and MgO molar percentages, proved to be consistent with the assumptions. The almost identical additive of alkaline metal oxides is also essential. It proves that the presence of barium oxide is the only variable in both groups of the materials.

The next step was aimed at investigating the influence of the barium oxide additive on the characteristic temperatures of the glass and crystalline materials in question. The tests were performed using the input material after grinding and drying. The measurement was taken until the total melt away of pastille was obtained (the height of the sample must be lower than one-third of the length of the base). The results of the measurements are presented in **Table 2**. The dilatometric characteristic temperatures, transition and softening, were conducted using the mechanical dilatometer Netzsch 402C. The samples, cuboid with dimensions of 10mm × 5mm × 5mm, were cut from the glazes fired at a temperature of 1230°C. The results dilatometric curves as a function of the temperature and values of the characteristic temperatures results are shown in **Figures 1**, **2** and in **Table 2**. By the analysis of the results obtained, it must be stated in the first place that the influence of the BaO additive on the characteristic temperatures depends on the type of the base material. In calcium materials, which contain CaO from alkaline earth oxides, the BaO additive reduces the values of individual characteristic tempera

Table 2. Characteristic temperatures of tested glass-ceramic composites.

Description of characteristic temperatures	I PORC 01-4 Ca				I PORC 01-4 Mg			
	Addition of Bao/mass%							
	0 %	4 %	9 %	14 %	0 %	4 %	9 %	14 %
	Temperature/°C							
Transition DL*	753	737	720	710	761	759	744	702
Softening DL*	835	801	784	1139	830	882	992	1138
Sphere—HSM**	1252	1247	1245	1230	1290	1299	1307	1298
Half-sphere—HSM**	1285	1270	1266	1250	1315	1324	1327	1318
Melting—HSM**	1324	1305	1296	1279	1335	1346	1344	1334

* Dilatometric
** Hot-stage microscope

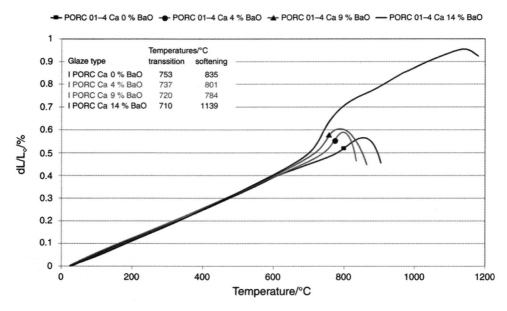

Figure 1. Dilatometric curves of glass-ceramic glaze PORC 01-4 Ca + BaO with values of characteristic temperatures.

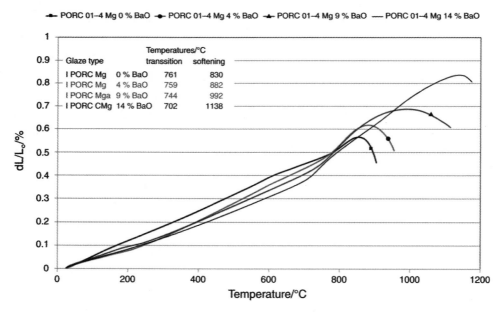

Figure 2. Dilatometric curves of glass-ceramic glaze PORC 01-4 Mg + BaO with values of characteristic temperatures.

-tures, compared to the input material. The reduction in all temperatures is proportional to the quantity of the barium oxide additive. Particularly, visible changes concern the transition temperature (reduction by 43°C) and the melting temperature (reduction by 45°C). In other cases, the differences are as follows: reduction by 22°C for the sphere temperature and by 35°C for the hemisphere temperature. The influence on the characteristic temperatures of barium oxide increases together with its quantity; the influence is more noticeable at high temperatures. Thus, it may be concluded that in total, the results are consistent with these data relating to the materials in the system of SiO_2–Al_2O_3–Na_2O–K_2O–CaO. A different behaviour of various BaO additives was identified in the other tested system of SiO_2–Al_2O_3–Na_2O–K_2O–MgO, wherein magnesium oxide (MgO) is the representative of alkaline earth oxides in the base material. For the transition temperature, the material behaves in a similar manner, as was the case of the previously discussed system with the CaO additive; that is successive proportional reduction in this temperature to the increase in the BaO additive can be observed. As for the temperatures defined as the sphere, hemisphere and melting, the behaviour of the material is completely different. The sphere and hemi-sphere temperatures increase slightly from 9 to 12°C for the 4mass% and the 9mass% additives of BaO. For the 14mass% additive of BaO, the temperatures are reduced to the base material temperature, without the additive of the oxide. These differences can be accounted for by various levels of refractoriness of calcium and magnesium oxides and a higher tendency for crystallisation in systems with magnesium oxide. Another deviation from the described phenomena is the influence of the barium oxide additive on the dilatometric softening temperature in both systems. With the 4mass% and 9mass% additive of BaO, changes in this temperature can be described exactly as for the sphere and hemisphere temperatures for systems with calcium and magnesium oxides, while for the 14mass% additive of BaO, a sharp increase in the dilatometric softening temperature (by approximately 300°C) is observed in both systems. The authors have not found an explanation for this effect yet. The most likely explanation is the large increase in the amount of crystalline phases in glazes containing 14% by mass, barium oxide, in both types of material in the presence of calcium oxide and magnesium oxide [**Figure 9(a), (b)**].

Based on measurements of the characteristic temperatures, a graphic relationship between viscosity and temperature was determined. The Vogel-Fulcher-Tammann (VFT) equation was used for the calculations[14].

$$\log \eta = A + B/(T - T_0)$$

where A, B and T_0 are constants. Graphically, the VFT equation is represented with a straight-line plot of logg versus $1/(T - T_0)$, where A is the y-intercept, B is the slope of the line and T_0 is the measure of deviation of the curve from the ideal straight line of nonassociated liquids. The melt viscosity can be estimated from dilatometric and HSM measurements based on three known reference points: $\eta = 10^{12}$Pa s as the dilatometric T_g; $\eta = 10^{9.25}$Pa s as the dilatometric T_s; and $\eta = 10^{3.55}$Pa s at the HSM $T_{1/2}$, where a sample forms a half-sphere shape during HSM analysis[14]. The results are presented in **Figures 3** and **4**.

The introduction of barium oxide into the SiO_2–Al_2O_3–Na_2O–K_2O–CaO system influences the viscosity of the aluminium-silicon-oxygen alloy only for the case of the 14mass% additive of BaO (**Figure 3**). It can be seen that the parabolic nature of viscosity changes and the viscosity abruptly decreases from approximately 1250°C. According to the authors, this change is mostly influenced by an unexplained rise in the value of the dilatometric softening temperature (T_s). For other amounts of this oxide, *i.e.* 4 and 9mass%, the viscosity changes slightly,

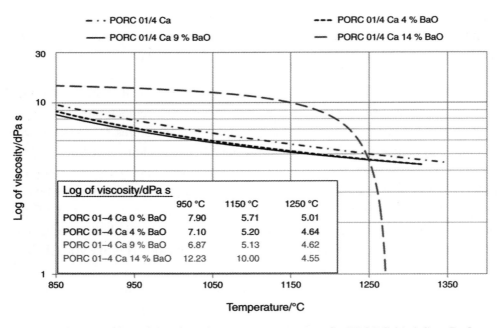

Figure 3. Curves of log of the viscosity versus temperature for I PORC 01-4 Ca + BaO.

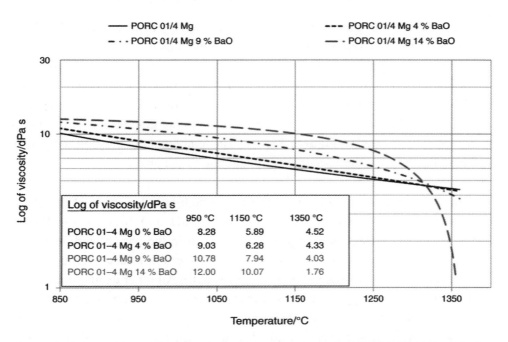

Figure 4. Curves of log of the viscosity versus temperature for I PORC 01-4 Mg + BaO. compared to the base material.

The BaO additive in the SiO_2–Al_2O_3–Na_2O–K_2O–MgO system exerts a considerably greater influence on the viscosity in the function of temperature. It is clear that the increase in the barium oxide additive results in the divergence of the viscosity logarithmic curves between 900°C and 1250°C (**Figure 4**); a higher BaO content results in a higher viscosity. Above 1250°C, the viscosity curves begin to converge and achieve the same values at a temperature of approximately 1300°C. An increase in the viscosity within the specified range of temperatures probably results from the occurrence of large amounts of crystalline phases in magnesium materials (**Figures 8**, **9**), as was the case of the changes in the characteristic temperatures. The logarithmic viscosity curve obtained for the 14 mass% additive shows a similar parabolic output to the analogous curve obtained for calcium materials. This similarity is caused by the same abrupt rise in the dilatometric softening temperatures as observed for the SiO_2–Al_2O_3–Na_2O–K_2O–CaO system.

In the next step, an attempt was made to explain the described phenomena by investigating the changes during thermal processing. For this purpose, thermal characteristics of the glass-crystalline materials were performed with the DSC method using the Netzsch STA 449 F1 Jupiter apparatus in a synthetic air atmosphere with the heating rate of 10°C min^{-1}. All measurements were taken while heating the materials to a temperature of 1230°C at a rate of 10°C min^{-1}. In order to facilitate the interpretation, the DSC curve obtained for barium carbonate, heated under the same conditions, was overlaid onto the graphs. The DSC sintering curves are presented in **Figures 5** and **6**. The temperatures shown in the graphs are ONSET points, *i.e.* they occur at the beginning of the process.

In the graphs of the DSC lines relating to both tested systems of the glass-crystalline materials—SiO_2–Al_2O_3–Na_2O–K_2O–CaO and SiO_2–Al_2O_3–Na_2O–K_2O–MgO with the BaO additive—the effects seen in the barium carbonate distribution curve are reflected, *i.e.* mostly the endothermic effects starting in the carbonate at 806°C, 960°C and 1000°C (**Figures 5, 6**). All other exothermic effects should correspond to crystallisation processes of new phases during the heating of the compositions with a various BaO content. For the SiO_2–Al_2O_3–Na_2O–K_2O–CaO system, these temperatures are equal to approximately 996°C and 1066°C for the 9mass% and 14mass% additives (**Figure 5**). These values should correspond

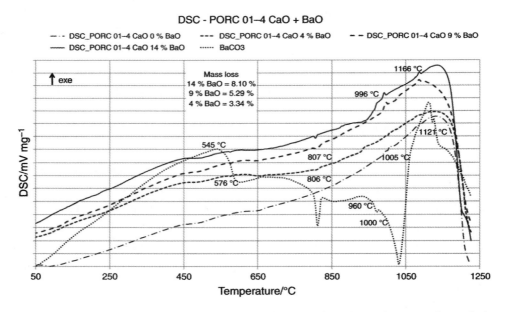

Figure 5. Curve of DSC thermal transformations during thermal processing of the SiO_2–Al_2O_3–Na_2O–K_2O–CaO + BaO system.

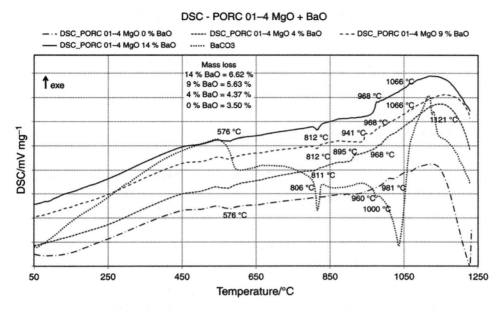

Figure 6. Curve of DSC thermal transformations during thermal processing of the SiO_2–Al_2O_3–Na_2O–K_2O–MgO + BaO system.

to areas in which two crystalline phases present in these systems—sanidine and hyalophane—should be formed (**Figures 7** and **8**). The point, in which it is likely for anorthite to occur (**Figure 7**) as regards the material without the BaO additive, should correspond to the peak on the DSC curve at approximately 1005°C (**Figure 5**).

The situation of the SiO_2–Al_2O_3–Na_2O–K_2O–MgO + BaO system is slightly different (**Figure 6**). Admittedly, the effects which can be associated with the changes occurring while heating $BaCO_3$, *i.e.* endothermic effects at 806°C, 960°C and 1000°C, can be observed along the lines corresponding to the individual amounts of barium oxide. The effect observed at 981°C should correspond to the occurrence of protoenstatite (**Figures 8** and **9**). The remaining effects observable at 941°C, 968°C and 1066°C should show temperatures at which forsterite, hyalophane and celsian occur (**Figures 6, 8, 9**).

Scanning electron microscopy images shows that the presence of BaO, in both types of materials, generally favours the crystallisation of crystalline phases, and additionally, it is in proportion to the increase in the content of barium oxide.

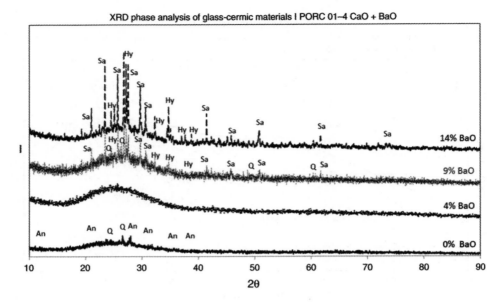

Figure 7. XRD phase composition XRD phase analysis of glass-cermic materials I PORC 01-4 CaO + BaO diagram for the glass-crystalline materials from the system composed of SiO_2–Al_2O_3–Na_2O–K_2O–CaO + BaO (Q—quartz, An—anorthite, Sa—sanidine, Hy—hyalophane).

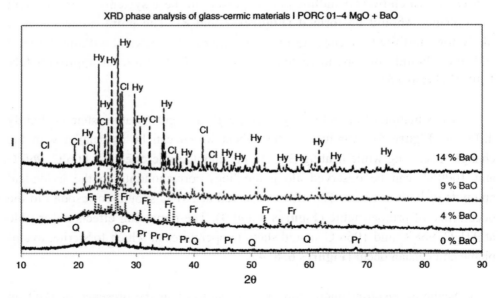

Figure 8. XRD phase composition diagram for the glass-crystalline materials from the system composed of SiO_2–Al_2O_3–Na_2O–K_2O–MgO + BaO (Q—quartz, Pr—protoenstatite, Fr—forsterite, Hy—hyalophane, Cl—celsian).

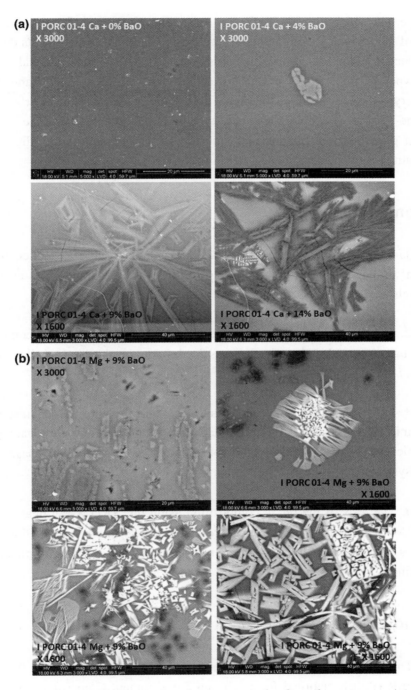

Figure 9. SEM images of the glass-ceramic glazes samples from the systems composed of $SiO_2–Al_2O_3–Na_2O–K_2O$—a CaO/b MgO with the BaO additive.

In materials which contain MgO, the crystalline phase is definitely more. Both types of glazes (calcium and magnesium) exhibit different crystal habits due to the difference in the phase composition (**Figures 7, 8**). The common crystal phase for both types of materials is only a barium alumino-silicate—hyalophane, and all other crystalline phases are different. The calcium glazes contain quartz, calcium alumino-silicate—anorthite—and potassium alumino-silicate—sanidine, and magnesium glazes contain magnesium silicate—forsterite, aluminium silicate—protoenstatite—and barium silicate—celsian

4. Conclusions

1. The additive of barium oxide has a considerable influence on both thermal properties and the phase composition of the obtained glass-crystalline materials.

2. The behaviour of barium oxide significantly depends on the composition of the base material and, as in the presented study, on the type of alkaline earth oxides, *i.e.* CaO and MgO.

3. In the thermal DSC analysis diagrams, it is possible to recognise the endothermic effects associated with the decomposition of barium carbonate and exothermic effects resulting from the formation of new crystalline phases in the tested systems.

4. The tests of the phase composition using the XRD X-ray phase analysis and SEM-EDX scanning electron microscopy confirm the results of the thermal analyses, and as such, the type of crystalline phases and the texture of materials are defined.

Acknowledgements

This research has been carried out thanks to financing under the framework of NCBiR (Polish National Research and Development Committee) programme Nos. N N508 477734 and PBS1/B5/17/2012.

Open Access: This article is distributed under the terms of the Creative Commons Attribution 4.0 International License (http://creativecommons.org/licenses/by/4.0/), which permits unrestricted use, distribution, and reproduction in any medium, provided you give appropriate credit to the original author(s) and the source, provide a link to the Creative Commons license, and indicate if changes were made.

Source: Janusz Partyka, Katarzyna Gasek. Effect of addition of BaO on sintering of glass-ceramic materials from SiO_2–Al_2O_2–Na_2O–K_2O–CaO/MgO system[J]. Journal of Thermal Analysis and Calorimetry, 2016(3):1–9.

References

[1] Holand W, Beall GH. Glass ceramic technology. 2nd ed. London: Wiley; 2012. ISBN 978-0-470-48787-7.

[2] Manfredini T, Pellacani GC, Rincon JM. Glass-ceramic materials. Fundamentals and applications. Modena: Mucchi Editore; 1997. ISBN-10: 887000287X.

[3] Brow RK. Inorganic glasses and glass-ceramics chapter 6 in characterization of ceramics. Stoneham: Butterworth-Heinemann; 1993. p. 103–18.

[4] Zachariasen WH. The atomic arrangement in glass. J Am Chem Soc. 1932;54:3841.

[5] Gunawardane RP, Glasser FP. The System K_2O–BaO–SiO_2. J Am Ceram Soc. 1976;59(5-6):233–6.

[6] Gunawardane RP, Glasser FP. Phase equilibria and crystallization of melts in the system Na_2O–BaO–SiO_2. J Am Ceram Soc. 1974;57(5):201–4.

[7] ArgyleJ F, Hummel FA. System BaO–MgO–SiO2; compatibility triangles. B = BaO; M = MgO; S = SiO_2. Glass Ind. 1965;46(12): 710–8.

[8] Chen J, Xiao X, Yang K, Wu H. Effect of Al_2O_3/SiO_2 Ratio on the viscosity and workability of high-alumina soda-lime-silicate glasses. J Chin Ceram Soc. 2012; 40(7):1001–7.

[9] Ahmed M, Earl A. Characterizing glaze melting behaviour via HSM Hot Stage Microscopy. Am Ceram Soc Bull. 2002;81(3): 47–51.

[10] Zhou W, Zhang L, Yang J. Preparation and properties of barium aluminosilicate glass-ceramics. J Mater Sci. 1997;32(18):4833–6.

[11] Lambrinou L, Van der Bies O. Study of the devitrification behaviour of a barium magnesium aluminosilicate glass-ceramic. J Eur Ceram Soc. 2007;27:1805–9.

[12] Szumera M. Charakterystyka wybranych metod termicznych. 2nd ed., LAB Labora-

toria, Aparatura, Badania. 2013. 18(1), p. 24–33.

[13] Qu LJ, Li B, Wang J, Gu YM. Application of DSC technique in study of glass ceramic. Adv Mater Res. 2010;105–106:743–5.

[14] Paganelli M, Sighinolfi D. Understanding the behaviour of glazes with automatic heating microscope. Ceram Forum Int CFI/Ber DKG. 2008;85(5).